Monographs in Theoretical Computer Science
An EATCS Series

For further volumes:
http://www.springer.com/series/776

Carlo A. Furia • Dino Mandrioli
Angelo Morzenti • Matteo Rossi

Modeling Time in Computing

 Springer

Carlo A. Furia
Department of Computer Science
Zürich, Switzerland

Dino Mandrioli
Angelo Morzenti
Matteo Rossi
Dipartimento di Elettronica e Informazione
Politecnico di Milano
Milan, Italy

ISSN 1431-2654
ISBN 978-3-642-32331-7   ISBN 978-3-642-32332-4 (eBook)
DOI 10.1007/978-3-642-32332-4
Springer Heidelberg New York Dordrecht London

Library of Congress Control Number: 2012947316

ACM codes: F.1, F.2, F.4

Preface

Time has become a part of computing, where it has claimed its own models. This book summarizes several decades of research on developing, analyzing, and applying time models to computing and, more generally, engineering.

Every researcher contributing to such a burgeoning field is forced to consider several available approaches, to understand what they have in common and how they differ. This continued effort prompts a systematic work of comparison, classification, and assessment of the diverse models and methods. This book picks up this call and tries to put it into practice.

Our survey paper with the same title[1] is a precursor work: while paper and book share the same fundamental themes, this book is aimed at a wider public than computer science researchers – including engineers, students, and practitioners – and is more ambitious in scope and number of topics and examples.

Scope and Topics

Models including a notion of time are ubiquitous in the natural sciences and engineering, and they have also received the attention of philosophers, linguists, and other scholars. It is in computing, however, that the abstractions provided by the traditional models of time in the physical sciences can be inadequate. To bridge this hiatus between abstractions, computer science research has been quite prolific in spawning novel models of time. This book is a thorough presentation of the state of the art in this area, including some historical perspective and not omitting the occasional controversial aspects.

This book is *not* meant to be an encyclopedic catalog of models of time, nor to be a reference handbook that collects independent chapters sharing a common theme.

---

[1] Furia, C.A., Mandrioli, D., Morzenti, A., Rossi, M.: Modeling time in computing: a taxonomy and a comparative survey. ACM Comput. Surv. **42**(2), 1–59 (2010).

Rather, it is an organized presentation of the issues related with timing modeling and analysis that recur in computer science and other branches of engineering and science, and of the most significant available solutions. The connections among the themes of different chapters and the usage of a taxonomy of fundamental issues help readers develop a solid understanding of the problems and their solutions, which is applicable outside the necessarily limited array of topics covered in the book, in their concrete research and practice. For each chapter, we selected examples and approaches that cover some fundamental aspects of time modeling, and presented them in a way that abstracts away the inessential low-level details and focuses on the novel conceptual contributions and their relations to other approaches. Each chapter is supplemented by a comprehensive collection of bibliographic remarks, where readers interested in knowing all the details about a specific approach, or looking for variants of a fundamental idea, can find up-to-date references to the scientific literature.

Audience

The book addresses an ambitiously broad audience. Graduate students and researchers in computer science – or with a computer science background – are the primary intended readership. Researchers and practitioners in other scientific and engineering disciplines interested in time modeling with a computational flavor will also find the book interesting and approachable: the comparative and conceptual approach of the book makes it a valuable introduction for non-experts, and a useful preparatory reading to more specialized technical references.

A large number of engineering disciplines target software-intensive systems whose correctness crucially depends on the timing behavior, for example, embedded control systems, complex manufacturing systems, avionics systems, railway signaling systems, and so on. The concepts, models, techniques, and methods developed by computer scientists to describe timing behavior have increasingly been applied beyond their original application domains to model and analyze such software-intensive systems. This requires a mutual knowledge exchange between traditionally distinct engineering areas: domain experts are usually unfamiliar with the concepts and notations specialized by computer science to model time, whereas computer scientists may be unprepared to apply their methods to heterogeneous systems operating in unfamiliar domains. This book may contribute to forming a shared vocabulary: the presentation uses a rigorous, yet not overly technical, style, which is approachable by readers with heterogeneous specialized backgrounds.

Addressing a wide public has two opposite risks: repeating material already known and making the presentation not self-contained. We structured the book to minimize these risks, by providing a flexible reading approach as we explain next.

Book Structure

The book has two parts and a total of 12 chapters. Chapters 1–3, as well as the concluding Chap. 12, belong to the frame; Part I includes Chaps. 4–6, Part II the remaining Chaps. 7–11.

Chapter 1 introduces the book, its contents, topics, structure, and goals. Chapter 2 introduces the notions of formalism and model in general terms, and some of their fundamental classification criteria; while doing so, it presents the fundamentals of propositional and predicate logic.

Chapter 3 is a cornerstone of the whole book: it introduces some essential issues that arise when modeling time across every type of system. The presentation of the numerous formalisms in the rest of the book constantly refers to these "dimensions" to put the different models on a common ground.

Part I is a concise summary of the models of time that are traditional in engineering and the natural sciences, including fundamental computer science. Chapter 4 targets dynamical systems and control theory; Chap. 5 takes the point of view of hardware design, and Chap. 6 of software algorithmic and complexity analysis. Part I is meant to provide heterogeneous readers with a homogeneous background; hence the reading will focus on the topics that are less familiar. For example, readers with expertise in control, electrical, or mechanical engineering will probably skim through Chap. 4 and spend more time on Chaps. 5 and 6 to acquaint themselves with some basic computer science models, in preparation for Part II. Conversely, computer scientists may find the content of Chap. 4 especially useful, whereas they will probably only use Chaps. 5 and 6 as references or reminders.

Part II covers advanced and specialized formalisms developed for dealing with specific issues of time modeling in heterogeneous software-intensive systems. Chapter 7 presents formalisms that all share finite state machines as common "ancestors"; Chap. 8 discusses Petri nets in many variants; Chap. 9 targets notations based on mathematical logic, such as temporal logic; Chap. 10 is about process algebras – widely used to model concurrency, but less prominent in timing analysis; Chap. 11 presents some "dual-language approaches" that combine two notations with different characteristics to model and verify complex systems (model checking frameworks are the most popular applications of the dual-language approach).

Chapter 12 concludes the book with summarizing remarks and hints towards future developments and open challenges.

Prerequisites, Dependencies, and Teaching

The book assumes a basic knowledge of the standard topics of engineering and science undergraduate curricula: calculus, probability theory, algorithms, and programming.

A more advanced knowledge of computer science topics, in particular automata, formal languages, and mathematical logic, is not required but will make the book easier to read and understand in full. Readers without a computer science background should, in any case, read Chaps. 2, 5, and 6 with attention and make sure they understand the basic notions presented there. Mathematical logic is a particularly critical topic, in that Chap. 2 cannot replace a detailed introduction to such a technical topic; consequently, readers totally unfamiliar with propositional and predicate logic may have to skip the more advanced topics covered in Chaps. 9 and 11.

Chapter 4 is in contrast required reading for readers with little scientific knowledge outside "traditional" computer science. More generally, many of the examples in the book refer to systems from various engineering domains; hence it is important to develop an interdisciplinary view of problems and solutions as the reading progresses.

This book can be used as a textbook in multiple ways. The typical usage could be with a one-semester graduate course, where instructors can focus on the topics and models from Part II that are closest to the course topics and goals, according to the students' background. In particular, every chapter of Part II starts by presenting formalisms in their basic form (e.g., transition systems in Chap. 7, and basic Petri nets in Chap. 8) and then continues with richer extensions (e.g., Statecharts in Chap. 7, and timed Petri nets in Chap. 8). The teaching may stop at the basic levels, or tackle advanced extensions only of certain formalisms. An aspect that is mostly independent of the others is modeling with probabilities; readers can safely skip these parts if uninterested or when lacking a background in probability theory.

Exercises

Exercises are an important device for taking full advantage of the book's contents. While they are optional in a first reading, some of them become necessary for assimilating every technical concept and for understanding the most subtle aspects of the presentation. This does not necessarily mean writing out every solution in full detail: checking that the requests are clear and making the first steps toward a solution may be enough to take advantage of most exercises.

The exercises appear close to the topics they target. They have varying difficulty and scope; those requiring a stronger technical background are marked with a "♦"; their headings may specifically indicate which notions are assumed (e.g., the theory of computation, or probability theory).

A few exercises, marked with a "♣", have a distinctively conceptual flavor, in that their main purpose is to be thought provoking and they may admit different legitimate solutions, possibly informal. This type of exercise mostly occurs in Chap. 3, given its foundational nature.

Acknowledgments

The people who supported, directly or indirectly, the development of the book are too many to acknowledge; hence we start by apologizing for the inevitable omissions.

We thank the colleagues of many years who worked with us in our research on time modeling and contributed to broadening and shaping our views. This includes people from academia – the "historical fellows" Carlo Ghezzi and Pierluigi San Pietro, as well as the other major coworkers Connie Heitmeyer, Dick Kemmerer, Alberto Coen-Porisini, Miguel Felder, Sandro Morasca, Mauro Pezzè, Angelo Gargantini, Matteo Pradella, Paul Kolano, Paola Spoletini, Marcello Bersani, and Luca Ferrucci – and also practitioners from industry, in particular Piergiorgio Mirandola and the whole ENEL-CRA group, which provided technical cooperation besides financial support. We also acknowledge the fruitful collaboration with Selenia-Alenia, Elsag, Agusta-Westland, and Parvis.

The presentation has benefitted from the suggestions made indirectly by Bertrand Meyer while working with the first author on different topics; the first author also thanks Scott West for the English tips he gave – knowingly and unknowingly – and Sebastian Nanz for suggesting bibliographic references for Chap. 10 and for putting up with lengthy and frequent phone meetings with the other authors.

Finally, we are very grateful to the editor Grzegorz Rozenberg, to the anonymous reviewer, and to the Springer staff for their support and careful reading of various drafts of the book. Uncaught errors and deficiencies are, of course, our responsibility.

Zürich, Switzerland   Carlo A. Furia
Milan, Italy   Dino Mandrioli
                                         Angelo Morzenti
                                          Matteo Rossi

Contents

Chapter 1
It's About Time

Imagine a world without time...

Can't do it? The egregious difficulty you have certainly experienced with such a Herculean stretch of imagination shows that the notion of time is so deeply entrenched in our mental models of the world that completely eliminating it is hardly possible.

Time plays a central role in everyday life, where thoughts, actions, and statements often include temporal references: "I'm going on vacation tomorrow", "I have to finish this job by the end of the week", "John is always late". Time is also a favorite object of philosophical inquiry, and a subject of religious speculation. The very notion of time pervades science and engineering.

Physics has dated the beginning of our universe to time $\mathcal{T}_0$ – approximately 13.7 billion years ago. Philosophy and religion, as well as physics itself, speculate about the state of affairs *before* $\mathcal{T}_0$. Past and future seem perfectly well defined notions at the level of intuition, but, whereas mathematics can treat the past as "future with reversed sign", physics establishes – with the second principle of thermodynamics – that the arrow of time does not go backward.

Engineering, concerned with the invention of systems that operate in the physical world and have some purpose, has to deal with several different notions of time: it affects human life, hopefully improving it; it applies knowledge about the laws of physics; it develops and analyzes mathematical models of the world; and, at least indirectly, it refers to speculations offered by philosophers, since philosophy is rooted in, is used to analyze, and in one way or another impacts our lives. For example, when engineering a subway system, the time saved by travelers using the transportation is a major driver in determining the routes and the frequency and speeds of trains. In addition, braking and acceleration times impact the passengers' safety and comfort, and are to be included in mathematical models of the cars.

Computer science has rapidly become pervasive in engineering and, as a consequence, in everyday life. Cars, for example, embed many electronic components, and also systems in many other different domains – banking, medical care,

C.A. Furia et al., *Modeling Time in Computing*, Monographs in Theoretical Computer Science. An EATCS Series, DOI 10.1007/978-3-642-32332-4_1,
© Springer-Verlag Berlin Heidelberg 2012

transportation, etc. – depend, directly or indirectly, on the assistance of computing devices.

The proliferation of computing devices in the modern world is evident everywhere. What is, instead, less apparent is how the relatively short history of computing has produced novel problems and challenges in dealing with time, and solutions to them when designing systems with computational components.

This book has precisely the goal of analyzing this situation: on the one hand, computers are devices subject to physical laws like every other physical object; therefore, they can be modeled in terms of the motion of electrons in semiconductors and electromagnetic waves. On the other hand, computers work in widely diverse application domains, which implies dramatically different notions of time and its flow. For instance, users of automated teller machines certainly do not need a model of the electrons that flow in the circuits controlling the withdrawal of money; on the contrary, the users are concerned with the machine's responsiveness and expect to be able to receive money within a few seconds. Similar dualities occur in many other situations where people interact with computer-controlled devices.

The fundamental conceptual tool for coping with heterogeneous concerns in complex situations is *abstraction*, which consists of focusing on what is relevant in a certain context and for a certain purpose, while neglecting irrelevant details. Abstraction pervades computer science, which often has to deal with interacting multifarious domains as suggested by the examples above. Unsurprisingly, the models applied in computer science are often more diverse and heterogeneous than those in other sectors of engineering such as electrical or mechanical engineering, whose application domains are fairly established and well understood.

Abstraction of time is a special, and crucial, case: the roles and perceptions of time are heterogeneous, spanning very different domains – sometimes including psychological aspects ("happy times flow faster") – and hence models of time follow a great variety of approaches and have spawned diverse notations and formalisms.

At one extreme, given that computers are physical objects, we could model and analyze their behavior according to the physical laws of electromagnetism, which describe the flow of electrons through semiconductors or even the evolution of their quantum states.

At the other extreme, the theory of computation is fundamentally based on drastic abstractions of time, almost to the point of removing it completely from the models: in many traditional applications, only the results of a computational process matter, not so much how long it takes to obtain them. This was true with the slow batch computer systems of the past, when users input a collection of punched cards and came back after 1 day to pick up the printout of the results; but it also happens with the fast interactive computers of the present, when users perceive only the overall responsiveness of the system, and the time of each individual operation is negligible. In these scenarios, computational processes are abstracted as *functions* from input to output data.

Between these two extremes there is a continuum of abstractions and models, based on application environments, design goals, and preferences of the designers. Let us sketch a few examples, developed in greater detail throughout the book.

When moving from the point of view of electronic circuit design to that of hardware architecture, we apply *discretization*, namely the change from continuous to discrete domains. This applies both to time and to other domains used in the formal models. Discretization can be seen as a tool for mathematical analysis via numerical computation, which has burgeoned also in domains where time and dynamics are not primary concerns, for instance in the *finite element methods* for static analysis of structures.

Computational complexity theory defined another major historical approach to modeling time in computing. In some sense, computational complexity fills the "abstraction gap" of purely functional models, as it describes *how long* computations take, independently of what output they produce. Take, for instance, the problem of sorting a sequence of elements. First, we can describe and implement a few algorithms that obtain the desired result. Then, we classify the algorithms according to their complexity, preferring the most efficient ones, which require, say, a time proportional to $n \cdot \log n$ for every sequence with n elements. The computational complexity abstraction of time sharply departs from the traditional approaches in other fields of engineering, where system behavior is modeled by the evolution of *state as a function of time*. For example, the laws of mechanics describe the position and velocity of masses as functions of time, from which one can compute the time and space required by, say, a car to reach a full stop from a given initial speed.

However, the traditional view of computation as a sequential process that starts from some initial state, reads some input, and produces an output after some time is inadequate to model systems where computational elements work in collaboration with modules of different kinds. This is the case with so-called "reactive systems", which are often *embedded*. Reactive systems include computing devices as parts of a more complex system where different processes, with different dynamics, interact and *coordinate* with one another towards a common goal, or *compete* to access limited shared resources. Also, when the computations must obey *quantitative timing constraints* (e.g., "the shared resource cannot be occupied for longer than 100 seconds", "as a consequence of an alarm the system must be shut down within ten seconds"), the systems are called "real time".

The structure of reactive systems can be highly complex and they may include heterogeneous components that require diverse mathematical models. Often, the external *environment*, whose behavior is only partially controllable or observable, plays a prominent role in interacting with the other system components. The environment often includes users and actors – human or otherwise. For example, a car is a complex system made of interacting mechanical and electronic components, which interacts with a much larger and complex environment consisting of other cars, drivers, pedestrians, roads, and so on. Modeling, analyzing, and designing such systems requires the ability to formalize quite different features and their mutual interactions.

If we focus on time modeling, we notice how many different notions of time belong to different levels of abstraction. In the example of cars in traffic, there are, among other notions of time, those of revolutions per minute of the engines, processor clocks in the electronic embedded components, reaction times of drivers, schedules of traffic lights, and so on. Such notions of time have quite different features and therefore require different mathematical models: the microseconds of electronic signals; the hours needed to go from city to city; the precisely determined time necessary to reach a full stop; the uncertain reaction time of average drivers from the instant an obstacle appears on the road to when the brake pedal is pushed. All these "times" belong to the same big picture; competent designers must be able to analyze their dynamics in isolation whenever possible, but also be able to understand their interactions when relevant – for instance, when documenting the behavior of brakes from the user's perspective.

Heterogeneity, however, is not always an issue: a special class of systems consists of collections of homogeneous components that cooperate towards a common goal. This is the case, for example, with multiple identical pistons and cylinders in a car, which together have more power than a single cylinder could have, or of the parallel processors in a multi-core machine. With homogeneous components, coordination and synchronization become the main modeling and design concerns.

In response to the advent and rapid ongoing evolution of heterogeneous reactive systems, the scientific community has developed a rich collection of formalisms, notations, and techniques to deal with the various aspects of timing analysis. The introduction and evolution of the modeling notations has inevitably often been haphazard and demand-driven, corresponding to the evolving needs of applications. As a result, publications describing specific approaches, methods, and tools abound, but there is a lack of comprehensive systematic analyses that investigate general issues and survey the peculiarities of the different contributions.

Filling this void is the main goal of this book, which aims at fostering the critical thinking of readers towards:

- Understanding the subtleties of system dynamics when analyzing problems and investigating possible solutions (we will see that time is often "hidden" in models that do not feature it explicitly);
- Evaluating and comparing models and approaches and selecting the most appropriate ones for the specific needs (we will see that, unlike in other fields of engineering, the "best" formalisms are not always evident; on the contrary, tailoring and integrating existing solutions may be necessary in some new cases).

To achieve these goals, the book develops in two main directions. It presents some fundamental categories useful for comparing and evaluating modeling notations encapsulating time. These categories include issues such as whether time is modeled as a discrete or a dense domain. The book's other, orthogonal, direction is historical, which starts with a review of the traditional time models in science and engineering in general, and in computer science in particular. The presentation continues with more recent models that address specifically the situation of complex systems where computing devices interact with subsystems of other types. In this respect, it is important to emphasize how an interdisciplinary approach is becoming more and more relevant in modern system design: with the exception of very few highly specialized fields, it is essential that software designers understand the application domain and, conversely, domain engineers have a working knowledge of the computing subsystem's behavior and of its interactions. The same interdisciplinary approach may be relevant also for the general public, beyond the technicians and engineers, since, as we emphasized before, human-computer interaction is a primary attribute of many complex systems.

Within this global picture, time plays a fundamental role, on the one hand being the unifying variable that spans the life of the whole universe, on the other hand showing itself in so many different ways and forms to the various actors of the universe's life, from subnuclear particles that exist for a few nanoseconds to stars that "die" billions of years after they "have been born", from the pace of a human heart to the time needed to obtain a university degree.

In correspondence with the above directions, the book is structured into three introductory chapters and two parts, and concluded by a short epilogue. After this introduction, Chap. 2 presents the notions of formalism and model in general terms, and some of their fundamental classification criteria; it also briefly discusses the fundamentals of propositional and predicate logic, which should help make the rest of the book self-contained for a reasonably large readership.

Chapter 3 is a cornerstone of the whole book, as it introduces a taxonomy of essential issues of modeling time in diverse systems. The presentation of the numerous formalisms in the rest of the book recurrently refers to these "dimensions" to compare and contrast different models on a common ground.

Part I contains a concise summary of the models of time that are traditional in engineering and the natural sciences, including traditional computer science. It is meant to provide heterogeneous readers with a homogeneous background.

Part II covers advanced and specialized formalisms specifically developed to support time modeling in heterogeneous software-intensive systems. The aim of Part II is not to offer an exhaustive list of the innumerable contributions available in the literature; this would be a Herculean task, but also probably of little value. On the contrary, the presentation privileges depth over exhaustiveness, and focuses on significant semantic subtleties of a few important formalisms and critical issues, rather than cataloging every minimal variation of the basic approaches. Readers interested in additional details will still find detailed, commented bibliographic references at the end of each chapter. We hope that this presentation style will help readers extend the analysis to other paradigms or approaches not included in the main text. Chapters 7–9 discuss three main and complementary families of formalisms: those based on finite state machines; Petri nets; and those extending mathematical logic. Chapter 10 is about process algebras – widely used to model concurrency, but less prominently so in timing analysis. Chapter 11 presents "dual-language approaches" which combine two notations with different characteristics to model and verify complex systems (model checking frameworks are the most popular applications of dual-language approaches).

Chapter 12 concludes the book with summarizing remarks and hints towards future developments and challenges.

The book's content focuses on the way formalisms can be used to model system behavior and properties and on their expressive power – also in the informal sense of naturalness and ease of use. Analysis and verification techniques and tools for the various formalisms have, in contrast, a more limited coverage, as the book is not meant to focus on verification techniques. Nevertheless, every chapter in Part II includes a section that mentions analysis and verification techniques and tools based on the notations introduced in the chapter.

1.1 Bibliographic Remarks

While it is arguable that *Homo sapiens* began thinking about time shortly after the development of natural language, ancient philosophers were the first whose observations have been recorded and preserved to this day, and have influenced the evolution of science and engineering. The rest of this section gives a very sketchy outline of some of these ideas [8, 10].

The Greek pre-Socratic (and pre-sophistic) naturalist philosophers of the fifth and fourth centuries B.C. suggested informal models of universal time. Some of them, most notably Heraclitus and his followers, predicated a notion of time that is "monotonic" (using modern terminology) in that it never repeats itself; others, most notably those from Parmenides's school, considered time an illusion devoid of physical reality. Among Parmenides's disciples, Zeno of Elea has become famous for his paradoxes on the advancement of time; some critical behaviors in the formal analysis of systems have been named after him (see Sect. 3.6). Many philosophers following the Greek naturalists have adopted, and refined, either Parmenides's or Heraclitus's ideas about time; some thinkers, such as Vico in the seventeenth century or Nietzsche in the nineteenth, have developed an intermediate view where time undergoes real progress but periodically repeats itself.

Kant's gnoseology describes time as an a priori concept, ingrained in the human mind and hence usable as a universal reference in describing the physical world. This view bolstered the development of classical Newtonian mechanics; when Einstein generalized Newton's models with his Theory of Relativity, he also had to perfect its philosophical underpinnings to account for the fact that different observers measure time differently according to their relative motion. Contemporary physics, with its experiments and speculations, keeps on questioning the traditional views of time and introduces new, original explanatory models.

Literature developed around original notions of time is also abundant, as it includes a large part of science-fiction books and movies that entertain the possibility of time-travel; since it is impossible to cite even a fraction of these many books, let us just mention the irresistible description of the problems of grammar related to time travel in Douglas Adams's "The Restaurant at the End of the Universe" [1]. The notion of time in science has also inspired some major literary masterpieces, such as some of Borges's short stories [3], and several of Calvino's novels [4, 5].

The rest of this book offers many specific technical references on time. More general examples of recent papers that discuss some "philosophical" aspects of time with technical rigor include Alur and Henzinger's well-known surveys [2, 7] (the second survey, [7], shares the title with this chapter), Koymans [9], and Schreiber [11]. Finally, the same basic motivations that spawned our survey paper [6] also guided us in developing this book.

References

1. Adams, D.: The Restaurant at the End of the Universe. Pan Macmillan, London (1980)
2. Alur, R., Henzinger, T.A.: Logics and models of real time: a survey. In: Real Time: Theory in Practice. Lecture Notes in Computer Science, vol. 600, pp. 74–106. Springer, Berlin/New york (1992)
3. Borges, J.L.: Collected Fictions. Penguin, New York (1969)
4. Calvino, I.: Cosmicomics. Harcourt Brace, New York (1968). Original Italian title: *Le cosmicomiche*
5. Calvino, I.: t Zero. Harcourt Brace, New York (1969). Original Italian title: *Ti con zero*
6. Furia, C.A., Mandrioli, D., Morzenti, A., Rossi, M.: Modeling time in computing: a taxonomy and a comparative survey. ACM Comput. Surv. **42**(2), 1–59 (2010). Article 6
7. Henzinger, T.A.: It's about time: real-time logics reviewed. In: Sangiorgi, D., de Simone, R. (eds.) Proceedings of the 9th International Conference on Concurrency Theory (CONCUR'98). Lecture Notes in Computer Science, vol. 1466, pp. 439–454. Springer, Berlin/New York (1998)
8. Hetherington, S. (ed.): Epistemology: The Key Thinkers. Continuum, London/New York (2012)
9. Koymans, R.: (Real) time: a philosophical perspective. In: de Bakker, J.W., Huizing, C., de Roever, W.P., Rozenberg, G. (eds.) Proceedings of the REX Workshop: "Real-Time: Theory in Practice". Lecture Notes in Computer Science, vol. 600, pp. 353–370. Springer, Berlin/New York (1992)
10. Russell, B.: A History of Western Philosophy. Simon and Schuster, New York (1967)
11. Schreiber, F.A.: Is time a real time? An overview of time ontology in informatics. In: Halang, W.A., Stoyenko, A.D. (eds.) Real Time Computing. NATO ASI, vol. F 127, pp. 283–307. Springer, Berlin/New York (1994)

Chapter 2
Languages and Interpretations

Everybody is familiar with the notion of language – at least to the extent that they can speak one. The scope of language, however, extends well beyond interhuman verbal exchange, and comprises any form of communication that takes place according to some rules. *Natural languages* originate in the natural world to support communication among people; the spoken languages – such as English, Italian, and Chinese – are obvious examples, but non-verbal natural idioms – such as gestures, pictures, and music – are also common. This book is about a different kind of language: *artificial languages* designed for the description and analysis of phenomena – in particular, their temporal aspects. The choice of artificial languages is also vast and varied, ranging from mathematical notation – algebra, graphs, mathematical logic – to programming languages – such as Java and Haskell – and communication protocols – such as Internet's TCP and HTTP.

Whether natural or artificial, every language is structured as a collection of *sentences*. Each sentence is an arrangement of elementary blocks from the language's *alphabet*. Often, the alphabet is finite and its elements are combined into linear sentences, such as the English alphabet and sentences, but infinite alphabets or more complex arrangements are also possible – for example, the alphabet of all sounds is uncountably infinite, and sentences develop bidimensionally in most visual graphic languages.

2.1 Syntax and Semantics

The definition of a language covers its *syntax* and its *semantics*. The rules of *syntax* define how to build correct sentences from the alphabet; in other words, syntax defines which sentences, from among all possible combinations of alphabet elements, are *well-formed* (acceptable sentences of the language) and which are not. Consider, for example, the English language. The English vocabulary defines

C.A. Furia et al., *Modeling Time in Computing*, Monographs in Theoretical Computer
Science. An EATCS Series, DOI 10.1007/978-3-642-32332-4_2,
© Springer-Verlag Berlin Heidelberg 2012

all valid combinations of letters into words; the English grammar describes the rules
to build sentences out of words, as in

> A *sentence* consists of a *noun phrase* followed by a *verb phrase*. The *noun phrase* is a *noun*
> (possibly preceded by a *determiner* such as an article) or a *pronoun*. The *verb phrase* is a
> *verb* (possibly followed by a *noun phrase* such as an object). The *number* of the noun in the
> noun phrase and of the verb in the verb phrase must agree.

The English grammar and vocabulary collectively describe the syntax of the
language. Another example is the syntax of programming languages, such as Java:

> A *block* is a sequence of *elements* within braces. Each *element* is a *statement*, a *local class
> declaration*, or a *local variable declaration*.

Semantics associates a *meaning* to every syntactically well-formed sentence of
a language; in other words, semantics connects the sentences – which are symbols
– to an *interpretation* – which is the content they express, and which can belong to
any domain, such as those of measurements, decisions, references to facts, and so
on. For example, the natural language sentence

> In case of fire, do not use the elevator.

expresses a suggestion (possibly, an order) about how to behave in the case of a
fire to minimize the risk of personal injury. The Java semantics associates with the
sentence

> **if** $(x > 3)$ { $x = x + 1$; } **else** { $x = x - 1$; }

a behavior that depends on the variable x: if it evaluates to a value greater than
3, increment it; otherwise decrement it.

The semantics of a sentence is *ambiguous* if the sentence may have multiple
different meanings. For example,

> Eats shoots and leaves.

may refer to a panda (whose diet consists of bamboo shoots and leaves) or to
a gunman (who fires his weapon after eating, and then abandons the place),
according to whether we interpret "shoots and leaves" as nouns or verbs. Conversely,
syntactically different sentences may convey the same meaning according to the
semantics. For example, the three sentences

> The gardener sprays water on the roses.
> The gardener sprays the roses with water.
> Water is sprayed on the roses by the gardener.

all essentially convey the same picture even if they combine words according to
different structures. In a formal setting, the two sets of linear equalities

$$\begin{cases} x = 5, \\ y = 6, \end{cases} \qquad \begin{cases} x = 5, \\ y = x + 1, \end{cases}$$

indicate the same pair of values for x and y, but with different syntax.

2.2 Language Features

Given that the notion of language is very broad, very different classification criteria for languages are possible, according to the features of interest within a certain scope. For instance, understandability is very relevant in a teaching context, whereas conciseness is important when space is a concern.

Chapter 3 presents the dimensions that are specific to the central topic of the book, namely time modeling. The current section illustrates more generic features, applicable to languages independently of their application domains, which will also recur in the book's presentation of modeling languages.

2.2.1 Formality

The syntax and semantics of most natural languages are *informal*: even when there are standardized vocabularies and grammars, they lack absolute precision, and their interpretation may be ambiguous or subjective, so that the well-formedness or the meaning of certain sentences depends on the context in which they are used. Imprecision is the price to pay for naturalness: the attribution of meaning to sentences and idioms evolve without regulation, according to social customs and recurring practices; hence syntactic and semantic rules must accommodate unpredictable changes and the specialization of a stable language into dialects, jargons, and slangs.

In contrast, *formal* languages such as mathematics have been defined with an unambiguous syntax and a very precise semantics, so that the fact that a sentence such as

$$\int_0^\pi \cos(x)\mathrm{d}x \ = \ 0$$

is well formed and true is not subject to dispute.

Between the two extremes of natural languages and mathematics, there are several intermediate degrees of formality. Even if mathematics ultimately is fully formal, the presentation of mathematical theories and results often embeds mathematical notation in natural language text, such as in the book you are reading and in every other scientific textbook. In other cases, a language may have a formal syntax but a semantics that omits or overlooks some details. Such notations are often called "semiformal" languages.

Example 2.1 (UML). The Unified Modeling Language (UML) graphical notation is a diagrammatic notation widely used in software engineering. The UML standard defines the syntax of diagrams quite precisely, but their semantics only in natural language. This increases the flexibility of the notation, but it also implies that the meaning of UML diagrams is not always unambiguous. Take, for example, the UML *sequence diagrams* of Fig. 2.1.

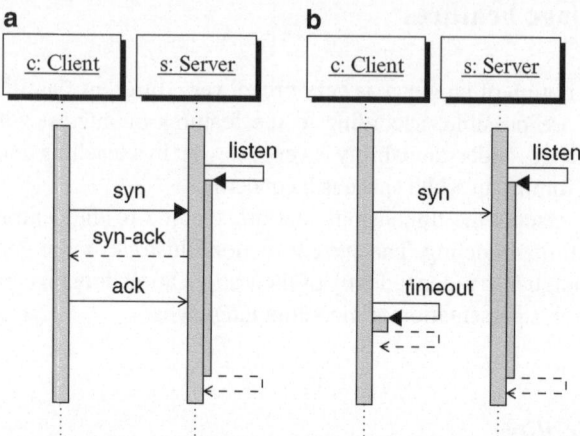

Fig. 2.1 UML sequence diagrams describing the "three-way handshake" of the TCP protocol

The UML diagrams are syntactically correct, and they arguably describe the "three-way handshake" of the TCP communication protocol in the cases (a) in which the communication is successfully established and (b) in which a time-out occurs after the syn message. This meaning is, however, only conveyed informally, and even within the level of abstraction provided by the diagrams several aspects have multiple possible interpretations. For example, the UML standard does not provide a precise meaning for the combination of the diagrams, so what happens if the syn-ack message is received after the time-out is triggered (is the communication successfully established or not?) is not well defined: our own intuition and knowledge of the protocol leads us to conclude that the communication is not successfully established, but this is not prescribed by the semantics of UML. ∎

Programming languages are also often defined only semiformally. While there are standard notations to formalize *syntax* – the Backus-Naur form or variants thereof – their semantics is often described only informally using natural language. This is not only a matter of form, because informality in the definition of programming languages has the same drawbacks as it does in natural languages: the recurring practices take the place of a formal semantics in defining the correct "meaning", with the result that different, possibly incompatible, "dialects" of a language develop according to the compilers or techniques programmers use. Even widely used and standardized languages such as C have had to face these problems. Indeed, a complete and completely formal semantics is available only for a few general-purpose programming languages.

2.2.2 Medium

A language *medium* is the usual means by which sentences of the language are expressed. The preferred media of natural languages are speech and writing – with syllabic alphabets for most Western languages and ideogrammatic notations for several Asian languages – but they often encompass other complementary media such as gestures and facial expressions.

For more formal languages such as those described in this book, the primary medium is textual, over a finite alphabet (typically including Latin and Greek alphanumerical characters). A textual syntax is often supplemented by a graphical one, whose semantics may have different degrees of formality. One of the recurring themes of the book is the analysis of the most-delicate aspects that hamper the definition of a sound formal semantics, or make the semantics counterintuitive.

2.3 Languages for System Modeling

A *system* is a collection of *components* that work together within an *environment* to perform one or more *functions*.[1] To analyze and design systems, it is fundamental to construct system *models*: abstract representations of systems, which include their essential features and support analysis and prediction of their behavior.

A system model has several aspects:

Structure: the components in the system, and how they are connected and communicate (with one another and with the environment).

Behavior: how the components work, and how they interact with one another and with the environment.

Requirements: the function and goals that the system should achieve (relative to the environment).

"Modeling languages" are suitable notations to describe system models. Some modeling languages are sufficiently rich to be applicable to every aspect of the model; others target a specific one.

This book describes many modeling languages, with particular emphasis on models of *dynamic* systems: systems characterized by their behavior over time. We will develop the theme of time modeling thoroughly in the book; Fig. 2.2 suggests a very preliminary and informal view of the activity. Two equally informal examples follow.

---

[1] In fact, the word "system" derives from the Greek words $\sigma\upsilon\nu$ ("together") and $\iota\sigma\tau\eta\mu\iota$ ("to put, to compose").

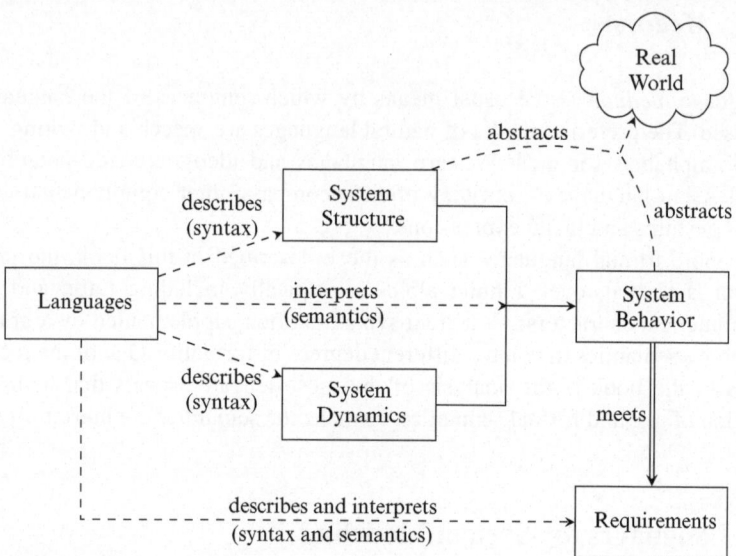

Fig. 2.2 Models, behaviors, languages, and requirements of dynamic systems

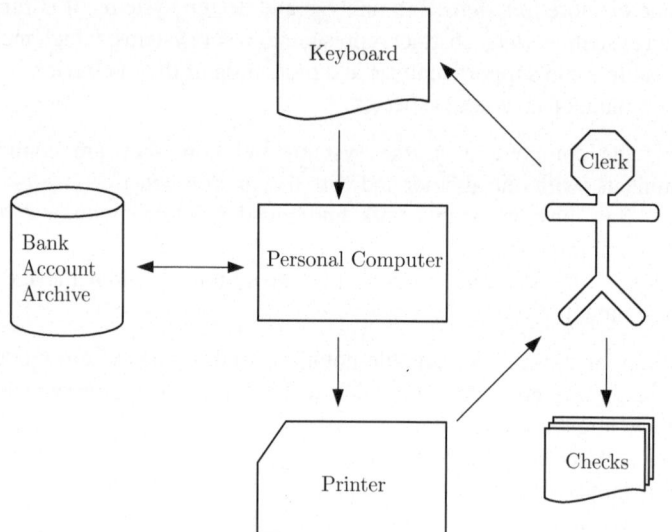

Fig. 2.3 The structural model of a simple banking system

Example 2.2. Figure 2.3 is an informal graphical model of the *structure* of a system where users interact with a personal computer through an input keyboard and an output printer; the computer can read and write a local file system.

The system *behavior* is also described informally, using natural language: the user is inserting the data about a collection of checks into the computer, where a bank management system runs. For every check, she types the bank account numbers of the drawer and the recipient on the keyboard; then she types the amount and commits. Upon commit, the banking system records the information in the archive stored in the file system and prints a receipt slip for auditing.

The system *requirements* specify that the user click commit only if the system signals that the drawer's account has enough money. Notice that the requirements refer to the system as a whole, including the user whose behavior they constrain to achieve a defined goal. ∎

Example 2.3. As an example of a system where timing is central, consider the dynamics of a braking car. A simple mathematical description models the car as a point mass of P kilograms, and its braking interaction with the road as a kinetic friction with friction coefficient of μm/s^2. Then, the system dynamic behavior specifies the car speed $v(t)$ at generic time t, assuming that the speed is v_0 when the braking starts, with the equation

$$v(t) = v_0 - \mu t. \tag{2.1}$$

An example of requirements for such a system is that if V denotes the maximum speed of the car, the breaking always takes no longer than T seconds and D meters. We can determine if the system behaves according to the requirements by analyzing (2.1) with the tools of elementary calculus and mechanics. ∎

2.4 Operational and Descriptive Languages

The principle of "separation of concerns" is an engineering cornerstone, as it enables the analysis of complex systems by separating different dimensions. A special instance of the general principle is the "what vs. how" prescription about how to structure the engineering of a system: first define *what* to achieve, and then detail *how* to achieve it. Figure 2.2 adheres to this prescription by separating the *requirements* (what the system has to achieve) from the structural and behavioral *design* (how the system meets the requirements). At a lower level of abstraction, the duality between *specification* and *implementation* mirrors the one between requirements and design.

Modeling languages often target only one of several concerns. Some languages are explicitly designed, or simply better suited, for describing system behavior as sequences of transitions between configurations; we call these languages "operational". Other notations lend themselves to describing and formalizing system requirements abstractly; we call these languages "descriptive".

Abstract machines and dynamical systems are paradigmatic examples of operational notations, which describe the evolution of the *state* as a reaction to the input stimuli; for this reason, *state-based* is a synonym of operational. Logic is instead a classic instance of descriptive language, which formalizes *properties* and

Fig. 2.4 A two-state machine describing the safe of Example 2.4 operationally

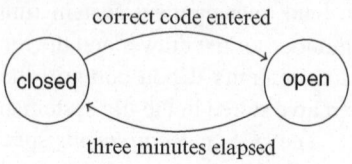

implications. The classification into operational and descriptive is, however, largely a matter of style and conventions, and the same language can often describe both transitions between configurations and system properties, as the following example demonstrates.

Example 2.4 (Natural language model of a safe).

Operational formulation: When the last digit of the correct security code is entered, the safe opens; then, if the safe remains open for three minutes, it automatically closes.

Descriptive formulation: The safe is open if and only if the correct security code has been entered no more than three minutes ago.   ∎

The remainder of the current chapter gives a few sketchy examples of operational and descriptive notations; the rest of the book gives a much more extensive and systematic presentation of several formalisms in both categories.

2.4.1 Operational Formalisms

Example 2.3 is distinctly operational, because Eq. (2.1) describes the evolution of the state (the speed v) over time from an initial value (v_0). Correspondingly, the derived behavior is also operational; for example, the displacement $x(t)$ of the car at time t is computed through elementary kinematics as

$$x(t) = \int_0^t (v_0 - \mu t)\, dt = v_0 t - \frac{\mu}{2}t^2, \tag{2.2}$$

which is valid only until the car reaches a full stop at time v_0/μ.

The possible states (that is, the values of x and v over time) in the car example range over a bounded interval of the real numbers. In contrast, the operational description of the safe in Example 2.4 suggests only two distinct states, corresponding to the safe being *open* or *closed*; the actions of *opening* the safe and *closing* it correspond to transitions between the two states. Figure 2.4 gives a graphical representation of the safe's operational model using a widely used graphical convention – which the rest of the book will use and improve on several occasions.

The operational model of Fig. 2.4 is very abstract, as the two transitions summarize actions that consist of sequences of simpler events. For example, entering the

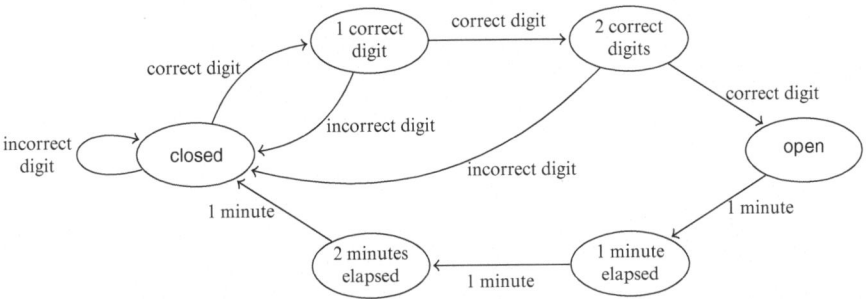

Fig. 2.5 A state machine refining the model of Fig. 2.4

three-digit code does not happen instantaneously, but one digit at a time; if one of the digits is incorrect, a new typing of the code must start over. Correspondingly, we can introduce two intermediate states to "count" the number of correct digits entered. Similarly, we can refine the elapsing of three minutes of time into a sequence of intermediate states, one per minute. Figure 2.5 represents this more detailed model with the same graphical notation of Fig. 2.4.

Exercise 2.5. Extend the safe operational model of Fig. 2.5 to accommodate:

- A command "stay open", which resets the time-out counter when the safe is open.
- A command "leave open", which stops the time-out counter until another command "close now" is issued.
- A security mechanism that, whenever an incorrect digit is entered, makes it impossible to enter a new code for the next two minutes.
- A further security mechanism that guarantees that the user is notified of the incorrectness of the code only after all digits have been entered.   ■

General-purpose imperative programming languages – C, Pascal, Java, Eiffel, etc. – essentially are operational notations, as their programs consist of *instructions* that change the state when executed. Indeed, a common approach to defining the formal semantics of programming languages is *operational*: the effect of each instruction is defined in terms of how it affects the state (memory) of an abstract machine (e.g., a formalization of the Java Virtual Machine for Java). In contrast, logic programming languages – such as Prolog and Curry – indirectly describe computations through the defining properties of their output; hence it is natural to formalize their semantics with some form of logic.

2.4.2 Descriptive Formalisms: Mathematical Logic

Mathematical logic originates from the efforts of ancient philosophers (Aristotle, in particular) to model the rules of reasoning rigorously. Logic is a very versatile language, suitable for describing facts and properties, and hence a fundamental descriptive notation. This section presents the basics of mathematical logic and gives an idea of the kind of descriptive models that logic can express.

2.4.2.1 Propositional Logic

Propositional logic (also called "propositional calculus") is the simplest variant of mathematical logic, at the core of every more expressive logic language. Propositional logic sentences are called "formulae". Formulae are built out of a countable infinite alphabet of "propositional letters" (or simply "propositions"), which are just character identifiers such as

$$A, B, \ldots, X, \ldots, \text{open, closed}, \ldots,$$

$$\text{1\_minute\_elapsed}, \ldots, \text{Train\_1\_faster\_than\_Train\_2}, \ldots .$$

Propositional letters are combined with "logical connectives" (also called logical "operators"), $\neg$ ("not"), $\wedge$ ("and"), $\vee$ ("or"), $\Rightarrow$ ("implication"), and $\Longleftrightarrow$ ("double implication", "co-implication", or "equivalence"), according to the following rules:

(i) Every propositional letter L is a well-formed formula;
(ii) If F, G are well-formed formulae, then the following are well-formed formulae:

    (a) $\neg F$ ("not F"),
    (b) $F \wedge G$ ("F and G"),
    (c) $F \vee G$ ("F or G"),
    (d) $F \Rightarrow G$ ("F implies G"),
    (e) $F \Longleftrightarrow G$ ("F if and only if G").

When multiple connectives are present in the same formula, the binding power decreases, in the same order as above, from $\neg$ down to $\Longleftrightarrow$; as in mathematics, parentheses are used to enforce a different order of application of connectives. Thus, for example,

$$A \wedge B \Rightarrow C \Longleftrightarrow \neg D$$

is the same as

$$((A \wedge B) \Rightarrow C) \Longleftrightarrow (\neg D).$$

Given a well-formed formula in propositional logic, an *interpretation* is an assignment of a value in the domain {**True**, **False**} to every predicate letter appearing in the formula. The assigned values are *truth values*, because they declare which propositions hold and which do not in the interpretation. The truth value of a formula follows from the interpretation of its propositions according to the following rules (which mirror the natural language meaning of logic connectives):

$\neg F$	is **True**	if and only if	F is **False**;
$F \wedge G$	is **True**	if and only if	both F and G are **True**;
$F \vee F$	is **True**	if and only if	F or G, or both, are **True**;
$F \Rightarrow G$	is **True**	if and only if	F is **False**, or else both F and G are **True**;
$F \Longleftrightarrow G$	is **True**	if and only if	F and G are both **True**, or both **False**.

Exercise 2.6. The rules show that some of the connectives are redundant, in that they are subsumed by the others. More precisely, two formulae are equivalent if, for every interpretation of propositions, they are always both **True** or both **False**. Show that:

- Every double implication $A \Longleftrightarrow B$ is equivalent to $(A \Rightarrow B) \wedge (B \Rightarrow A)$.
- Every implication $A \Rightarrow B$ is equivalent to $\neg A \vee B$.
- De Morgan's laws: Every disjunction $A \vee B$ is equivalent to $\neg(\neg A \wedge \neg B)$, and every conjunction $A \wedge B$ is equivalent to $\neg(\neg A \vee \neg B)$. ∎

A propositional logic formula is "valid" when it evaluates to **True** for every interpretation of the propositions; it is "satisfiable" when it evaluates to **True** for some interpretation (at least one). A valid formula is **True** entirely on the basis of its propositional structure, independently of any interpretation; for example, $A \vee \neg A$ is valid because every propositional letter evaluates to **True** or **False**.

Example 2.7 (Propositional logic model of the safe). The simple descriptive model of the safe (Example 2.4) translates into a propositional formula over the propositions open, closed, and correct_code_entered_within_three_minutes:

$$(\text{open} \Rightarrow \neg \text{closed}) \quad \wedge$$
$$(\text{open} \Rightarrow \text{correct\_code\_entered\_within\_three\_minutes}) \tag{2.3}$$

Consider an extension of the safe model, where a command "stay open" forces the safe to stay open indefinitely. The propositional logic model would accommodate this extended behavior with new propositions to represent the command and whether the safe was opened in the past:

$$\text{open} \Longleftrightarrow \left(\begin{array}{c} \text{correct\_code\_entered\_within\_three\_minutes} \\ \vee \left(\begin{array}{c} \text{safe\_opened\_in\_the\_past} \\ \wedge \\ \text{stay\_open\_issued\_since\_last\_opening} \end{array} \right) \end{array} \right). \tag{2.4}$$

∎

The semantics of propositional logic defines the truth value of a composite formula from the truth value of its simpler components, down to the propositional letters. "Deduction systems" leverage this feature to derive the truth value of every formula from a set of simpler ones, in a calculational fashion. The basic formulae are called "axioms" and define some a priori knowledge about the system modeled: every axiom is a formula assumed to be true. The deduction system's *inference rules* describe how to derive new true formulae from the axioms. The inference rules are typically independent of the modeled system, which only affects the axioms, and are universal for the chosen logic. True formulae derived by applying some inference rules are called "theorems" in mathematical logic jargon.

Deduction systems can also provide an alternative definition of the semantics of propositional logic, where we can "calculate" properties of systems specified in propositional logic by applying inference rules to a set of standard axioms in addition to those specific to the system description. It is sufficient to use the universal inference rule of "Modus Ponens",

If a formula F holds and the implication $F \Rightarrow G$ holds, then the formula G holds,

and consider for any formulae F, G, and H the axioms

(AX1) $F \Rightarrow (G \Rightarrow F)$;
(AX2) $(F \Rightarrow (G \Rightarrow H)) \Rightarrow ((F \Rightarrow G) \Rightarrow (F \Rightarrow H))$;
(AX3) $(\neg F \Rightarrow \neg G) \Rightarrow (G \Rightarrow F)$

in addition to the system description axioms. In the case of the safe of Example 2.7, the formula

$$\neg \, \mathsf{correct\_code\_entered\_within\_three\_minutes} \Rightarrow \mathsf{closed} \qquad (2.5)$$

specifying that the safe is closed if the correct code has not been entered in the last three minutes is a system property. We derive (2.5) from the axioms of propositional logic and the system-specific axiom (2.3), with repeated applications of Modus Ponens (we abbreviate open, closed, and correct_code_entered_within_three_minutes with O, C, and E).

(D1)	$O \Rightarrow E$	special case of (2.3)
(D2)	$\neg(\neg O) \Rightarrow \neg(\neg E)$	equivalent form[2] of (D1)
(D3)	$\neg O \Rightarrow C$	special case of (2.3)
(D4)	$(\neg(\neg O) \Rightarrow \neg(\neg E)) \Rightarrow (\neg E \Rightarrow \neg O)$	instance of (AX3)
(D5)	$\neg E \Rightarrow \neg O$	Modus Ponens with (D2), (D4)
(D6)	$(\neg E \Rightarrow (\neg O \Rightarrow C))$	
	$\Rightarrow ((\neg E \Rightarrow \neg O) \Rightarrow (\neg E \Rightarrow C))$	instance of (AX2)
(D7)	$(\neg O \Rightarrow C) \Rightarrow (\neg E \Rightarrow (\neg O \Rightarrow C))$	instance of (AX1)
(D8)	$\neg E \Rightarrow (\neg O \Rightarrow C)$	Modus Ponens with (D3), (D7)
(D9)	$(\neg E \Rightarrow \neg O) \Rightarrow (\neg E \Rightarrow C)$	Modus Ponens with (D8), (D6)
(D10)	$\neg E \Rightarrow C$	Modus Ponens with (D5), (D9)
		QED

---

[2]The equivalence is formally derivable from the same axioms.

The simplicity of propositional logic stems from the usage of atomic items (the propositional letters) to indicate arbitrarily complex facts and actions through their truth values. This very abstract view, however, restricts the expressiveness of the logic, and makes it often too limited to model and reason about entities with complex structures and behavior. Even in the simple Example 2.7, we had to coalesce different types of information into the simple propositional letters: for example, the proposition correct_code_entered_within_three_minutes conflates information about the entering of the code with the fact that the code was correct, and with timing information about when it last happened. This is not very flexible and generalizes poorly. What happens, for example, if we want to formalize the fact that the safe is open at a generic time? If we associate instants of time with the natural numbers $0, 1, 2, \ldots$, we can introduce a denumerable sequence of propositions open_0, open_1, open_2, ... Nonetheless, it is impossible to write formulae that mention all such propositions, and hence explicitly define the state of the safe at every instant. Things are even worse if we assume a continuum of time instants, for example, corresponding to the nonnegative real numbers; in this case, the required number of propositions is not available even in principle.

These observations call for an extension of propositional logic that supports atomic elements more complex than propositions. In particular, it should support *parametric* elements, and the ability to express properties about elements evaluated for infinitely many different values of the parameters. Such parametric elements are called "predicates"; predicate logic extends propositional logics to accommodate them. The rest of this chapter describes the basics of predicate logic, which underpins many other formal languages discussed in the rest of the book.

2.4.2.2   Predicate Logic

Predicate logic (also called "predicate calculus", or "first-order logic") extends propositional logic with variables, functions, Boolean predicates, and quantifiers. More precisely, the alphabet of predicate logic includes:

- Symbols for *constants $a, b, c, \ldots$*;
- Symbols for *variables $t, u, v, w, x, y, z, \ldots$*;
- Symbols for *functions $f, g, h, \ldots$*;
- Symbols for *predicates $P, Q, R, \ldots$*;
- The logic quantifiers $\forall$ ("for all", universal quantifier) and $\exists$ ("there exists", existential quantifier);
- The same logic connectives as propositional logic.

For notational convenience, the constant, function, and predicate symbols can include standard mathematical symbols such as $+$, $-$, sin, $<$, $=$, $>$, and π, with their usual syntax (e.g., infix binary operators) whenever useful.

Predicate logic builds well-formed formulae incrementally from the alphabet as follows. First, constants, variables, and functions are combined into *terms*:

(i) A constant c and a variable x are terms;

(ii) If f is an n-ary function and $t_1, t_2, \ldots, t_n$ are n terms, then the function application $f(t_1, t_2, \ldots, t_n)$ is a term.

Second, terms and predicates are combined into *atomic formulae*:

(iii) If P is an n-ary predicate and $t_1, t_2, \ldots, t_n$ are n terms, then $P(t_1, t_2, \ldots, t_n)$ is an atomic formula.

Finally, atomic formulae are combined with quantifiers and logic connectives into well-formed formulae of arbitrary complexity:

(iv) Every atomic formula is a well-formed formula;

(v) If F is a well-formed formula and x is a variable, then both $\forall x(F)$ and $\exists x(F)$ are well-formed formulae;

(vi) If F is a well-formed formula, then $\neg F$ is a well-formed formula;

(vii) If F and G are well-formed formulae and $\star$ is any binary connective of propositional logic ($\wedge, \vee, \Rightarrow, \Longleftrightarrow$), then $F \star G$ is a well-formed formula.

Example 2.8. Let us show a few examples of well-formed formulae of predicate calculus.

- Every well-formed propositional formula is also a well-formed predicate formula, because propositional letters correspond to argumentless predicates, and hence to atomic formulae.
- Every mathematical equality or inequality is a well-formed formula, because equality and other relational operators are binary predicates (usually written in the infix form). For example:

  - $x = x + 1$,
  - $\sin^2(x) + \cos^2(x) = 1$,
  - Bernoulli's inequality: $(1 + x)^r \geq 1 + rx$.

- More generally, mathematical statements are expressible in the language of predicate logic. For example:

  No natural number equals its successor: $\forall n(n \neq n + 1)$,
  The sine of π is 4 or 2: $\sin(\pi) = 4 \vee \sin(\pi) = 2$,
  Bernoulli's inequality holds for every nonnegative integer r and   nonnegative real x: $\forall r(\forall x(r \in \mathbb{Z} \wedge r \geq 0 \wedge x \in \mathbb{R} \wedge r \geq 0 \Rightarrow (1 + x)^r \geq 1 + rx))$. ∎

While reading the examples above, you have probably tried to figure out which formulae express true facts and which are false. Intuition based on mathematical knowledge is sufficient for these simple examples, but being able to do it systematically for every formula requires the precise *semantics* of predicate logic. Given a well-formed first-order formula, an "interpretation" associates a value of a suitable domain with every variable and constant symbol, and a concrete function and relation with every function and predicate symbol. In practice, mathematical symbols and relations will be interpreted to convey the usual meaning adopted in mathematics. From the interpretation of the basic symbols, we know the value of every term, and the truth value of every atomic formula. Finally, the truth value of a generic formula follows compositionally from the same rules of the propositional calculus semantics. The only new operators are the quantifiers, whose treatment requires a few more details:

- In every quantified formula $\forall x(F)$ or $\exists x(F)$, F is the quantifier's "scope".
- Every occurrence of a variable x within the scope of a quantifier $\forall x$ or $\exists x$ is called "bound"; a variable occurrence that is not bound is "free".[3]
- A universally quantified formula $\forall x(F)$ evaluates to **True** if F evaluates to true for *every* possible interpretation of x.
- An existentially quantified formula $\exists x(F)$ evaluates to **True** if F evaluates to true for *some* possible interpretation of x.

In general, the truth value of a formula depends on the interpretation given to constants, variables, functions, and predicate symbols. With the same definition as in propositional logic, a well-formed predicate formula is "valid" if it evaluates to **True** for *every* possible interpretation of constants, variables, predicates, and functions; it is "satisfiable" if it evaluates to **True** for *some* interpretation.

Example 2.9. Let us consider the truth value of some of the formulae in Example 2.8.

- Under the standard interpretation of mathematical functions and predicates over numerical domains, logic truth coincides with mathematical truth (of course!),

  - $x = x + 1$ evaluates to **False**, regardless of the value assigned to x by the interpretation;
  - Correspondingly, $\forall n(n \neq n+1)$ evaluates to **True** under every interpretation of n.

- However, under nonstandard interpretations of constants, functions, and predicates, the semantics may become counterintuitive:

  - $x = x + 1$ evaluates to **True** under the interpretation where $+$ is not an ordinary sum, but is the function that returns its first argument: $x + 1$ evaluates to x, and hence the equality holds;
  - $\sin(\pi) = 4 \ \vee \ \sin(\pi) = 2$ is **False** with the standard interpretation for 2, 4, π, and sin, but it is **True** in other interpretations, such as the one when sin denotes a constant function with value 4.

- A well-formed formula F is "closed" if no variable appearing in F is free. The truth value of a closed formula is independent of the interpretation of its variables (but it may still depend on the domain of variables and on the interpretation of constants, functions, and predicates). For example, $\forall x(\exists y(y = x + 2))$ is **True**

---

[3] For simplicity, assume that different quantifier instances bind variables with different names. This is without loss of generality, and it helps keep the presentation plain.

over the domain of natural numbers for the standard interpretation of equality and sum, regardless of the values chosen for x, y.   ∎

If functions and predicates can have arbitrary interpretations, the semantics of predicate formulae may seem detached from the practice of mathematics, and of limited practical utility. In fact, predicate calculus is a very powerful modeling language, mainly when used in combination with some "first-order theories" that enforce the intended semantics of functions and predicates. A first-order theory supplements predicate calculus with a set of *axioms*. As in the deduction systems mentioned in the context of propositional logic, axioms are well-formed formulae that are assumed to be **True**; in other words, in a first-order theory we only consider interpretations that satisfy the axioms. Validity and satisfiability are redefined for a theory T accordingly: a formula is "T-valid" if it evaluates to **True** in every interpretation where T's axioms also evaluate to **True**; it is "T-satisfiable" if it evaluates to **True** in some interpretation where T's axioms evaluate to **True**. The details of how to describe the characteristic properties of arithmetic and other mathematics are quite involved and outside the book's scope. The rest of the book will simply assume they are available whenever needed to reconcile intuition with strictly formal reasoning.

Example 2.10 (Predicate logic model of the safe). With predicate calculus, we can make the descriptive model of the safe more detailed than in Example 2.7, which used only propositional logic. To this end, we introduce the two variables t, u to denote time; for simplicity, we can assume that they vary over the integers, but other options could be accommodated along the same lines. The propositions for the safe being open or closed and correct code being entered – used in the propositional model of Example 2.7 – become predicates with t or u as a parameter, so that their truth is time-dependent. A predicate formula that specifies the behavior of the safe is the following:

$$\forall t \left(\begin{array}{l} (\text{open}(t) \iff \neg\text{closed}(t)) \\ \wedge \\ (\text{open}(t) \iff \exists u \, ((t - 3 \leq u < t) \wedge \text{correct\_code\_entered}(u))) \end{array} \right).$$

(2.6)

∎

Exercise 2.11. Provide a predicate logic extension of the propositional formula (2.4), along the lines of Example 2.10.   ∎

The various examples have demonstrated the versatility of logic as a descriptive formalism: the formulae describe systems through their characterizing *properties* of interest, rather than as explicit sequences of transitions and reached states. Anyway, as remarked at the beginning of the current section, the distinction between operational and descriptive is largely a matter of style. The following example provides evidence of this fact by sketching an operational model of the safe formalized with predicate calculus. Similarly, we will discuss the converse correspondence between state-based formalisms as models of logic formulae in the context of temporal logic (Chap. 9) and dual-language approaches (Chap. 11).

Example 2.12 (An operational model with logic). Using the predicates of Example 2.10, the following formula translates the state-based operational model of Fig. 2.4; the first line is the usual mutual exclusion between states (implicit in the model of Fig. 2.4); the second line describes the transition from closed to open and the permanence of open for three minutes; the last line specifies that the state closed does not change unless a correct code is entered (also implicit in the model of Fig. 2.4):

$$\forall t \left(\begin{array}{l} (\text{open}(t) \iff \neg\text{closed}(t)) \\ \land\ (\text{correct\_code\_entered}(t) \Rightarrow \forall u((t+1 \le u \le t+3) \Rightarrow \text{open}(u))) \\ \land\ \left(\forall u \left(\begin{array}{l} (t \le u \le t+2) \\ \Rightarrow (\text{open}(u) \land \neg\text{correct\_code\_entered}(u)) \end{array} \right) \Rightarrow \text{closed}(t+3) \right) \\ \land\ (\text{closed}(t) \land \neg\text{correct\_code\_entered}(t) \Rightarrow \text{closed}(t+1)) \end{array} \right) . \tag{2.7}$$

The rest of the book will provide plenty of examples of operational and descriptive formalisms; their strong connections with the basic "universal" languages introduced in the present chapter will be apparent. ∎

2.5   Bibliographic Remarks

Pinker wrote several fascinating books about the origins, role, and evolution of natural languages [14–16].

Version 2 of the Unified Modeling Language (UML) is a standard of the Object Management Group [4, 8, 21]. Stevens discusses TCP and other Internet protocols in depth [20]. Every compiler construction book discusses the Backus-Naur form or extensions thereof [1,2,6,22]; Knuth traces back the origins of the notation [10]. The formalization of the semantics of programming languages now has a rich history and many comprehensive texts [12, 13, 17, 18]. These books typically abstract away some of the low-level details of real programming languages, but others have tried to formalize a specific language in its entirety, as Stärk et al. [19] did for Java.

For more specific references on operational and logic formalisms, see the bibliographic remarks at the end of other chapters – in particular, Chaps. 4, 5, 7, 8, and 11 for state-based notations and Chaps. 9 and 11 for logic-based formalisms. Mendelson [11], Enderton [7], and Kleene [9] are classic general introductions to mathematical logic; Ben-Ari [3] and Bradley and Manna [5] present the same basic topics from a computer science perspective.

References

1. Aho, A.V., Lam, M.S., Sethi, R., Ullman, J.D.: Compilers: Principles, Techniques, and Tools, 2nd edn. Addison-Wesley, Boston/London (2006)
2. Appel, A.W.: Modern Compiler Implementation in ML. Cambridge University Press, Cambridge/New York (2004)
3. Ben-Ari, M.: Mathematical Logic for Computer Science, 2nd edn. Springer, London/New York (2003)
4. Booch, G., Rumbaugh, J., Jacobson, I.: Unified Modeling Language User Guide, 2nd edn. Addison-Wesley, Upper Saddle River/Boston (2005)
5. Bradley, A.R., Manna, Z.: The Calculus of Computation: Decision Procedures with Applications to Verification. Springer, Berlin (2007)
6. Cooper, K., Torczon, L.: Engineering a Compiler, 2nd edn. Morgan Kaufmann, Burlington (2011)
7. Enderton, H.B.: A Mathematical Introduction to Logic, 2nd edn. Academic, San Diego (2001)
8. Fowler, M.: UML Distilled: A Brief Guide to the Standard Object Modeling Language, 3rd edn. Addison-Wesley, Boston (2003)
9. Kleene, S.C.: Mathematical Logic. Dover, Mineola (2002)
10. Knuth, D.E.: Backus normal form vs. Backus Naur form. Commun. ACM 7, 735–736 (1964)
11. Mendelson, E.: Introduction to Mathematical Logic, 5th edn. Chapman and Hall, London (2009)
12. Meyer, B.: Introduction to the Theory of Programming Languages. Prentice Hall, New York (1990)
13. Mitchell, J.C.: Concepts in Programming Languages. Cambridge University Press, New York (2002)
14. Pinker, S.: The Language Instinct: How the Mind Creates Language. William Morrow, New York (1994)
15. Pinker, S.: Words and Rules: The Ingredients of Language. Basic Books, New York (1999)
16. Pinker, S.: The Stuff of Thought: Language as a Window into Human Nature. Viking Adult, New York (2007)
17. Reynolds, J.C.: Theories of Programming Languages. Cambridge University Press, Cambridge/New York (1998)
18. Riis Nielson, H., Nielson, F.: Semantics with Applications: An Appetizer. Springer, New York/London (2007)
19. Stärk, R.F., Schmid, J., Börger, E.: Java and the Java Virtual Machine: Definition, Verification, Validation. Springer, Berlin/New York (2001)
20. Stevens, W.R.: TCP/IP Illustrated, vol. 1–3. Addison-Wesley, Reading (1994)
21. UML 2.0. http://www.omg.org/spec/UML/2.0/ (2005)
22. Wirth, N.: Compiler Construction. Addison-Wesley, Harlow/Reading (1996)

Chapter 3
Dimensions of the Time Modeling Problem

Modeling is all about abstraction: select which aspects should be included in the model, the details of their descriptions, and the form the descriptions should take. Models of *time*, in particular, must consider several distinctive issues that have to do with the nature of time and how it is represented. This chapter presents these *dimensions* of the time modeling problem within the general framework of the book.

Some of the dimensions denote issues that are pervasive in the modeling of time in the literature, for example, the use of discrete or continuous time domains. Others shed light on aspects specific to some classes of formalisms, for example, the presence of an explicit or implicit reference to time.

The dimensions will guide the presentation and comparison in the following chapters of how the various notations model time; they will focus the presentation on the most significant instances and equip readers with references and skills useful for analyzing any other formalism that includes some notion of time, beyond those detailed in this book. Correspondingly, the dimensions can guide a knowledgeable choice – and possibly a tailoring – of the notation most appropriate to specific modeling needs (as will be briefly discussed in the epilogue).

The "dimensions" of this chapter, however, informally refer to aspects that are neither necessarily exhaustive nor independent. Unlike the dimensions in an orthonormal mathematical basis, some dimensions of time modeling depend on each other, with the result that only certain combinations are sometimes possible, reasonable, or relevant in practice. The following chapters will illustrate the dependencies among different dimensions in several concrete examples.

3.1 Discrete Versus Dense Time Domains

A first natural categorization of the formalisms dealing with time-dependent systems is between the use of discrete and dense sets as domains for the "time variable".

C.A. Furia et al., *Modeling Time in Computing*, Monographs in Theoretical Computer Science. An EATCS Series, DOI 10.1007/978-3-642-32332-4_3,
© Springer-Verlag Berlin Heidelberg 2012

Recall that a discrete set consists of isolated points whereas a dense set, ordered by "$<$", is such that for every two points t_1, t_2, with $t_1 < t_2$, there is always another point t_3 in between: $t_1 < t_3 < t_2$. In the scientific literature and applications, the most widely adopted discrete time models are natural and integer numbers – denoted by $\mathbb{N}$ and $\mathbb{Z}$, respectively – whereas the typical dense models are rational and real numbers – denoted by $\mathbb{Q}$ and $\mathbb{R}$, respectively. For instance, differential equations normally assume the real (or even the complex) numbers as variable domains, whereas difference equations are defined over integers. Computing devices are formalized through discrete models when their behavior is paced by a clock, so that it is natural to measure time by counting clock ticks.

In addition to the well-known classification into discrete and dense domains, a few more accurate distinctions are useful for better evaluating and comparing the various formalisms available in the literature and those that will be proposed in the future.

3.1.1   Continuous Versus Non-continuous Time Models

Dense domains include both continuous and non-continuous sets. For some models, the distinction is relevant and must be considered.

The notion of continuous domain originated from the observation that there exist *incommensurable* physical quantities: two values v_1, v_2 are incommensurable if there exist no integers n, m such that $n \cdot v_1 = m \cdot v_2$; hence the ratio v_1/v_2 is not a rational number in the dense non-continuous set $\mathbb{Q}$. For example, the diameter and the circumference of every circle are incommensurable, and the *irrational* number π denotes their constant ratio. Other irrational numbers are introduced to denote the results of operations naturally applicable to every rational number whose results are not rational, such as the square root of 2. The extension of a dense non-continuous set such as $\mathbb{Q}$ with all irrational numbers gives a *continuous* domain; the real and complex numbers are the most widely known and used continuous domains.

The problem with incommensurable quantities is relevant also when measuring time: the periods of two clocks that are not perfectly synchronous are likely incommensurable. We do not have to look for contrived examples of this phenomenon: the solar day and year are indeed incommensurable time spans. Adopting a continuous set as the time domain makes it possible to model incommensurable times precisely, thus making the analysis more general and uniform; for example, showing that a model has behaviors with certain characteristics may be simpler under the assumption of a continuous time domain.

On the other hand, the greater generality of continuous domains becomes an obstacle when performing numerical and algorithmic analyses of the models, because irrational numbers (which are the overwhelming majority in a continuous set) have no finite representation as series of digits; hence a digital computer can only rely on *approximations* of their exact values in terms of rational numbers. The finite precision of the approximations must allow for the computation of solutions with an error that is acceptable for the application domain. For example, the incommensurability of day and year requires an approximation to construct calendars. The simple convention of approximating 1 year to 365 days introduces a considerable drift, which accumulates and becomes unacceptable after only a few years; the Julian calendar introduced a more precise approximation using a leap year every 4 years; the Gregorian calendar further refined the approximation (years that are exactly divisible by 100 are not leap years unless they are exactly divisible by 400) but still introduces an error of 1 day every few thousand; in general, every approximation introduces a drift between calendar and astronomical day after a sufficiently long period of time.

Another, more specific, context in which the distinction between continuous and merely dense time domains is relevant is the algorithmic analysis of timed models: some sophisticated time analysis algorithms work correctly only under the restriction that certain time parameters of the model are rational. We will mention examples of such algorithms when discussing timed automata in Chap. 7 and Petri nets in Chap. 8.

3.1.2 Bounded, Finite, and Periodic Time Models

System modeling often assumes behaviors that may proceed indefinitely in the future (and maybe in the past), so it is natural to model time as an unbounded set. This typically complicates the analysis of system properties, which may become undecidable[1] in the general case because no observation over a finite amount of time can be conclusive about the longer-term behavior (see Sect. 3.8.2 for more comments on the aspect of decidability).

There are significant cases, however, where all relevant system behavior can be a priori enclosed within a *bounded* "time window". For instance, braking a car to a full stop requires at most a few seconds; thus, if we want to model and analyze the behavior of an antilock braking system, there is no loss of generality if we assume as a temporal domain, say, the real range [0, 60] seconds. In many cases, a restriction to bounded time highly simplifies algorithmic analysis and simulation.

When a domain is not only bounded but also discrete, it becomes *finite*. For systems where the time domain and every other domain are finite, all system properties are, in principle, decidable, because behavioral analysis reduces to the enumeration of a finite number of system configurations. If a domain is not discrete but only bounded, its *discretization* – that is, the discrete approximation of its values – may support an exhaustive analysis of system behavior that is precise

---

[1] A property is decidable if there exists an algorithmic procedure that can determine, in finite time, whether the property holds in any given system model; otherwise, it is undecidable. Chapter 6 introduces the notion with more precision for readers unfamiliar with the theory of computability.

enough to replace the exact analysis on dense domains. Section 3.1.3 discusses the widely used *sampling* technique to achieve discretization.

A special case of unbounded behavior occurs when a system is *periodic*, that is, recurrently returns to certain states during its evolution. The time between two consecutive visits to a repeated state is called *period*. Since the evolution over an unbounded time domain consists entirely of infinite repetitions of the period, the analysis of a periodic system's behavior reduces to the analysis over bounded time: the properties holding over the whole time domain follow from the behavior over a single finite period.

The well-known problem of studying the termination of computer programs illustrates how periodicity can simplify timing analysis. Determining whether a generic program halts for a given input boils down to timing analysis: "determine if there exists a time t such that the program, run with the given input, stops after t time units". Termination is undecidable in the general case: the best we can do is run the program with the input, but then we can never conclude that it will not halt in the future if it has not halted after a finite (arbitrarily large) amount of time. If, however, we observe that the system behavior is periodic, then there exists a period Δ such that the state of the computation – which comprises the memory and the input – is the same at all times $t_1, t_1 + \Delta, t_1 + 2\Delta, t_1 + 3\Delta$, and so on indefinitely. If termination only depends on the state, nontermination over a single period entails nontermination everywhere, because the behavior over a period characterizes the overall behavior.

Section 3.8.3 mentions several analysis techniques that rely on periodicity and finiteness of behaviors and domains to achieve automation by means of exhaustive enumeration. Bounded model checking, presented in Chap. 11, is a prime example of such techniques with significant practical impact.

3.1.3 Hybrid Systems

The discussion about dense vs. discrete domains of the present section focuses on time, but system models must select discrete or dense domains also for other state components and variables. Chapter 2, for example, presented some models of physical systems where all domains – e.g., time, speed, and position – are continuous. Computing systems, in contrast, are usually modeled with discrete time (paced by the clock) and state (sequences of digital bits).

Combining discrete time with discrete state variables and dense time with dense state domains is a common choice, but alternatives exist: *hybrid* system models combine discrete and dense domains. All combinations are possible: discrete time and dense state space, dense time and discrete state space, and even cases where the time model integrates discrete and dense components of time, or discrete and dense state domains coexist. Indeed, there are several circumstances in which hybrid models are the natural choice; they are mainly, but not exclusively, related to the problem of integrating heterogeneous components.

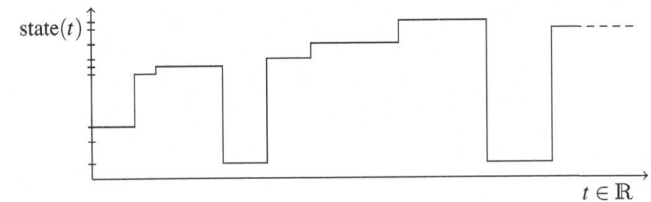

Fig. 3.1 A square-wave form over dense time

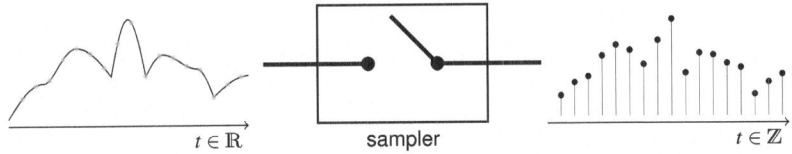

Fig. 3.2 A continuous behavior sampled

A typical example is a system consisting of a power plant controlled by some digital computing device. Differential equations on continuous time are a natural model for the physical process (for example, a chemical reaction or the production of electric power). The digital computer has a discrete, possibly finite, state space over a discrete time domain. The two components interact through devices such as samplers and holders.

More generally, dense and discrete domains coexist in hybrid models in different ways:

- Components with a discrete – possibly finite – set of states evolve over a dense time domain. In such cases, behaviors are graphically described as *square wave* forms and evolve as piecewise constant functions of time, as shown in Fig. 3.1.
- Sampling the state at regular (discrete) intervals provides an approximation of the behavior of state variables over a continuous time domain, as pictured in Fig. 3.2. A classic problem of control and information theory is how to guarantee that the approximation introduced by sampling does not lose any relevant information about the continuous-time behavior. Some sections of Chaps. 7 and 9 discuss the role of sampling techniques for operational and descriptive temporal models.
- The *time domain* is hybrid when it consists of a discrete sequence of "macro-steps", but between each pair of discrete steps there exist finer-grain dynamics modeled over dense time. This setting accommodates, for example, the abstraction of electronic components into logic gates – discussed in Chap. 5 – as well as discrete-time systems with time-outs that can occur asynchronously. Finite-state automata augmented with dense-timed clock variables, such as the timed and hybrid automata of Chap. 7, are more general examples of this type of hybridism.

Example 3.1. The braking system mentioned in Example 2.3 is a hybrid model: the system senses the information coming from the wheels in the form of variables varying over continuous time and state domains (e.g., angular speed and friction) and processes it through a digital embedded device that computes, in real time, the ideal pressure to be applied to the calipers. Processing the continuous components with a digital system – with finite precision and synchronized to a discrete clock – requires sampling and approximating the information coming from the sensors; then, the actuators translate the discrete series of values output by the computer to the calipers, which operate over "physical" continuous time.   ■

We will go back to the issue of hybrid models in Sect. 3.7, when discussing composition of modules.

Exercise 3.2 (♣). Which of the following systems or processes are naturally described by a hybrid model? What are the discrete and the dense/continuous components? To which of the three aforementioned classes of hybrid models do the systems naturally belong?

- A thermostat controlling the temperature of a room by turning on and off heating.
- The controller of a railroad junction.
- The emission of light from a heated chemical element.   ■

3.2 Ordering Versus Metric

A formalism may permit the expression of metric constraints on time, or, equivalently, of constraints that exploit the metric structure of the underlying time model (if it has any).

A domain (possibly a time domain) has a *metric* when it is equipped with a notion of *distance*, that is a *measure* function $d(t_1, t_2)$ associated with pairs of points t_1, t_2 of the domain that satisfies the properties of

(i) Nonnegativity: $d(t_1, t_2) \geq 0$;
(ii) Identity of indiscernibles: $d(t_1, t_2) = 0$ if and only if $t_1 = t_2$;
(iii) Symmetry: $d(t_1, t_2) = d(t_2, t_1)$;
(iv) Subadditivity (also called triangle inequality): $d(t_1, t_3) \leq d(t_1, t_2) + d(t_2, t_3)$.

The typical time domains – the usual discrete and dense numerical sets $\mathbb{N}$, $\mathbb{Z}$, $\mathbb{Q}$, $\mathbb{R}$ – all have a "natural" metric in terms of Euclidean distance between two points: $d(t_1, t_2) = |t_1 - t_2|$.

Although all common choices for time domains possess a metric, an issue is whether the *language* in which the system is described permits using the same form of metric information as that embedded in the underlying time domain. For instance, some languages allow for stating that an event p (e.g., "push button") must precede temporally another event q (e.g., "take picture"), but do not include constructs to specify how much time elapses between the occurrence of p and that of q; thus, they cannot distinguish between the case in which the delay between p and q is one time unit from the case in which the delay is 100 time units. Languages where the relative *ordering* of events is expressible, but metric constraints are not, support a purely *qualitative* notion of time, as opposed to the *quantitative* time expressible with metric languages.

Example 3.3 (Parallel and real-time systems). In purely parallel systems, the correctness of a computation only depends on the *relative ordering* of computational steps, irrespective of their absolute distances. Reactive systems, where a controller component evolves concurrently with the controlled environment, are often purely parallel in this sense. For the formal description of such systems, a purely qualitative language is sufficient. Real-time systems also usually perform in parallel, but their correctness depends as well on the *time distance* between events; thus, the complete model of real-time systems requires quantitative languages, supporting the expression of metric constraints.

As a simple example of purely parallel system, consider two tasks T_1 and T_2 that exchange messages. T_1 can perform an action a only after receiving a datum from T_2; and T_2 produces the datum by performing another action b. This data dependency forces the ordering of actions a and b: a follows one or more occurrences of b, independently of the relative speed of the two tasks or of the transmission channel.

The same two-task system becomes real time if T_2 produces data at a fixed rate of n actions b per second and puts them in a buffer composed of a single slot, and we want to avoid the situation where T_2 tries to put a datum in a full buffer. T_1 then has to be fast enough: a specification of correct behavior may require, in addition to actions a following actions b, that no execution of a takes more than $1/n$ seconds. ∎

Following the difference between purely parallel and real-time systems, the research in the field of formal languages for system description has evolved from dealing with purely qualitative models to the more difficult task of expressing and reasoning about metric constraints. Consider, for instance, two sequences σ_1 and σ_2 of events p and q, where exactly one event per time step occurs,

$$\sigma_1 = p\,q\,p\,q\,p\,q\cdots,$$

$$\sigma_2 = p\,p\,q\,q\,p\,p\,q\,q\cdots,$$

that share the following property, expressible without referencing any metric information: "every occurrence of p is eventually followed by an occurrence of q"; in contrast, "p occurs in every instant" is a qualitative property that is false for both behaviors. Some metric properties, instead, discriminate between σ_1 and σ_2, as in "every occurrence of q is followed by another occurrence of q after two time steps", which holds for σ_1 but not for σ_2.

The notion of *invariance under stuttering* is an alternative characterization of the properties expressible qualitatively. Consider, for example, the discrete-time behavior σ_3 consisting of the following sequence of states, one per time step:

$$\sigma_3 = s_1\,s_2\,s_3\,s_4\,s_5\,s_6\,s_7\,s_8\cdots.$$

A time step i such that $s_i = s_{i+1}$ is called a "stuttering" step; for example, the first step in sequence σ_2 is stuttering (if we interpret it as a sequence of states rather than as a sequence of events). Adding or removing stuttering steps from a behavior does not affect the qualitative properties it satisfies. If two behaviors are identical up to the addition or removal of any number of stuttering steps, the two behaviors are called *stutter-equivalent* or *equivalent under stuttering*. For example, behaviors σ_1, σ_2 above are stutter-equivalent: every odd instant of time corresponds to a stuttering step in σ_2, and σ_1 equals σ_2 with all stuttering steps removed.

Stutter equivalence is an equivalence relation; the equivalence classes it induces precisely identify classes of behaviors that share identical qualitative properties. Note that stutter invariance is defined for discrete time models only.

Exercise 3.4 (♣). Argue that the property

$$\text{"sequences of events } a \text{ and sequences of events } b \text{ alternate"} \qquad (3.1)$$

is qualitative, whereas the property

$$\text{"events } a \text{ and } b \textit{ strictly} \text{ alternate"} \qquad (3.2)$$

is quantitative.   ∎

Exercise 3.5. With reference to Exercise 3.4, characterize the set of behaviors corresponding to (3.1) and show that any two members of the set are stutter-equivalent. Then, characterize the set of behaviors corresponding to (3.2) and show that there exist behaviors which are stutter-equivalent to elements of the set but are not in the set.   ∎

3.2.1 Total Versus Partial Ordering

The discussion so far assumed time and other domains with a *total* ordering: for every pair of distinct elements x, y in the domain, either x precedes y in the ordering (usually written $x < y$), or y precedes x ($y < x$). The definition of dense sets, in particular, is simpler for totally ordered domains, and so is the definition of a metric. There are circumstances, however, where sets with only partial ordering – where neither $x < y$ nor $y < x$ for some distinct elements x, y – are the best choice for the temporal domain in a system model.

Example 3.6. Modern cars implement several functions on their on-board embedded electronics as software. The antilock braking system mentioned in Sect. 3.1.2 is a common example; another subsystem electronically controlled is the one responsible for moving the car windows. Each subsystem must meet its timing requirements: among other things, the braking system must release the breaks within, say, $1/10$ second whenever the wheels are blocked and the vehicle is moving, and the motorized windows must shut completely within, say, 7 s whenever a passenger clicks the button. In an overall model of the car, the events "wheels become blocked" and "breaks released" are strictly ordered, and so are the two other events "button clicked" and "windows become closed". However, there is no reason to define an order between events of the braking subsystem and events of the window control: the events in the overall system are only partially ordered, and so are the instants of time when they may occur.   ■

The above example suggests that partial orderings arise naturally when composing the behavior of subsystems into composite systems with heterogeneous components: the events happening in different subsystems are usually unrelated, and synchronization among subsystems relies on explicit "messages" sent at the subsystems' interface. Part II of the book will present some notations that introduce partial orders when composing unrelated events, as well as others that always define a total order among events. Section 3.3 discusses another dimension that relies on the notion of total and partial ordering.

3.2.2 Time Granularity

System models with metric time usually possess a "natural" time scale, corresponding to the abstraction level of the temporal behavior in the model. In Example 3.6, the braking system operates within fractions of seconds, whereas the window system is paced by an order of magnitude slower time scale. The notion of *time granularity* captures this idea of "time scale", and different components in a composite system have different time granularities when their natural time scales differ, possibly by orders of magnitude.

In some sense, time granularity is a form of hybridism (see Sect. 3.1.3), which is frequent in complex composite systems where processes that evolve in the order of seconds or minutes – or even days or months (such as a chemical process, or a process at a hydroelectric power plant) – are controlled by fast digital electronic devices. In principle, a continuous time domain, such as the real numbers, can accommodate system models with arbitrarily heterogeneous time granularities: conversion among different time units is always possible, with possibly an arbitrarily small loss of precision if some units happen to be incommensurable (see Sect. 3.1.1).

If, however, the underlying time domain is discrete, the approximation error introduced when converting the coarser time units can be non-negligible and raise subtle semantic issues. Consider, for instance, the sentences

Every month, if an employee works, then she gets her salary.

and

Whenever an employee is assigned a job, this job should be completed within three days.

If the sentences are part of the same specification of an office system, we have to find a way to reconcile their time units. A discrete temporal domain with the day as time unit seems a natural choice, because the other time unit, the month, is of coarser granularity. However, a simple change of time units from months to days alters the meaning of the quantification "every": the specification "every month, if an employee works, then she gets her salary" has a different meaning than "every day, if an employee works, then she gets her salary", because working for 1 month means working for 22 variable days during the month, whereas getting a monthly salary means that there is one fixed customary day of every month when salaries for the whole month get paid. A change in the time unit (from months to days) is insufficient to capture the correct meaning of the original sentence.

In the other example, your boss states that "this job has to be finished within three days from now" at 4 P.M. on 16 June 2012. What does she mean exactly? "This job has to be finished within $3 \cdot 24 \cdot 60 \cdot 60$ seconds counting from now", or "this job has to be finished by 6 P.M. on 19 June 2012", or even "this job has to be finished by midnight on 19 June 2012"? Each interpretation may be valid, depending on the context of the claim.

Chapter 9 presents an approach to deal rigorously with different time granularities in the context of temporal logics.

Example 3.7. Consider the following structurally similar sentences:

- Tomorrow, I will eat.
- Tomorrow, I will work.
- Tomorrow, I will go to the bank to pay my monthly bills.
- Tomorrow, I will stay in the city.

Depending on the time unit used to interpret the sentences, the meaning of "Tomorrow, I will..." changes from sentence to sentence. In particular, if we introduce the finer granularity of hours, the first two sentences read as "Tomorrow there will be *some* (few) hours when I will be eating" and "Tomorrow there will be *some* hours (say, eight) when I will be working"; the third sentence probably translates to "Tomorrow there will be *one* hour during which I will pay my bills"; the fourth one likely refers to the fact that "Tomorrow, during *all* hours of the day I will be somewhere in the city". The different meanings of the verbs ("eat", "work", "go", "stay") hint at different scopes ("some", "all", "one") in terms of hours during the day.   ∎

Exercise 3.8 (♣). Determine the most appropriate time units to interpret the following sentences:

- Tomorrow, I will work, and then I will go out.
- Tomorrow, I will have two classes, separated by a short break.
- Tomorrow, I will be on vacation.   ∎

Exercise 3.9 (♣). A hydroelectric power production system consists of a reservoir, an electric production station, and pipes connecting the dam of the reservoir to the power station; sluice gates to control the amount of water to be sent to the power

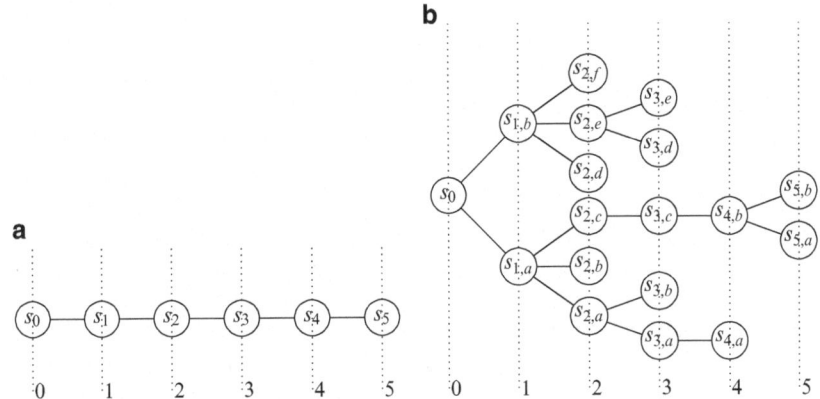

Fig. 3.3 A linear (**a**) and a branching (**b**) time model

station; and controlling devices which manage the sluice gates and the turbines that generate the power (e.g., to keep the frequency constant).

Consider a model of the global behavior of the system including the dynamics of the reservoir in terms of the amount of water coming in (from incoming rivers and the rain) and flowing out (from the pipes), the amount of power delivered by the plant, the control goals for the power supplied and water consumed, and the behavior of the digital controllers.

Which time unit would you use for each component of such a model? How would you combine them into a unique model?   ∎

3.3   Linear Versus Branching Time Models

The terms *linear* and *branching* refer to the structures on which formal languages are interpreted: *linear*-time formalisms are interpreted over *linear* sequences of states, whereas *branching*-time formalisms are interpreted over *trees* of states. In other words, a system description adopting a linear notion of time refers to linear behaviors, where the future evolution, from a given state at a given time, is always unique. Conversely, branching-time interpretations refer to behaviors structured in trees, where each "present state" may evolve into different "possible futures". Assuming discrete time, Fig. 3.3 pictures a linear sequence of states and a tree of states, over six time instants.

Linear behaviors are special cases of trees. Conversely, trees represent sets of linear behaviors that share common prefixes (i.e., that are prefix-closed). Under this duality, linear and branching models can be put on a common ground and compared; this has been extensively done in the literature.

In Fig. 3.3, the linear model (a) defines a total ordering of the symbols $s_0, s_1, \ldots, s_5$, whereas the branching model (b) induces a partial ordering of $s_0, s_{1,a}$, $\ldots, s_{5,b}$. Incidentally, the figure also suggests a metric on the time domain, so that symbols such as $s_{1,a}$ and $s_{1,b}$ are, in principle, not ordered but mark the same absolute time. If we ignore the metric information, we have a genuinely partial order where pairs such as $s_{1,a}$ and $s_{1,b}$ or $s_{1,a}$ and $s_{3,d}$ are unordered and it is undefined whether one occurs before or after the other.

The meaning of the branches in a branching-time model depends on the context and the system modeled. For example, the branch from s_0 to $s_{1,a}$ and $s_{1,b}$ in Fig. 3.3b may capture the fact that the system spawns two new parallel processes, whose behavior is described in each branch independently of the other. In a different interpretation, the same branch may describe a nondeterministic choice between two alternatives: each path in the tree is a totally ordered sequence of events or states that may happen in one of the possible computations, but elements in different branches have no order because they belong to mutually exclusive alternatives. Section 3.4 gives more details on such interpretations of branching time in the presence of nondeterministic computations.

Linear or branching semantic structures are then matched in the formal languages by corresponding syntactic elements that can express properties of specific features of the interpretation. This is possible, in principle, with all formal languages, but it is especially relevant for logic languages, and for temporal logics in particular. Linear-time temporal logics are interpreted over linear structures, and express properties of behaviors with unique futures, such as "if event p happens, then event q will happen eventually in the future". On the other hand, branching-time temporal logics are interpreted over tree structures, and state properties of branching futures, such as "if event p occurs, event q will occur along *some* of the originating branches". Chapter 9 discusses similar examples in greater depth with reference to temporal logics.

It is also possible to have semantic structures that are *branching in the past*, where different pasts merge into a unique present. Branching-in-the-past models are, however, uncommon in practical applications, so we will not deal with them.

Exercise 3.10 (♣). Consider a formalization of the game of chess in which a *state* is a given configuration of the pieces on the board, time instants coincide with moves of the players, and at every time instant there is a transition from the current state to the next one, determined by the move. Discuss whether linear or branching-time models adequately represent the following:

- A single match;
- The set of all matches starting with a given opening (e.g., all matches starting with the "Danish gambit" opening);
- A chess problem, such as: "starting from the given configuration, white to move and checkmate in three moves".   ∎

3.4 Deterministic, Nondeterministic, and Probabilistic Models

Linear time and branching time are features of languages and of the structures on which those are interpreted, whereas deterministic, nondeterministic, and probabilistic behaviors are attributes of the systems modeled or analyzed.

3.4.1 Deterministic Versus Nondeterministic Models

Consider systems including a notion of *input*, which evolve over time by reading the input, and changing the current state accordingly. A system is *deterministic* whenever the current value of state and input *uniquely* determine the future state. For instance, a light switch is a deterministic system, where pressing the button (input) when the light is in state off yields the unique possible future state of light on. Notice that, for a given input sequence, the initial state completely determines the behavior of a deterministic system.

Conversely, systems are *nondeterministic* if they can evolve to different future states from the same present state and input by making arbitrary "choices". For example, a resource arbiter is nondeterministic if it responds to two requests happening at the same time by "choosing" arbitrarily whom to grant the resource first, and the same pair of simultaneous requests may result in a different choice every time.

Example 3.11 (Ada's rendezvous). The rendezvous mechanism of the Ada programming language is a significant example of nondeterminism applied to the synchronization of parallel tasks. Consider two client tasks *Producer* and *Consumer* that depend on a server task *Buffer* to perform operations *Put* and *Get* respectively. In the Ada framework, when *Buffer* is ready to execute both *Put* and *Get* (i.e., it is neither full nor empty and there are pending requests – entry calls in Ada terminology), it chooses any of them nondeterministically. This behavior is embodied in the semantics of the **accept** statement. Figure 3.4 has a typical example of Ada code using this feature. Whenever the execution of *Buffer* reaches the **select** statement, all conditions ("guards") expressed by the **when** clauses are checked to determine which ones are enabled. If multiple **when** conditions hold and there are pending requests from other tasks for the corresponding entry, the system arbitrarily selects, in a nondeterministic fashion, one of the enclosed **accept** statements to be executed. The actual choice is resolved either by the compiler or by the operating system. Programmers must build programs which behave correctly independently of how the nondeterministic choices are actually resolved. ∎

We will further discuss the semantics of Ada's rendezvous mechanism in Sect. 3.7.

A notation with nondeterministic features supports a high level of abstraction, where details that pertain solely to the implementation, such as the precise order of

Fig. 3.4 The rendezvous
mechanism in Ada

```
task Buffer is
    entry Put (Item:  in  Integer );
    entry Get (Item:  out  Integer );
end;

task body Buffer is
begin loop
    select
        when Count < MAX => -- buffer not full
            accept Put (Item:  in  Integer ) do
                ...
            end; -- accept
        or
        when Count > 0 =>    -- buffer not empty
            accept Get (Item:  out  Integer ) do
                ...
            end; -- accept
    end select ;
end loop;
end Buffer ;
```

execution of tasks, are hidden as nondeterministic choices. At the same time, the lack of control over such lower-level details introduces subtleties which may make a full understanding of system behavior more difficult. For instance, in the loop of the Fig. 3.4 there is no control over the choice of pending request served at each iteration: the *Buffer* may always favor *Put* over *Get*, leaving the *Consumer* task waiting idly for a time that is related to the capacity of the buffer. In extreme cases – for instance, if the **accept** statements are not under the control of suitable **when** conditions or such conditions are always true – a process may even "starve" while waiting on an entry call because the server always selects other processes (we will discuss the notion of a *starving process* in Chap. 8).

More generally, nondeterminism is a powerful abstraction mechanism for incomplete knowledge in the description of systems. The sources of incompleteness and the semantics of the nondeterminism abstraction may vary with the application context.

In the example of Fig. 3.4, it is undetermined which **accept** statement should be executed when both are enabled. Similarly, system requirements may not commit to selecting from possible acceptable alternatives, with the objective of not over-constraining the implementation policies. For instance, a requirement of the type

When signal S occurs, the system must react by signaling T no later than ten seconds and no sooner than five seconds.

is met by systems that always produce T after 8 s, by other systems that react after 6 s in certain conditions and after 9 s in others, as well as by many other implementations.

Another context in which nondeterminism can formalize incomplete knowledge occurs in search problems in unstructured spaces or, more generally, where there is no a priori criterion for selecting from different options. Think, for example, of searching a generic graph or tree for nodes with certain properties.

Systems embedded in an external environment whose behavior is only partially known – power plants, distributed social groups, and so on – may also avail of nondeterministic abstractions to model imperfect knowledge of the environment.

The semantics of nondeterministic choice also depends on the application context and on the nature of knowledge incompleteness to be dealt with. For instance, in Example 3.11, the programmer may be oblivious to how the runtime system will choose from among enabled **accept** statements; hence he must guarantee that the program will meet its requirements *for all* possible operational conditions encompassed by its nondeterministic choice. Symmetrically, many generic search algorithms are modeled as nondeterministic visits of data structures: a visit is successful when *there exists some* sequence of nondeterministic choices that leads to the searched element.

We refer to the first type of nondeterminism – where every nondeterministic choice must produce an acceptable behavior – as *universal nondeterminism* (in analogy with universal logic quantification). Conversely, we will call the second type of nondeterminism – where it is sufficient that one nondeterministic choice leads to a valid solution – as *existential nondeterminism*. The following chapters will show various application contexts of nondeterministic abstractions and models, with references to the classification just introduced.

3.4.2 Nondeterministic Versus Probabilistic Models

A nondeterministic system can evolve into different future states from the same current state and input. The choice of which future state to take is arbitrary, and all possible alternatives are considered. In other words, nondeterminism is a convenient abstraction for sets of alternatives that can happen over different runs.

Probabilistic systems (also called "stochastic systems") can also choose from among different future states for a given current state and input. Unlike nondeterministic systems, however, the choice relies on probability distributions: the system selects the next state by drawing a value from a distribution and proceeding accordingly. An unbiased coin is an obvious example of a probabilistic system without input. Each state corresponds to the coin showing heads or tails. Flipping the coin moves the system to the next state, which is heads with probability $1/2$ and tails with probability $1/2$. The probability distribution associated with each transition induces a probability distribution on sets of behaviors. For example, the set of sequences of coin flips such that the first two draws are both head has a probability of $1/2 \cdot 1/2 = 1/4$.

Nondeterministic and probabilistic models are both concerned with representing incomplete information about the system's behavior: in the nondeterministic case there is no information at all about how choices are resolved, whereas in the probabilistic case the probability distribution gives a measure of the partial information available. This quantitative difference results in a sharp conceptual difference in the types of questions addressed in the analysis of the two families of models and, consequently, in the type of mathematics needed to perform such an analysis. The analysis of nondeterministic models addresses "yes/no" questions about whether every possible behavior meets some requirements (such as "all pending requests are served within three seconds") or, symmetrically, whether there exists any behavior that achieves a certain goal. The analysis of probabilistic models addresses typically quantitative questions about the "likelihood" of certain events happening or not happening. For example, we can ask if "pending requests are served within three seconds 90 % of the time" – or, equivalently, with a probability of at least 90 %.

The conceptual difference between nondeterministic and probabilistic models persists even if we compare nondeterministic transitions to probabilistic ones with uniform probability. For example, the unbiased coin described above could also be modeled as a series of nondeterministic transitions that arbitrarily choose between heads and tails. The difference between the probabilistic and the nondeterministic model of the coin lies in the different weights given to possible sequences of flips. In the nondeterministic model, every sequence is on equal ground with all the others; an unbounded sequence of heads, for example, is perfectly legitimate behavior. In the probabilistic model, in contrast, different sequences have different probabilities and hence different likelihoods of happening. An unbounded sequence of heads, in particular, has zero probability of happening; hence it is essentially ruled out by the abstraction of the model.

Probabilistic modeling is a natural choice for systems whose dynamics are known only partially and empirically or are too complex to model exactly. Most physical phenomena happening in the natural world are of this type. For example, geological data may suggest the probability of an earthquake of a certain magnitude happening in a certain region during 1 year. Systems including human users are also often conveniently modeled with probabilities, for example, to quantify the chance that an operator performs a sequence of events in an incorrect, unsafe order, or to model the accesses to a Web server resulting from users visiting a certain HTML page.

The design of probabilistic algorithms is another, more sophisticated, application area within computing, where probabilistic models of timing properties are widely deployed. It turns out that several computationally complex algorithmic tasks can be sped up significantly by randomizing certain choices during the computation. Probabilistic algorithms give correct answers only with certain finite probabilities; in practice, however, this probability can often be made arbitrarily close to 1, so as to favorably leverage the trade-off between time spent computing and correctness of the result. Section 6.4 describes some examples of probabilistic computational models and algorithms.

The examples of nondeterministic and probabilistic behavior discussed so far focused on the choice from among different future *states*, but the same abstractions apply to the choices of waiting times, delays, and other time intervals. For example, in a communication protocol for handshaking (such as TCP's three-way handshake in Example 2.1), the event "acknowledge" always follows deterministically the event "start connection", but the time elapsing between an occurrence of "start connect" and the corresponding occurrence of "acknowledge" is nondeterministic and varies according to the conditions of the communication network. In general, purely nondeterministic models consider arbitrary delays between a minimum T_{min} and a maximum T_{max} (which can be 0 and ∞ if every delay is possible), whereas probabilistic models associate a probability distribution T with the interval $[T_{min}, T_{max}]$ such that the delay is $t \in [T_{min}, T_{max}]$ with probability $T(t)$.

The term "stochastic" sometimes specifically refers to the application of probabilistic models to delays and timing information, as opposed to "probabilistic" models where the choice is of different future states. This terminology is, however, not universally accepted, and different research areas often use different conventions. This book uses the attributes "probabilistic" and "stochastic" as synonyms.

Finally, notice that, as with nondeterminism, probabilistic behavior may abstract incomplete knowledge of different origins: the input provided by the environment to the system may be known only statistically (the uncertainty may be in the input values or in its timing), and the system itself may react according to a deterministic or a probabilistic policy. When both system and environment have stochastic behavior, the overall model of systems embedded in the environment follows probabilities that depend on those of each component – this corresponds to the notion of *conditional probability*. Chapter 6 and the following ones describe various forms of probabilistic models, for different sources of uncertainty.

Exercise 3.12. Consider multiple consecutive iterations of the *Buffer* loop in Fig. 3.4 such that, at every iteration, both guards *Count* > 0 and *Count* < *MAX* are true and there are pending requests for both *Get* and *Put*. Assume that *Buffer* always spends one time unit to perform a *Get* and two time units to perform a *Put*.

- Build a branching-time behavior consisting of a tree that summarizes all possible sequences of events over four iterations.
- Describe the set of linear-time behaviors representing the same sequences of events as the tree.
- Under the same conditions, what are the minimum and maximum times *Buffer* takes to execute ten consecutive loop iterations?
- Assume that there are always some pending requests for both *Get* and *Put* by some client; how much time, at most, must elapse before a call for a *Get* is certainly served? How much time for a call for a *Put*? ∎

Exercise 3.13. Consider again multiple consecutive iterations of the *Buffer* loop in Fig. 3.4 such that, at every iteration, both guards *Count* > 0 and *Count* < *MAX* are true and there are pending requests for both *Get* and *Put*. Assume that, at every iteration, the *Buffer* chooses to execute *Put* with probability $p = 40\%$ and *Get* with probability $q = 60\%$.

- What is the probability that *Put* has never been executed after ten loop iterations?
- How many iterations are needed to guarantee that both operations have been executed at least once with probability greater than 95%?

- What values for the probabilities p and q minimize the number of iterations to guarantee that both operations have been executed at least once with probability greater than 90 %? ∎

Exercise 3.14 (♦).   Consider again multiple consecutive iterations of the *Buffer* loop in Fig. 3.4 such that, at every iteration, both guards *Count* > 0 and *Count* < *MAX* are true and there are pending requests for both *Get* and *Put*. Assume that, at every iteration, the *Buffer* chooses to execute *Put* and *Get* deterministically in strict alternation. Every execution of *Put* and *Get* takes time whose probability is uniformly distributed in the interval [10, 20] milliseconds.

- What is the average duration of ten loop iterations?
- What is the probability that executing ten loop iterations takes more than 120 ms?

(*Hint*: the exercise is simpler if the interval [10, 20] is taken to be discrete rather than continuous). ∎

3.4.3   Deterministic, Probabilistic, and Nondeterministic Versus Linear- and Branching-Time Models

There is a natural coupling between, on one side, deterministic systems and linear models, and, on the other side, nondeterministic or probabilistic systems and branching models. In linear-time models the future of any instant is unique, and hence the modeled system is deterministic, whereas in branching-time models each instant branches into different futures, corresponding to possible nondeterministic choices.

This natural correspondence notwithstanding, determinism and linearity of time are distinct concepts, which target different concerns. For instance, linear-time models are often preferred – even for nondeterministic systems – for their intuitiveness and simplicity. The discussion of Petri nets in Chap. 8 will provide examples of linear time domains expressing the semantics of nondeterministic formalisms. On the other hand, branching-time models can describe sets of computations of deterministic systems for different input values. For instance, the branches of a tree can describe all possible computations of an array sorting algorithm, where each branch corresponds to a choice made by the algorithm on the basis of comparisons between array elements. Analyzing the tree gives measures of the minimum, maximum, and average execution times.

3.5   Implicit Versus Explicit Time Reference

Some languages for the description of temporal properties make explicit reference to temporal items (attributes or entities of "type time", such as the occurrence times of events and the durations of states or actions), whereas other formalisms leave such references implicit in their syntax.

To illustrate, consider the case of pure first-order predicate calculus to specify system behavior and its properties, as done in some examples of Chap. 2. Formulae explicitly refer to terms ranging over the time domain and combine them with quantifiers; such formulae give properties where explicit time references are frequent, such as in the sentence

> For every instant of time t, the safe is open if and only if there exists another time instant u, smaller than t and at least as large as $t - 3$, such that the correct code has been entered at u.

which corresponds to a part of formula (2.6) in Chap. 2. On the contrary, formulae of classic temporal logic, despite its name, do not mention any temporal quantities explicitly, and express temporal properties in terms of an implicit "current time" and the ordering of events with respect to it; for example, a simple sentence in this style reads

> If the correct code is entered [implicitly assuming the adverb *now*], then the safe will open sometime in the future, and then it will close again.

Most formalisms adopt some kind of intermediate approach between the extremes of purely explicit and purely implicit references. For instance, many types of abstract machines can specify explicitly the duration of activities with implicit reference to their starting time (Statecharts, discussed in Chap. 7, and Petri Nets, presented in Chap. 8, are two representative examples). Other languages inspired by temporal logic (such as MTL, presented in Chap. 9) keep its basic approach of referring any formula to an implicit current instant (the *now* time) but can explicitly express time distances with respect to it. Such logics can express properties such as

> If the correct code is entered [*now*], then the safe will open immediately, and then it will close again after exactly three time units.

Using implicit references to time instants – in particular an implicit *now* – is quite natural and convenient when modeling so-called "time-invariant systems", which are the majority of real-life systems: in most cases, in fact, the system behavior does not depend on the absolute value of time but only on the relative time distances. Therefore, expressing explicitly where the *now* is located along the time axis is irrelevant for such system models.

Example 3.15 (Explicit and implicit time). Sentences stating historical facts typically use explicit time references[2]:

- During the year 1625, a dramatic famine struck Europe; the famine lasted until the beginning of the year 1630.
- The starving population was an easy target for an epidemic of plague, which began in 1629 and lasted until 1631.
- During the years 1625–1631, life expectancy dropped from 50 to 37 years.

---

[2]The following three sentences refer to some real historical facts mentioned in Alessandro Manzoni's *The Betrothed*; the dates and figures are plausible but not necessarily accurate.

Observations about morals usually use implicit time references to convey timelessness:

- Every lie is eventually uncovered.
- You can fool some of the people all of the time, and all of the people some of the time, but you cannot fool all of the people all of the time.[3]

Engineering artifacts are often time-invariant systems, naturally described with an implicit "now":

- The speed of a braking car decreases proportionally relative to the time since when braking starts (see Example 2.3).
- The discharge time of a capacitor attached to a resistor depends only on the resistor's resistance, the capacitor's capacity, and the initial charge accumulated, irrespective of the absolute time when discharging starts (the example is developed further in Chap. 4).   ∎

Exercise 3.16 (♣). Analyze the following sentences in natural language, and determine the kind of implicit or explicit time references they contain.

- World War II lasted 6 years from 1939.
- The last death of a US president in office occurred in 1963.
- The final agreement must be signed within 30 days from the subscription of the letter of intent.
- After he reached the age of 60, he was never in good health for more than 3 months.
- A vast majority of the "baby boomers" will not be able to retire before the age of 65.
- Life expectancy has steadily increased in the last three centuries, and it is now over 80 years in a few countries.
- You tried your best, and you failed miserably. The lesson is, never try.   ∎

3.6 The Time Advancement Problem

The problem of time advancement arises when the model of a timed system exhibits behaviors that do not progress past some instant. Usually, such standstill behaviors do not correspond to any physical "real" phenomena; they may be the consequence of some incompleteness and inconsistency in the formalization of the system, and must thus be ruled out.

The simplest manifestation of the time advancement problem arises when transitions that occur in null time are possible. For instance, several automata-based formalisms such as Statecharts and timed versions of Petri nets support such

---

[3]This quotation is usually attributed to Abraham Lincoln, but this is allegedly apocryphal.

abstract zero-time transitions (see Chaps. 7 and 8). Although truly instantaneous actions are physically unfeasible, they nonetheless are useful abstractions for events that take an amount of time which is negligible with respect to the overall dynamics of the system; pushing a button is an example of an action whose actual duration can usually be ignored and that can thus be represented abstractly as a zero-time event. When zero-time transitions are allowed, an infinite number of such transitions may accumulate in an arbitrarily small interval, thus modeling a fictitious infinite computation where time does not advance past the interval. Behaviors where time does not advance are usually called "Zeno" behaviors, from the ancient philosopher Zeno of Elea[4] and his paradoxes on time advancement. From a rigorous point of view, even the notion of behavior as a function – whose domain is time and whose range is the system state (see Chap. 4) – is ill-defined with zero-time transitions: if the transition is instantaneous, the system is both at the source state and at the target state in the same instant.

Even if actions are non-instantaneous, Zeno behaviors can still occur if time advances only by *arbitrarily small* amounts. Consider, for instance, a system that produces an unbounded sequence of events p_k, for $k \in \mathbb{N}$; each event p_k happens exactly t_k time units after the previous one (i.e., p_{k-1}). If the series of the relative times t_k (that is, the infinite sum $\sum_k t_k$ of the time distances between consecutive events) converges to a finite limit t, then the absolute time never surpasses t; in other words, *time stops* at t, while an infinite number of events occur in the finite time between any t_k and t.

Zeno behaviors exist also for continuous-valued time-dependent functions of time that vary smoothly. Take, for instance, the real-valued function of time

$$b(t) \quad = \quad \begin{cases} \exp\left(-\frac{1}{(t-t_0)^2}\right) \sin\left(\frac{1}{t-t_0}\right) & t \neq t_0 , \\ 0 & t = t_0 . \end{cases} \qquad (3.3)$$

$b(t)$ is very smooth, as it possesses continuous derivatives of all orders. Nonetheless, its sign changes an infinite number of times in any interval containing the time instant t_0; therefore, if we consider the event of function $b(t)$ changing its sign, an unbounded sequence of such events takes place before t_0, without time advancing past t_0; natural notions such as "the last or next instant at which the sign of b changes" are not defined at time t_0, and, consequently, we cannot describe the system by relating its behavior to such – otherwise well-defined – notions. Indeed, as will be explained precisely in Chap. 9 when discussing temporal logics, absence of Zenoness may be obtained through the mathematical notion of *analyticity*, which is even stronger than infinite derivability.

Even when Zeno behaviors are ruled out, and hence time progresses, the occurrence of an unbounded number of events in intervals of fixed finite length may lead to "irregular" behaviors that complicate the analysis. For example, the distance

---

[4]Circa 490–425 B.C.

between consecutive events may get indefinitely smaller while time diverges, such as in the harmonic sequence defined by $t_{k+1} = t_k + 1/k$. These behaviors are called "Berkeley", after the philosopher George Berkeley[5] and his investigations arguing against the notion of infinitesimal. Systems with Berkeley behaviors cannot be controlled by digital controllers operating with a fixed sampling rate since the behaviors cannot be suitably discretized. On the other hand, several real-life systems cannot guarantee an a priori bound on the "speed" of events; hence their model must include Berkeley behaviors.

Some well-known problems of – possibly – concurrent computation such as *termination, deadlocks*, and *fairness* can be considered as *dual* problems to time advancement, because they describe processes that fail to advance their *states*, while time keeps on flowing. Examples of these problems and their solutions are discussed with reference to a variety of formalisms in Part II of the book.

Two different approaches manage the time advancement problem: we refer to them as "a priori" and "a posteriori" methods. In a priori methods, the syntax or the semantics of the formal notation is restricted beforehand, in order to guarantee that every system model is exempt from time advancement problems by construction. For instance, in some notations every transition must necessarily take a positive time greater than some fixed value c. A less restrictive assumption, which guarantees a good level of abstraction while still avoiding a priori the risk of Zeno and even Berkeley behaviors, allows for only finite sequences of zero-time transitions that are followed by an event that takes a minimum fixed time. This view does not explicitly restrict the number of events occurring in any finite time, but ensures that no infinite sequence ever accumulates. It is well suited, for instance, for expressing a sequence of logic gate switches in a hardware processor that occur within a single clock interval.

A *posteriori* methods, in contrast, deal with time advancement issues only *after* the system specification has been built; the specification is analyzed against a formal definition of time advancement, in order to check that all of its actual behaviors do not run into the time advancement problem. A posteriori methods may be particularly useful for detecting possible criticalities in the behavior of real systems already built. For instance, the oscillations exhibited by a mathematical model with a frequency that goes to infinity within a finite time interval, as in the function $b(t)$ mentioned in (3.3) above, may be the symptom of some instability in the modeled physical system, in the same way a physical quantity – say, a temperature or a pressure – that tends to infinity within a finite time in the model is the symptom of a serious possible failure in the real system.

The same "duality" – a priori avoidance vs. a posteriori verification – is often assumed to deal with the symmetric problem of *process* advancement.

---

[5] Kilkenny, 1685–Oxford, 1753.

Exercise 3.17. Consider a system whose state s evolves according to the function of time $s(t) = \sin(\omega t^2)$. How would you classify such a behavior? A Zeno behavior? A Berkeley behavior? None of them?   ∎

Exercise 3.18. An unbounded sequence of events occur each at time $t_1, t_2, \ldots, t_i, t_{i+1}, \ldots$, where

$$t_k = \begin{cases} 0 & k = 1, \\ t_{k-1} + d_{k-1} & k > 1. \end{cases}$$

Define, if possible, a sequence of values $d_1, d_2, d_3, \ldots$ such that the resulting sequence of events is:

1. Zeno and all events but the first occur at irrational times;
2. Zeno and all events occur at integer times;
3. Non-Zeno and Berkeley;
4. Zeno and Non-Berkeley;
5. Non-Berkeley and all events but the first occur at irrational times;
6. Non-Berkeley and all events occur at integer times.   ∎

3.7   Concurrency and Composition

Most real *systems* – as the term itself suggests – are complex enough that it is useful, if not outright unavoidable, to model, analyze, and synthesize them as the composition of several subsystems. Such a composition/decomposition process may be iterated until each component is simple enough to be analyzed directly.

Composition and decomposition, also referred to as *modularization*, are general and powerful design principles in any field of engineering. In particular, in the case of – mostly sequential – software design, they have originated a rich collection of techniques and language constructs, from subroutines to abstract data types and object orientation.

The application of the same principles of modularity to concurrent and timed systems is definitely less mature, and in fact only a few programming languages deal explicitly with concurrency. From a programming language viewpoint, the central issue with the modeling of concurrency is the *synchronization* of activities (embodied in different constructs such as processes, tasks, and threads) when they access shared resources or exchange messages. The etymology of the word "synchronization" is quite descriptive of the timing issues that are at stake: "synchronization" combines the Greek words συν (which means "together") and χρονοσ (which means "time"), and in fact, concurrent activities evolve in parallel independently, until they must synchronize and meet at the same time. Synchronization may require that faster activities slow down and wait for the slower activities to meet themselves at the "same time".

When the concurrent activities of the modules are heterogeneous in nature, formally modeling the synchronization of components becomes even more intricate because of the difficulty of coming up with a uniform time model. For instance, a plant, a vehicle, and a group of people can each be one module, interacting with other modules for monitoring and control implemented in hardware and software. Consequently, time references can be implicit for some activities and explicit for others; also, the overall system model might include parts in which time is represented simply as an ordering of events and parts that are described through a metric notion of time; finally, the system may even be hybrid, with different components referring to time domains of different natures (discrete or continuous).

It is often convenient to distinguish, within concurrent components, between the *environment* and the system *embedded* into it, which typically monitors, controls, or manages the environment. We will see that the models of, and the roles attached to, system and environment significantly vary with notations and application domains. In some cases, the environment models an independent external entity that only supplies stimuli to the system, which inputs them; the system's inputs from the environment may be modeled as nondeterministic sequences of events. In other cases, the environment is just one of the components of an overall global system, and it forms a feedback loop with the other modules both by providing them with input and by reacting to their output. Models of the first type, where the environment is an independent external module, are called *open systems* (that is, open to the external environment), whereas models of the second type are called *closed systems*.

The following subsections provide a basic classification of the approaches dealing with the concurrent composition of timed units.

3.7.1   *Synchronous Versus Asynchronous Composition*

Synchronous and asynchronous compositions are two paradigms for combining the temporal evolution of concurrent modules.

Synchronous composition constrains state changes of the various units to occur at the very same time, or at time instants that are strictly and rigidly related. Time models with synchronous composition naturally refer to discrete time domains, although exceptions are possible where the overall system synchronizes over a continuous time domain.

Conversely, in *asynchronous* composition, each unit can progress independently of the others. In this view, there is no need to know in which state each unit is at every instant; in some cases this is even impossible: for instance, if we are dealing with a system that is geographically distributed over a wide area and the state of a given component changes in a time period that is shorter than that needed to send information about the state to other components. A similar situation occurs in totally different realms, such as the global stock market, where the differences in local times at locations all over the world make it impossible to define certain states about the market, such as when it is "closed".

While units progress independently most of the time, the "real" synchronization of asynchronously composed systems occurs with dedicated events at special "meeting points", and according to specific rules. The *rendezvous* mechanism of the Ada programming language, which we mentioned in Sect. 3.4.1 for its nondeterministic features, is a typical example of synchronization between asynchronous tasks: a task owning a resource (the *Buffer* in Example 3.11) waits to grant it until it receives a request thereof; symmetrically, a task that needs to access the resources raises a request (an *entry call*) and waits until the owner is ready to accept it. When both conditions are verified (an entry call is issued and the owner is ready to accept it) the rendezvous occurs, and the two tasks are synchronized. At the end of the entry execution by the owner, the tasks split again and continue their asynchronous execution. As we saw in Sect. 3.4, Ada combines this mechanism with a nondeterministic choice in case two or more different rendezvous are possible between the task owning the resources and those asking for their use at a given time.

Many formalisms feature some kind of asynchronous composition. Among these, Petri nets (described in depth in Chap. 8) exhibit similarities with the Ada task system.

Unsurprisingly, asynchronous composition is usually more complex to formalize precisely than synchronous composition. Chapters 7, 8, and 10 present several representative approaches to this problem.

Exercise 3.19. Consider the standard concurrency libraries of the following general-purpose programming languages:

- C (processes);
- Java (threads);
- C# (threads);
- Eiffel (processors and SCOOP).

Are the models of parallelism they implement synchronous or asynchronous? What kinds of synchronization mechanisms do they offer? Do you know any programming language or modeling notation that features a purely synchronous concurrency model? ∎

3.7.2 Message Passing Versus Resource Sharing

Another major classification of the mechanisms to compose and coordinate concurrent system components is into message passing and resource sharing. The two terms are rather self-explanatory: in message-passing coordination, components exchange messages over communication channels (such as buffers or pipes) to synchronize; in resource-sharing coordination, different components have access to the same resources (such as memory locations), and communication takes place when a unit reads from the shared space what another unit has written. "Google Docs" is an example of Internet-based application with shared concurrent access by multiple users, whereas email is a typical message-passing mechanism.

At lower levels of abstraction, communication will ultimately involve concurrent access to some shared resource. For example, in a high-level programming language with routine parameters passed by reference, the passage of actual parameters is akin to a message-passing communication mechanism even when it is implemented by means of sharing of global variables, accessed in a disciplined way. Even email communication involves several implementation steps where buffers, communication channels, and other resources are shared in the various stages of the transmission.

At higher abstraction levels of applications, however, message passing and resource sharing feature different and peculiar properties, and involve some clear trade-offs. Both types of coordination support asynchronous interaction among activities (processes, threads, and so on), though, in principle, they could both be deployed in fully synchronous systems. Message passing, on the one hand, decouples the timing of concurrent activities almost completely; an email message, for instance, might never be received. Resource sharing, on the other hand, usually requires stricter coordination rules, in particular to manage read and write access rights; this makes it more likely that some activities have to wait explicitly before accessing the shared resources.

Many modeling formalisms with some notion of module composition feature coordination modeling primitives that correspond to message-passing mechanisms (for example, channels and send/receive message primitives), to memory-sharing ones (for example, global variables), or to both. Chapters 4 and 10 describe some relevant examples.

In terms of other dimensions of time modeling, both message-passing and resource-sharing synchronization influence the *ordering* of events. For instance, reading an email message is possible only after it has been sent and subsequently delivered; the order is partial because messages need not be read in the same order in which they were sent. On the other hand, synchronization strategies may also depend on *metric* aspects of time in parallel systems with real-time requirements.

Exercise 3.20. Consider the following functional programming languages:

- Haskell;
- Erlang;
- Scala.

What kind of coordination mechanism do they offer: message passing or memory sharing? ∎

3.8 Analysis and Verification Issues

Formal models must be amenable to analysis to be useful, so that probing the models can determine whether the systems will behave as expected and will possess the desired features. The characterizing properties that the model (and then the system) must exhibit are often called *requirements*; hence the task of checking that a given model satisfies a set of requirements is called *verification*. Although this book is not about verification, the discussion and comparison of formalisms must refer to several notions related to verification and, more generally, formal analysis. The rest of the current section presents these notions, which have broad scope, focusing the discussion on their relevance to timed models.

3.8.1 Expressiveness

The notion of *expressiveness* refers to the possibility of characterizing extensive classes of properties; it is a fundamental criterion for the classification of formal languages. A language L_1 is more expressive than another language L_2 if the sentences of L_1 can define sets of behaviors that no sentence of L_2 can identify precisely, whereas everything definable with L_2 is definable with L_1 as well. This informal definition implies that the expressiveness relation among languages is a partial order, as there are pairs of formal languages whose expressive power is incomparable: for each language, there exist properties that can be expressed only with the other language. In other cases, different formalisms have the same expressive power; hence they can express the very same properties with different syntax. Expressiveness only deals with the logical possibility of expressing properties; hence it differs from other – somewhat subjective, but nonetheless very relevant – characterizations such as conciseness, readability, naturalness, and ease of use.

Several of the other dimensions of time modeling often mutually influence the expressiveness of the formalisms. For example, a language that can only constrain the ordering of events is less expressive, by definition, than a similar language that includes primitives to declare the temporal distance between consecutive events. In other cases, the possibility of expressing metric constraints depends on other dimensions; for example, classic temporal logic, presented in Chap. 9, can express time distances – even if somehow cumbersomely – only over discrete time domains.

3.8.2 Decidability and Complexity

Although in principle one might prefer the "most expressive" formalism, in order not to be restrained in what can be expressed, there is a fundamental trade-off between expressiveness and another important characteristic of a formal notation: *decidability*. A certain property is *decidable* for a formal language if there exists an algorithmic procedure that is capable of determining, for any model formalized in that language, whether the property holds or not in the model. Therefore, the verification of decidable properties is – at least in principle – a totally automated process. The trade-off between expressiveness and decidability arises because properties may become undecidable with more expressive languages. The verification of undecidable properties can only resort to semi-automated or manual methods, or to *partial* techniques such as testing and simulation. "Partial" means that the results of the analysis may be incomplete or incorrect for a subset of all possible behaviors of the models.

Let us consider the property of *termination* to illustrate the trade-off between expressiveness and decidability. The verification problem reads as follows: given any program expressed in a programming language L, determine if it will terminate for every possible input. Termination is a temporal property, where time can feature implicitly ("The program will *eventually* halt") or explicitly ("There exists a future time t such that the program will halt at t"). Whether termination is decidable depends on the expressiveness of the programming language L.

As discussed in Chap. 6, general-purpose programming languages, such as C and Lisp, achieve maximum expressiveness, and consequently termination is undecidable for programs in such languages. If, however, the expressive power is sufficiently restricted, termination becomes decidable. For example, the termination of programs written in a subset of C where dynamic memory allocation, recursion, and the preprocessor are disabled is decidable, because such programs use an a priori bounded amount of memory. This subset is, however, less expressive than the full language, and many C programs cannot be encoded under these restrictions.

While decidability is just a "yes/no" property, *complexity* analysis provides, in the case where a given property is decidable, a measure of the computational effort required by an algorithm to decide whether the property holds or not for a model. The computational effort is typically measured in terms of the amount of memory or time required to perform the computation, as a function of the length of the input (that is, the size of its encoding). Chapter 6 presents more details about this classical view of computational complexity.

3.8.3 Analysis and Verification Techniques

There exist two broad families of verification techniques: those based on *exhaustive enumeration* procedures, and those based on *syntactic transformations* like deduction or rewriting, typically in the context of some axiomatic description. Although large, these two classes do not cover the whole spectrum of verification algorithms, which comprises very different techniques and methods; here, however, we limit ourselves to sketching a minimal definition of these two basic techniques.

Exhaustive enumeration techniques are mostly automated, and are based on the exploration of graphs or other structures representing an operational model of the system, or of the space of all possible interpretations for formulae expressing the required properties. A typical example of this kind of technique is model checking, illustrated in Chap. 11.

Techniques based on *syntactic transformations* typically address the verification problem by means of logic deduction. These techniques can be applied when the model, its requirements, or both are in descriptive form; then the verification may consist of successive applications of deduction schemata, until the requirements are shown to be a logical consequence of the system model. Mathematical logic is a classic example of formalism focused on syntactic transformations, as shown in Chap. 2. Chapter 2 also exemplified how applying deduction schemata incurs in a trade-off between expressiveness and decidability: simple propositional logic supports deduction schemata where every expressible property is decidable, but its expressive power is limited to simple behaviors. The much more expressive predicate logic is therefore preferable for complex system specification, but no verification technique can decide the validity of every sentence of predicate logic. Chapter 9 will discuss similar trade-offs for logics supporting a notion of time.

3.8.3.1 Summing Up

This chapter presented some dimensions that characterize languages and methods for the modeling and analysis of timed systems. The dimensions will support the presentation of the languages in the rest of the book, and will help readers classify, compare, and evaluate other similar notations, and possibly even derive new ones if needed. The dimensions of this chapter are not orthogonal, and indeed we discussed many examples of mutual influence and dependence among them. The dimensions are also often qualitative, in that a rigid classification of every language against every dimension would often be vacuous, unsubstantiated, or even misleading for widely different notations with heterogeneous scopes. The rest of the book explicitly discusses the most relevant dimensions for each notation; when doing so in Part II of the book, the keywords referring to the dimensions discussed are graphically **EMPHASIZED** and referenced in the index with sub-entries corresponding to the formalisms under discussion. Dealing with the dimensions not mentioned explicitly is a useful exercise that will improve your understanding of the book's content.

3.9 Bibliographic Remarks

Koymans discusses the nature of time domains for real-time modeling [18]. Some textbooks consider hybrid systems in a general setting [22, 29]. A few authors address the issue of different time granularities [6, 8, 28]. The classical theory of sampling considers the equivalence between continuous-time signals and their discrete-time samplings [4, 9]. Wirth [30] first pointed out the difference between purely parallel and real-time systems.

The concept of invariance under stuttering was introduced by Lamport [21] and characterized for temporal logics [19, 24]. The differences between linear- and branching-time models were discussed extensively in classic references on temporal logics [10, 11, 18, 20] and real-time logics [1]; Koymans [18], in particular, mentioned branching-in-the-past time models. Classic temporal logic with the notion of an implicit current time was introduced by philosophers [17]. Pnueli pioneered its usage as a modeling tool in computing [26]. Abadi and Lamport introduced the attribute "Zeno" to describe behaviors where time stops; by analogy, we suggested the attribute "Berkeley" to characterize a different category of time advancement problems [13, 14]. Notions of nonstandard real analysis allow for a simpler approach to model and reason about Zenoness in the presence of zero-time transitions [15].

Fairness and other concurrency problems are described in various texts on parallel programming [2, 3, 12]. Most software engineering books elucidate the notions of requirements and verification [16, 25, 27].

The main features of the Ada programming language are described in many texts, such as Booch and Bryan [5]. A first critical semantic analysis of the rendezvous mechanism and a possible formalization by means of Petri nets was published by Mandrioli et al. [7, 23].

References

1. Alur, R., Henzinger, T.A.: Logics and models of real time: a survey. In: Real Time: Theory in Practice. Lecture Notes in Computer Science, vol. 600, pp. 74–106. Springer, Berlin/New York (1992)
2. Andrews, G.R.: Foundations of Multithreaded, Parallel, and Distributed Programming. Addison-Wesley, Reading (2000)
3. Ben-Ari, M.: Principles of Concurrent and Distributed Programming, 2nd edn. Addison-Wesley, Harlow/New York (2006)
4. Benedetto, J.J., Ferreira, P.J. (eds.): Modern Sampling Theory. Birkhäuser, Boston (2001)
5. Booch, G., Bryan, D.: Software Engineering with ADA. Addison-Wesley, Boston (1994)
6. Burns, A., Hayes, I.J.: A timeband framework for modelling real-time systems. Real-Time Syst. **45**(1–2), 106–142 (2010)
7. Cocco, N., Mandrioli, D., Milanese, V.: The Ada task system and real-time applications: an implementation schema. J. Comput. Lang. **10**(3/4), 189–209 (1985)
8. Corsetti, E., Crivelli, E., Mandrioli, D., Morzenti, A., Montanari, A., San Pietro, P., Ratto, E.: Dealing with different time scales in formal specifications. In: Proceedings of the 6th International Workshop on Software Specification and Design, pp. 92–101. IEEE, Los Alamitos (1991)
9. Deming, W.E.: Some Theory of Sampling. Dover, New York (2010)
10. Emerson, E.A.: Temporal and modal logic. In: van Leeuwen, J. (ed.) Handbook of Theoretical Computer Science, vol. B, pp. 996–1072. Elsevier, Amsterdam/New York (1990)
11. Emerson, E.A., Halpern, J.Y.: "Sometimes" and "not never" revisited: on branching versus linear time temporal logic. J. ACM **33**(1), 151–178 (1986)
12. Francez, N.: Fairness. Monographs in Computer Science. Springer, New York (1986)
13. Furia, C.A., Rossi, M.: A theory of sampling for continuous-time metric temporal logic. ACM Trans. Comput. Log. **12**(1), 1–40 (2010). Article 8
14. Furia, C.A., Pradella, M., Rossi, M.: Automated verification of dense-time MTL specifications via discrete-time approximation. In: Cuéllar, J., Maibaum, T., Sere, K. (eds.) Proceedings of the 15th International Symposium on Formal Methods (FM'08). Lecture Notes in Computer Science, vol. 5014, pp. 132–147. Springer, Berlin/New York (2008)
15. Gargantini, A., Mandrioli, D., Morzenti, A.: Dealing with zero-time transitions in axiom systems. Inf. Comput. **150**(2), 119–131 (1999)
16. Ghezzi, C., Jazayeri, M., Mandrioli, D.: Fundamentals of Software Engineering, 2nd edn. Prentice Hall, Harlow (2002)

17. Kamp, J.A.W.: Tense logic and the theory of linear order. Ph.D. thesis, University of California at Los Angeles (1968)
18. Koymans, R.: (Real) time: a philosophical perspective. In: de Bakker, J.W., Huizing, C., de Roever, W.P., Rozenberg, G. (eds.) Proceedings of the REX Workshop: "Real-Time: Theory in Practice". Lecture Notes in Computer Science. vol. 600, pp. 353–370. Springer, Berlin/New York (1992)
19. Kučera, A., Strejček, J.: The stuttering principle revisited. Acta Inform. **41**(7/8), 415–434 (2005)
20. Lamport, L.: "Sometime" is sometimes "not never": on the temporal logic of programs. In: Proceedings of the 7th ACM Symposium on Principles of Programming Languages (SIGPLAN-SIGACT), pp. 174–185. ACM, New York (1980)
21. Lamport, L.: What good is temporal logic? In: Mason, R.E.A. (ed.) Proceedings of the 9th IFIP World Congress. Information Processing, vol. 83, pp. 657–668. North-Holland, Amsterdam/New York/Oxford (1983)
22. Lygeros, J., Tomlin, C., Sastry, S.: Hybrid Systems: modeling, Analysis and Control. Available online https://inst.eecs.berkeley.edu/~ee291e/sp09/handouts/book.pdf (2008)
23. Mandrioli, D., Zicari, R., Ghezzi, C., Tisato, F.: Modeling the Ada task system by Petri nets. J. Comput. Lang. **10**(1), 43–61 (1985)
24. Peled, D., Wilke, T.: Stutter-invariant temporal properties are expressible without the next-time operator. Inf. Process. Lett. **63**(5), 243–246 (1997)
25. Pfleeger, S.L., Atlee, J.: Software Engineering: Theory and Practice, 3rd edn. Prentice Hall, Upper Saddle River (2005)
26. Pnueli, A.: The temporal logic of programs. In: Proceedings of the 18th IEEE Symposium on Foundations of Computer Science (FOCS'77), pp. 46–67. IEEE, New York/Long Beach (1977)
27. Pressman, R.: Software Engineering: A Practitioner's Approach, 7th edn. McGraw-Hill, Dubuque (2009)
28. Roman, G.C.: Formal specification of geographic data processing requirements. IEEE Trans. Knowl. Data Eng. **2**(4), 370–380 (1990)
29. van der Schaft, A., Schumacher, H.: An Introduction to Hybrid Dynamical Systems. Lecture Notes in Control and Information Sciences, vol. 251. Springer, London/New York (2000)
30. Wirth, N.: Toward a discipline of real-time programming. Commun. ACM **20**(8), 577–583 (1977)

Part I
Historical Approaches

The examples in the previous chapters indicate that time has had a prominent role in the mathematical models used in the natural sciences and in engineering for quite a long time, well before the development of modern computer science. This part of the book is a concise account of such mathematical models, including a notion of time that preceded – and influenced – the modern approaches.

The latter are the focus of the second part of the book. This first part of the text follows instead a historical approach, from the notion of time in the classical dynamical models used in physics and major branches of engineering, to the fundamental models of computation adopted in computer science – where the role of time often deviates from that in the classical models.

The content in this part makes the book self-contained by offering a uniform view on the essential foundations of time modeling familiar to scientists and engineers. It is also a preparation for the second part, which discusses how the modern approach to embedding computing devices in physical systems has called for a revision and expansion of the models of time in both realms (physics and computing).

The presentation in this part, as well as in Part II, targets the dimensions introduced in Chap. 3 in order to center such broad themes on the core subject of the book: the role of time and how it is modeled. The dimensions also have a didactic purpose: they help readers generalize the analysis and apply it to other approaches not discussed in the book.

Part I consists of three chapters. Chapter 4 discusses how time features in dynamical systems, the most traditional family of models used for physical phenomena and engineering systems. Chapter 5 shows the traditional view of electronic devices, where the state space and time are continuous, and relates it with the notion of digital computing devices with Boolean state space and discrete time. Chapter 6 presents "algorithmic" models of computation and the abstract notion of time adopted for the analysis of computational complexity. These concepts will also be useful for presenting the characteristics of verification algorithms for the models of Part II.

Part I
Historical Approaches

Chapter 4
Dynamical Systems

Every inhabitant of the earth can count on two natural clocks offered by the passing of days and years. It is no surprise, then, that measuring time has been a common activity since the dawn of humanity, and the notion of time features prominently in so many inventions, from natural language to modern science.

Time provides a uniform background for describing phenomena: given a clock, natural or otherwise, match every relevant event with the time when it happens. This produces descriptions where the evolution of a system consists of a sequence of "snapshots" of the visible state, each associated with a time measure that positions the snapshot against the temporal background.

Mathematics can formalize this intuitive notion of system evolution, so that it becomes rigorous and analyzable. To this end, we associate each element that belongs to the description of a phenomenon, including time, with a suitable mathematical entity – typically, elements of a set of numbers. The entities are called *variables* because they characterize how the system evolves. Every "snapshot" is then completely described by a collection of values, one for each variable plus a distinct value of time. The nontemporal variables define the *state of the system*. The system's evolution is then a function s that maps every time t to the state value $s(t)$ recorded at that time. Time is an independent variable in this view, and the other variables change according to it. The systems described in this way, with the system state as a function of time, are called *dynamical systems*.[1] In a sense, they are the quintessential operational formalism, as almost every operational model ultimately defines a dynamical system.

The present chapter is a succinct introduction to dynamical systems through a number of simple and well-known examples. We will often discuss the examples according to the dimensions introduced in Chap. 3, also referenced in the following chapters. This will help us understand the influence of classical dynamical systems

---

[1] The attribute "dynamical" comes from the Greek word for "force" ($\delta \upsilon \nu \alpha \mu o \sigma$), which is the source of system evolution in mechanical phenomena.

C.A. Furia et al., *Modeling Time in Computing*, Monographs in Theoretical Computer Science. An EATCS Series, DOI 10.1007/978-3-642-32332-4_4, © Springer-Verlag Berlin Heidelberg 2012

on the various models discussed in Chap. 5 and, in particular, the connection between physical models of electronic devices and abstract models of computation.

4.1 Discrete-Time Dynamical Systems

Natural numbers were invented or, as a genuine Platonist would put it, discovered before rational and real numbers. Correspondingly, our historical account of dynamical systems begins with a renowned example involving only discrete numbers, introduced in the early thirteenth century by the Italian mathematician Leonardo Pisano – commonly known by his patronymic Fibonacci. A man of eclectic interests, his most influential work remains the development of mathematical models for the analysis of the birth rates of rabbits and other animal species.

Example 4.1 (Fibonacci's rabbits). Fibonacci's rabbits reproduce according to a few simple rules (an abstraction of what happens in reality):

1. A rabbit's pregnancy lasts exactly 1 month;
2. A birth produces exactly two siblings, a male and a female;
3. A newborn rabbit becomes fertile when it turns 1-month old and it remains fertile for its whole life;
4. A rabbit's life is indefinitely long (within the time scales considered).

Under these rules, a farmer who starts breeding a couple of adult rabbits will have two couples after 1 month. By the end of month 2, the older couple has another pair of siblings, for a total of three pairs. By the end of month 3, the couples in the first two generations will each generate a pair of rabbits, which gives $2 + 3 = 5$ couples in total. One month later, five couples are fertile and three deliver, so the rabbit population grows to $3 + 5 = 8$ couples, and so on.   ∎

We formalize the behavior described by Example 4.1 with a single state variable R counting the number of rabbit *couples*: the dynamics of the system is completely described by the function $R : \mathbb{N} \to \mathbb{N}$ which counts the number of couples $R(t)$ alive at the end of the tth month (the time unit is the month, modeled by the set of natural numbers). According to the rules for Fibonacci's rabbits, $R(t)$ is the sum of the couples $R(t-1)$ who were around at the end of the previous month (the rabbits never die, rule (4)) plus the newly born, denoted $newR(t)$:

$$R(t) \;=\; R(t-1) \;+\; newR(r) \,. \tag{4.1}$$

We can express the number $newR(t)$ of new couples born at the end of month t by the difference $R(t-1) - newR(t-1)$: one for each couple of the previous generation, except for those born at month $t-1$, who are exactly 1 month old at the end of month t and become fertile only then (rule (3)). Finally, the newly born couples $newR(t-1)$ at $t-1$ are also expressible as $R(t-1) - R(t-2)$ according to (4.1). Thus, we can eliminate $newR$ by rewriting (4.1) as:

$$R(t) = R(t - 1) + newR(t)$$
$$= R(t - 1) + R(t - 1) - newR(t - 1)$$
$$= R(t - 1) + R(t - 1) - (R(t - 1) - R(t - 2)) \qquad (4.2)$$
$$= R(t - 1) + R(t - 2).$$

The equation

$$R(t) = R(t - 1) + R(t - 2) \qquad (4.3)$$

is sufficient to compute the dynamics of the system – that is, the concrete number of rabbits at every generation – from the initial conditions – that is, the rabbits that are bred initially. For example, if a farmer starts out with a fertile pair of rabbits at time 0, we have the sequence:

$$R(0) = 1, \quad R(1) = 2, \quad R(2) = 3, \quad R(3) = 5, \quad R(4) = 8, \quad \dots.$$

If there are no rabbits at time 0, and a pair of newly born rabbits are introduced at the end of month 1, we have the sequence called *Fibonacci numbers*:

$$R(0) = 0, \quad R(1) = 1, \quad R(2) = 1, \quad R(3) = 2, \quad R(4) = 3, \quad \dots.$$

Equation (4.3) expresses the value of R at t implicitly with a *difference equation* that recursively relates the values of R at different times. The difference equation uniquely defines the value of $R(t)$ for every $t \geq 0$, given the *initial value* of R at 0.

Similar models of the evolution of a dynamical system in terms of implicit difference (or, for continuous time and state domains, differential) equations are widely used in many branches of science and engineering. As the rest of the chapter will demonstrate, such models support powerful analysis techniques and can accommodate the description of widely different systems. In the case of Fibonacci's numbers, it is possible to turn the implicit Eq. (4.3) into a *closed form* where an *explicit* function of t gives the value of the tth Fibonacci number for the initialization $R(0) = 0, R(1) = 1$:

$$R(t) \quad = \quad \frac{1}{\sqrt{5}}\left(\left(\frac{1 + \sqrt{5}}{2}\right)^t - \left(\frac{1 - \sqrt{5}}{2}\right)^t\right). \qquad (4.4)$$

4.2 Continuous-Time Dynamical Systems

With the birth of modern science in the seventeenth century, calculus and the mathematics of continuous functions over real numbers gained a dominant role as the mathematical tools par excellence for formalizing and analyzing a vast number of systems modeling natural phenomena or engineering applications. Along these

Fig. 4.1 The diagrammatic
representation of a capacitor

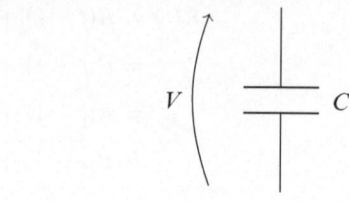

Fig. 4.2 An RC electric
circuit

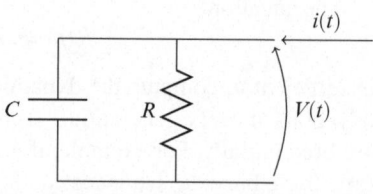

lines, the next example shows a dynamical system specified using the mathematics
of the continuum that models a common component of electric circuits.

Example 4.2 (The capacitor). Capacitors are standard components of many electric
circuits, in whose diagrammatic representations they take the form of the icon in
Fig. 4.1.

The variable that characterizes a capacitor's state is the *charge Q* accumulated
between the two plates and measured in coulomb, or the *voltage V* measured in
volt. The two representations are equivalent because charge and voltage are directly
proportional according to the equation $Q = C \cdot V$, where the *capacity C* is a
constant parameter in every ideal capacitor.

The charge Q varies over time in response to the current i that flows through the
capacitor's terminals. More precisely, i is the amount of charge flowing per time
unit, modeled as the first-order derivative of the continuous function $Q : \mathbb{R} \to \mathbb{R}$:

$$\frac{\mathrm{d}}{\mathrm{d}t} Q(t) = i(t). \tag{4.5}$$

If we integrate both sides of (4.5), we obtain the value of the state variable Q as an
explicit function of time t, which is a complete characterization of the dynamics of
our system given the initial conditions:

$$Q(t) = Q(0) + \int_0^t i(t)\mathrm{d}t. \qquad \blacksquare$$

Example 4.2 becomes more interesting if we make it a bit more complex:
consider a system made of a capacitor, characterized by a capacity C, in parallel
with a resistor with resistance R, as in Fig. 4.2. The state variable is now the voltage
V measured between the terminals, which is linked to the value of the current i
entering the circuit according to the elementary laws of electric circuits:

- The current i splits into the two parallel branches occupied by the resistor and
  the capacitor: $i(t) = i_R(t) + i_C(t)$;

- The voltage V and the current i_R at the resistor's terminals are proportional according to Ohm's law: $V(t) = R \cdot i_R(t)$;
- The change per time unit of the voltage and the current i_C flowing through the capacitor are proportional as discussed previously: $C \frac{d}{dt} V(t) = i_C(t)$.

With a little elementary calculus, we can derive the dynamics of V as an explicit function of time from the implicit dynamics defined by the differential equations just described: in the special case where no current i is ever supplied and the voltage is 1 V initially (conventionally, at time 0), V varies over time according to the exponential function $V(t) = \exp(-\frac{t}{R \cdot C})$.

4.3 The State-Space Representation of Dynamical Systems

The examples just discussed introduce the fundamental elements common to a large family of dynamical systems and to a standard approach to model them, called "state-space representation". In this approach, the model of the *state* of a system is a vector $\mathbf{x}$ of n variables:

$$\mathbf{x} = x_1 \, x_2 \, \ldots \, x_n.$$

Each variable formalizes a component of the system state, and ranges over a suitable numerical domain – discrete or continuous according to the "nature" of the item it models.

Two other vectors, $\mathbf{u}$ and $\mathbf{y}$, respectively describe the *input* stimuli the system is subject to and the *output*, which models the part of the system directly observable from the outside. $\mathbf{u}$ has m components

$$\mathbf{u} = u_1 \, u_2 \, \ldots \, u_m$$

and changes over time independently of the system evolution. $\mathbf{y}$ has l components

$$\mathbf{y} = y_1 \, y_2 \, \ldots \, y_l,$$

often just the visible components of the state; in general, it is a function of the state and the input.

Every time t in the evolution of the system has an associated value of state $\mathbf{x}(t)$, input $\mathbf{u}(t)$, and output $\mathbf{y}(t)$ variables. The value of the state $\mathbf{x}$ changes over time in response to the input supplied and depending on the history of the system, recorded by the most recent value of the state. If the time domain is discrete – typically the integers $\mathbb{Z}$ – a difference equation is the classical tool for formalizing the evolution of $\mathbf{x}$ over time:

$$\mathbf{x}(t+1) = f(\mathbf{x}(t), \mathbf{u}(t), t), \tag{4.6}$$

where the function f combines the "current" state $\mathbf{x}(t)$ and input $\mathbf{u}(t)$ to determine the "next" state $\mathbf{x}(t+1)$. If the domains of state, input, and time are instead

Fig. 4.3 A parallel RLC
electric circuit

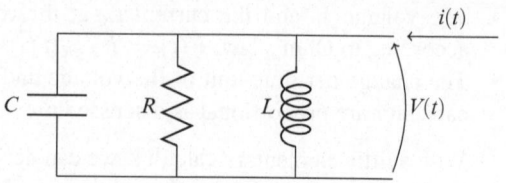

continuous sets – typically the reals $\mathbb{R}$ – a differential equation expresses the state's instantaneous rate of change as a function of the value of the state and input:

$$\frac{d}{dt}\mathbf{x}(t) = f(\mathbf{x}(t), \mathbf{u}(t), t). \tag{4.7}$$

Since in the models (4.6) and (4.7) the state changes over time according to its "previous" values, every complete evolution of the system must start from an "initial" time, customarily assumed to be 0. Correspondingly, the computation of a complete evolution depends on the initial value $\mathbf{x}(0)$ of the state. The choice of the "initialization time" is often immaterial because the Eqs. (4.6) and (4.7) are typically restricted not to depend directly on variable t, but only indirectly through references to $\mathbf{x}$ and $\mathbf{u}$:

$$\mathbf{x}(t+1) = f(\mathbf{x}(t), \mathbf{u}(t)), \tag{4.8}$$

$$\frac{d}{dt}\mathbf{x}(t) = f(\mathbf{x}(t), \mathbf{u}(t)). \tag{4.9}$$

Dynamic equations in the forms (4.8) and (4.9) are called *time invariant*, for reasons better explained in Sect. 4.6.4, as they mention only *differences* or *rates of change* over time of the states, not the value of any absolute time instant.

The model of the output is purely algebraic for both discrete and continuous time domains:

$$y(t) = g(\mathbf{x}(t), \mathbf{u}(t), t), \tag{4.10}$$

that is, the value of the state and the input at t determine the value of the output *instantaneously* at the same t. The following example discusses a continuous-time dynamical system with two state variables and one input variable.

Example 4.3 (The oscillator). Figure 4.3 shows a circuit which adds an inductor – characterized by its constant inductance L – in parallel to the circuit in Fig. 4.2. The resulting model is familiar to many engineers: it captures, in terms of ideal electric components, the fundamental behavior of every system with oscillatory phenomena, that is, *waves* in different means and forms – electromagnetic, acoustic, mechanical, and so on – but sharing the same mathematical modeling framework.

Consider a state system vector $\mathbf{x}$ with two component variables: the voltage V measured between the circuit's terminals and the current i_L flowing through the inductor (unlike the simpler system of Example 4.2, the RLC circuit requires two

Fig. 4.4 A harmonic forced
dampened oscillator

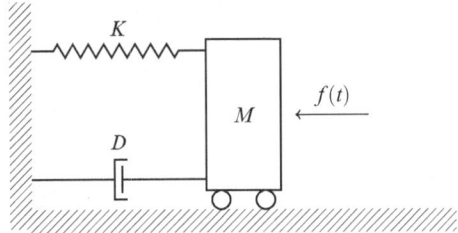

state variables to be characterized completely). The current i entering the circuit
is the (only) input variable, while the measured state component V coincides with
the output. The equation that governs the dynamics of **x** instantiates the generic
schema (4.7) according to the elementary laws of electromagnetism:

$$\frac{\mathrm{d}}{\mathrm{d}t} i_L(t) = \frac{1}{L} V(t), \tag{4.11}$$

$$\frac{\mathrm{d}}{\mathrm{d}t} V(t) = \frac{1}{c} \left(i(t) - i_L(t) - \frac{V(t)}{R} \right). \qquad \blacksquare$$

Even if the Eq. (4.11) adopt concepts and notation from electrical engineering
(voltage, current, resistance), the dynamics of every ideal oscillator follows equations structurally identical to (4.11). Consider, for example, the mechanically driven
damped oscillator in Fig. 4.4. The mono-dimensional position ("displacement") p
and the speed v of the mass correspond to current i_L and voltage V; the input force
f applied to the system is the equivalent of i in the circuit; the constant parameters
are instead the mass M of the oscillator, the viscous damping coefficient D of the
dampener, and the elasticity K of the spring, analogues of C, $1/R$, and $1/L$ in the
RLC circuit. Applying these substitutions to (4.11) we obtain the formalization of
the mechanical oscillator as a dynamical system (K and $1/L$ appear in different
positions according to their measurement units):

$$\frac{\mathrm{d}}{\mathrm{d}t} p(t) = v(t), \tag{4.12}$$

$$\frac{\mathrm{d}}{\mathrm{d}t} v(t) = \frac{1}{M} \left(f(t) - K \cdot p(t) - D \cdot v(t) \right).$$

The branch of mathematics known as harmonic analysis provides powerful tools
for computing the explicit state dynamics for systems described by differential
equations such as (4.11) or (4.12): the state components have a behavior, as an
explicit function of time, that is a linear combination of sinusoidal functions. For
example, the state variables of a simplified RLC circuit – where there is no resistor
(R is ∞), no input current ($i(t) = 0$ for every time t), and the system variables

voltage V and current i_L are initially 0 V and 1 A respectively – oscillate like the sinusoidal functions:

$$i_L(t) = \cos\left(\frac{t}{\sqrt{LC}}\right),$$

$$V(t) = -\sqrt{\frac{L}{C}} \sin\left(\frac{t}{\sqrt{LC}}\right). \qquad (4.13)$$

Exercise 4.4. Generalize Fibonacci's rabbits (Example 4.1) to the case where, at the end of every month, the farmer can buy an arbitrary number of new rabbit couples, or get rid of some of the couples already in the breeding farm. Formalize this generalization with a suitable input and revise the dynamical system model accordingly.

(*Hint*: using two state variables, $m(t)$, for the number of mature (fertile) couples, and $i(t)$, for the number of immature (infertile) couples, simplifies the model. You can then introduce an output variable $P(t)$ which counts the total number of couples, mature and immature). ■

4.4 Dynamical Systems as Models of Computation

The introductory examples of this chapter consider dynamical systems that model natural phenomena, but it is easy to notice the analogies between, on the one hand, some of the concepts used in the dynamical system framework and, on the other hand, abstract models of computation such as finite state automata and Turing machines (presented in Chap. 6): the notions of input and output, the state as a memory of the past evolution of the system, and the operational update of the state according to its previous values and the input. The analogies are stronger for discrete-time dynamical systems; Chap. 6 describes in greater detail some classic computational models and their connection with discrete-time dynamical systems. The following example discusses *cellular automata*, an unconventional type of discrete-time dynamical systems which can model some natural phenomena as computational processes, thus further emphasizing the proximity between the theory of dynamical systems and the operational models of computation.

Example 4.5 (Cellular automata). The term *cellular automata* designates a family of varied unconventional discrete-time dynamical systems where the state is represented by a spatial configuration of *cells*, each of which represents a component of the state. The update rule for the state of cellular automata, corresponding to Eq. (4.6), typically takes the spatial configuration into account, in that the next state of a cell depends only on the state of the *neighbor* cells. Such distributed models lend themselves to the formalization of biological phenomena, such as the evolution of populations or the development of skin patterns in some animal species, as well as to the analysis of distributed computation models.

In this example, we consider a cellular automaton known as "rule 110" from its number in Wolfram's catalog. *Rule 110* defines the evolution of a mono-dimensional spatial configuration of cells; the time domain is the set of natural numbers and the state is an infinite string of binary variables $s_i(t) \in \{0, 1\}$, for every $i \in \mathbb{Z}$:

$$s(t) = \ldots s_{i-2}(t)\, s_{i-1}(t)\, s_i(t)\, s_{i+1}(t)\, s_{i+2}(t) \ldots .$$

The following equation – again, a special case of (4.6) – defines the state dynamics of any cell according to its and its neighbors' values at the previous instant:

$$s_i(t+1) = \begin{cases} 1 & \text{if } s_{i-1}(t)s_i(t)s_{i+1}(t) \in \{110, 101, 011, 010, 001\}, \\ 0 & \text{otherwise.} \end{cases} \quad (4.14)$$

The output coincides with the whole state, and there is no input: the system is *closed*; correspondingly, the evolution of the system is completely determined given the initial state $s_i(0)$ of every cell s_i, for $i \in \mathbb{Z}$.

The system dynamics allowed by *rule 110* is highly complex in spite of the simplicity of the update rule: with an appropriate choice of the initial state, the evolution of the cellular automaton can emulate the computation performed by any Turing machine (see Chap. 6). In other words, the system defined by *rule 110* can "compute" any generic computable function. ∎

4.5 From Continuous to Discrete

In pure mathematics, real numbers are constructed from rationals, and rational numbers are defined as pairs of integers. Nonetheless, in spite of their common roots, the mathematics of the continuum and of the discrete have long been two largely separate branches, sharing little by way of common vocabulary, models, and techniques. For example, the toolset developed to solve differential equations over continuous domains has included, for a long time, mostly symbolic techniques, and only a few primitive numerical algorithms based on **DISCRETIZATION**.

The field of *numerical analysis* has bridged these two branches, originally separate. Numerical analysis is concerned with computing approximate numerical solutions to the problems of the mathematics of the continuum; a major branch in numerical analysis targets numerical solutions to differential equations such as those defining dynamical systems. While numerical analysis has existed for a long time – and the inventors of differential calculus, such as Newton, also introduced some numerical methods – it is only with the advent of the digital computer that numerical methods have greatly developed and prospered. Today, efficient implementations of numerical methods are indispensable to every branch of science and engineering dealing with differential calculus and dynamical systems.

Applying numerical methods requires a process of *discretization*, which approximates real numbers to within a finite number of discrete digits, so that they become amenable to manipulation by digital computers – with finite memory and in finite time. In this section, we give an idea of the discretization of dynamical system models with continuous time and state space that is performed to numerically compute their evolution. The presentation assumes a simple discretization with *fixed-point* arithmetic, where the precision of the approximation is fixed and cannot change during the computation. State-of-the-art numerical analysis uses the more flexible – and more complicated – *floating-point* arithmetic, but discussing it is out of the scope of this book.

To discretize the dynamics of a system, choose a *precision* $p_v \in \mathbb{Z}$ for every variable v in the system – including all state, input, output, and time variables. The precision denotes the granularity of the discretization: assuming, for simplicity, a decimal representation, a variable v with "real" value $\bar{v}$ has the discretized approximate value $\lfloor \bar{v} \cdot 10^{p_v} \rfloor / 10^{p_v}$; hence, the domain of every discretized variable is discrete by construction. Then, transform the differential equations defining the continuous dynamics of the original system into difference equations describing the dynamics of the discretized state space. In particular, the precision used to represent the t time variable determines the width of the discrete unit time step.

Take, for instance, the dynamics of the RC circuit in Example 4.2, assuming no input current is applied:

$$\frac{\mathrm{d}}{\mathrm{d}t} V(t) = -\frac{1}{RC} V(t). \tag{4.15}$$

If p and q respectively denote the precision of the voltage's and the time's approximation, (4.15) is approximated by the difference equation

$$\frac{\left\lfloor V\left(\frac{\lfloor t \cdot 10^q \rfloor + 1}{10^q}\right) \cdot 10^p \right\rfloor}{10^p} = -\frac{1}{RC} \frac{\left\lfloor V\left(\frac{\lfloor t \cdot 10^q \rfloor}{10^q}\right) \cdot 10^p \right\rfloor}{10^p}, \tag{4.16}$$

or, equivalently for $k \in \mathbb{N}$ and $\hat{V} = \lfloor V(k) \cdot 10^p \rfloor / 10^p$,

$$\hat{V}(k+1) = -\frac{1}{RC} \hat{V}(k). \tag{4.17}$$

We can determine the approximate dynamics of the system in an interval $[0..T]$ by computing recurrence Eq. (4.17) for K steps, with $K = \lfloor T \cdot 10^q \rfloor$, from an initial value of tension $\hat{V}_0 = \hat{V}(0)^2$:

---

[2]For simplicity, we assume that the value of constant $1/(RC)$ does not impact the precision of the calculation.

$$\hat{V}(0) = \hat{V}_0,$$

$$\hat{V}(1) = (-RC)^{-1}\hat{V}(0) = (-RC)^{-1}\hat{V}_0,$$

$$\hat{V}(2) = (-RC)^{-1}\hat{V}(1) = (-RC)^{-2}\hat{V}_0,$$

$$\vdots$$

$$\hat{V}(K) = (-RC)^{-K}\hat{V}_0.$$

The toolset of mathematical analysis can determine *closed-form* solutions, in the form of explicit functions of the independent variables, for only a few restricted classes of differential equations. In many cases, analytically deriving an explicit solution is provably impossible; in others, it is possible but extremely hard and hence impractical; often, even a partial analysis of the properties of the solution, such as its existence or its uniqueness, is unattainable. These limitations entail that explicitly computing the state dynamics $\mathbf{x}(t)$ for most but the simplest continuous-time dynamical systems is unfeasible, and the process of discretization and numerical analysis is the only viable approach to studying the dynamics of the complex systems that are considered in the natural sciences and engineering; in this respect, the widespread success of numerical analysis methods is unsurprising. Even in the simpler cases where closed-form solutions are computable, such as in our simple example of the RC circuit, numerical approximation can be a valuable analysis tool: the discretized difference equations provide a straightforward and effective algorithm to simulate and analyze the dynamics without the need for sophisticated mathematical machinery.

The power of numerical analysis does come with its own baggage of technical challenges, such as estimating and limiting the accumulation of approximation errors introduced by the finite discretization, and having efficient and robust simulation algorithms for many classes of systems. Our simplified presentation did not touch on any of these aspects, which called for the development of a rich set of techniques and methods at the boundary of mathematics and computer science.

4.6 Dynamical Systems and the Dimensions of Temporal Modeling

This section analyzes how time is modeled in dynamical systems such as those discussed in the previous part of the chapter. The analysis follows Chap. 3. It focuses on a few significant aspects, while the exercises and examples stimulate further analysis and help readers independently extend similar observations to other models.

4.6.1 Discrete and Continuous (Time) Domains

The examples of dynamical systems seen so far feature a somewhat natural correspondence between the sets used to modeling the state (as well as input and output) variables and the independent variable "time": **DISCRETE** time usually comes with state variables over discrete domains (such as in Fibonacci's rabbits), whereas **CONTINUOUS** time matches a state space of continuous, or complex, numbers (as in the RLC circuit). This correspondence is even more evident for models using differential equations, which rely on the notion of continuous function – usually defined for functions with continuous domain and range.

While this correspondence is common, significant exceptions exist where the time domain and the state domain have different characteristics, giving rise to **HYBRID** models. The first case is a dynamical system with continuous (or merely dense) state space which evolves in steps over discrete time. For example, a model different from the one in Example 4.1 could formalize the evolutionary history (over discrete generations) of a genetic feature that varies over a continuum, such as the average length of the rabbits' ears or the fraction of their fur that is black.

The opposite case of discrete state variables varying over continuous time is also possible: the evolution of such systems is a discontinuous function of time, and the discontinuities coincide with the instantaneous "jumps" from a discrete state to another (see Fig. 3.1 for an example). Such piecewise constant functions are, among other things, models for hardware digital components; Chap. 5 shows some examples and discusses the pivotal role of these hybrid models in raising the level of abstraction from the physical details of electrical circuits to a higher-level "functional" view based on the notions of abstract computation.

4.6.2 Irregular and Zeno Behaviors

The general form (4.6) and (4.7) of the state-space representation of dynamical systems allows for the definition of very complex behaviors, including some that, even if they are perfectly defined in mathematical terms, do not seem to correspond to any "real" physical phenomenon of everyday experience. Several examples have a simple definition but very **IRREGULAR** behavior.

- A continuous-time system with state x following the *Dirichlet function*,

$$x(t) = \begin{cases} 1 & \text{if } t \text{ is rational,} \\ 0 & \text{if } t \text{ is irrational,} \end{cases}$$

  which is discontinuous everywhere and cannot even be properly plotted.
- A continuous-time system with state x described by the trigonometric tangent function $\tan(t)$, undefined for every value of $t = \pi/2 + k\pi$, for $k \in \mathbb{Z}$.

- The discrete-time *logistic map* defined by the difference equation

$$x(t + 1) = r x(t) \cdot (1 - x(t)),$$

for constant $r > 0$, which defines highly irregular *chaotic* behavior difficult to predict (see also Exercise 4.10).

A particular but frequent case is that of differential or difference equations that admit both "regular" and "irregular" (in some sense) solutions. To rule out irregular solutions that are not physically feasible, without constraining a priori the generality and flexibility of the form (4.6) and (4.7), the theory of dynamical systems usually includes some a posteriori conditions that every solution representing the temporal evolution of a system must satisfy in order to be considered acceptable. Typically, the state must be defined for every value of time (after initialization); the state of continuous-time systems over continuous domains must also usually be a continuous function of time, possibly with its derivatives up to some order. Other requirements on the regularity of a solution may include the **BOUNDEDNESS** of the state by a constant (e.g., $\|\mathbf{x}(t)\| < k$ for some constants k and for all t, where $\|\mathbf{x}(t)\|$ represents some suitable *norm*, or measure, of the state space domain). These conditions may still allow for Zeno behaviors; for instance, Sect. 3.6 introduced the function $b(t)$ in (3.3) which is defined everywhere, is continuous with all its derivatives, and is bounded by 1, but is nonetheless Zeno. It is not difficult to find dynamical systems whose solutions include Zeno behaviors such as b's, for instance, in the case of unstable oscillatory phenomena.

The dynamical systems approach, which favors versatility and the power to accurately model widely different types of systems at the price of complicating the analysis and requiring sophisticated mathematical machinery, is conducive to dealing with Zeno, or in any way irregular, solutions a posteriori. In fact, this is frequently the case with irregular behaviors which correspond to phenomena that are physically feasible but undesirable because they drive the state into unstable (and possibly catastrophic) configurations. A recurrent example is systems with unimpeded positive feedback, such as the density $N(t)$ of a population of bacteria growing with unconstrained resources:

$$\frac{\mathrm{d}}{\mathrm{d}t} N(t) = r \cdot N(t), \tag{4.18}$$

where r is a constant proportional to the growth rate of the population. The general solution of (4.18) is $N(t) = N(0) \exp(r\,t)$, which diverges exponentially for every positive growth rate. This exponential divergence corresponds to an "explosive" growth of the population, which would deplete any amount of *finite* resources in a short time.

Other engineered systems include behaviors with a model such as that of (4.18), for example, nuclear reactions or resonant oscillations in electrical or mechanical devices. From a designer's viewpoint, the process of **VERIFICATION** of such systems must ascertain that no uncontrolled phenomena arise in operational conditions. Enforcing this may require a specific design consideration, such as the introduction of an input force that neutralizes the explosive growth or filters out Zeno phenomena.

Exercise 4.6. Consider the following types of irregular behavior: discontinuous, continuous with discontinuous derivative, Zeno, Berkeley, unbounded. For each of the following choices of state and time domains, which types of irregular behavior may occur?

1. Continuous and unbounded state space and time (say, $\mathbb{R}$);
2. Continuous and bounded state space and time (say, the real interval $[0, 1]$);
3. Dense state space and time (say, $\mathbb{Q}$);
4. Continuous time (say, $\mathbb{R}$) and discrete state space (say, $\mathbb{Z}$);
5. Discrete time (say, $\mathbb{Z}$) and continuous state space (say, $\mathbb{R}$);
6. Discrete state space and time (say, $\mathbb{Z}$).

(♣) Can you think of real systems where such irregular behaviors can arise?   ■

4.6.3 Determinism in Dynamical Systems

Are dynamical systems **DETERMINISTIC**? The discussion of the present chapter seems to suggest a resounding "yes": modeling the evolution of the state as a *function* of time entails that the state $\mathbf{x}(t)$ is *unique* for every time t; hence the evolution is uniquely determined by the initial state and by the input function. The evolution, however, is usually described *implicitly* by means of difference or differential equations of the form of (4.6) or (4.7); hence the system is deterministic only if the implicit equations have a *unique solution*. The classical theory of dynamical systems usually targets models with unique solutions, and all the examples presented in this chapter are indeed perfectly deterministic.

While commonplace, this is not required, and modern system theory deals with models that admit multiple feasible solutions and are therefore not deterministic. Chapter 3 discussed the difference between purely nondeterministic behavior – where every alternative choice is possible – and probabilistic (stochastic) processes – where a probability distribution determines the likelihood of different alternatives. According to this distinction, dynamical system theory typically targets **PROBABILISTIC** systems, where input, state, and output are stochastic processes and are analyzed with the tools of probability theory. The abstraction of pure nondeterminism, in contrast, is usually out of the scope of classical dynamical system theory. The presentation of major examples of operational models where *time* is a stochastic variable belongs to Part II of the book.

4.6.4 Other Dimensions of Temporal Modeling in Dynamical Systems

This section concludes with a sketchy presentation of some other dimensions of temporal modeling from Chap. 3 applied to dynamical systems. As a useful exercise, you are encouraged to comment and expand these capsule presentations to the dimensions that are most relevant to your interests.

4.6.4.1 Implicit Versus Explicit Time Reference

Time is distinctly **EXPLICIT** in the general model of dynamical systems presented by Eqs. (4.6) and (4.7), where the future values of the state may depend on the precise value of t. Time is also explicit in the solutions of the state-space equations (4.6) and (4.7), where the state vector $\mathbf{x}$ is an explicit function of the variable "time".

On the other hand, the most common instances of dynamical systems – such as those considered in this chapter – feature **IMPLICIT** time, because the dynamic equations do not refer directly to the value of t, but only to the state (and input) at the same implicit time. As mentioned in Sect. 4.3, systems with such an implicit usage of time references are usually *time invariant*: if the system reaches the same state at two different times t_1, t_2 (that is, $\mathbf{x}(t_1) = \mathbf{x}(t_2)$) and it is subject to the same input function in the future (that is, $\mathbf{u}(d + t_1) = \mathbf{u}(d + t_2)$ for all positive d), then the future values of the state are the same at corresponding instants of time (that is, $\mathbf{x}(d + t_1) = \mathbf{x}(d + t_2)$ for all positive d).

4.6.4.2 Concurrency and Composition

Being able to describe complex systems as **COMPOSITIONS** of simpler standard components is a central concern for the practical application of dynamical system models to engineering. Correspondingly, the literature on dynamical systems offers a rich set of notations and methods to describe and analyze complex systems. Figure 4.5 shows the abstract example of a block diagram where the subsystems S_1 and S_2 work in parallel on the same input; their outputs are summed up and fed to subsystem S_3 as input; and S_3's output forms a negative feedback loop that is combined with S_1's and S_2's outputs. Correspondingly, there exist techniques to compute the overall dynamics of such composite systems from the equations describing each elementary block and how they are composed.

From the point of view of time modeling, composition of dynamical system modules is typically **SYNCHRONOUS** because the same time variable t is referenced absolutely in every subsystem. This suggests that, in spite of their flexibility, dynamical system models are often unsuitable for describing systems with a high degree of distribution – where it is difficult or even impossible to have a "global"

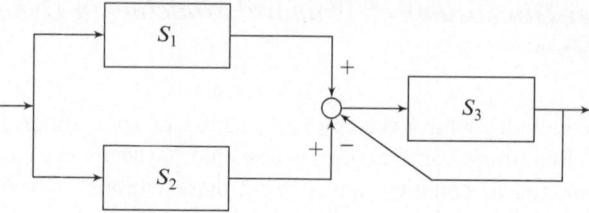

Fig. 4.5 A dynamical system block diagram

shared time totally ordered – or systems where the timing behavior is known only partially. Some of the models described in Part II have evolved to accommodate such special modeling requirements.

Exercise 4.7 (♦,♣). (*For readers familiar with the theory of differential equations*). Consider a class of differential equations you are familiar with (for example, linear ordinary differential equations or second-order partial differential equations). Its theory includes some general results about the properties of solutions, such as existence, uniqueness, regularity, dependence on initial conditions, and so on. What role do these results have from the point of view of *verification* (as presented in Sect. 3.8) of a dynamical system specified with differential equations from the class? For example, does a sufficient condition on the existence of a solution imply the *decidability* of the verification problem? ∎

Exercise 4.8. Describe a "realistic" dynamical system where the state function is necessarily *partial*, that is, not defined for some instant of time after initialization. How could you adapt the model to deal with such cases? ∎

Exercise 4.9 (♦). (*For readers with some knowledge of Fourier or Laplace transforms*). An alternative model of dynamical systems is based on the notion of *transfer function*: a function that relates the input and output of a system in terms of their frequency. In its most common form, the transfer function maps the *Fourier transform* of the input function **u** to the transform of the output function **y**. Intuitively, the transform is a representation of a function of time in its frequency domain.

- Is time implicit or explicit in dynamical system models in terms of the transfer functions?
- Are there non-Zeno behaviors of the input as a function of time whose Fourier transform is a Zeno function of frequency? What about vice versa? ∎

Exercise 4.10 (♦,♣). (*Readers unfamiliar with the basics of computation theory should postpone this exercise to after reading Chap. 6*). A dynamical system has *chaotic behavior* when its dynamics is difficult to predict because of certain characteristics, including in particular *sensitivity to initial conditions*. Informally, sensitivity to initial conditions is a form of *instability*, where a tiny change in the initial value of the state variables may result in a conspicuous change in the evolution of the state. In terms of predictability, this means that if the initial state is known only with finite precision, the state dynamics becomes completely unpredictable after some time. The logistic map (mentioned in Sect. 4.6.2) is an example of a discrete-time system with chaotic behavior.

Consider now the notion of *undecidability* applied to the dynamics of a class of discrete-time systems C: a property of C's dynamics (e.g., "Does the state of every system in C ever become positive?", "Does the state reach equilibrium?") is *undecidable* if its yes/no answer cannot be computed by any algorithmic procedure.

- Is the dynamics of a dynamical system with chaotic behavior always undecidable?
- Conversely, is a dynamical system whose dynamics is undecidable always chaotic? ∎

4.7 Notations and Tools for Dynamical System Analysis

As we will see in the following chapters, and somehow anticipated in the exercises of the previous section, the models and techniques of dynamical system theory have historically influenced the models used in computer science and buttressed their conceptual foundations. The influence has become, however, reciprocal, in that the modern models used in software engineering and programming languages are useful to structure the description and analysis of complex dynamical systems in a way that enhances reusability and modularity.

A significant example of reciprocal influence occurs in the Modelica language. Modelica supports the formalization of dynamical systems according to the notions of object-oriented design, a modern paradigm to develop complex software. Modelica integrates the mathematics of dynamical systems (difference and differential equations) with concepts closer to computer science and logic. The formal model of a system can include *descriptive* components and *partial* specifications, thus going beyond the distinctively operational attitude of dynamical system models. The Modelica language also supports the inclusion of different types of information about the system specified, such as the unit of measure of the variables, or other form of documentation usually left implicit in plain dynamical system models. Finally, the encapsulation mechanisms provided by object orientation support the definition of hierarchical systems and the reuse of template components in many different application domains of science and engineering.

MATLAB is an industrial-strength commercial environment for numerical computing, including a high-level programming language, especially suited for matrix manipulations, plotting of functions and data, implementation of numerical algorithms, and creation of user interfaces. The MATLAB environment includes Simulink, a tool for modeling, simulating and analyzing dynamic systems, offering a customizable set of block libraries and a block-diagramming graphical interface.

The MATLAB/Simulink tool environment is widely employed in control theory, digital signal processing, and "model-based design".

4.8 Bibliographic Remarks

There is a wide choice of books on modern dynamical system theory and its applications; readers are referred to Khalil [14] and Skogestad and Postlethwaite [22] as examples of such textbooks and for further reference.

Every book on the history of mathematics includes a section on Fibonacci's work; see, for example, Boyer [4]. Chong [7] describes Fibonacci's life and work; Horadam [13] discusses his influence on the development of mathematics. Mathematical Platonism [2, 10] is a (widespread) philosophy of mathematics that attributes absolute existence to abstract mathematical entities such as points and numbers.

Chua et al.'s textbook [8] is a comprehensive introduction to the classical theory of electric circuits and their dynamics.

Von Neumann introduced cellular automata in the late 1940s [24, 25], influenced by Ulam's models of crystal growth. Wolfram [26] and Schiff [21] catalog several variants of cellular automata and their modeling capabilities. Cook [9] provided the first proof that "rule 110" is Turing-complete.

Leader [16] and Burden and Faires [5] are among the most recent textbooks with extensive presentation of numerical analysis. Hamming [11] is a less up-to-date, but still classic, reference.

Signal processing theory deals with stochastic processes in the context of dynamical systems [3, 6, 15, 20].

Arnold [1] provides a classic nontechnical review of the history and applications of chaotic behavior; for technical presentations of the same subject, see Strogatz [23] and Hilborn [12]. Symbolic dynamics [17] applies methods from the theory of formal languages to the dynamics of complex nonlinear systems; Exercise 4.6 is inspired by Moore [19].

Modelica [18] is a nonproprietary standard developed by the Modelica Association.

References

1. Arnold, V.: Catastrophe Theory. Springer, Berlin (2004)
2. Benacerraf, P., Putnam, H. (eds.): Philosophy of Mathematics: Selected Readings. Cambridge University Press, Cambridge (1983)
3. Bittanti, S., Colaneri, P.: Periodic Systems: Filtering and Control. Springer, London (2009)
4. Boyer, C.B.: A History of Mathematics. Wiley, New York (1991). 2nd edn, revised by Uta C. Merzbach
5. Burden, R.L., Faires, J.D.: Numerical Analysis. Brooks Cole, Pacific Grove (2010)

6. Byrne, C.L.: Signal Processing: A Mathematical Approach. A. K. Peters, Wellesley (2005)
7. Chong, P.K.: The life and work of Leonardo of Pisa. Menemui Mat. **4**(2), 60–66 (1982)
8. Chua, L.O., Desoer, C.A., Kuh, E.S.: Linear and Nonlinear Circuits. McGraw-Hill, New York (1987)
9. Cook, M.: Universality in elementary cellular automata. Complex Syst. **15**, 1–40 (2004)
10. Ernest, P.: New angles on old rules. Times Higher Educational Supplement. http://people. exeter.ac.uk/PErnest/pome12/article2.htm (1996). Accessed 6 Sept 1996
11. Hamming, R.: Numerical Methods for Scientists and Engineers. Dover, New York (1987)
12. Hilborn, R.: Chaos and Nonlinear Dynamics. Oxford University Press, Oxford (2001)
13. Horadam, A.F.: Eight hundred years young. Aust. Math. Teach. **31**, 123–134 (1975)
14. Khalil, H.: Nonlinear Systems, 2nd edn. Prentice-Hall, Upper Saddle River (1995)
15. Lathi, B.P.: Signal Processing and Linear Systems. Oxford University Press, Oxford (2000)
16. Leader, J.J.: Numerical Analysis and Scientific Computation. Addison-Wesley, Boston (2004)
17. Lind, D., Marcus, B.: An Introduction to Symbolic Dynamics and Coding. Cambridge University Press, Cambridge (1995)
18. Modelica and the Modelica Association. http://www.modelica.org
19. Moore, C.: Unpredictability and undecidability in dynamical systems. Phys. Rev. Lett. **64**(20), 2354–2357 (1990)
20. Proakis, J.G., Manolakis, D.K.: Digital Signal Processing, 4th edn. Prentice Hall, Upper Saddle River (2006)
21. Schiff, J.L.: Cellular Automata: A Discrete View of the World. Wiley, Hoboken (2008)
22. Skogestad, S., Postlethwaite, I.: Multivariable Feedback Control: Analysis and Design, 2nd edn. Wiley, Chichester (2005)
23. Strogatz, S.H.: Nonlinear Dynamics and Chaos. Westview Press, Cambridge (2001)
24. von Neumann, J.: The general and logical theory of automata. In: Jeffress L.A. (ed.) Cerebral Mechanisms in Behavior—The Hixon Symposium, pp. 1–31. Wiley, New York (1951)
25. von Neumann, J.: The Theory of Self-reproducing Automata. University of Illinois, Urbana (1966)
26. Wolfram, S.: Cellular Automata and Complexity. Perseus Books Group, Oxford (1994)

Chapter 5
Time in Hardware Modeling and Design

The design of digital computers hides myriad abstractions that tame the formidable complexity of harnessing the unreliable dynamics of continuous physical systems to build reliable digital components with programmable behavior. From the perspective of this book, the successive abstractions of time adopted in hardware design are a cogent illustration of the role and importance of deploying widely diverse models of time and of understanding their differences and relationships. This chapter presents – along the dimensions introduced in Chap. 3 – the fundamental abstractions used in hardware design, from the "micro" elementary behavior of transistors and other electronic components, to their combination to form elementary logic gates and sequential logic, to the "macro" description of hardware architectures.

5.1 From Transistors to Sequential Logic Circuits

The ascent from the solid-state physics of semiconductors to general-purpose digital devices is too long to be accounted for, even concisely, within the scope of this book. The presentation of this chapter skips the first levels of abstraction, involving mostly complex dynamical systems based on partial differential equations to describe semiconductor physics, and starts from the *transistor* as a component providing the basic functionality of a *switch*. Figure 5.1 shows the icon of an NMOS transistor, with its three terminals: source, drain, and gate. NMOS (N-type Metal-Oxide Semiconductor) refers to a particular implementation in a doped semiconductive medium, whose details are not relevant to our discussion.

5.1.1 Logic Devices from Transistors

A few transistors and resistors are sufficient to implement the simplest digital devices: logic gates. A logic gate transforms input signals into output signals. Each signal corresponds to an electrical quantity – typically, a voltage. The signal varies

C.A. Furia et al., *Modeling Time in Computing*, Monographs in Theoretical Computer Science. An EATCS Series, DOI 10.1007/978-3-642-32332-4_5,
© Springer-Verlag Berlin Heidelberg 2012

Fig. 5.1 An NMOS
transistor

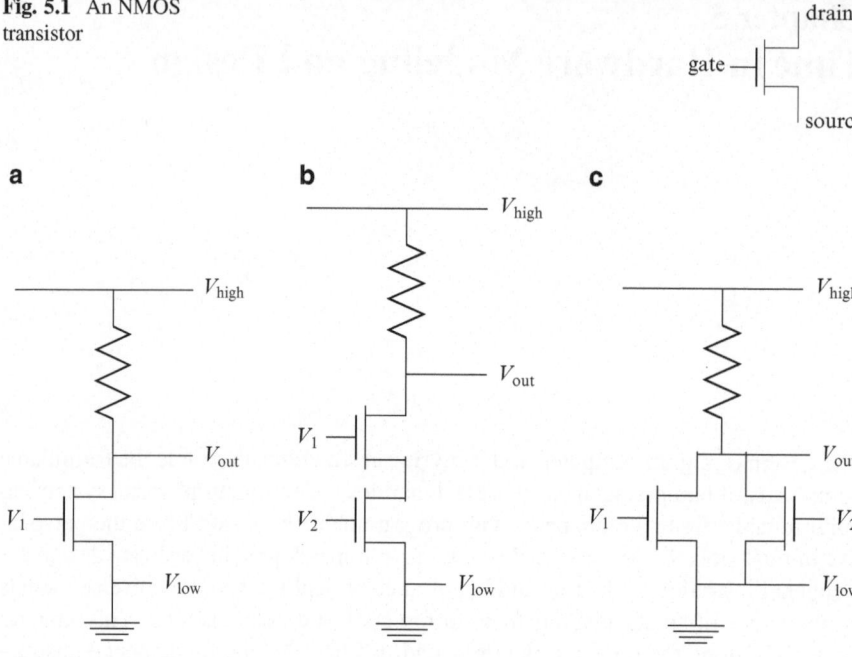

Fig. 5.2 Transistor devices implementing logic gates: NOT (**a**), NAND (**b**), NOR (**c**)

over a continuous domain, but is normally constrained between a minimum ("low")
and a maximum ("high"). Figure 5.2 shows three logic gates: a NOT gate (with one
input and one output), a NAND gate, and a NOR gate (each with two inputs and one
output). In each gate, the terminal at the bottom retains the "low" voltage V_{low} and
the terminal at the top retains the "high" voltage V_{high}.

Let us demonstrate the basic behavior of the transistor as a digital device with
the example of the NOT gate. Figure 5.3 depicts the output signal V_{out} in a NOT
gate when the input commutes from "low" to "high". Initially, before time t_1,
the input V_1 is close to the "low" value. In this configuration, the properties of
the underlying semiconductor materials make the transistor behave as a very high
resistance between the drain (maintained at V_{low}) and the output (signal V_{out}); hence
the value of V_{out} stays close to the "high" value V_{high}, possibly with a few oscillations
due to noise and imprecise readings. At time t_1, V_1 starts growing and reaches the
value "high" at t_2. After a little delay, at time t_3, the transistor reacts to this new
configuration and starts behaving as a very low resistance between drain and output.
Correspondingly, a current starts to flow between the terminal at the top and the one
at the bottom, and the voltage V_{out} – at the bottom end of the resistance – decreases
until it reaches a value very close to V_{low} at time t_4. The average delay $\Delta = t_4 - t_2$
characterizes the "reaction speed" of the transistor and is called *switching delay*.

The NAND and NOR gates combine two transistors to obtain a more complex
behavior. In the NAND gate, the transistors are put in series, so that the output is

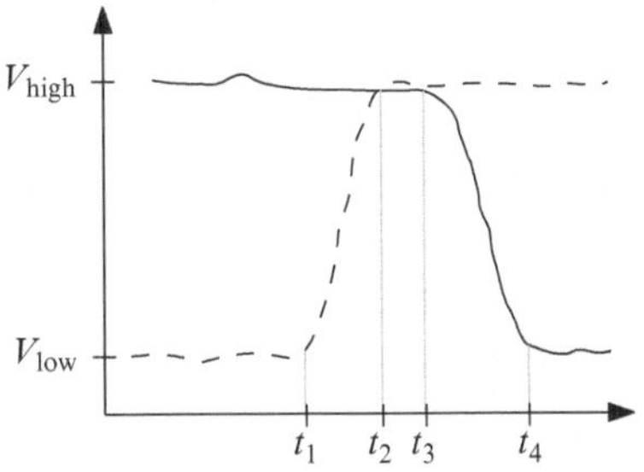

Fig. 5.3 The switching of a NOT gate: input V_1 (*dashed*) and output V_{out} (*solid*) signals

Fig. 5.4 Truth tables of logic functions: NOT (**a**), NAND (**b**), NOR (**c**)

a

NOT

V_1	V_{out}
0	1
1	0

b

NAND

V_1	V_2	V_{out}
0	0	1
0	1	1
1	0	1
1	1	0

c

NOR

V_1	V_2	V_{out}
0	0	1
0	1	0
1	0	0
1	1	0

"low" if and only if both inputs are "high"; in the NOR gate, the transistors are in parallel and the output is "high" if and only if both inputs are "low".

5.1.2   An Abstract View of Logic Gates

Figure 5.3 shows that real voltages may jiggle about the reference values "low" and "high", but tend to stick close to them before and after every transition. Correspondingly, the first abstraction of the behavior of a transistor ignores the precise values of V_{high} and V_{low} and models the inputs and outputs as varying over the continuous bounded interval $[0, 1]$. It also ignores the oscillations of a signal around the maximum or the minimum, and approximates the signal by a straight line, respectively at 1 or 0. We call this view the *logic-gate abstraction*.

From the timing analysis viewpoint, the logic-gate abstraction, which ignores the oscillatory behavior due to noise, focuses on the *transition times* such as the switching delay between input and output. The input-output behavior is instead abstracted by a function mapping input to output that ignores intermediate transition states; these abstract behavioral functions are described with *truth tables* such as those in Fig. 5.4. The graphical representation also evolves to match this more abstract view of the logic gates: Fig. 5.5 shows the standard symbols for NOT, NAND, and NOR gates whose behavior is characterized only by switching delays and truth tables, independently of the underlying transistor implementation.

Under the logic-gate abstraction, it is easier to analyze the transition delays of more complex devices that combine several elementary logic gates. For example, connecting two NOT gates in series gives a device whose output copies the input with an average delay of 2Δ time units, where Δ denotes the switching delay of a

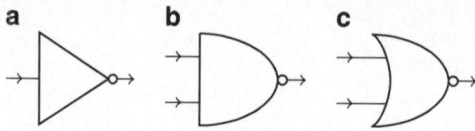

Fig. 5.5 Icons of logic gates: NOT (**a**), NAND (**b**), NOR (**c**)

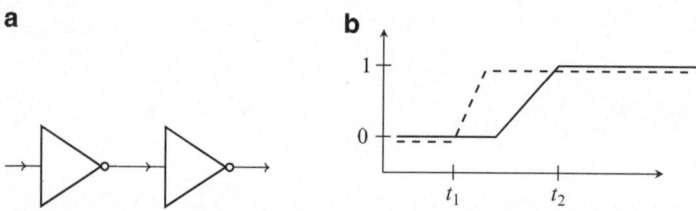

Fig. 5.6 Two NOT gates in series (**a**); a possible input/output graph (**b**), where $t_2 - t_2 = 2\Delta$ (for readability, the input signal is *dashed* and slightly shifted down, while the output signal is *solid*)

single NOT element. Figure 5.6 shows the diagram of this simple device and the corresponding "abstract" input-output graph when the input undergoes a transition from 0 to 1.

Exercise 5.1. Describe each of the following logic gates under the logic-gate abstraction, and illustrate how you can build each of them from the logic gates already presented.

- AND, whose output is 1 iff both inputs are 1; the icon for the AND gate is like the NAND's but without the disc on the output signal.
- OR, whose output is 0 iff both inputs are 0; the icon for the OR gate is like the NOR's but without the disc on the output signal.
- XOR, whose output is 1 iff either input is 1, but not both are 1.   ∎

5.1.3 From Combinatorial to Sequential Logic Circuits

The temporal behavior of logic gates is essentially characterized by the switching delay: if we combine layers of logic gates in arbitrarily complex configurations but without feeding back any output to any input, the overall output is essentially just a delayed function of the input, as the circuit "forgets" about the previous configurations of the input after the cumulative switching delay elapses. These "memoryless" networks of logic gates, where there is no feedback from output to input, are called *combinatorial logic*.

Circuits with feedback among logic gates form *sequential logic*, whose simplest components are elementary memory elements known as *latches*. Figure 5.7 shows a latch of type $\overline{\text{SR}}$ obtained from two NAND gates. In every stable state, the output

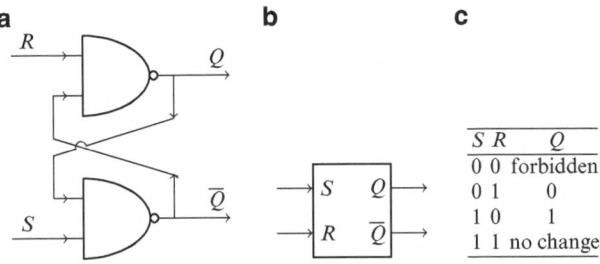

Fig. 5.7   A $\overline{SR}$ NAND latch: implementation (**a**), icon (**b**), functional behavior (**c**)

signals Q and $\overline{Q}$ are always opposite and represent the value stored by the latch (Q) and its complement ($\overline{Q}$). The input signals S and R (for "set" and "reset") are used to change the value of the output. Consider an initial configuration where $S = R = 1$: the NAND element whose inputs are Q and S outputs the opposite of Q after one switching delay Δ; similarly, the other NAND element outputs Q; therefore, the stable output value does not change as long as $S = R = 1$. If we force the input to $S = 0$ and $R = 1$, we *reset* the output to 0: the NAND gate with S as input must output $\overline{Q} = 1$ after Δ; this value of $\overline{Q}$ is passed to the other NAND gate together with $R = 1$; hence $Q = 0$ after an average cumulative delay of 2Δ. If, instead, we select the input $S = 1$ and $R = 0$, a similar analysis shows that we *set* the output Q to 1 after 2Δ.

What happens to the $\overline{SR}$ latch output when $S = R = 0$? In this case, both NAND gates will output 1, hence violating the assumption that Q and $\overline{Q}$ always store opposite values. This configuration is called "forbidden state" because it invalidates the abstract view of the latch as a programmable memory element with a defined functional behavior. In fact, consider a transition of the input from $S = R = 0$ to $S = R = 1$, where both inputs become high simultaneously. If the NAND gates have the very same switching delay Δ, the outputs will be 0 after Δ, 1 after 2Δ, 0 again after 3Δ, and so on. If, instead, the NAND gate connected to the output Q switches much faster than the other gate, the output will be $Q = 0$, $\overline{Q} = 1$, and the latch will resume normal behavior. In yet other cases, where the switching delays are close but not perfectly equal, the output will oscillate between 0 and 1 for some time, until it stabilizes to normal behavior. Such scenarios – where the switching delays affect the functional behavior of a component – are called *race conditions*, and break the logic-gate abstraction that separates time and functional behavior: the race conditions reveal details of the underlying implementation.

Exercise 5.2.   Add combinatorial components to the input of an $\overline{SR}$ latch to prevent it from going into a forbidden state. (*Hint*: filter the inputs S, R through a NOR gate which signals when $S = R = 0$; then, use an XOR gate, which outputs 1 iff either input is 1, but not both are, to change the input of the latch to "hold"). ∎

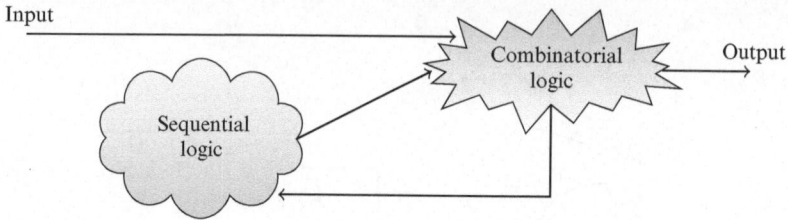

Fig. 5.8 The general structure of a sequential machine

5.2 From Two to Many States: Raising the Level of Abstraction

Elementary memory components such as latches are also called *bistables*, since their output has exactly two stable states, 0 and 1 (also represented as **False** and **True**). We can compose several bistables to obtain larger state spaces, that is, larger memory components: the composition of two bistables gives a "four-stable", and the composition of n bistables gives a "2^n-stable" – n-bit memory which can take 2^n configurations.

5.2.1 Sequential Machines and Zero-Time Transitions

More generally, complex circuits mix combinatorial and sequential components as in Fig. 5.8. The sequential part stores the current state of the system; the combinatorial components compute the output and the "next" state of the system as a function of the current state and the input; the new state is fed back to the sequential part to be stored as the new current state. Circuits following this distinctly **OPERATIONAL** model are called (digital) "sequential *machines*".

Exercise 5.3 (♣). Show the correspondence between the notions of state, input, and output in the model of sequential machines and in the model of dynamical systems introduced in Chap. 4.   ∎

Example 5.4 (Sequential machine). The circuit of Fig. 5.9 is a simple sequential machine where a NAND logic gate connects two $\overline{\text{SR}}$ latches. The overall circuit has three inputs (S_1, R_1, R_2) and two outputs (O_1, O_2) with their complements ($\overline{O}_1$, $\overline{O}_2$), which in this case are simply the outputs of the two latches. The cumulative switching delay of the circuit is the sum of the switching delays of the two latches and the NAND gate. The combinatorial part of the machine is such that we can reset the output O_2 to 0 only when $R_1 = \overline{O}_1 = 1$, that is, when the latch #1 is reset or held at $O_1 = 0$. Correspondingly, the following Fig. 5.10 describes the input/output behavior of the circuit, where "*" denotes a "don't care" value (which does not

Fig. 5.9 A sequential machine with two $\overline{SR}$ latches and a NAND gate

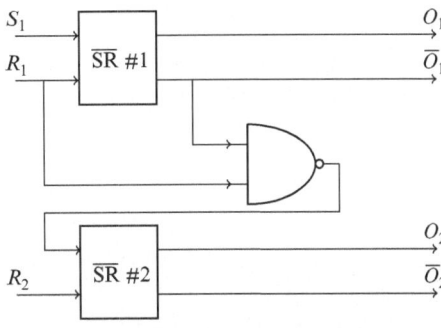

Fig. 5.10 The input/output behavior of the machine in Fig. 5.9

S_1	R_1	O_1	new O_1	R_2	S_2	O_2	new O_2
1	1	0	0	1	0	*	0
1	1	1	1	1	1	1	1
1	1	1	1	1	1	0	0
1	1	1	1	0	1	*	1
1	0	*	1	1	1	1	1
1	0	*	1	1	1	0	0
0	1	*	0	1	0	*	0

Fig. 5.11 A behavior of the machine in Fig. 5.9

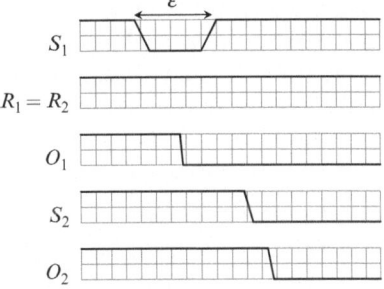

affect the output) and configurations leading to "forbidden states" of the latches are omitted.

Figure 5.11 depicts a behavior of the machine for an initial output $O_1 = O_2 = 1$, where the two inputs R_1 and R_2 hold the value 1 while S_1 switches to value 0 for ε time units. The behavior in the picture is consistent with the logic-gate abstraction as long as ε is greater than the switching delay of the $\overline{SR}$ latch; otherwise, S_1's transitions may not have a visible effect on the other parts of the circuit and, in particular, on O_1 and $\overline{O}_1$.   ■

In the logic-gate abstraction used so far, transitions among the states 0 and 1 take a variable non-null amount of time (see Figs. 5.6b and 5.11). This is consistent with the "microscopic" behavior of electronic components (Fig. 5.3): the speed of change of any physical quantity is finite. The previous Example 5.4, however, has shown that only the time between consecutive stable states affects the functional behavior of the machine, while the precise speed of each transition does not. This

Fig. 5.12 The state Q of an
$\overline{\text{SR}}$ latch set to 1 with
zero-time transitions (Δ is the
switching delay of the latch)

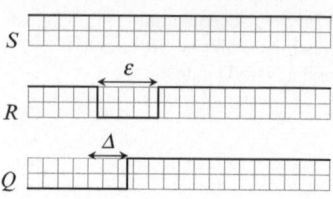

encourages our introducing the additional abstraction of **ZERO-TIME** *transitions*:
given that the detailed transition dynamics is irrelevant, we abstractly represent it as
if it were instantaneous. Correspondingly, the behaviors become piecewise constant
functions with domain time and range $\{0, 1\}$, such as the behavior of an $\overline{\text{SR}}$ latch
changing the state Q from 0 to 1 depicted in Fig. 5.12.

5.2.2 Finite-State Machines

Finite-state machines are a convenient and widely used notation to describe
the functional behavior of sequential machines under the zero-time abstraction.
Innumerable operational models used in various application domains are variants
or extensions of finite-state machines. This section continues with the description
of the basic model as used in the context of hardware design, while the rest of the
book discusses several important variants, especially in Chap. 7.

A finite-state machine consists of:

- A finite set Q of states;
- A finite set I of input values;
- A transition function $\delta : Q \times I \rightarrow Q$.

The transition function δ describes how the state changes in response to every
input: if the machine is in some state $q \in Q$ and receives some input $i \in I$, it
changes its state to $q' = \delta(q, i)$ in zero time. The abstraction introduced by the
finite-state machine model only considers stable inputs – the input states held long
enough to have visible effects on the rest of the circuit.

Consider again the $\overline{\text{SR}}$ latch (Fig. 5.7) under the zero-time transition abstraction.
To model its functional behavior, we consider a finite-state machine with the set of
states $\{0, 1\}$, modeling the value stored by the output signal Q, and the set of input
values $\{s, r\}$ for set (inputs $S = 1$, $R = 0$) and reset ($S = 0$, $R = 1$) signals. The
transition function does not allow for transitions to the forbidden state, and models
the "hold" state $S = R = 1$ implicitly by considering only the sequence of visible
set and reset inputs:

$$\begin{cases} \delta(0, s) = \delta(1, s) = 1 \,, \\ \delta(0, r) = \delta(1, r) = 0 \,. \end{cases}$$

Fig. 5.13 Graphical representation of the finite-state machine for an $\overline{SR}$ latch

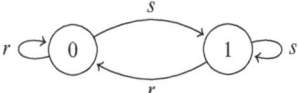

The widespread usage of the finite-state model and its variants is due not only to its simplicity but also to its standard intuitive graphical representation: a directed graph with the nodes corresponding to the states in Q and, for every transition $\delta(q, i) = q'$, an arc with label i connecting node q to node q'. Figure 5.13 represents the finite-state model of the $\overline{SR}$ latch described in the previous paragraph.

Exercise 5.5. Extend the finite-state machine model of the $\overline{SR}$ latch to include transitions to the "forbidden state".

(*Hint*: no transition exits the forbidden state).   ■

Exercise 5.6. Model the sequential machine of Fig. 5.9 with a finite-state machine.
                                                                                    ■

5.2.3   Finite-State Machines with Output: Moore and Mealy Machines

In dynamical system models (discussed in Chap. 4), it is customary to include the notion of *output* to describe the quantities directly observable as a function of the current state and, possibly, of the input. The basic finite-state machine model of Sect. 5.2.2 can also be extended with a similar notion of output. In hardware design, the two standard models of finite-state machines with output are called "Moore machines" and "Mealy machines", after their inventors.

A *Moore machine* models the output with a function $\eta : Q \to O$ that assigns an output value, from a finite set O, to every *state*. Hence, the current state univocally determines the output, but different states may produce the same output. The graphical representation of Moore machines appends each node with an output label (next to the state label).

A *Mealy machine* models the output with a function $\eta : Q \times I \to O$ that assigns an output value, from a finite set O, to every *transition*. Hence, the current state and input determine the output. The graphical representation of Mealy machines appends each arc with an output label (next to the input it corresponds to).

Consider again the finite-state machine of Fig. 5.13, modeling an $\overline{SR}$ latch. Add an output value $o \in O$ whenever the state is set to 1 from 0. Figure 5.14 shows a Mealy machine with this behavior, where the special output value ε is used for transitions that do not correspond to any visible output. There are various ways to accommodate such "silent" outputs in the formalization of Moore or Mealy machines:

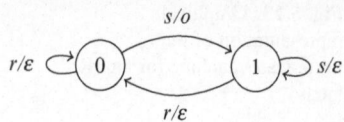

Fig. 5.14 A Mealy machine modeling an $\overline{SR}$ latch with output

1. ε is an output symbol (that is, $\varepsilon \in O$) with special meaning.
2. The output function η is partial.
3. The codomain of the output function η is O^*: every output value is in general a finite (possibly empty) string of output values. (This model is compatible with the underlying physical behavior only if the emission of the output strings can be performed within the time abstracted by zero-time transitions).

Exercise 5.7. Model the behavior of the Mealy machine in Fig. 5.14 with a Moore machine that produces the same output sequence in response to the same input sequence. ∎

The notion of output as a function of the current state and input, adopted by Mealy machines, is closer to the general definition of output in the state-space representation of dynamical systems (Sect. 4.3), but whenever the output is independent of the current input the abstraction of Moore machines may be more practical. In any case, Exercise 5.8 shows that Moore and Mealy machines are equivalent formalisms.

Exercise 5.8. Generalizing Exercise 5.7, prove that Moore and Mealy machines are equivalent formalisms:

1. Show that a generic machine M with output (Moore or Mealy) defines a translation function $\tau^M : I^* \to O^*$ from strings of input symbols to strings of output symbols.
    (*Hint*: τ^M has an inductive definition in terms of δ and η).
2. Given a generic Moore machine M, build a Mealy machine M' such that $\tau^{M'} = \tau^M$.
3. Given a generic Mealy machine M, build a Moore machine M' such that $\tau^{M'} = \tau^M$. ∎

5.3   From Asynchronous to Synchronous Logic Circuits

Let us revise the abstractions introduced so far in the description of logic components. The modeled physical quantities (voltages, currents, etc.) are *continuous* but *bounded* by a maximum and a minimum; they vary over a **CONTINUOUS TIME** domain. In the logic-gate abstraction, behaviors become continuous functions of time, the bounded domain is normalized to the real interval $[0, 1]$, and the switching delays represent time **IMPLICITLY**. With the introduction of zero-time transitions,

Fig. 5.15 The clock as a
square wave (*dotted line*) or
as a sequence of impulses
(*solid line*, slightly shifted for
readability)

$$\overset{\longleftrightarrow}{1/f}$$

$$\overset{\longleftrightarrow}{1/(2f)}$$

behaviors become piecewise-constant functions of time into the discrete finite set
of Boolean values $\{0, 1\}$. Time is always **METRIC** and **ASYNCHRONOUS**: the state
(and the output) of a sequential machine reacts, with a delay, to transitions occurring
at *any* time.

5.3.1   Logic Synchronized by a Clock:
The Complete-Synchrony Abstraction

Asynchronous logic is used "analogically", mostly for specialized hardware components; the preferred technology to build general-purpose digital computers uses,
instead, another family of logic components where state changes are **SYNCHRO-
NIZED** with a *clock*: a periodic signal that is shared by all components in the circuit
to achieve global synchrony. The clock typically is a square wave with constant
frequency f and a 50 % duty cycle, which holds the value 1 for $1/(2f)$ (half the
period) and the value 0 for the other half. If the period is long enough with respect
to the switching delays of the circuit synchronized by the clock, we can further
abstract the clock signal as a sequence of zero-time impulses every $1/f$ time units.
Figure 5.15 shows the two equivalent representations.

Elementary memory elements with a clock are called "flip-flops"; they are
synchronous variants of latches. Figure 5.16a shows a simple way to synchronize the
transitions of an $\overline{\text{SR}}$ latch with a clock: combine each input S' and R' with the clock
through an OR gate (see Exercise 5.1), so that the latch's inputs are relevant only
when the clock signal is 0. The synchronous behavior forced by the clock justifies an
input-output diagram even more abstract than the one in Fig. 5.12: the precise time at
which the "real" input changes asynchronously is irrelevant, because the device will
read the input only synchronously with the clock; correspondingly, we can replace
the synchronized input S, R for the real input S', R'. Figure 5.16b shows such a
complete-synchrony abstraction applied to a transition of the sequential machine. In
the practice of digital circuit design, the synchronous level of abstraction is called
the "register-transfer" level.

Example 5.9 (T-type flip-flop). Figure 5.17a shows the icon for a flip-flop known as
"T-type" (or just "T"); the input line with a triangle is reserved for the clock signal.
If the input signal T is 0, the output Q (and its complement $\overline{Q}$) holds its previous
value. If the input T is 1, the output toggles – that is, changes from 1 to 0 or vice
versa – synchronously with the *rising* edge of the clock signal. Figure 5.17b shows
the behavior of a T flip-flop under the complete-synchrony abstraction.   ∎

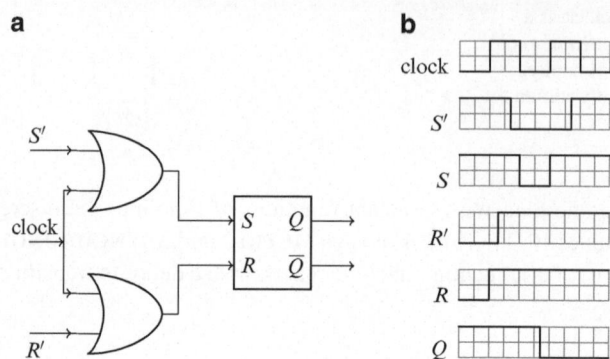

Fig. 5.16 The $\overline{\text{SR}}$ latch synchronized with the clock (**a**) and a synchronized reset transition (**b**)

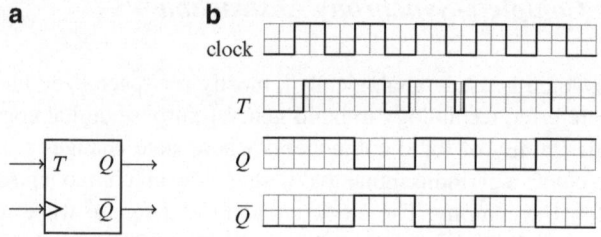

Fig. 5.17 The symbol for a T flip-flop (**a**) and its input/output diagram (**b**)

Fig. 5.18 The symbol for a
JK flip-flop

Exercise 5.10. Modify the circuit of Fig. 5.16a so that the latch reads the input only when the clock is 1.   ■

Exercise 5.11. According to Fig. 5.17b, the output of a T flip-flop toggles synchronously with the rising edges of the clock.

- Describe how to use a T flip-flop with this property to transform a clock signal with frequency f to a clock signal with frequency $f/2$.
- Add a combinatorial component to the T flip-flop to build a component that toggles at the *falling* edges of the clock.   ■

Exercise 5.12 (JK flip-flop). The "JK flip-flop" (Fig. 5.18) is another synchronous variant of the $\overline{\text{SR}}$ latch, with the inputs S and R respectively renamed to J and K. When $J = 1$ and $K = 0$, the output Q is set to 1 at the next rising edge of the clock; when $J = 0$ and $K = 1$, the output Q is reset to 0; when $J = 1$ and $K = 1$, the output Q toggles; and when $J = 0$ and $K = 0$, the output is held.

- Draw an input-output diagram for the JK flip-flop.
- Modify the JK flip-flop so that it behaves as a T flip-flop.
- Modify the JK flip-flop so that it behaves as an asynchronous $\overline{SR}$ latch (ignore the behavior in the forbidden state).

    (*Hint*: replace the clock with a constant signal).
- Model the JK flip-flop as a Mealy machine.   ■

Exercise 5.13 (♣). This section applied the notion of synchronous and asynchronous behavior to sequential logic. Consider now purely combinatorial logic circuits.

- Would you classify their behavior as synchronous or asynchronous?
- Can you describe the behavior of a synchronous NOT gate and design a component that implements it? Is the resulting circuit still purely combinatorial?   ■

5.3.2 From Continuous to Discrete Time

Under the complete-synchrony abstraction, the transitions of every component take place simultaneously with the clock transitions, and the effect of each transition becomes visible exactly when the next clock cycle begins. In other words, all the visible behavior of the signals occurs at isolated instants of time, corresponding to the clock's periodic impulsive "ticks" (Figure 5.15), whereas the value of the signals between any two consecutive transitions of the clock is irrelevant. The notion of **EVENT** captures this idea of *instantaneous* occurrence, and it provides a further abstraction of the behavior of a synchronized circuit: a sequence of isolated events with constant frequency (determined by the clock). Since the events take no time and are evenly spaced out in time, it is reasonable to measure time globally by *counting* the discrete tick events of the clock. Hence, a **DISCRETE** *set* (such as the natural numbers) becomes an appropriate domain for time.

Finite-state machines, introduced with reference to asynchronous behavior in Sect. 5.2.2, can model synchronous behaviors as well, under a different interpretation of the input as *events*. Take, for example, the finite-state machine in Fig. 5.13, which models an asynchronous $\overline{SR}$ latch. The input event s is triggered whenever the latch's signal S is set to 1 while the other signal R is 0 (vice versa for the other input event r). Since such events happen asynchronously, the finite-state machine can sit in a state for an indefinitely long amount of time, and change state asynchronously whenever the input is triggered. Now, interpret the same finite-state machine as a synchronous device. At the beginning of every clock cycle, the input event s is triggered if S is 1 and R is 0, and the input event r is triggered if S is 0 and R is 1: the finite-state machine can change state only synchronously with the clock.

The finite-state model of Fig. 5.13 represents the "hold" state of the $\overline{SR}$ latch ($S = R = 1$) only implicitly, by retaining the same state unless a set or reset event occurs. In the synchronous interpretation, we can model this aspect explicitly:

Fig. 5.19 A finite-state
machine for a synchronized
$\overline{SR}$ latch, including "no
change" events

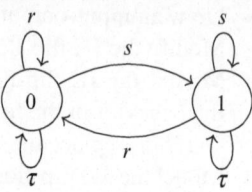

introduce a new event $\tau \in I$ corresponding to $S = R = 1$ when the clock
ticks; the state is held whenever τ occurs (see Fig. 5.19). Then, an event occurs
(and a transition is taken) at the beginning of *every* clock cycle, and counting the
number of consecutive events measures the elapsed time.[1] The machine is fully
synchronized with the rest of the system, and time is genuinely **METRIC**, as it is in
the asynchronous model.

Exercise 5.14. Model a T-type flip-flop with a finite-state machine in the synchronous interpretation:

- With implicit hold of the state when $T = 0$;
- With explicit "no change" events when $T = 0$.   ■

Exercise 5.15. A serial-in-serial-out n-bit shift register is a synchronous sequential
machine with an n-variable state $v_1, \ldots, v_n$ that reads a sequence $b_1 b_2 \ldots b_i \ldots$ of
bits from the input, where v_i and b_i are in $\{0, 1\}$ for every i. The machine updates
the state according to the following rules. At the ith clock tick:

- The machine inputs bit b_i and the state variable v_1 takes the value b_i;
- For every other $2 \leq j \leq n$, the state variable v_j takes the value of the state
  variable v_{j-1} at the previous clock tick; that is, the state *shifts* by one position.

For example, for $n = 3$ and an initial state $\langle v_1, v_2, v_3 \rangle = \langle 0, 0, 0 \rangle$, an input sequence
1 0 0 1 0 1 1 0 determines the sequence of states:

$$\langle 0,0,0 \rangle \; \langle 1,0,0 \rangle \; \langle 0,1,0 \rangle \; \langle 0,0,1 \rangle \; \langle 1,0,0 \rangle \; \langle 0,1,0 \rangle \; \langle 1,0,1 \rangle \; \langle 1,1,0 \rangle \; \langle 0,1,1 \rangle .$$

1. Model a 3-bit shift register as a discrete-time dynamical system.
2. Model a 3-bit shift register as a finite-state machine.
     (*Hint*: consider how many states and transitions you need).
3. Describe how to implement a 3-bit shift register using three $\overline{SR}$ latches synchronized by the same clock.   ■

Exercise 5.16. Consider the circuit in Fig. 5.20: the signals $Q, \overline{Q}$ output the state,
whereas D and E are inputs.

1. Describe its functional behavior under the logic-gate abstraction and asynchronous time.

---

[1] Under the assumption that the forbidden input $S = R = 0$ never occurs.

Fig. 5.20 A D-type flip-flop

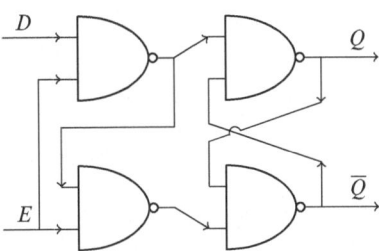

2. Show a finite-state machine formalizing the behavior of (1).
3. Assume now that a clock signal feeds the input E; in this configuration, the circuit implements a so-called "D-type flip-flop". Describe the flip-flop's behavior under the complete-synchrony abstraction.
4. Show a finite-state machine formalizing the behavior described in (3). Can you reuse the same machine as the one in (2)? ∎

5.4 Modular Abstractions in Hardware Design

The complete-synchrony abstraction and the models introduced in the previous section are in principle sufficient to design any synchronous digital machine: even the most powerful supercomputer can be modeled by a state machine with a finite – albeit huge – set of states. It is clear, however, that the notations described in Sect. 5.3 are practical only for small components with well-defined behavior. To manage the complexity of general-purpose hardware, we need a higher level of abstraction and models suitable for describing large modular architectures of components.

5.4.1 The Behavioral Abstraction

The new abstraction needed is called "behavioral" in hardware engineering terminology. Behavioral abstraction focuses on the *functional* behavior of the individual components and on the flow of data between them. It also provides mechanisms to encapsulate components into modules, and to compose modules hierarchically. Temporal aspects are entirely **IMPLICIT** in behavioral abstraction, because the clock is relied upon to guarantee the appropriate synchronization.

Consider, for example, the informal behavioral model of an adder in Fig. 5.21. The ingoing and outgoing arrows denote the adder's inputs (A_1, A_2) and outputs (SUM, $OVERFLOW$); the box contains a description of the function performed by the adder: SUM contains the addition of the values A_1 and A_2, and $OVERFLOW$ is a Boolean flag which is true if and only if the sum is above a certain maximum

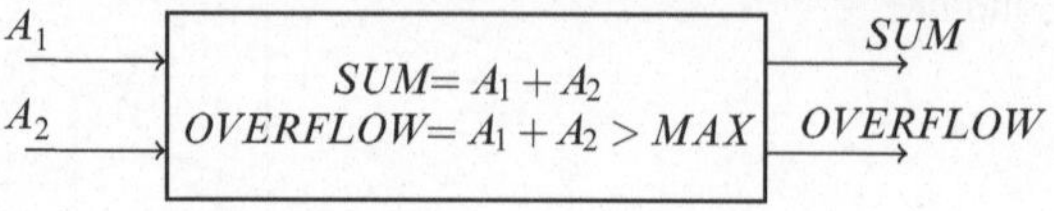

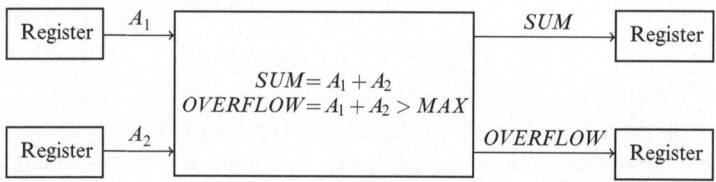

Fig. 5.21 The dataflow diagram of an adder

Register —A_1→ | SUM = $A_1 + A_2$ |—SUM→ Register
Register —A_2→ | OVERFLOW = $A_1 + A_2 > MAX$ |—OVERFLOW→ Register

Fig. 5.22 An adder with input/output memory registers

MAX. In Fig. 5.22, we add four memory components and connect them to the adder; the result is a circuit that, at every clock cycle t, reads A_1 and A_2 from the registers, computes their sum (and overflow bit), and stores the result in the two other registers, where it is available at the next clock cycle $t + 1$. These temporal aspects of the behavior are entirely implicit in Figs. 5.21 and 5.22, which explicitly model only the data flow and the components involved.

5.4.2 Hardware Description Languages

The industrial practice of hardware design uses languages that are more formal and expressive than the informal dataflow diagrams shown in Figs. 5.21 and 5.22. They are typically textual, and sufficiently rich to support multiple levels of design abstraction. They also typically come with a collection of supporting tools to perform various types of analyses, including extraction of graphical representation from the textual description, simulation, and synthesis (see Sect. 5.5 for more details on these aspects).

The rest of this section gives a short informal overview of VHDL, a widely used hardware description language. VHDL originated from a project started in the 1980s by the US Department of Defense to build "very high speed integrated circuits" (that is what the "V" in VHDL stands for, while "HDL" is an acronym for "Hardware Description Language"). The Ada programming language was also launched by the same institution during the same years, and as a result some concepts and syntax in VHDL borrow from Ada (although the two languages mostly have different scopes).

Let us provide a behavioral description of the adder in Fig. 5.21 using VHDL. We introduce two separate declarations. First, in Fig. 5.23, an **entity** describes the interface of the component: its inputs and outputs, and their types (integers of generic maximum size N, except for the $OVERFLOW$ bit).

Fig. 5.23 An **entity** declaration in VHDL

entity *ADDER* **is**
    generic (*N*: *integer*)
        port (
                A_1, *A_2*:   **in** *integer* **range** 0 **to** *N*;
                SUM: **out** *integer* **range** 0 **to** *N*;
                OVERFLOW: **out** *bit* ;
        );
    end *ADDER*;

Fig. 5.24 An **architecture** declaration in VHDL

architecture *BEHAVIOR* **of** *ADDER* **is**
    constant *MAX*: *integer* := *N* + 1;
  begin
    SUM <= *A_1* + *A_2*;
    OVERFLOW <= 0 **when** *SUM* < *MAX* **else** 1;
  end *BEHAVIOR*;

Fig. 5.25 An **entity** declaration of a compound element

entity *DOUBLER* **is**
    generic (*N*: *integer*)
        port (
                I_1, *I_2* :   **in** *integer* **range** 0 **to** *N*;
                TWICE: **out** *integer* **range** 0 **to** *N*;
                OVER: **out** *bit* ;
        );
    end *DOUBLER*;

Then, in Fig. 5.24, an **architecture** defines the functional behavior of the adder: *SUM* receives the result of the sum *A_1* + *A_2*, whereas the *OVERFLOW* bit is set to 0 if the sum is not greater than *N* and to 1 otherwise. The two "signal assignments" to *SUM* and *OVERFLOW*, denoted by the "<=" operator, are executed concurrently.

We now build a more complex component by connecting multiple instances of the adder. Precisely, the system will compute *TWICE* = 2 * (*I_1* + *I_2*) by having two adders each computing *I_1* + *I_2* in parallel, and then adding up the results with a third adder. The interface of the overall component, shown in Fig. 5.25, also includes an overflow bit.

The architecture built, shown in Fig. 5.26, instantiates the three adders and connects them by *signals*, which may correspond to actual physical connections in a real implementation. The mapping between the components' inputs and outputs by means of signals is defined, through the "=>" operator, in the **port map** section. The cumulative overflow bit is set to 1 if at least one of the adders overflows. Figure 5.27 shows the data flow of the *DOUBLER* component.

The examples introduced so far are at the behavioral level of abstraction, but VHDL also supports lower-level descriptions. For example, the definition of a T-type flip-flop (introduced in Example 5.9) at the register-transfer level under the complete synchrony assumption can mention the *CLOCK* signal explicitly and describe how the state is updated, by means of the sequential assignment operator ":=", at the rising edge of the clock (that is, **on** *CLOCK* and when *CLOCK* = '1'). The result is shown in Fig. 5.28. The semantics of a VHDL **process** ensures that the

```
architecture STRUCT of DOUBLER is
    signal SUM_1, SUM_2: integer range 0 to N;
    signal OVER_1, OVER_2, OVER_3: bit;
begin
    ADDER_1: ADDER
            generic map (N => N);
            port map (A_1 => I_1,      A_2 => I_2,
                        SUM => SUM_1,  OVERFLOW => OVER_1);
    ADDER_2: ADDER
            generic map (N => N);
            port map (A_1 => I_1,      A_2 => I_2,
                        SUM => SUM_2,  OVERFLOW => OVER_2);
    ADDER_3: ADDER
            generic map (N => N);
            port map (A_1 => SUM_1,  A_2 => SUM_2,
                        SUM => TWICE,  OVERFLOW => OVER_3);
    OVER <= OVER_1 or OVER_2 or OVER_3;
end STRUCT;
```

Fig. 5.26 An **architecture** declaration of a compound element

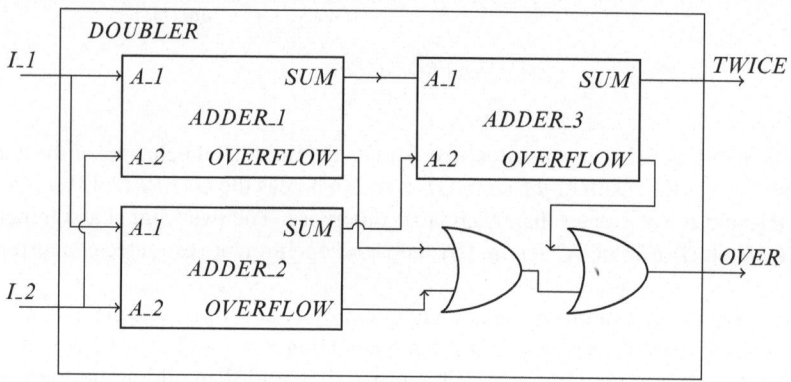

Fig. 5.27 Dataflow diagram of component *DOUBLER*

process statements are executed *sequentially*. Therefore, the process model must specify in detail how the various data components are updated and in what order. Processes in different components execute in parallel and synchronize with shared signals.

5.5   Methods and Tools for Hardware Analysis

In the last decades, hardware design techniques have made tremendous progress towards *automation*. Hardware description languages, such as VHDL, presented in

```
architecture RTL of T_FLIP_FLOP is
    variable new_Q: bit ;
begin
    process (T, CLOCK, Q)
    begin
            wait on CLOCK;
            new_Q := Q;
            if (CLOCK = '1' and T = '1') then
                new_Q := not new_Q;
            end if ;
            Q <= new_Q;
            Q' <= not new_Q;
    end process;
end RTL;
```

Fig. 5.28 A T-type flip-flop description in VHDL

the previous section, are supported by a wide array of tools that facilitate design at several levels of abstraction by taking care of the most tedious aspects.

The following activities are especially relevant for hardware design:

Testing, which consists of subjecting a physical device to certain inputs and observing the outputs to verify that the device behaves as expected.

Simulation, which is essentially testing performed on an executable model of the device, rather than on the real implementation.

Formal verification, where the model is analyzed rigorously by means of mathematical techniques to ensure that it satisfies the properties expected.

Synthesis, that is, the automatic construction of models at a lower level of abstraction that implement a higher-level design. The lower level may be detailed enough to be physically implementable and then testable.

At the levels of abstraction where transistors and other electronic components are modeled explicitly, testing of the physical implementation is the traditional verification approach. Testing is, however, time consuming and expensive because it requires a physical implementation that can be checked for correctness. Therefore, powerful and accurate simulation tools have been developed; SPICE and its successors are among the most used software tools in this area. These programs can simulate, for example, the behavior of transistors and produce graphical displays such as that of Fig. 5.3, which accurately represent the behavior of real transistors but are built without an actual implementation.

Automated synthesis is possible, and widely used, for models at the logic-gate abstraction level: the synthesizer automatically produces a description in terms of transistors that behave as in the logic-gate model. Simulation is also commonly used to demonstrate the behavior of the network in critical conditions, to ensure that race conditions do not arise (see Sect. 5.1.3), and – for synchronous circuits – to guarantee that the cumulative switching delays of the various components are small enough to react to the inputs before the end of the clock cycle, so that there are no "missed" transitions when the circuits operate in synchrony with a clock of known frequency. There are also simple analytical techniques to perform this analysis of the switching delays exhaustively on the critical parts of the system. Simulation can also produce behavioral graphs, such as the one in Fig. 5.6b, from any given input signals.

At the register-transfer or behavioral level, under the complete synchrony abstraction, automated synthesis into logic-gate models is often possible. Hardware description languages such as VHDL, however, offer an expressive collection of constructs, but only a much smaller subset is directly synthesizable. Simulation, on the other hand, is commonplace even for complex behavioral descriptions; simulators can produce graphs such as those in Figs. 5.12 and 5.7c, and the simulator input languages often offer features to program sophisticated simulation sessions and define complex input stimuli to be fed to the system.

Formal verification is also quite developed for hardware design at a high level of abstraction, and it has even become part of industrial practice. The increasing availability of computational power has made some of these techniques practical even for very large state spaces. Formal verification usually works on models such as finite-state machines and variants thereof. Part II of this book (Chaps. 7, 9, and especially 11) contains extensive examples of formalisms also widely used to model and verify hardware designs at a high level of abstraction.

5.6 Bibliographic Remarks

Von Neumann first established solid foundations to the construction of reliable systems from unreliable components [29]; his work evolved into the modern theory of error correction [3, 21].

There are numerous books on electronic system design, with widely varying breadth and targets. A number of them mostly focus on integrated circuit design and digital electronics [5,12,23,32,44]; some treat analog electronics as well [2,11]. Other textbooks start from the logic-gate abstraction level to introduce the design of more complex components at a higher level of abstraction [18, 40, 45]. The books targeting computer systems describe the highest levels of abstraction in the construction of digital programmable computers [10, 30, 38].

Finite state machines are extensively studied as abstract computing devices in the theory of formal languages and computability [7, 8, 13, 22, 25, 34]. Booth [4] and Kohavi [19] show the application of finite-state machine models to hardware design; Kam et al. [16, 17] present analysis and synthesis techniques for the same models. The bibliographic notes in Part II of this book cite many more models based on finite-state machines. Moore and Mealy, respectively, introduced the machines named after them [24, 26].

VHDL [43], Verilog [39], and SystemC [37] are among the most established hardware description languages. There are many other such languages; some of them are developed as libraries and extensions of general-purpose programming languages, such as SystemC, which is a collection of C++ [36] classes.

Electronic design automation techniques and methods [14, 15, 33], prominently including testing, simulation, and automated synthesis, have reached industrial strength and wide distribution. SPICE [35] is an open-source circuit simulation program that has evolved through several versions [27, 28, 31]. Abarbanel-Vinov et al. [1], Kurshan [20], Fix [6], and Vardi [41, 42] document the gain of acceptance of formal verification techniques (especially those based on finite-state models and model checking) in industry.

References

1. Abarbanel-Vinov, Y., Aizenbud-Reshef, N., Beer, I., Eisner, C., Geist, D., Heyman, T., Reuveni, I., Rippel, E., Shitsevalov, I., Wolfsthal, Y., Yatzkar-Haham, T.: On the effective deployment of functional formal verification. Form. Method Syst. Des. **19**, 35–44 (2001)
2. Agarwal, A., Lang, J.: Foundations of Analog and Digital Electronic Circuits. Morgan Kaufmann, Amsterdam (2005)
3. Baylis, J.: Error Correcting Codes: A Mathematical Introduction. Chapman and Hall, London/New York (1997)
4. Booth, T.L.: Sequential Machines and Automata Theory. Wiley, New York (1967)
5. DeMassa, T.A., Ciccone, Z.: Digital Integrated Circuits. Wiley (2008)
6. Fix, L.: Fifteen years of formal property verification in Intel. In: Grumberg, O., Veith, H. (eds.) 25 Years of Model Checking-History, Achievements, Perspectives. Lecture Notes in Computer Science, vol. 5000, pp. 139–144. Springer, Berlin (2008)
7. Gill, A.: Introduction to the Theory of Finite-State Machines. McGraw-Hill, New York (1962)
8. Ginsburg, S.: Introduction to Mathematical Machine Theory. Addison-Wesley, Reading (1982)
9. Grumberg, O., Veith, H. (eds.): 25 Years of Model Checking—History, Achievements, Perspectives. Lecture Notes in Computer Science, vol. 5000. Springer, Berlin (2008)
10. Heuring, V.P., Jordan, H.F.: Computer Systems Design and Architecture, 2nd edn. Prentice Hall, Harlow (2003)
11. Hickman, I.: Analog Electronics, 2nd edn. Newnes, Oxford/Boston (1999)
12. Hodges, D., Jackson, H., Saleh, R.: Analysis and Design of Digital Integrated Circuits. McGraw-Hill, Boston (2003)
13. Hopcroft, J.E., Motwani, R., Ullman, J.D.: Introduction to Automata Theory, Languages, and Computation, 3rd edn. Addison-Wesley, Boston (2006)
14. Jansen, D.: The Electronic Design Automation Handbook. Springer, Dordrecht (2003)
15. Jansen, D.: Electronic Design Automation for Integrated Circuits Handbook. CRC, Boca Raton (2006)
16. Kam, T., Villa, T., Brayton, R.K., Sangiovanni-Vincentelli, A.L.: Synthesis of Finite State Machines: Logic Optimization. Springer, Boston (1997)
17. Kam, T., Villa, T., Brayton, R.K., Sangiovanni-Vincentelli, A.L.: Synthesis of Finite State Machines: Functional Optimization. Springer, Boston (1997)
18. Katz, R.H., Borriello, G.: Contemporary Logic Design, 2nd edn. Prentice Hall, Upper Saddle River (2004)
19. Kohavi, Z.: Switching and Finite Automata Theory, 2nd edn. McGraw-Hill, New York (1978)
20. Kurshan, R.P.: Verification technology transfer. In: Grumberg, O., Veith, H. (eds.) 25 Years of Model Checking-History, Achievements, Perspectives. Lecture Notes in Computer Science, vol. 5000, pp. 46–64. Springer, Berlin (2008)

21. MacWilliams, F.J., Sloane, N.J.A.: The Theory of Error-Correcting Codes. North Holland, Amsterdam (1998)
22. Mandrioli, D., Ghezzi, C.: Theoretical Foundations of Computer Sciences. Wiley, New York (1987)
23. Martin, K.: Digital Integrated Circuit Design. Oxford University Press, New York (1999)
24. Mealy, G.H.: A method for synthesizing sequential circuits. Bell Syst. Tech. J. **34**, 1045–1079 (1955)
25. Minsky, M.L.: Computation: Finite and Infinite Machines. Prentice Hall, Englewood Cliffs (1967)
26. Moore, E.F.: Gedanken-experiments on sequential machines. In: Automata Studies. Annals of Mathematical Studies, vol. 34, pp. 129–153. Princeton University Press, Princeton (1956)
27. Nagel, L.W.: Spice2: a computer program to simulate semiconductor circuits. Ph.D. thesis, EECS Department, University of California, Berkeley (1975)
28. Nagel, L.W., Pederson, D.: SPICE (simulation program with integrated circuit emphasis). Tech. Rep. UCB/ERL M382, EECS Department, University of California, Berkeley (1973)
29. von Neumann, J.: Probabilistic logics and the synthesis of reliable organisms from unreliable components. In: Automata Studies, pp. 329–378. Princeton University Press, Princeton (1956)
30. Patterson, D.A., Hennessy, J.L.: Computer Organization and Design, 4th edn. Morgan Kaufmann, Burlington (2008)
31. Quarles, T.L.: Analysis of performance and convergence issues for circuit simulation. Ph.D. thesis, EECS Department, University of California, Berkeley (1989)
32. Rabaey, J.M., Chandrakasan, A., Nikolic, B.: Digital Integrated Circuits, 2nd edn. Prentice Hall, Upper Saddle River (2003)
33. Rubin, S.M.: Computer Aids for VLSI Design. Addison-Wesley, Reading (1987). Available online at http://www.rulabinsky.com/cavd/
34. Sipser, M.: Introduction to the Theory of Computation, 2nd edn. Course Technology, Boston (2005)
35. SPICE: Simulation program with integrated circuit emphasis. http://bwrc.eecs.berkeley.edu/Classes/IcBook/SPICE/
36. Stroustrup, B.: The C++ Programming Language, 3rd edn. Addison-Wesley, Boston (2000)
37. The open SystemC initiative. http://www.systemc.org
38. Tanenbaum, A.S.: Structured Computer Organization, 5th edn. Prentice Hall, Upper Saddle River (2005)
39. Thomas, D.E., Moorby, P.R.: The Verilog Hardware Description Language, 5th edn. Springer, New York (2002)
40. Tocci, R.J., Widmer, N., Moss, G.: Digital Systems: Principles and Applications, 11th edn. Prentice Hall, Harlow (2010)
41. Vardi, M.Y.: From church and prior to PSL. In: Grumberg, O., Veith, H. (eds.) 25 Years of Model Checking-History, Achievements, Perspectives. Lecture Notes in Computer Science, vol. 5000, pp. 150–171. Springer, Berlin (2008)
42. Vardi, M.Y.: From philosophical to industrial logics. In: R. Ramanujam, S. Sarukkai (eds.) ICLA. Lecture Notes in Computer Science, vol. 5378, pp. 89–115. Springer, Berlin (2009)
43. IEEE P1076—VHDL analysis and standardization group. http://www.eda.org/twiki/bin/view.cgi/P1076/
44. Weste, N., Harris, D.: CMOS VLSI Design: A Circuits and Systems Perspective, 4th edn. Addison-Wesley, Boston (2010)
45. Wirth, N.: Digital Circuit Design for Computer Science Students: An Introductory Textbook. Springer, Berlin (1995)

Chapter 6
Time in the Analysis of Algorithms

Compared to its standing in many other disciplines, the function of time in traditional software modeling seems modest. Undergirding software theory and practice is the notion of *algorithm* – the most abstract view of computational process, in terms of elementary abstract operations. An algorithm is an effective procedure to solve a computational task that can be carried out with resources precisely defined; each algorithm defines a way to implement the computation of a *function* of some input into some output. In this very abstract view, time is mostly irrelevant, and the models that combine individual algorithms into larger pieces of software focus on purely functional behavior. On the other hand, *computational complexity* considers an abstract notion of running time of each algorithm as its *time complexity*, and considers also the algorithm's usage of other computational resources (such as memory). Time complexity analysis "counts" the number of elementary operations performed when executing the algorithms on an abstract processor; time still is an uncomplicated entity compared to its complex attributes in other disciplines such as dynamical system theory (discussed in Chap. 4). Representing a significant departure from the unified view of dynamical systems, the traditional models of algorithms are double-edged, as they complement a *functional* view with aspects of time *complexity* analysis.

This twofold view of time in software – *functional* and *complexity-theoretical* – has historical origins and a theoretical basis. Historically, the functional/complexity-theoretical duality mirrors, and further abstracts, the macro/micro views used in hardware synchronous models and described in Chap. 5: the macro behavior of a hardware circuit is the sequence of actions performed at every tick of the clock, but each macro-step is just a summary of the behavior of individual signals that propagate through the elementary circuit elements with their asynchronous delays. The functional/complexity-theoretical duality and its elementary notion of time are also justified by the theoretical results described in Sects. 6.8 and 6.2.7, which suggest that, within the framework of algorithmic computation universally used, more complex models of time would not affect the results of the analyses performed.

C.A. Furia et al., *Modeling Time in Computing*, Monographs in Theoretical Computer Science. An EATCS Series, DOI 10.1007/978-3-642-32332-4_6,
© Springer-Verlag Berlin Heidelberg 2012

This chapter introduces the fundamental notions used in the computational complexity analysis of algorithms, and illustrates some of the most important results. The purpose of this chapter is twofold. First, it completes the historical account of Part I of the book by presenting the characteristics of time in the traditional models used to describe software and computational processes. The presentation focuses on the aspects that are most relevant for the formalisms introduced, whereas other dimensions mentioned in Chap. 3 (such as concurrency and composition) are discussed in Part II of the book with reference to more specific notations derived from the basic models of the present chapter. Second, it introduces some notions of computational complexity that will be referenced in Part II of the book to discuss analysis and verification algorithms for the various models and notations presented. In this way, the book is self-contained also for readers with no prior knowledge of computational complexity.

6.1   Models of Algorithms and Computational Complexity

Algorithms run on digital machines, and every digital machine is, in principle, just a finite-state machine – with a possibly huge but finite number of states. Therefore, it seems natural to found the notion of algorithm and computational process on abstract operational models such as finite state machines – in the synchronous interpretation with discrete time that abstracts away from implementation details. Chapter 5, however, already discussed how a finite-state model is impractical when the number of states involved is too large, as in the case of modern digital computers with large memories and complex processors.

In addition to having this practical limitation, the general notion of algorithm calls for a more abstract view of computation, which finite-state models fail to capture satisfactorily even in principle. An algorithm describes, in an abstract yet implementable setting, a general procedure to solve a certain computational *problem*; sorting a list of elements, computing the product of two integer numbers in their decimal representation, and finding the shortest path between two given nodes on a graph are all examples of computational problems with well-known algorithmic solutions. The definition of computational problem – and of the algorithms that solve it – must encompass inputs of *any* size: we are not interested in sorting a list of elements of length up to 100 (or even 100^{100}), but *of any length*; we want to devise algorithms that compute the product of two integers *of any size*, and that find the shortest path on a graph *with any number of nodes*. Correspondingly, it is natural to express the complexity of an algorithm – an abstract measure of the "time" taken by the computation – as a function of the input *size*. This setting requires models that extend abstract finite-state machines with unbounded resources, so that they can perform computations involving arbitrarily large numbers of steps and amounts of memory, and can summarize the behavior of any computing device irrespective of its configuration and resources. Such *infinite-state abstractions* capture the notion of algorithm and have yielded general results that are directly applicable to the design and analysis of *real* programs running on digital computers with finite resources.

The rest of the present chapter reviews the fundamental notions of computational complexity in two contexts:

- Section 6.2 deals with automata models. "Automata" is the standard name given to abstract machines in the context of algorithmic analysis. The section focuses on two fundamental automata classes: finite-state automata and *Turing machines*, which directly extend finite-state machine models with unbounded computational resources. (Other, more specialized, automata classes are out of this book's scope; interested readers can find references in the bibliographic remarks at the end of the chapter). Automata models underpin the most fundamental results of computability and computational complexity theory; they introduce a very abstract view of computational processes that enables the analysis of the fundamental capabilities and limits of computation, such as the properties shared by every conceivable algorithm that solves a certain computational problem.
- Section 6.3 presents computational models based on computing architectures à la von Neumann, and in particular the "random access machine". These models introduce, still in an abstract setting, some finer-grained details of real computer architectures and organization. They support the analysis of properties of specific algorithms and of some of their implementation details on real computers.
- Section 6.4 discusses extensions of the fundamental abstract models of Sect. 6.2 with probabilistic features.

The presentation of this chapter is concerned with the historical approach of the current Part I; hence it mainly focuses on the traditional models of computation as a strictly sequential process that starts from the input, performs a sequence of elementary computational steps, and produces, if it terminates, the output. In these models, the time domain is **DISCRETE** (the natural numbers) and so are all the domains included in the computational model (such as the state). The evolution over time is sequential and **DETERMINISTIC**. However, the present chapter will also discuss the role of nondeterministic and probabilistic models of computation when relevant and instrumental to the general themes of the chapter.

6.2 Computational Complexity and Automata Models

The theory of computational complexity primarily uses models derived from the finite-state machines used in hardware design (see Chap. 5), called finite-state automata in complexity theory parlance. The next Sect. 6.2.1 introduces the general approach to defining computational complexity measures; then, Sect. 6.2.2 discusses the complexity analysis of finite-state automata. The rest of Sect. 6.2 introduces Turing machines, an extension of the finite-state model with unbounded resources. Turing machines support the most general definition of computational process (Sects. 6.2.3 and 6.2.4) and allow for a uniform and abstract analysis of the computational complexity of algorithmic problems (Sects. 6.2.5 and 6.2.7). The closing Sects. 6.2.8–6.2.10 discuss the role of nondeterminism in the context of computational complexity, and present some fundamental results based on the comparison of these models with the standard deterministic ones.

6.2.1 Measures of Computational Complexity

In the functional view, an algorithm A transforms an input x into an output $f_A(x)$. Computational complexity measures the resources used in A's computation as a function of the input *size* (also called "length"). More precisely, it associates two complexity measures with every algorithm A: the time complexity T_A and the space complexity S_A; both are functions from $\mathbb{N}$ to $\mathbb{N}$, following an inherently DISCRETE presentation of time and other state domains. Different inputs of the same size will, in general, produce computations of different length or memory usage; the most common definition of complexity – and the one we use throughout the chapter – assumes a *worst-case* scenario: take the maximum number of steps and the maximum amount of memory used for all possible inputs of a given size. Then, for every n, the time complexity $T_A(n)$ measures the maximum number of elementary operations ("steps") performed by any computation with input of size n; and the space complexity $S_A(n)$ measures the maximum amount of elementary memory elements ("cells") used during any computation with input of size n. For example, a time complexity $2n$ is associated with an algorithm that takes up to twice as many steps as the input size; a space complexity $n^2 + 3^n$ characterizes an algorithm that may use up to 13 (that is, $2^2 + 3^2$) memory cells for an input of size 2.

The time and space complexities are in general partial functions; we write $T_A(n) = \infty$ if A does not terminate for some input of length n, and $S_A(n) = \infty$ if some input of size n generates a computation that uses an unbounded amount of memory.

The intuition behind the notions of time and space complexities should be clear, but their precise definition seems to require fixing many subtle details, such as the encoding of the input (and how we measure its size) and the elementary steps allowed by the computational model. Somehow surprisingly, Sect. 6.8 will show that these details are ultimately negligible with respect to other factors, and a sound and robust definition of complexity is possible at a very high level of abstraction, independently of most implementation details. Until then, the presentation relies on the readers' intuition to deploy the complexity measures appropriately. The presentation focuses on the time complexity, given the general theme of the book, and discusses space complexity occasionally, mostly to show its relationship with time complexity.

6.2.2 Finite-State Automata

Let us apply the notion of complexity measures to the abstract computational model defined by finite-state automata in the perfect synchrony abstraction (Chap. 5). The

Fig. 6.1 A k-tape Turing machine

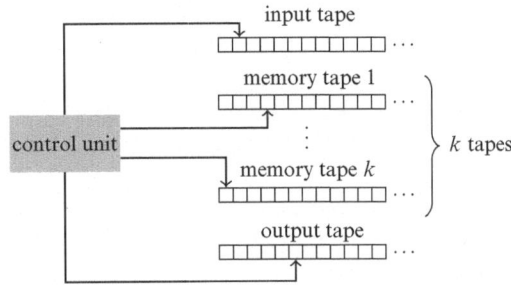

input of a finite-state automaton is the sequence of input events; an input of n events has length n. It is irrelevant whether the automaton has an explicit output (such as in Moore and Mealy machines) or the sequence of states visited implicitly determines a trace of the computation. In both cases, the computation of an input of length n consists of exactly n computational steps because each transition exactly takes one input event. Therefore, $T_A(n) = n$ for every finite-state automaton A. The memory of a finite-state automaton is its finite set of states. Hence, $S_A(n) = |Q|$ for every finite-state automaton A with $|Q|$ states, and the space complexity of finite-state automata does not depend on the input size.

The literature on automata and formal language theory usually calls *real-time machine* any automaton that requires no more steps than the input length to process the input. Finite-state automata are real-time machines according to this terminology. The correspondence between this formal definition of "real time" and the informal usage to denote "predictable timing behavior" – discussed in Chaps. 1 and 3 – is loose. More precisely, the real time of automata implies the "informal" real time, but the latter is usually meant to include a broader class of behaviors and systems.

6.2.3   Turing Machines

Section 6.1 argued for a formal definition of algorithm and computational process based on an abstract computing machine more general than simple finite-state automata. This section presents "Turing machines", a model that extends finite-state automata with unbounded scratch memory to perform arbitrarily long computations. Turing machines are named after the British mathematician Alan M. Turing, who introduced them in a groundbreaking article published in 1936.

Definition 6.1 (Turing machine). A k-tape Turing machine, pictured in Fig. 6.1, consists of:

- A *control unit*, which is essentially a finite-state machine with a finite set Q of states managing the access to the input, output, and memory devices.
- $k + 2$ *tapes*: one for the input, one for the output, and k for scratch memory. Each tape consists of a sequence of cells, bounded to the left and unbounded (infinite)

to the right; cells on each tape are numbered with natural numbers. Each cell stores an element from a finite set Σ of symbols (the *alphabet*). Σ includes a special *blank* symbol "$\square$". The memory tapes are rewritable, whereas the input is read-only and the output write-only.

- $k+2$ moving *heads*, one for each tape; the control unit accesses the tapes through the heads.
- A *transition function δ*, whose details are described below, which defines how the control unit operates on the other components of the machine and on its own state.

At the beginning of every computation:

- The input tape stores an encoding of the input.
- All memory cells in the tapes other than the input are blank.
- Each head occupies position 0 on its memory tape.
- The control unit is in an *initial state $q_0 \in Q$*.

A computation consists of a sequence of steps from the initial setup. After every step during the computation:

- The control unit is in one of the control states.
- The tapes store only a finite number of non-blank symbols.
- Every head occupies exactly one cell on its tape (the "current cell").

The state of the control unit, the non-blank portions of the tapes, and the positions of the heads determine a *configuration* of the Turing machine.

The transition function

$$\delta : Q \times \Sigma^{k+1} \to Q \times \Sigma^{k+1} \times \{R, L, S\}^{k+1} \times \{R, S\}$$

defines every step (also called "move" or "transition") from a configuration to the next one in a computation as follows. Whenever

- The control unit is in state $q \in Q$;
- The current cell under the input head stores the symbol $i \in \Sigma$;
- For every $1 \leq j \leq k$, the current cell under the jth tape head stores the symbol $m_j \in \Sigma$;
- $\delta(q, i, m_1, \ldots, m_k)$ is defined and equal to

$$\langle q', o, m'_1, \ldots, m'_k, h_I, h_1, \ldots, h_k, h_O \rangle$$

for some $q' \in Q$, $o \in \Sigma$, $m'_1, \ldots, m'_k \in \Sigma$, $h_I \in \{R, L, S\}$, $h_1, \ldots, h_k \in \{R, L, S\}$, $h_O \in \{R, S\}$,

the Turing machine changes configuration in the following way:

- The control unit switches to the state q';
- The current cell under the output head is written with the symbol o;

- For every $1 \leq k \leq k$, the current cell under the jth tape head is rewritten with the symbol m'_j;
- The input head moves right (by one position) if h_I is R, moves left if h_I is L, and does not move ("stays") if h_I is S;
- For every $1 \leq k \leq k$, the jth tape head moves right (by one position) if h_j is R, moves left if h_j is L, and does not move if h_j is S;
- The output head moves right (by one position) if h_O is R, and does not move if h_O is S (it cannot move left).

A computation stops when no move is possible; this occurs when $\delta(q, i, m_1, \ldots, m_k)$ is undefined, or when $\delta(q, i, m_1, \ldots, m_k)$ requires the head of the input tape or of one of the memory tapes to move left when it is already at the beginning of the tape.

It is customary to extend the graphical representation of finite-state automata to Turing machines: a graph with nodes corresponding to the control states Q and an arc between every pair of nodes q, q' with label

$$i, m_1, \ldots, m_k / o, m'_1, \ldots, m'_k / h_I, h_1, \ldots, h_k, h_O$$

whenever the transition function defines

$$\delta(q, i, m_1, \ldots, m_k) = \langle q', o, m'_1, \ldots, m'_k, h_I, h_1, \ldots, h_k, h_O \rangle. \quad \blacksquare$$

The k-tape Turing machine in Definition 6.1 is different from Turing's original model, which has a single tape used for input, output, and scratch memory. We preferred a slightly more complex model, because it better fits the traditional dynamical system view (Chap. 4), with a clear separation between input, output, and state. The rest of the current Sect. 6.2 will also show how the k-tape machine supports a more realistic complexity analysis than the single-tape machine, where significant time may be spent simply accessing the remote input or output portions of the tape.

Example 6.2. Consider the problem of computing the successor $succ(x) = x + 1$ of natural numbers encoded in binary; encode the input as a sequence of 0/1 characters, one per cell, with the most significant digit occupying the second position of the input tape, after a leading blank. With this setup, a Turing machine M_{succ} with one memory tape that solves the problem scans the input backward (from right to left) while writing on the memory tape (from left to right):

1. Before detecting the first 0 digit, it rewrites every 1 digit as 0 on the memory tape with an implicit "carry";
2. When it detects the first 0 digit, it rewrites it as 1 on the memory tape;
3. After detecting the first 0 digit, it copies all the remaining input characters to the memory tape;
4. If there are no 0 digits, the machine adds an extra 1 digit after writing to the memory tape as many 0's as 1's are in the input;

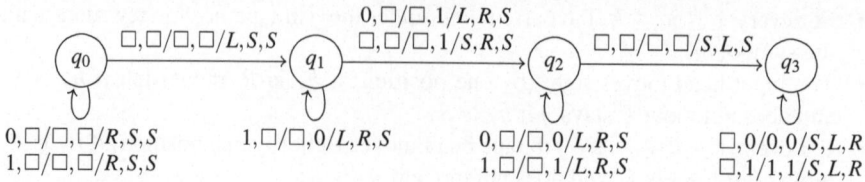

Fig. 6.2 A 1-tape Turing machine M_{succ} that computes the successor of a binary number (Multiple transitions between the same pairs of states are represented by multiple labels attached to the same edge, a customary convention that we will use whenever convenient)

5. After processing all the input (that is, when the input head reads a blank "□"), the machine reverses the content of the memory tape to the output and stops.

Figure 6.2 pictures such a Turing machine for the successor function, where q_0 is the initial state, q_1 is reached when the input head is on the rightmost input cell, q_2 denotes that the first 0 (if any) has been detected, and q_3 is the state where the machine halts after writing the output.   ■

Exercise 6.3. Define a Turing machine M_{double} that computes the function $double(x) = 2x$ of the input encoded in binary as in Example 6.2.

(*Hint*: doubling amounts to a shifting of the digits in binary).   ■

Example 6.4. A 2-tape Turing machine M_{square} that computes the function $square(x) = x^2$ of natural numbers encoded in unary (that is, the encoding of x is a sequence of x symbols 1) works as follows:

• Scan the input and copy it on both memory tapes; after copying all the input, the input head does not move anymore.
• For each of the x cells of the first memory tape that stores a 1:

– Scan and copy to the output the whole non-blank content of the second memory tape;
– Move the head on the second memory tape back to the first position.   ■

Exercise 6.5. Complete the formalization of the Turing machine M_{square} in Example 6.4.   ■

6.2.4   Universal Computation and the Church-Turing Thesis

When computing, a Turing machine goes through a sequence of configurations. The sequence may be finite – the machine halts because it reaches a configuration where no further steps are defined – or infinite. The configuration completely determines the future behavior of a Turing machine: it is the equivalent of the *state* in dynamical system models (Chap. 4), but the standard terminology of automata models reserves the word "state" for the state of the control unit, which varies over a finite domain, whereas the domain of all possible configurations is *infinite*.

An infinite configuration space endows Turing machines with a conspicuous *computational (expressive) power*, their primitive memory model notwithstanding. In particular, it is clear that Turing machines can define computations inexpressible with finite-state automata: a finite-state automaton can only process the input "on line" as it reads it, but it has no long-term memory other than that of the finite states. Turing machines, in contrast, can process the input "off-line" after storing it on a memory tape. For example, no finite-state automaton can compute the function x^2 (encoded in unary as in Example 6.4) because an output of length x^2 cannot be generated "in real time": finite-state automata produce the output synchronously with every input character, and only a finite number of characters can be output with every input event.

The realization that Turing machines achieve greater computational power than finite-state automata immediately prompts the question of whether there exist computing devices even more powerful than Turing machines. More precisely, we are interested in computational models that are implementable, that is, we abstract away inessential details of real physical processes, which can be engineered in real computing devices. The Turing machine model is implementable: the unbounded memory tapes are only an abstraction for resources that are not fixed a priori but can grow as the computation progresses.

Even with this qualification about implementations, the best answer to the question about the existence of "the most powerful" computing device is only a conjecture known as the "Church-Turing thesis":

Every implementable computational process can be computed by a Turing machine.

The evidence corroborating the validity of the Church-Turing thesis is overwhelming, as very diverse computational models, defined in extremely different contexts and with heterogeneous scopes, have always been proved to be no more powerful than the simple Turing machine. Section 6.3 explains, in particular, that the RAM – a model much closer to the architectures of digital computers – has the very same computational power as the Turing machine.

The Church-Turing thesis and its experimental validation justify the choice of the Turing machine model to present *general* results about the properties of computational processes. It also implies that there exist "programmable" Turing machines that can simulate every other Turing machine by encoding its transition function in the input, similarly to what is done in general-purpose digital computers where instructions and data are both stored in the central memory (Sect. 6.3 discusses these models in more detail). This property is called *universality*: some Turing machines can simulate every other Turing machine.

The computational power of Turing machines also poses ultimate limits on their analyzability. There are many questions about the behavior of Turing machines that, while perfectly well defined, cannot be computed by any Turing machine and therefore, according to the Church-Turing thesis, by any algorithm or computing device. In particular, every question about the long-term dynamics of a generic Turing machine, seen as an operational model of a dynamical system, is *undecidable*: there exists no algorithm that can answer such a question reliably and in finite time.

The computational complexity and timing analysis cannot be completely automated: with great power comes great undecidability.

Example 6.6. The following questions about the long-term dynamics of a generic Turing machine M are undecidable:

- The *halting problem*: does M halt for every input?
- What is the maximum number of memory cells used by M during any computation? Is it always finite?
- What is the maximum k such that M halts for every input of size less than or equal to k?
- The *busy-beaver problem*: what is the maximum number of moves made by M in a halting computation? ∎

6.2.5 Complexity of Algorithms and Problems

According to the Church-Turing thesis, the Turing machine provides the most general notion of computational process. This section introduces the timing analysis of Turing machines with the approach of computational complexity discussed in Sect. 6.2.1.

A computation consists of a sequence of configurations; the timing details of the configuration changes are abstracted away, and the only measure of time is the count of steps, each considered atomic. The time complexity $T_M(n)$ of a Turing machine M is then the function that counts the maximum number of M's moves with an input of size n.

Example 6.7. Consider the 2-tape Turing machine M_{square} – outlined in Example 6.4 – that computes the square of the input encoded in unary form. For an input of size n, M_{square} performs the following moves:

- $n + 1$ steps to scan the input until the input head reaches the first blank, while copying it over the two memory tapes in parallel;
- $1 + 2(n + 1)$ steps for each input character: one step to move the head on the first tape to the next cell, $n + 1$ steps to scan the characters stored on the second tape until the first blank character, and another $n + 1$ steps to move the head back to the beginning of the tape just before the first position with a stored character.

In total, M_{square} performs

$$(n + 1) + n \cdot (1 + 2(n + 1)) = 2n^2 + 4n + 1 = T_{M_{square}}(n)$$

moves for every input of size n. ∎

There are two different, but tightly connected, branches that deal with computational complexity along the lines of Example 6.7. On the one hand, the "analysis of algorithms" is concerned with studying the complexity of individual Turing machines that formalize a certain algorithm; Example 6.7 is an instance of analysis of an algorithm to compute the square of numbers in unary form. On the other hand, "computational complexity theory" analyzes the inherent complexity of *every* algorithm for a certain computational *problem*. Example 6.7 considers the complexity of a particular solution to the problem of "computing the function *square*", whereas computational complexity theory would investigate the complexity that *any* algorithm that correctly computes the function *square* must have, no matter how different the algorithms that compute the function are. The algorithmic analysis in Example 6.7, in particular, shows that it is possible to compute *square* with a time complexity $T_{M_{square}}$, but it leaves open the possibility of devising more ingenious algorithms that achieve the same goal with fewer computational steps.

Computational complexity theory classifies the computational problems into *complexity classes* according to their inherent complexities: two problems in the same class can be solved with a similar amount of resources. The rest of the present section gives a more rigorous definition of complexity classes, and briefly presents some of the most important ones. As for every other topic discussed in Part I, providing a comprehensive overview of such a complex and broad topic as computational complexity theory is impossible in this book – and out of its scope. The presentation focuses instead on the nitty-gritty with as few technical details as possible, and it privileges the results that are directly applicable to the presentation of the modern computational models that include a notion of time, introduced in Part II.

6.2.6 The Linear Speed-Up Theorem

Consider again the simple time complexity analysis of the Turing machine in Example 6.7. The exact form of $T_{M_{square}}$ depends on details of how the Turing machine works; for example, if the machine could detect when a head is on the first character written at the left end of a tape, it would perform n fewer steps for the input n. Remarkably, the linear speed-up theorem shows that these details are ultimately unimportant, and seemingly brittle analyses such as the one for $T_{M_{square}}$ turn out to be the basis of a robust and representative characterization of the time complexity of real algorithms, irrespective of their implementation details. Even more interestingly, a similar speed-up theorem holds for other models of computations, such as the RAM model of von Neumann architectures discussed in Sect. 6.3 in which the linear speed-ups correspond to an increase in the bit size of memory words in a real computer. This wide applicability of the notion of speed-up validates the generality of time complexity analysis.

Theorem 6.8 (Linear speed-up). *Given any Turing machine M solving a problem with time complexity $T_M(n)$ and any rational number $c > 0$, it is possible to build another Turing machine m' that, after reading the whole input string, solves the same problem as M with time complexity $c \cdot T_M(n)$, that is,*

$$T_{M'}(n) = \max(n, c \cdot T_M(n)).$$ ∎

The fundamental idea behind the proof of the linear speed-up theorem is the trade-off between complexity of the input (and memory) encoding and the resources needed to solve a problem: if the alphabet of M' has an element for every k-tuple of characters of the alphabet of M (for a suitable choice of k according to c's value), M' can coalesce multiple sequential steps of M into one of its steps, thus achieving a faster running time for the same problem.

The linear speed-up theorem entails the robustness of the complexity analysis of Turing machines – according to the principles of Sect. 6.2.1 – with respect to changes in the details of how the machines work. More precisely, every specific that does not affect the asymptotic value of the complexity measures, but accounts for at most a constant factor, can be overlooked. This suggests ignoring multiplicative factors completely and partitioning complexity measures according to their asymptotic behavior, using the following notation.

Definition 6.9 (Asymptotic notation).   Consider two functions $f, g : \mathbb{N} \to \mathbb{N}$.

- f is $O(g)$ ("big oh of g") if there exist positive constants c, k (k integer) such that $f(n) \leq c \cdot g(n)$ for all $n > k$;
- f is $\Omega(g)$ ("big omega of g") if there exist positive constants c, k (k integer) such that $f(n) \geq c \cdot g(n)$ for all $n > k$;
- f is $\Theta(g)$ ("big theta of g") if f is $O(g)$ and $\Omega(g)$.   ∎

Example 6.10. The function $2n^2 + 4n + 1 = T_{M_{square}}(n)$ is:

- $O(n^2)$, $\Omega(n^2)$, and $\Theta(n^2)$;
- $\Theta(n^2 + n + 1)$, $\Theta(n^2 + n)$, $\Theta(3n^2)$, $\Theta(5n^2)$, $\Theta(100^{1000}n^2)$;
- $O(n^4)$, $O(4^n)$, $O(\exp(\exp(\exp(\exp(n^{1/2})))))$.   ∎

Exercise 6.11. Show the following relation between exponential functions: for every $1 < b < c$, b^n is $O(c^n)$.   ∎

The Θ relation is an equivalence relation among complexity functions. The equivalence classes in the partition induced by the equivalence relation are robust with respect to constant multiplicative factors and additive factors with asymptotically slower growth; machines with complexity measures in the same class have the same *asymptotic complexity*. Correspondingly, we can overload the notation and let $\Theta(g)$ denote the *set* of all functions f that are $\Theta(g)$. Customarily, $\Theta(g)$ is presented with g in the simplest form with unit constants. For example, $T_{M_{square}}(n)$ in Example 6.4 induces the *set* of functions $\Theta(n^2)$, and M_{square} has asymptotic complexity in $\Theta(n^2)$.

The asymptotic notation supports the analysis of the computational complexity of algorithms, and the invention of new ones, based on their *asymptotic complexity*: improvements in the hardware can achieve a linear speed-up of the algorithms currently available, but a new algorithm with asymptotically faster behavior will overwhelmingly outperform the other algorithms for inputs of increasingly large size, irrespective of how optimized the machines that run them are.

Example 6.12. Consider the two well-known sorting algorithms implemented in a high-level language on a von Neumann architecture:

- Bubble Sort, with asymptotic time complexity in $\Theta(n^2)$;
- Merge Sort, with asymptotic time complexity in $\Theta(n \log n)$.

The performance of Merge Sort on a slow computer will inevitably surpass the performance of Bubble Sort on a much faster computer as the input size becomes sufficiently large. ∎

6.2.7 Complexity Classes and Polynomial Correlation

Let us now move our focus from the analysis of individual *algorithms* to the classification of computational *problems* according to the complexity of their algorithmic solutions. A complexity measure induces the class of problems that have solutions with that complexity.

Definition 6.13 (TIME and SPACE complexity classes). A time complexity measure[1] $T(n)$ defines the *complexity class* $\text{TIME}(T(n))$ of all *problems* that can be solved by some Turing machine with time complexity in $O(T(n))$; a space complexity measure $S(n)$ defines the *complexity class* $\text{SPACE}(S(n))$ of all *problems* that can be solved by some Turing machine with space complexity in $O(S(n))$. (DTIME and DSPACE are other names used for TIME and SPACE that emphasize the deterministic nature of the computational models). ∎

Since complexity classes are *sets* of problems, it is natural to compare them by means of set-theoretic relations, such as "$\subseteq$", "$\neq$", and "$\supset$". When each of $m > 1$ classes $C_1, \ldots, C_m$ is included in the following one, $C_1 \subseteq \cdots \subseteq C_m$, we say that they define a "hierarchy". Exercise 6.14 presents a simple hierarchy between pairs of time complexity classes.

Exercise 6.14. Show that if $f(n)$ is $O(g(n))$ then $\text{TIME}(f(n)) \subseteq \text{TIME}(g(n))$. ∎

Example 6.15. Consider the following computational problems.

PALINDROME: determine if the input sequence is a *palindrome*. (A sequence is a palindrome if it reads the same left to right and right to left; for example, a, $b\,b$, $a\,b\,c\,c\,b\,a$, and $a\,b\,c\,b\,a$ are palindromes, but $a\,b$ and $a\,b\,b\,c\,a$ are not).
PALINDROME is in $\text{TIME}(n)$ for a 1-tape Turing machine that works as follows: copy the input on the memory tape, and compare the input with itself read backward.
SORT: sort a sequence of n integer elements.

---

[1] A rigorous definition should include some regularity constraints on $T(n)$ and $S(n)$ as functions from $\mathbb{N}$ to $\mathbb{N}$; see the bibliographic remarks for references with all the omitted details.

Example 6.12 implies that *SORT*, when implemented on a von Neumann architecture, is in TIME($n \log n$); for algorithms based on comparison, $\Theta(n \log n)$ is also the best asymptotic time complexity possible.

MATRIX_MULTIPLY: compute the product of two $n \times n$ matrices.

The algorithm that implements the definition of matrix multiplication on a von Neumann architecture shows that *MATRIX_MULTIPLY* is in TIME(n^3). Every algorithm for *MATRIX_MULTIPLY* necessarily takes at least n^2 steps because it must generate the n^2 elements of the result matrix; hence *MATRIX_MULTIPLY* is not in TIME(n^c) for any $c < 2$. At the time of this writing, the asymptotically fastest algorithm known for *MATRIX_MULTIPLY* runs in time $\Theta(n^{2.373})$; hence *MATRIX_MULTIPLY* is in TIME($n^{2.373}$).

ENUMERATE: generate all n-digit numbers in base b.

ENUMERATE when implemented on k-tape Turing machines with alphabet of cardinality greater than or equal to b is in TIME($n \cdot b^n$): there are exactly b^n n-digit base-b numbers, and generating an n-digit number with a finite alphabet takes time proportional to n.   ∎

With the linear speed-up theorem, the notion of asymptotic behavior fosters an analysis of algorithms that is robust with respect to the *details* of the particular machine, or program, chosen to implement the given algorithm. Can we similarly define complexity classes that are robust with respect to the choice of *computational model*?

In general, in fact, whether a problem is in a class TIME($f(n)$) may depend on the choice of computational model. Consider again the problem *PALINDROME*: determine if the input string is a palindrome. Example 6.15 outlines a 1-tape Turing machine solving *PALINDROME* that runs in time $\Theta(n)$. If, however, we adopt the original and simpler variant of Turing machines with a single tape – used for input, output, and scratch memory – then we could prove that even the best algorithm for *PALINDROME* requires time $\Theta(n^2)$. In all, the problem of computing *PALINDROME* is certainly in TIME(n^2), but it is in TIME(n) for some (universal) computational models and not for others.

Exercise 6.16. Consider the problem of sorting with a Turing machine.

1. Describe a k-tape Turing machine B that sorts a sequence of natural numbers, represented in binary form and separated by "#" symbols, using the Bubble Sort algorithm; what is the complexity $T_B(n)$ of the machine (with n the length of the whole input string)? Compare it with the complexity of a corresponding program running on a von Neumann architecture.

2. Consider the sorting problem of (1), but this time assume that the natural numbers considered have at most K binary digits (with K a given constant). Describe a Turing machine M that uses the Merge Sort algorithm to sort the sequence of input numbers, and determine its complexity $T_M(n)$, with n again the length of the whole input string.   ∎

It is possible to introduce the notion of a complexity class being robust with respect to the choice of computational model. It rests on the notion of *polynomial correlation*, which extends the Θ notation to accommodate polynomial transformations.

Definition 6.17 (Polynomial correlation). Two functions $f, g : \mathbb{N} \to \mathbb{N}$ are *polynomially correlated* if there exist two polynomial functions $p, q : \mathbb{N} \to \mathbb{N}$ of any degree such that f is $O(p(g(n)))$ and g is $O(q(f(n)))$. ∎

Example 6.18.

1. The functions n and n^2 are polynomially correlated: $n^2 = n \cdot n$.
2. The functions n and $n \log n$ are polynomially correlated: n is $O(n \log n)$ and $n \log n$ is $O(n^2)$.
3. The functions 2^n and n^n are *not* polynomially correlated:

$$\frac{n^n}{(2^n)^k} = \frac{n^n}{2^{k \cdot n}} = \frac{n^n}{\left(2^k\right)^n} \xrightarrow{n \to \infty} \infty$$

for every constant k. ∎

Consider two classes C_1, C_2 of abstract machines as powerful as Turing machines; for example, C_1 and C_2 are two variants of Turing machines, or a Turing machine and a markedly different model based on hardware architectures (such as the RAM described in Sect. 6.3). We say that C_1 can *efficiently simulate* C_2 if for every machine M_2 in C_2 that runs in time $T_2(n)$ there exists a machine M_1 in C_1 that simulates M_2 and runs in time $T_1(n)$ such that $T_1(n)$ and $T_2(n)$ are polynomially correlated. The notion of simulation refers to a way of "reproducing" the results of every algorithm defined in a computational model within another computational model, with at most a polynomial slowdown per step.

Of course, a polynomial slowdown may be quite conspicuous in practice, and in fact the analysis of individual algorithms is finer grained and does not group algorithms together only because they have polynomially correlated running times. The identification of "efficient" with "polynomial" is, however, apt for the study of the computational complexity of problems, where it provides a way to define classes of computational problems independently of the computational model chosen. This practice underlies a refinement of the Church-Turing thesis that takes computational complexity into account. This refinement usually goes by the name of "strong Church-Turing thesis", although it is not due to Church or Turing:

> Every implementable computational process can be computed by a Turing machine with at most a polynomial slowdown with respect to the complexity of any other abstract machine computing the same process.

As in the original Church-Turing thesis, the qualification "implementable" means that "unreasonable" computational models, which fail to capture features of real implementations, are ruled out.

We are finally ready to define a few standard complexity classes. They are robust with respect to any computational model that is efficiently (i.e., polynomially) simulated by a Turing machine (for space complexity classes, the polynomial correlation is between space complexity measures).

Definition 6.19 (Deterministic complexity classes).

- P (also PTIME) is the class of problems that can be solved in polynomial time (and unlimited space), that is,

$$P = \bigcup_{k \in \mathbb{N}} \text{TIME}(n^k).$$

P is usually considered the class of problems that are computationally tractable, according to the intuition developed above.
- EXP (also EXPTIME) is the class of problems that can be solved in exponential time (and unlimited space), that is,

$$EXP = \bigcup_{k \in \mathbb{N}} \text{TIME}(\exp(n^k)).$$

- PSPACE is the class of problems that can be solved in polynomial space (and unlimited time), that is,

$$PSPACE = \bigcup_{k \in \mathbb{N}} \text{SPACE}(n^k).$$

- EXPSPACE is the class of problems that can be solved in exponential space (and unlimited time), that is,

$$EXPSPACE = \bigcup_{k \in \mathbb{N}} \text{SPACE}(\exp(n^k)).$$

- ELEMENTARY is the class of problems that can be solved in iterated exponential time (and unlimited space), that is,

$$ELEMENTARY = \bigcup_{k \in \mathbb{N}} \text{TIME}(\overbrace{\exp(\exp(\cdots \exp(n) \cdots)))}^{k}.$$

Correspondingly, a decidable problem is *nonelementary* if its time complexity grows faster than any iterated exponential function. Chapter 9 will mention an algorithm with nonelementary complexity. ∎

A computation cannot use more time than space, because writing a memory cell requires at least one step and each cell can be rewritten multiple times. Therefore, $\text{TIME}(f(n)) \subseteq \text{SPACE}(f(n))$ for every complexity measure $f(n)$. The next section introduces other fundamental complexity classes, for nondeterministic models of computation.

6.2.8   *Nondeterministic Models of Computation*

All the models of computation discussed so far in the present chapter are deterministic. Chapter 3, however, mentioned the usefulness (or necessity) of nondeterministic and probabilistic models for the formalization of certain classes of processes and systems. Section 6.4 will introduce probabilistic computational models, whereas the remaining parts of Sect. 6.2 will introduce nondeterministic models of computation and present their features in the context of computational complexity timing analysis.

A nondeterministic model of computation can produce several distinct computations starting from the same initial input. Hence, a nondeterministic process in the functional model associates with every input a *set* of outputs – all those that the computation can produce starting from the input given. Correspondingly, whereas the "state" uniquely determines the next step taken in a deterministic computation, a nondeterministic process allows multiple steps to be taken from the same "state" – for an appropriate notion of state. Concretely, the transition function of a nondeterministic automaton defines a *set* of transitions for each current "state" and input. The next definition details this idea for finite-state automata and Turing machines; Chaps. 7 and 8 will present different extensions of the finite-state automaton model that feature a notion of nondeterminism.

Definition 6.20 (Nondeterministic finite-state automaton and Turing machine).
A *nondeterministic finite-state automaton* consists of the same components as a deterministic automaton (Sect. 5.2.2), but the transition function has signature

$$\delta : Q \times I \to \wp(Q),$$

where $\wp(Q)$ is the powerset of Q. For current state $q \in Q$ and input $i \in I$, the nondeterministic automaton can change its state to any of the states in the set $\delta(q, i)$.

A *nondeterministic Turing machine* consists of the same components as a deterministic Turing machine (Definition 6.1), but the transition function has signature

$$\delta : Q \times \Sigma^{k+1} \to \wp\left(Q \times \Sigma^{k+1} \times \{R, L, S\}^{k+1} \times \{R, S\}\right).$$

For current state $a \in Q$, input symbol $i \in \Sigma$, and memory symbols $m_1, \dots, m_k \in \Sigma$, the nondeterministic machine can change its configuration according to any of the tuples in $\delta(q, i, m_1, \dots, m_k)$.   ∎

The interpretation of nondeterministic features with respect to the process modeled depends on the kinds of questions that we want the model to address. Automata, in particular, model the computation of some function of the input. Given that nondeterministic automata express a *set* of possible behaviors for a given input and initial state, which of these behaviors is expected to produce the *correct* output? If an external "hostile" environment can influence the nondeterministic alternatives, the computational process has to guarantee a correct output for *every* nondeterministic behavior, of any length. This assumption is called "external" or "demonic" nondeterminism, and is adopted in certain notations such as process algebras (see Chap. 10); it corresponds to the general notion of UNIVERSAL NON-DETERMINISM discussed in Sect. 3.4.1. If, instead, the nondeterministic choices represent the possibility of selecting the "best" alternative whenever convenient, it is sufficient that *one* of the nondeterministic behaviors yields a correct output. This semantics is called "internal" or "angelic" nondeterminism; it corresponds to the notion of EXISTENTIAL NONDETERMINISM of Sect. 3.4.1, and it is the standard assumption in the context of computational complexity analysis.

The rest of the chapter adopts the latter *existential* view of nondeterminism, of which there are two intuitive interpretations in the context of computational complexity. In the first view, a nondeterministic machine has a somehow uncanny power to "choose" the *best* alternative at each step – the choices leading to a successful computation, if one exists, with as few steps as possible. In the other view of nondeterminism the machine spawns *parallel computations* at every step; each computation examines the effects of a particular choice from among the alternatives, and the machine combines the results returned by the parallel threads. This notion of parallelism is *unbounded*, because even if the choices available at every step are finite, the length of computations is, in general, unbounded, and the machine may spawn parallel processes at every step. In both views – nondeterministic choice and unbounded parallelism – a BRANCHING-TIME domain often is the natural model of nondeterministic computations: all the possible computations originating from the same initial state are represented as a tree (see Chap. 3).

Remark 6.21. Definition 6.20 conservatively considers automata and Turing machines with a unique initial state; but nondeterminism can easily accommodate multiple initial states. We discuss the issue with reference to finite-state automata; extending the same concepts to Turing machines is straightforward. If $Q_0 \subseteq Q$ is a set of initial states of automaton A, every computation of A starts from a nondeterministically chosen state $q \in Q_0$ and then continues as in Definition 6.20. The additional source of nondeterminism is entirely reducible to the case of unique initial states, because the initial nondeterministic choice is expressible by adding transitions that exit a unique initial state: let us build an automaton A' with unique initial state q_0 equivalent to A. A' includes all states and transitions in A, plus a fresh state $q_0 \notin Q$ that is its unique initial state. For every initial state $q \in Q_0$ of A, input symbol $i \in I$, and state $q' \in \delta(q, i)$ directly reachable from the initial state, A' also includes a transition from q_0 to q' with input i. A' can make precisely the same first transitions as A, and it behaves identically to A after leaving q_0; hence the two automata are equivalent. Figure 6.3 shows the equivalence construction on a simple automaton with two initial states.   ■

The introduction of nondeterministic models brings forth the question of their impact on computational power and resource usage. The following sections investigate these fundamental questions, focusing on Turing machines (Chap. 7 will discuss nondeterministic extensions of finite-state automata in more detail).

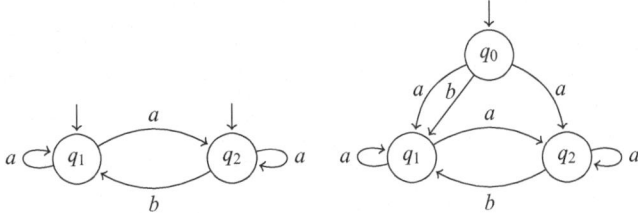

Fig. 6.3 Two equivalent nondeterministic finite-state automata

- Are nondeterministic computational models more powerful, in terms of express-ible computations, than deterministic models?
- Within models with the same expressive power, do nondeterministic models achieve a better (that is, lower) computational complexity? In other words, is nondeterministic computation more efficient?
- Is the nondeterministic abstraction a suitable model of "real" computational processes? Is it readily implementable? What is the impact of nondeterministic computation on the Church-Turing thesis (Sect. 6.2.4) and on the strong Church-Turing thesis (Sect. 6.2.7)?

To approach these questions rigorously, we first introduce complexity measures and classes for nondeterministic models.

Definition 6.22 (Nondeterministic complexity classes). A nondeterministic Tur-ing machine N runs in $\#_N$ steps for an input x if the shortest sequence of steps that is allowed by the nondeterministic choices and correctly computes the result for input x has length $\#_N$. Correspondingly:

- A nondeterministic Turing machine has time complexity $T(n)$ if the longest computation with input of size n runs in $T(n)$.
- A time complexity measure $T(n)$ defines the *complexity class* NTIME($T(n)$) of all *problems* that can be solved by some *nondeterministic* Turing machine with time complexity in $O(T(n))$.
- NP (also NPTIME) is the class of problems that can be solved in polynomial time (and unlimited space) using nondeterminism, that is,

$$\text{NP} = \bigcup_{k \in \mathbb{N}} \text{NTIME}(n^k).$$

- NEXP (also NEXPTIME) is the class of problems that can be solved in exponential time (and unlimited space) using nondeterminism, that is,

$$\text{NEXP} = \bigcup_{k \in \mathbb{N}} \text{NTIME}(\exp(n^k)). \qquad \blacksquare$$

To simplify the presentation, and to focus on the central issues, we do not analyze the consequences of nondeterminism for space complexity. It is useful to remark, however, that PSPACE and EXPSPACE define classes of complexity even higher than NP: it is known that

$$P \subseteq NP \subseteq PSPACE = NPSPACE \subseteq EXP \subseteq NEXP \subseteq EXPSPACE$$
$$= NEXPSPACE.$$

Also, notice that each deterministic class is included in its nondeterministic counterpart (e.g., P is included in NP) because determinism is a special case of nondeterminism where there is a unique choice at every step.

6.2.9   Nondeterministic Turing Machines

Nondeterminism does not increase the computational power of Turing machines. Therefore, the Church-Turing thesis applies to nondeterministic models too:

> Every (implementable) nondeterministic computation can be computed by a deterministic Turing machine.

The intuition behind a deterministic simulation of a nondeterministic Turing machine rests on the idea of nondeterministic choice: whenever the nondeterministic computation can "choose" from among multiple next steps, the deterministic Turing machine will sequentially try each alternative, exhaustively enumerating all possible computations originating from the current configuration (the complete branching-time tree). This simulation scheme is not difficult to implement, but it introduces in general an exponential blowup in the running time. Precisely, consider a nondeterministic Turing machine N that runs in time $T_N(n)$. The nondeterministic complexity measure factors in the power of choice, that is, it does not consider the alternative steps that are possible but useless for the computation. A deterministic Turing machine D simulating N by enumerating the computations will, however, have to include the "unfruitful" choices as well in the enumeration; if N's transition function allows for up to b alternatives at every step, D runs in time polynomially correlated to $b^{T_N(n)}$ in the worst case.

Example 6.23. The "map 3-coloring problem" (3-*COL*) is the problem of deciding if you can color with green, blue, or red each region on a generic map in such a way that no two adjacent regions have the same color.

A nondeterministic algorithm – expressible with a nondeterministic Turing machine – can solve 3-*COL* as follows: pick a color; use it to color one of the regions not colored yet; repeat until all regions are colored; when the coloring is complete, check if no two adjacent regions have the same color and report success or failure. The time complexity of this nondeterministic algorithm is $O(n^3)$ in the number of regions: n steps to nondeterministically produce a coloring, and then $O(n^3)$ steps to check if the coloring satisfies the adjacency property.

To this end, a memory tape stores the regions that are adjacent as sequences with separators; for example, with reference to US states, the sequence AZ NV UT CO NM CA encodes the fact that Arizona (AZ) is adjacent to Nevada (NV), Utah (UT), Colorado (CO), New Mexico (NM), and California (CA). If US states are encoded in unary notation respecting the alphabetic order (Alabama = 1, Alaska = 11, Arizona = 111, etc.), the adjacency list has length $O(n^3)$. Another tape encodes the colors given to each state in a fixed order, say alphabetical; for instance, a list beginning with **red green blue** means that Alabama is colored red, Alaska green, and Arizona blue. With this setting, since every state has $O(n)$ neighbors, and checking the color of each neighbor takes $O(n)$ steps to look up the list of colors, the complete check has cubic complexity.

A deterministic Turing machine simulating the nondeterministic algorithm must enumerate, in the worst case, all possible 3^n colorings. Therefore, it runs in time $O(n^3 \cdot 3^n)$, exponential in the nondeterministic time complexity.   ∎

6.2.10   NP-Completeness

Is the exponential blowup that deterministic Turing machines incur in simulating nondeterministic computations unavoidable in the worst case? In other words, can we *efficiently implement* the nondeterministic model of computation? These are outstanding open questions, whose complete answers have so far eluded the best scientists – and not for lack of trying!

If the blowup is unavoidable, it means that nondeterminism is not a model of efficiently implementable computational processes; the strong Church-Turing thesis does not apply to nondeterministic processes; unbounded parallelism is incommensurable with the bounded amount of parallelism that deterministic machines with multiple processors can achieve; and the complexity class P is a strict subset of the class NP.

The prevailing conjecture, supported by overwhelming theoretical and practical evidence, is that nondeterminism cannot be efficiently implemented in the general case: every attempt at designing an efficient (i.e., polynomially correlated) deterministic simulation of nondeterministic computations has failed, but it is extremely hard to prove that such an achievement is impossible, even in principle.

If we set aside the theoretical aspects and focus on practical evidence, it seems that nondeterminism is not a reasonable abstraction of "physical" computational processes. Nonetheless, it is still a convenient abstraction because it captures the essence shared by a huge family of heterogeneous problems with enormous practical relevance, known as NP-complete problems. A precise definition of NP-complete problem appears below. Intuitively, an NP-complete problem is one for which checking if a candidate solution is correct is computationally tractable, but constructing a correct solution from scratch seems intractable. As explained in the previous sections, the notion of "tractable" means that there exists a polynomial-time algorithm; hence a tractable problem is in the class P. In contrast, an intractable problem has only solutions with super-polynomial (typically, exponential) running times. Nondeterministic computations can use the power of "magic choice" and construct the correct solution in the same running time needed to check it, and hence efficiently with respect to the use of nondeterministic resources; or, equivalently, they can explore in parallel all the exponentially many possible solutions and select the correct ones.

The complexity class NP includes all these NP-complete problems that are "easy to check and (seemingly) hard to solve" for deterministic algorithms, but which nondeterministic algorithms can process efficiently. The NP-complete problems are then the "hardest problems in NP". Some examples of NP-complete problems:

- The "map 3-coloring" problem (3-*COL*) (introduced in Example 6.23).
- The "traveling salesman" problem (*TSP*) (also called "shortest Hamiltonian cycle" problem): given a list of cities and their distances from each other, find the shortest path that goes through every city exactly once.
- The "Boolean satisfiability" problem (*SAT*): given a combinatorial circuit of Boolean logic gates (see Chap. 5) with a single output, find if there is an input that makes the output evaluate to true. An equivalent presentation of the same problem is in terms of Boolean formulae in propositional logic.
- The "register allocation problem": given a program in a high-level language, find an assignment of the program variables to a minimum number of registers in such a way that no two variables use the same register at the same time.

All these problems have enormous practical relevance. What is remarkable is that, in spite of their widely different domains, they all are intimately connected by being NP-complete: the existence of an efficient implementable algorithm for *any* NP-complete problem would imply that *every* NP-complete problem has a similarly efficient solution. In complexity-theoretical terms, if there exists an NP-complete problem which is also in P, then $P = NP$ and every NP problem has a deterministic tractable solution. Conversely, if we succeeded in proving that a specific NP-complete problem has no efficient solution, then *no* NP-complete problem would have an efficient solution, and $P \subset NP$.

Definition 6.24 (NP-completeness).   For a problem p, $P(x)$ denotes the solution of p for input x; similarly, $M(x)$ denotes the (unique) output of the deterministic Turing machine M with input x. Then, a problem c in NP is "NP-complete" if, for any other problem p in NP, there exist two deterministic Turing machines $R_{p \to c}$ and $R_{c \to p}$ with polynomial time complexities such that $R_{c \to p}(C(R_{p \to c}(x))) = P(x)$ holds for every input x. ∎

Thus, a problem c is NP-complete if we can use it to solve every other problem p in NP with the same time complexity up to a polynomial factor: transform the input to p into an "equivalent" input to c in polynomial time with $R_{p \to c}$; solve c for the transformed input; transform the solution back in p's form in polynomial time with $R_{c \to p}$. The bottleneck in this process is solving c: if it can be done efficiently, then every other problem in NP is solvable efficiently (with at most a polynomial slowdown) by piggybacking on c's solution.

Conventional wisdom, supported by the results outlined above, considers the border between tractable and intractable problems in any "reasonable" computational model to lie close to the P/NP (or, more precisely, P/NP-complete) frontier. However, the failure to come up with efficient solutions for NP-complete problems has not prevented the development of algorithms that use sophisticated heuristics and highly optimized data structures to achieve performances that are acceptable in practice for inputs of reasonable size. Part II of the book will mention several examples of verification tools that exploit these results to analyze complex properties of systems automatically with acceptable performance, in spite of the problems being NP-complete or of even harder complexity classes.

6.3 Computational Complexity and Architecture Models

The automata-based models analyzed in Sect. 6.2 are suitable for investigating fundamental questions about the limits of computing and the inherent computational complexities of problems. Part II will show that automata are also appropriate abstractions for domain-specific modeling, especially when they include a notion of time. Automata-based models usually are, however, of limited usefulness for modeling general-purpose digital computing devices. The von Neumann hardware architecture typically underlies the internal organization of these devices. Therefore, the fine-grained analysis of the running time (and memory used) by digital computer programs must use formal models that represent the principles of the von Neumann architecture accurately, with suitable abstractions.

6.3.1 Random Access Machines

We already discussed the inadequacy of finite-state automata for providing a reasonable abstraction of general-purpose computers. Turing machines provide the "right" expressive power for representing general-purpose computing, but they are imprecise models of the computational complexities of real programs. In particular, the assumption that every move of a Turing machine counts for one unit of time is too simplifying for general-purpose computers, where different elementary operations performed by the CPU can have quite different running times. For example, memory access operations normally take more clock cycles than other operations accessing only internal CPU registers, and input/output operations are even orders of magnitude slower.

There is a simple partial solution to this inadequacy of the Turing machine model: assign different costs to different moves of the Turing machine. This may somehow complicate the complexity analysis, but it makes the model more detailed. Such a more detailed model, however, would be superseded by some linear speed-up, and it would still fail to capture a more fundamental discrepancy between automata-based models and architectures à la von Neumann: the *memory model*. Turing machines

Fig. 6.4 A random access
machine (RAM)

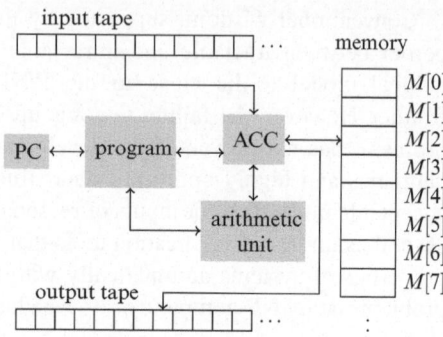

access their memory tapes strictly sequentially; therefore the position of a datum
affects the number of steps required to access it. Computers with von Neumann
architectures have direct access memory, and an elementary operation such as load
or store can atomically transfer a block of bits from one memory location to any
other location, regardless of their relative positions in the memory.

This section develops computational models that, while still abstract compared to
the architectures of real computers, include memory with direct access, and therefore support an accurate analysis of algorithmic complexity. The analysis results
are largely extensible to general-purpose computers with similar architectures. One
classic abstract model of computer with direct access memory is the "Random
Access Machine" (RAM).

Definition 6.25 (Random access machine). A random access machine (RAM),
pictured in Fig. 6.4, consists of:

- Input and output tapes, each a sequence of cells accessed sequentially with one
  head (similarly to Turing machines).
- A direct access memory of unbounded size; the memory is a sequence of cells,
  each addressed by a natural number.
- A control unit with:

  - A hardwired program consisting of a sequence of instructions, each numbered
    by a natural number;
  - A register called the "accumulator" (ACC);
  - The "program counter" (PC), another register storing the address of the next
    instruction to be executed;
  - A unit capable of performing arithmetic operations.

Every memory cell, including the registers and those on input/output tapes and in
the direct access memory, can store any integer value, or a character from a finite
alphabet.

The RAM executes the instructions programmed sequentially, except for jump
instructions that explicitly modify the PC and change the instruction to be executed
next. Figure 6.5 lists the instructions available and their semantics; $M[n]$ denotes
the content of memory at address n.   ∎

Instruction	Semantics	Comments
READ x	$M[x] \leftarrow$ current input value; input head advances by one position.	Copy the value in the current input cell to the memory at address x.
READ@ x	$M[M[x]] \leftarrow$ current input value; input head advances by one position.	"@" denotes *indirect* addressing: copy the input to the memory cell whose address is stored in memory at address x.
WRITE x	$M[x] \rightarrow$ current output cell; output head advances by one position.	
WRITE@ x	$M[M[x]] \rightarrow$ current output cell; output head advances by one position.	
WRITE= x	$x \rightarrow$ current output cell; output head advances by one position.	"=" denotes *immediate* addressing: write the operand's value x.
LOAD x	$ACC \leftarrow M[x]$	Copy the content of memory at address x into the accumulator.
LOAD@ x	$ACC \leftarrow M[M[x]]$	
LOAD= x	$ACC \leftarrow x$	
STORE x	$M[x] \rightarrow ACC$	
STORE@ x	$M[M[x]] \rightarrow ACC$	
ADD x	$ACC \leftarrow [ACC] + M[x]$	Add the contents of the accumulator to memory content at address x and store the result back in the accumulator.
ADD@ x	$ACC \leftarrow [ACC] + M[M[x]]$	
ADD= x	$ACC \leftarrow [ACC] + x$	
SUB	[...]	Subtraction, multiplication, and divi-
MULT		sion arithmetic operations are defined
DIV		similarly to ADD.
JUMP lab	$PC \leftarrow$ instruction (lab)	Set the program counter to the address of the instruction with label "lab"; this instruction is executed next.
JZ lab	**if** $[ACC] = 0$ **then** $\quad PC \leftarrow$ instruction (lab) **else** $\quad PC \leftarrow [PC] + 1$ **end**	Conditional jump: jump to "lab" if the accumulator stores 0; otherwise continue sequentially.
HALT		Execution stops.

Fig. 6.5 The instruction set of the RAM

Example 6.26. Figure 6.6 shows a RAM program that checks if the input is a prime number; it implements the trivial algorithm that tries all possible divisors. In the comments, n denotes the input value read. ∎

6.3.2 Algorithmic Complexity Analysis with Random Access Machines

Let us analyze the (time) complexity of the RAM program in Fig. 6.6 with the same principles used for Turing machines: count the maximum number of instructions

Label	Instruction	Comment
	READ 1	Store n into $M[1]$.
	LOAD= 1	If $n = 1$ then it is trivially prime
	SUB 1	and execution ends immediately
	JZ *yes*	with a positive answer.
	LOAD= 2	Initialize $M[2]$ to 2
	STORE 2	to start a loop of tests.
loop:	LOAD 1	When $M[2] = n$ exit the loop:
	SUB 2	all possible divisors have been tested
	JZ *yes*	hence the number is prime.
	LOAD 1	Assuming integer arithmetic, if
	DIV 2	$M[1]$ equals $(M[1] / M[2]) \times M[2]$ then
	MULT 2	$M[2]$ contains a divisor of n,
	SUB 1	hence n is not prime:
	JZ *no*	exit loop and report negative answer.
	LOAD 2	Increment $M[2]$ by 1
	ADD=1	to test the next divisor.
	STORE 2	
	JUMP *loop*	Repeat loop.
yes:	WRITE= 1	Output positive answer.
	HALT	
no:	WRITE= 0	Output negative answer.
	HALT	

Fig. 6.6 A RAM program for primality testing

executed in a run with input n. Every integer n has n possible divisors tested by the program, and the loop consists of 12 instructions; this gives a time complexity function $\Theta(n)$. It is clear that the implementation of the same algorithm with a Turing machine would not have the same asymptotic complexity: just multiplying two numbers requires a number of steps that increases with the number of *digits* of the numbers.

This significant discrepancy suggests the need for more detailed inspection to understand the scope and limits of the RAM's abstraction. The assumption that every memory cell in the RAM can store an *integer* implies that a single instruction can process an infinite amount of information: there is no a priori bound on the maximum integer storable. On the one hand, the architectures of digital computers do offer powerful instruction sets, which can manipulate and perform arithmetic on the content of any register with a fixed number of clock cycles. On the other hand, every architecture has memory cells and registers of *finite size*, usually measured in bits. In a 64-bit architecture, for example, the RAM's abstraction of unbounded integers storable in every cell is appropriate as long as the program deals only with integers between -2^{63} and $+2^{63} - 1$ that fit a "real" memory cell. When this hypothesis does not hold, the complexity analysis on the RAM may give results that are not generalizable to a real architecture. The following Example 6.27 makes a compelling case.

Fig. 6.7 A RAM program
that computes the double
exponential of the input

Label	Instruction	Comment
	READ 1	Store the input n into $M[1]$.
	LOAD= 2	Initialize $M[2]$ to 2.
	STORE 2	
	LOAD= 1	$M[3]$ is used as counter,
	STORE 3	initialized to the value 1.
loop:	LOAD 1	When the counter reaches n,
	SUB 3	$M[2]$ contains the result:
	JZ *result*	print it and stop.
	LOAD 2	Square $M[2]$, that is
	MULT 2	$M[2]$ receives
	STORE 2	$M[2] \times M[2]$.
	LOAD 3	Increment the counter.
	ADD= 1	
	STORE 3	
	JUMP *loop*	
result :	WRITE 2	
	HALT	

Example 6.27. Figure 6.7 shows a RAM program that computes the double expo-
nential 2^{2^n} of the input n by successive squaring: $2^{2^n} = 2^{\overbrace{2 \cdot 2 \cdots \cdots 2}^{n}}$. The loop is
executed n times, which gives again a time complexity $\Theta(n)$ relative to the input
value. ∎

The time complexity analysis of Example 6.27 is clearly unacceptable: in a
k-bit architecture, the intermediate result would overflow after k loop iterations,
and it would occupy 2^{i-k} k-bit cells at the ith iteration of the loop. Any reasonable
complexity analysis must take the extra elementary operations used to manipulate
multi-cell numbers into account.

We combine these observations in the "logarithmic cost criterion", which evaluates the time complexity of a RAM program according to the following principles.

- An integer value n occupies $\log_2 n$ bits, split into $(\log_2 n)/k$ cells in a real
  k-bit architecture; then, accessing the content of a memory cell that stores
  an integer n requires $(\log_2 n)/k$ elementary operations. The logarithmic cost
  criterion incorporates these estimates by modeling the access to a memory cell
  storing a number n as taking time $\Theta(\log n)$; the precise value of k is abstracted
  away in the asymptotic notation.
- Similarly, accessing a memory cell at address n requires $\Theta(\log n)$ elementary
  resources in a real architecture, to send the address over the memory bus and to
  receive back the content of the cell. The logarithmic cost criterion thus counts
  $\Theta(\log n)$ time units for every access to memory at address n.
- The measure of the input length is also proportional to the value of the input
  cells: an input consisting of the sequence of integer values $i_1 i_2 \ldots i_n$ has length
  $\Theta(\log i_1 + \log i_2 + \cdots \log i_n)$, the parameter to be used in the complexity
  measures.

Fig. 6.8 The cost of some
RAM instructions with the
logarithmic criterion

Instruction	SemanticsCost
READ x	$\ell(\text{current input value}) + \ell(x)$
READ@ x	$\ell(\text{current input value}) + \ell(x) + \ell(M[x])$
WRITE x	$\ell(x) + \ell(M[x])$
WRITE@ x	$\ell(x) + \ell(M[x]) + \ell(M[M[x]])$
WRITE= x	$\ell(x)$
LOAD x	$\ell(x) + \ell(M[x])$
LOAD@ x	$\ell(x) + \ell(M[x]) + \ell(M[M[x]])$
LOAD= x	$\ell(x)$
STORE x	$\ell([ACC]) + \ell(x)$
STORE@ x	$\ell([ACC]) + \ell(x) + \ell(M[x])$
ADD x	$\ell([ACC]) + \ell(x) + \ell(M[x])$
ADD@ x	$\ell([ACC]) + \ell(x) + \ell(M[x]) + \ell(M[M[x]])$
ADD= x	$\ell([ACC]) + \ell(x)$
JUMP lab	1
JZ lab	$\ell([ACC])$
HALT	1

- Characters are from a finite alphabet; hence accessing a cell storing a character only costs constant $\Theta(1)$ time units.

The base of the logarithms is irrelevant because $\log_a(n) = \log_b(n)/\log_b(a) = \Theta(\log_b(n))$: every encoding other than unary (in base 2, 3, or more) achieves the same asymptotic complexity.

The simplistic approach to complexity measures used before is called *uniform cost criterion* for comparison with the logarithmic cost. The uniform cost criterion is perfectly adequate for algorithms that only manipulate integers of absolutely bounded size and use an amount of memory absolutely bounded (that is, bounded for every input). Otherwise, only the logarithmic cost criterion achieves a reliable complexity analysis.

Figure 6.8 shows the cost of the basic RAM instructions computed according to the logarithmic cost criterion; in the table, $\ell(x)$ is a shorthand for the expression

$$\ell(x) \;=\; \begin{cases} 1 & x = 0 \text{ or } x \text{ is a character,} \\ \lfloor \log_2 |x| \rfloor + 1 & \text{otherwise.} \end{cases}$$

Intuitively, each operation "costs" something proportional to the logarithmic size of its operands and result. For example, the sum of two integers n_1 and n_2 requires us to manipulate approximately $\lfloor \log_2(n_1) \rfloor$ and $\lfloor \log_2(n_2) \rfloor$ bits for the operands and $\lfloor \log_2(n_1 + n_2) \rfloor$ bits for the result; this affects the memory cells used and is also proportional to the number of truly elementary sum operations required by the RAM's addition.

For example, each iteration of the loop for the program in Fig. 6.6 corresponds to a number of elementary steps proportional to the logarithm of the input n and the current divisor, that is $\Theta(\log n)$. Repeated for n iterations, this gives a total number of steps $\Theta(n \log n)$. Finally, we express this complexity with respect to the input size $m = \ell(n)$ – which is $\Theta(\log n)$ – to get a time complexity measure $\Theta(m \cdot 2^m)$.

Exercise 6.28. Complete Fig. 6.8 with the logarithmic cost of the missing instructions.

- How would you measure the cost of mult? Does it affect the complexity result for the program in Fig. 6.6?
- Determine the time complexity of the RAM program in Fig. 6.7 using the logarithmic cost criterion. ∎

Exercise 6.29. Consider the algorithm that performs binary search of an integer in a sequence of integers.

- Write a RAM program that performs binary search of a sequence stored in memory.
- Determine the time complexity of the program developed in (6.29), both with the uniform cost and with the logarithmic cost criteria; compare the two measures.
- Outline (without details) a Turing machine that performs binary search; determine its time complexity measure; compare it to the RAM's in (6.29). What is the reason for the difference? ∎

Exercise 6.30. Consider the problem *PALINDROME* (Example 6.15), determining if a sequence of *characters* is a palindrome or not.

- Write a RAM program that determines if the input sequence is a palindrome; determine its time complexity (both with uniform and with logarithmic cost criteria).
- Write a Turing machine that determines if the input sequence is a palindrome; determine its time complexity.
- Compare the time complexities of the RAM and Turing machine models. ∎

Exercise 6.31. The "Random Access Stored Program machine" (RASP) is similar to the RAM, but stores its program in the direct access memory together with the data. It is the equivalent of a universal Turing machine for RAMs. The RASP execution model consists of fetching each instruction and copying it in a special register before executing it; the value of the program counter determines the next instruction to fetch. Extend the cost measures in Fig. 6.8 for the RASP machine model.

6.3.3 Random Access Machines and Complexity Classes

The examples and exercises in the previous parts of Sect. 6.3 show that the same problems may have different complexities in the RAM model and the Turing machine model. In most cases, the RAM's direct access memory allows it to achieve asymptotically faster running times than the Turing machine, but there are problems for which the humble Turing machine model can be faster.

Consider, for example, the problem of generating an output sequence that equals the concatenation of the input with itself (e.g., if the input sequence is 17, the output is 1717), and assume that the elements in the sequence are from a finite set. The Turing machine can solve the task in $\Theta(n)$ steps, by copying the input to a memory tape and then copying it twice onto the output tape. The RAM must copy the elements of the sequence in memory using $\Theta(n)$ memory addresses; this causes a logarithmic slowdown with the logarithmic cost criterion and a time complexity $\Theta(n \log n)$ asymptotically worse than the Turing machine's.

Section 6.2.7 anticipated the differences in analyzing single algorithms versus problems and justified them. After presenting the RAM model, we can better understand the different scope of complexity analysis for algorithms and for problems.

We base the analysis of the optimal implementation of specific algorithms on a machine model that, while abstract, captures the fundamental characteristics of the architecture of the real computers used. The RAM is a suitable model for investigating the complexities of algorithms for sequential programs running on architectures à la von Neumann. The bibliographic remarks at the end of the chapter mention other models of different architectures (e.g., parallel architectures) with the same level of abstraction as the RAM. The linear speed-up theorem applied to RAMs guarantees that details such as the bit size of memory cells or the execution speed of individual instructions do not affect the algorithmic complexity analysis performed with RAM models.

In the more detailed abstraction of RAMs and similar machines, it is sometimes useful to consider finer-grained complexity classes that are not robust with respect to polynomial correlations but are with respect to constant-factor (i.e., linear) speed-up. In particular, the two deterministic complexity classes LIN (problems solvable in "linear" time $O(n)$), also called LTIME, and L (problems solvable in logarithmic space $O(\log n)$), also called LSPACE, characterize problems with very efficient solutions that easily scale up to very large input sizes.

On the other hand, we use Turing machines (and other very abstract models) to determine the inherent complexities of computational problems. The results are valid beyond Turing machines, because every algorithm formalized in a "reasonable" abstract (or real) machine will have complexity polynomially correlated with the Turing machine's; hence the complexity class of a problem is universal with respect to any "implementable" computational model. In particular, a Turing machine can simulate the computation of every RAM with a polynomial slowdown[2] – with respect to the "realistic" logarithmic cost criterion; and a RAM can simulate the computation of every k-tape Turing machine with a polynomial (actually, logarithmic) slowdown – measured with the logarithmic cost criterion.

---

[2] Merely quadratic, under "reasonable" hypotheses.

6.4 Randomized Models of Computation

A computational process is *randomized* if its evolution over time depends on chance. Whereas the behavior of a deterministic process is completely determined by the initial state and the input, and a nondeterministic process can "choose" among possible next states and will explore *all* choices, a randomized (or *stochastic*) process can "flip a coin" or "throw a die" to select the next state from the current one. Chapter 3 outlined the basic motivation for introducing randomness in computational models: stochastic behavior is common in many physical phenomena. More precisely, it is inherent in some (as in the microscopic world of quantum mechanics) and an extremely convenient abstraction of others (such as in statistical mechanics for modeling the collective behavior of large populations of simple particles). The theme of the present chapter gives an additional justification for considering randomized models: since randomized models are clearly implementable – because random processes exist in nature – does using randomness as a resource increase computational power?

The present section discusses stochastic extensions of finite-state automata (Sect. 6.4.1) and Turing machines (Sect. 6.4.2), their impact on the results discussed in the previous sections, and their connections with some of the models presented in Part II of the book. In accordance with the outline of Sect. 3.4, the rest of this section presents notations incorporating randomness with two different roles: in some cases, the choice of which transition to make is randomized; in others, the *timing* of transitions is subject to chance. Whereas it is obvious that the latter case – of randomized timing – has a direct and **EXPLICIT** impact on timed behavior, the former case – of randomized transition choices – has an indirect impact, since different transitions may produce different response times.

6.4.1 Probabilistic Finite-State Automata

Finite-state automata extended with a notion of probability are usually called "Markov chains". The name is after Andrey Markov, who first suggested using the word "chain" to designate a family of discrete-time stochastic processes with memory. Even if the origins of Markov chains and finite-state automata are quite different, it is natural to present Markov chains as probabilistic extensions of finite-state automata. More precisely, Markov chains coincide with probabilistic finite-state automata when they possess a finite number of states, but Markov chains over a denumerable number of states are also possible. This section uses the two nomenclatures interchangeably, and presents three variants of the probabilistic finite-state automaton model: vanilla discrete-time Markov chains (Sect. 6.4.1.1) with probabilities associated with transitions; Markov decision processes (Sect. 6.4.1.2), which include input events; and continuous-time Markov chains (Sect. 6.4.1.3), where the *timing* of transitions is randomized.

Fig. 6.9 A probabilistic
finite-state automaton
modeling a student taking
exams

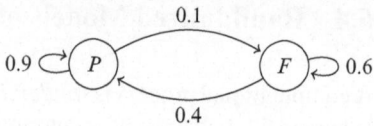

6.4.1.1 Discrete-Time Markov Chains

Probabilistic finite-state automata (also, "discrete-time Markov chains") associate a probability with each transition that determines the chance that the transition is taken.

Definition 6.32 (Probabilistic finite-state automaton/Discrete-time Markov chain). A *probabilistic finite-state automaton* is a finite-state automaton without an input alphabet, extended with a probability function $\pi : Q \times Q \to [0, 1]$ which determines which transitions are taken. The probability function is normalized:

$$\sum_{q' \in Q} \pi(q, q') = 1$$

for every $q \in Q$. When the automaton is in some state $q \in Q$, it can make a transition to any state $q' \in Q$ with probability $\pi(q, q')$. At every step, the automaton moves to the drawn state q', where it gets ready for a new transition.

   Discrete-time Markov chains generalize discrete-time probabilistic finite-state automata to any *countable* set Q of states.   ∎

   Probabilistic automata include no explicit notion of *input*, as transitions are chosen at random solely according to the probability distribution defined by π for the present state. This feature makes probabilistic automata suitable models of systems where a finite number of "events" can happen with certain probability but without direct control.

Example 6.33. Figure 6.9 shows a probabilistic finite-state automaton that models the following behavior of a student who takes a series of exams: when the student passes an exam (state P), there is a 90 % chance that she will pass the next exam as well (transition to P); when the student fails an exam (state F), there is a 60 % chance that she will fail the next exam (transition to F). The figure shows the probabilities associated with each transition and the corresponding next states.   ∎

   The analysis of deterministic models focuses on the properties of individual runs with a given input; that of nondeterministic models captures the behavior of every nondeterministic run on a given input. Analyses of probabilistic finite-state models typically take a different angle, and target the *probability* of certain runs or the *average* long-term behavior of the automaton.

Example 6.34. Let us go back to Example 6.33 and analyze a few properties of the model, under the assumption that the probability of passing the very first exam is 0.9 (that is, P is the initial state).

- The probability that the student passes the first k consecutive exams is 0.9^k.
- The probability that the student passes *every other* exam for $2k$ exams is $(0.1 \cdot 0.4)^k = 0.04^k$.
- The probability that the student *always* fails goes to 0 as the number of exams increases.   ∎

Computing the probability of specific runs is simple because the probabilistic choices depend only on the current state and are otherwise independent of the outcomes of the previous stochastic events. This property – which characterizes Markov processes – is in fact called "Markov property".

Another significant feature of probabilistic models à la Markov is the decrease of probability of every *specific* behavior as the length of the behavior increases: whereas in nondeterministic models *every* nondeterministic choice is possible and must be considered, assigning probabilities to different choices entails that the long-term behavior is more and more likely to asymptotically approach the average behavior. Precisely, the *steady-state probability* gives the likelihood that an automaton will be in a certain finite state after an arbitrarily long number of steps. To compute the steady-state probability, represent the probability function π as a $|Q| \times |Q|$ *probability matrix M*, whose element in row i, column j is the probability of transitioning from the ith to the jth state. The steady-state probability has the characterizing property of being independent of the initial and current states: a row vector $\mathbf{p} = p_1 \ldots p_{|Q|}$ of nonnegative elements, whose ith element denotes the probability of being in the ith state, is the steady-state probability if it satisfies $\mathbf{p} \cdot M = \mathbf{p}$ (it does not change after one iteration) and $\sum_{1 \le i \le |Q|} p_i = 1$ (it is a probability distribution on states). In other words, the steady-state probability is a *left eigenvector* of the probability matrix with unit eigenvalue, whose elements are nonnegative and sum to 1.

Example 6.35. The probability function in Example 6.33 determines the probability matrix

$$M = \begin{bmatrix} 0.9 \; 0.1 \\ 0.4 \; 0.6 \end{bmatrix},$$

with left eigenvector

$$[p_1 p_2] = [0.8 \; 0.2]$$

that satisfies $[p_1 p_2] \cdot M = [p_1 p_2]$. This means that, in the long term, the student passes 80 % of the exams she attempts.   ∎

Exercise 6.36. A mono-dimensional "random walk" is a process that visits the sequence of integer numbers as follows. The process starts from the position 0; at any step, it flips an unbiased coin, increases its position by 1 if the coin lands on tails, and decreases it by 1 otherwise.

- Model the random walk with an infinite-state Markov chain.
- Compute the probability $P_{k,t}$ that the process is at integer position k after t steps, for every $k \in \mathbb{Z}, t \in \mathbb{N}$.
- Compute the average position after k steps. Does it depend on k?   ∎

Time is **DISCRETE, SYNCHRONOUS,** and **METRIC** in finite-state automata regardless of whether they are in their deterministic, nondeterministic, or probabilistic version. Probabilities, in particular, only determine the transition taken at every step, but time progresses uniformly and synchronously with every transition. Time is also **INFINITE** in probabilistic finite-state automata: the time domain is isomorphic to the natural numbers, and all runs are indefinitely long because the probability that *some* transition is taken is 1 in every state. This notion of infinite behavior further justifies the focus on long-term steady-state behavior rather than on properties of individual finite runs.

6.4.1.2 Markov Decision Processes

Probabilistic finite-state automata with input (also, "Markov decision processes") extend probabilistic automata (and standard Markov chains) with input events.

Definition 6.37 (Probabilistic finite-state automaton with input/Discrete-time Markov decision process). A *probabilistic finite-state automaton with input* is a probabilistic finite-state automaton with probability function $\pi : Q \times I \times Q \to [0, 1]$ over transitions. The probability function is normalized with respect to the next states: for every $q \in Q$ and $i \in I$,

$$\sum_{q' \in Q} \pi(q, i, q') = 1.$$

When the automaton is in state $q \in Q$ and inputs an event $i \in I$, it can make a transition to any state $q' \in Q$ with probability $\pi(q, i, q')$. At every step, the automaton moves to the drawn state q', where it gets ready for a new transition.

Discrete-time Markov decision processes generalize discrete-time probabilistic finite-state automata with input to any *countable* sets Q and I of states and input events.   ∎

Input events are useful for modeling external influences on the system, such as control actions or changes in the environment; input does not, in general, deterministically control the system evolution – which remains stochastic – but it does affect the probability of individual transitions. According to the real-world entity it models, the input of Markov decision processes is identified by one of many names: "environment", "scheduler", "controller", "adversary", or "policy". These terms present decision processes as **OPEN** systems, to which an external abstract entity supplies input (see also Sect. 3.7). The composition of decision process and environment is a **CLOSED** system that characterizes an embedded process operating in specific conditions.

Example 6.38. Figure 6.10 shows a probabilistic finite-state automaton with input that models a student taking a series of exams; it is an extension of Example 6.33. While preparing for an exam, the student may attend classes on the exam's topic

Fig. 6.10 A probabilistic
finite-state automaton with
input modeling a student
taking exams

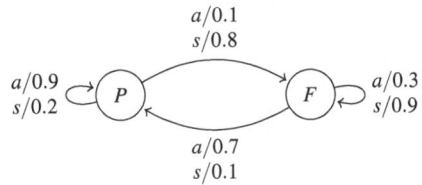

(event a) or skip them (event s). While attending classes is no guarantee of passing, it significantly affects the probability of success: after the student has passed an exam (state P), there is a 90 % chance that she will also pass the next one if she attends classes (transition to P with event a); if she has not, the probability shrinks to only 20 % (transition to P with event s). Conversely, after the student has failed an exam (state F), there is a 70 % chance that she will pass the next one if she attends classes (transition to P with event a), but only 10 % if she skips them (transition to P with event s). The figure shows the probabilities associated with each transition and input event.   ∎

As with Markov chains, the properties of interest with probabilistic finite-state models *with input* include the probabilities of sets of runs or average long-term behavior. Probabilities depend, however, on the supplied input sequences, which corresponds to the notion of *conditional probability* (mentioned in Sect. 3.4.2), as demonstrated in the following example.

Example 6.39. Let us derive some probabilistic properties of the model of Example 6.38, generalizing Example 6.34 for the presence of input events. Assume that P is the initial state.

- The probability that the student passes the first k consecutive exams when she always attends classes is 0.9^k.
- The probability that she passes the first $2k$ consecutive exams when she attends classes of every other exam is $(0.9 \cdot 0.2)^k = 0.18^k$.
- The probability $pass_k$ that the student passes the first k consecutive exams is, in general, a function of the input sequence $i_1 i_2 \ldots i_k$:

$$pass_k(i_1, i_2, \ldots, i_k) = \prod_{1 \le j \le k} p(i_j),$$

where

$$p(i_j) = \begin{cases} 0.9 & i_j = a, \\ 0.2 & i_j = s. \end{cases}$$ ∎

Since the input modifies the behavior of Markov decision processes, some interesting analysis problems consist of *computing the input* that ensures behaviors with certain properties. Using the terminology introduced before, these problems are presented as "controller (or scheduler) *synthesis*". A synthesis problem for Example 6.38 is finding the input sequence that maximizes the probability of

Fig. 6.11 A Mealy machine
controlling the probabilistic
finite-state automaton with
input of Fig. 6.10

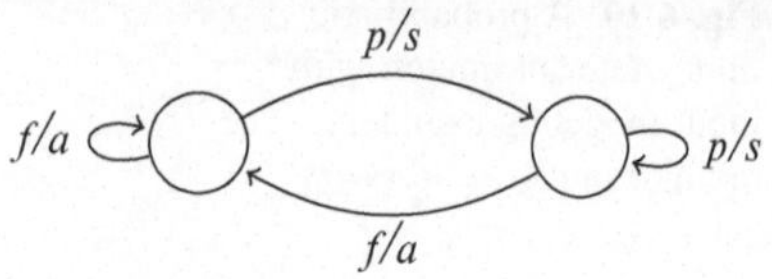

passing all the exams; it is clear that the solution is the input consisting of only
events a. The following example illustrates less trivial synthesis problems.

Example 6.40. Consider the following synthesis problem for the model of Example 6.38: "determine the input that maximizes the probability of *alternating* between
passing and failing exams".

The obvious candidate – the input sequence $a\,s\,a\,s\,\ldots$ that strictly alternates a
and s – does not make for a satisfactory solution: if, for example, the student fails
the first exam even if a was selected, we should input a again as the second event,
to maximize the probability that she passes the second exam, hence alternating
between passing and failing. The proper solution to the synthesis problem is thus not
simply a fixed input sequence, or even a set of inputs. It is, instead, a *computational
process* that responds to the probabilistic choices of the Markov decision process
with suitable inputs, according to the strategy: "when the student passes an exam,
generate input s; otherwise, generate input a".

We can formalize this *controller process* as the *deterministic Mealy machine*
in Fig. 6.11 (Sect. 5.2.3 defines Mealy machines), with input symbols p and f
corresponding to "passing" and "failing" an exam, and output symbols a and s to
be fed to the Markov decision process.   ∎

Exercise 6.41. For the probabilistic finite-state automaton of Example 6.38, informally describe controllers that:

1. *Minimize* the probability of *alternating* between passing and failing exams.
2. *Maximize* the probability of passing *exactly two* exams consecutively after every
   failed exam, and otherwise alternating between passing and failing.
3. (♦) *Maximize* the probability of passing exactly $k + 1$ exams consecutively after
   the kth sequence of consecutive failed exams, and otherwise alternating between
   passing and failing. For example, if the first exam is failing, the probability of
   having exactly two passing exams next should be maximized; if the first exams
   are $f\,p\,f\,f\,f\,p\,f\,p\,f\,f$, the probability of having exactly five passing exams
   next should be maximized (because there have been four sequences, of lengths
   one, three, one, and two, each consisting of consecutive failed exams).

Formalize the controllers described in items (1–3):

4. Formalize the controllers in (1) and (2) as finite-state Mealy machines.
5. (♦) Argue that the controller in (3) is not expressible as a finite-state Mealy
   machine, and suggest a more expressive computational model that can formalize
   the controller.   ∎

6.4.1.3 Continuous-Time Markov Chains

In contrast with the discrete-time probabilistic automata presented in the two previous subsections, CONTINUOUS-TIME probabilistic finite-state automata (also, continuous-time Markov chains) model time as a continuous set (the real numbers) and assign a probability distribution to the *residence* (also, "sojourn") *time* in every state. This produces ASYNCHRONOUS behavior, where an automaton sits in a state and moves to a different state after an amount of time drawn from a probability distribution that may depend on the current state. In this respect, continuous-time probabilistic finite-state automata extend with probabilities the asynchronous interpretation of finite-state machines discussed in Sect. 5.2.2. For continuous-time probabilistic finite-state automata (and Markov chains), the probability distributions of residence times ensure that the probability of remaining in the current state for the next t time units does not depend on the previous states traversed by the automaton, but only on the current one, in the same way as the probability of making a transition only depends on the current state in discrete-time probabilistic automata (that is, the Markov property). The only probability distribution that satisfies the Markov property is the *exponential* one: the probability that an automaton waits for t time units decreases exponentially with t. Chapter 7 will present more general examples of finite-state automata with probabilities on time that allow for probability distributions other than the exponential.

Definition 6.42 (Continuous-time probabilistic finite-state automaton/Continuous-time Markov chain).
    A *continuous-time probabilistic finite-state automaton* extends a (discrete-time) probabilistic finite-state automaton with a rate function $\rho : Q \to \mathbb{R}_{>0}$. Whenever the automaton enters state $q \in Q$, it waits a time given by an exponential distribution with parameter $\rho(q)$ and probability density function

$$p(t) = \begin{cases} \rho(q)e^{-\rho(q)t} & t \geq 0, \\ 0 & t < 0, \end{cases}$$

whose corresponding cumulative distribution function is

$$P(t) = \int_{-\infty}^{t} p(x)dx = \begin{cases} 1 - e^{-\rho(q)t} & t \geq 0, \\ 0 & t < 0. \end{cases}$$

When it leaves q, the next state is determined as in the underlying discrete-time probabilistic finite-state automaton.
    Continuous-time Markov chains generalize continuous-time probabilistic finite-state automata to any *countable* set Q of states. ∎

Example 6.43. A lamp with a lightbulb can be in one of three states: on, off, and broken. When it is off, it is turned on after 100 s on average; while turning on, the lightbulb breaks in 1 % of the cases. When the lamp is on, it is turned off after 60 s on

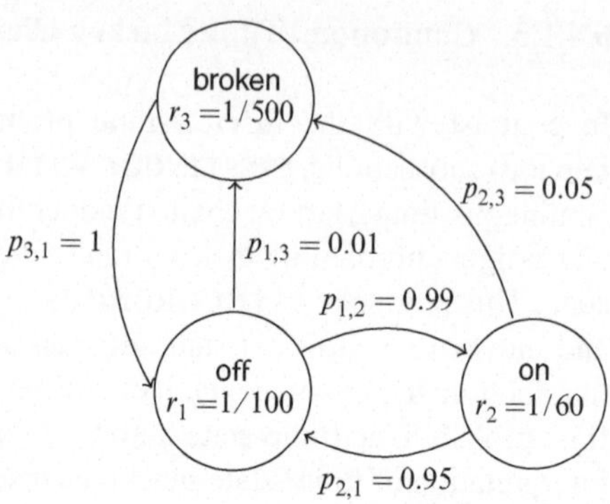

Fig. 6.12 A continuous-time
probabilistic finite-state
automaton modeling a lamp
that may break

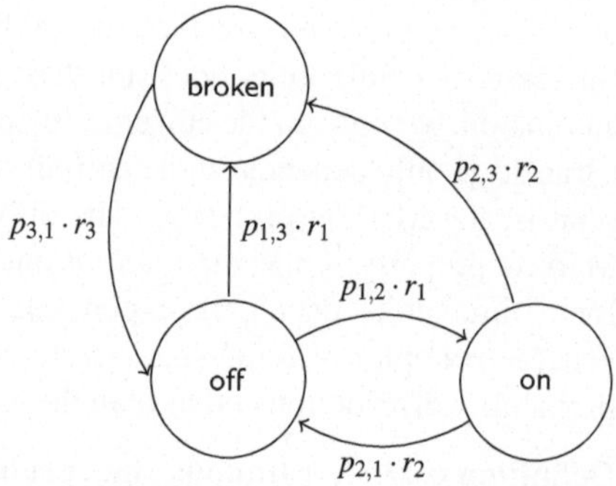

Fig. 6.13 An equivalent
presentation of the
continuous-time probabilistic
finite-state automaton of
Fig. 6.12

average; while turning off, the lightbulb breaks in 5 % of the cases. Finally, it takes
500 s on average before a broken lightbulb is replaced with a new one. Figure 6.12
models the behavior of this lamp with a continuous-time probabilistic finite-state
automaton, assuming seconds as time unit. Every state q is labeled with its rate
$\rho(q)$: since an exponential distribution with rate r has mean $1/r$, the rates are the
reciprocals of the average residence times.   ∎

Probabilistic automata underpin several computational models that include a
distinctive notion of continuous time. Chapter 8 will present, in particular, stochastic
Petri nets, whose semantics is based on continuous-time Markov chains and variants
thereof. Chapter 7 will present other variants and generalizations of probabilistic
finite-state automata.

Some of those applications naturally refer to an alternative definition of continuous-time Markov chains, which we first illustrate in Example 6.44, then present
formally in Definition 6.45, and finally claim equivalent to Definition 6.42 in
Exercise 6.46.

Example 6.44. Figure 6.13 shows a continuous-time probabilistic finite-state
automaton with the same states and transitions as that in Fig. 6.12, but with a positive
rate associated with every *transition* rather than state. The rates are parameters of
the exponential distributions associated with each transition, which determine the
time before each transition is taken.

Suppose, for example, that the automaton enters state off; it will stay there until it moves either to state on or to state broken. Transitions from off to on happen after time that follows an exponential distribution with rate $p_{1,2}r_1$; those from off to broken follow instead an exponential distribution with rate $p_{1,3}r_1$. Correspondingly, the overall residence time in off follows the distribution of the *minimum* of the times before a transition to on or to broken: whichever transition happens first preempts the other one. The minimum of two stochastic variables with exponential distribution is also exponentially distributed with rate given by the sum of the distributions' rates; hence, state off has rate $p_{1,2}r_1 + p_{1,3}r_1 = r_1$, as for the automaton in Fig. 6.12. Finally, a transition outgoing from off reaches on in $p_{1,2}r_1/(p_{1,2}r_1 + p_{1,3}r_1) = p_{1,2}$ of the cases, and broken in the other $p_{1,3}r_1/(p_{1,2}r_1 + p_{1,3}r_1) = p_{1,3} = 1 - p_{1,2}$ fraction of the cases. These numbers show that the models in Figs. 6.12 and 6.13 have identical behavior. ∎

Definition 6.45 (Continuous-time probabilistic finite-state automaton with transition rates/Continuous-time Markov chain with transition rates). A *continuous-time probabilistic finite-state automaton* (also, "continuous-time Markov chain") with transition rates includes transition function $\delta : Q \to \wp(Q)$ and rate function $\rho : Q \times Q \to \mathbb{R}_{>0}$ for transitions between different states. Whenever the automaton enters state $q \in Q$, it draws, for every state $q_k \in \delta(q)$ directly reachable from q, a time t_k exponentially distributed with parameter $\rho(q, q_k)$. It will then make a transition from q to q' after exactly t' time units, where $t' = \min_k\{t_k\}$ is the shortest drawn time and $q' = q_h$ is the corresponding state with $h = \arg\min_k\{t_k\}$. ∎

Exercise 6.46 (♦). Following Example 6.44, show that Definitions 6.42 and 6.45 are equivalent. Namely, show how to construct a continuous-time probabilistic finite-state automaton with transition rates with behavior identical to that of any given continuous-time probabilistic finite-state automaton (with state rates), and, conversely, how to construct a continuous-time probabilistic finite-state automaton (with state rates) with behavior identical to that of any given continuous-time probabilistic finite-state automaton with transition rates. ∎

Continuous-time probabilistic automata augmented *with input* are the continuous-time counterparts of Markov decision processes, where input determines the transition rates of states or transitions. The following exercise asks for a more precise definition.

Exercise 6.47. Define continuous-time Markov decision processes by combining Definitions 6.44 and 6.37. ∎

6.4.2 Probabilistic Turing Machines and Complexity Classes

A probabilistic Turing machine can randomly choose which transition to take from among those offered by the transition function. Therefore, the transition function of a probabilistic Turing machine has the same signature as that of a nondeterministic Turing machine (Sect. 6.2.8). An alternative, but equivalent, model of a probabilistic Turing machine consists of a deterministic machine augmented with an unbounded tape of random bits; the transition function can take the value of the current bit into account to determine the next configuration.

It should be apparent that the use of randomness does not increase expressive power: everything computable by a probabilistic Turing machine is computable by a nondeterministic Turing machine that simply includes every transition that has nonzero probability of taking place. A deterministic Turing machine can then simulate the probabilistic Turing machine, thus confirming the Church-Turing thesis for probabilistic models.

The computational complexity analysis is more interesting – and much more challenging. The output of a Turing machine is in fact a random variable: on a given input, the machine can return completely different output values at random, or it can even randomly fail to terminate. This makes it difficult to compare the behavior of probabilistic Turing machines with that of deterministic Turing machines – described in purely functional terms – as well as that of nondeterministic machines – where the analysis simply focuses on the "best" possible outcome from among all those possible. To avoid these problems, the computational complexity analysis of probabilistic Turing machines usually focuses on *bounded-error* computations.

Definition 6.48 (Bounded-error probabilistic Turing machine). A "bounded-error probabilistic Turing machine" is a probabilistic Turing machine M that computes a function $F(x)$ of the input. For all inputs x, M halts; upon termination, it outputs the correct value $F(x)$ with probability greater than or equal to $2/3$.   ∎

The intuition behind Definition 6.48 is that the probability of a bounded-error Turing machine returning inconsistent results by chance can be made as small as needed with repeated runs of the machine on the same input. More precisely, consider n consecutive runs of a bounded-error probabilistic Turing machine M on an input x. Since the random choices made in every run are independent of those of the previous runs, one can derive, by applying suitable results of probability theory, an upper bound on the probability that a majority of them are incorrect as

$$\exp\left(-2n \cdot \frac{1}{6^2}\right),$$

which decreases exponentially with the number of runs n. Therefore, it is sufficient to run the Turing machine for, say, 100 runs, and the behavior in the majority of runs is almost certain to be the expected one.

Exercise 6.49 (♦). Show that any probability strictly greater than $1/2$ can replace $2/3$ in Definition 6.48 without changing the asymptotic behavior of bounded-error probabilistic Turing machines.   ∎

The notion of bounded error makes for a sound definition of computational complexity in probabilistic models based on *average running time*. Take any bounded-error probabilistic Turing machine, and let $T_1(x), T_2(x), \ldots, T_j(x)$ be the j independent, identically distributed random variables counting the number of steps in each of j runs on the same input x. Their average

$$\frac{T_1(x) + T_2(x) + \cdots + T_j(x)}{j}$$

may either converge to a constant $\mu(T(x))$ for increasing j or diverge to $+\infty$. $\mu(T(x))$ equals the expected value $\mathbb{E}[T(x)]$ of the random variables' distribution. The average k-run time complexity of the machine is then the expected value of the total running time:

$$\mu(T(x), k) = \mathbb{E}[T_1(x) + T_2(x) + \cdots T_k(x)]$$
$$= \mathbb{E}[T_1(x)] + \mathbb{E}[T_2(x)] + \cdots \mathbb{E}[T_k(x)]$$
$$= k \cdot \mu(T(x)),$$

which is $\Theta(\mu(T(x)))$ for every constant k. Thus, the asymptotic average running time is independent of how many finite runs are needed to achieve the desired probability of error and only depends on the probability distribution induced by the probabilistic model. Finally, define $T(n)$ as the maximum $\mu(T(x))$ from among all inputs x of length n.

$T(n)$ is the time complexity function for bounded-error probabilistic Turing machines; a space complexity function can be defined similarly. Such complexity functions are on a par with those used for deterministic and nondeterministic Turing machines; hence we can compare the (time) complexity across computational models. To this end, we introduce the complexity class BPP, which represents all problems that are tractable for probabilistic computational models with bounded error, in the same way as the class P collects all tractable problems for deterministic models.

Definition 6.50 (Probabilistic complexity classes).

- A time complexity measure $T(n)$ defines the *complexity class* BPTIME($T(n)$) of all *problems* that can be solved by some *bounded-error probabilistic* Turing machine with time complexity in O($T(n)$).
- BPP ("bounded-error probabilistic polynomial") is the class of problems that can be solved in polynomial time (and unlimited space) by a bounded-error probabilistic Turing machine, that is,

$$\text{BPP} = \bigcup_{k \in \mathbb{N}} \text{BPTIME}(n^k). \qquad \blacksquare$$

Using the complexity classes we know, the question of whether deterministic Turing machines can efficiently simulate (bounded-error) probabilistic Turing machines can be expressed as: "does P equal BPP?" If P = BPP then randomization does not make intractable problems tractable, and the strong Church-Turing thesis applies to probabilistic Turing machines that would be efficiently simulated by deterministic Turing machines with at most a polynomial slowdown. If instead $P \subset BPP$, then there exist problems that can be solved in polynomial time only with the aid of randomization.

The P/BPP question is currently open, as is the other outstanding question on the relation between the classes P and NP. Unlike for the latter, for the former there is currently a large amount of evidence, both theoretical and experimental, suggesting that P *equals* BPP. In particular, efficient deterministic algorithms have been found for various important problems whose initial solutions required randomization. Even if the consensus is that "derandomizing" polynomial-time algorithms is always possible, it may be arduous to devise deterministic algorithms for certain problems first approached with probabilistic computational models. The most famous example is the problem *PRIME* of determining whether an integer is prime or composite (without necessarily computing a factorization). Polynomial-time probabilistic algorithms for *PRIME* have been known since the 1970s, but only about 30 years later was the first deterministic polynomial-time algorithm developed.

On the basis of the likely equivalence between P and BPP, nontechnical presentations usually do not detail the role of randomization and simply assume a probabilistic model whenever it makes for a practically convenient implementation by means of pseudo-random number generators.

6.5 Bibliographic Remarks

There is a rich choice of textbooks on the general topics of this chapter, such as Hopcroft et al. [30], Sipser [68], Lewis and Papadimitriou [43], Manna [46], Linz [44], Rosenberg [60], Jones [32], and Mandrioli and Ghezzi [45].

The chapter discussed the difference in scope between computational complexity theory, which focuses on complexity classes and their relations, and analysis of algorithms, which is concerned with devising efficient algorithms for specific problems. There are several comprehensive references in both areas: Papadimitriou [54], Arora and Barak [8], Bovet and Crescenzi [12], Goldreich [25], and Kozen [40] take the complexity theory viewpoint; Cormen et al. [16], Aho et al. [5], Kleinberg and Tardos [37], Mehlhorn and Sanders [48], Sedgewick and Wayne [63], Skiena [69], Papadimitriou and Steiglitz [55], and the encyclopedic Knuth [39] focus on algorithm design.

The theory of computation does not stop at Turing machines; books on computability theory – such as Rogers [59], Cutland [18], Soare [72], and Odifreddi [53] – explore the rich mathematics behind universal computation. At the other end of the spectrum, the theory of finite automata and formal languages studies computational models of lesser power than Turing machines, such as finite-state and pushdown automata. For example, deterministic pushdown automata have the same $\Theta(n)$ asymptotic time complexity as finite-state automata, whereas nondeterministic pushdown automata can produce computation trees of size $\Theta(2^{kn})$ for inputs of length n. Anderson [7], Lawson [41], Simon [66], Sakarovitch [62], Shallit [64], Ito [31], Khoussainov and Nerode [35], and Perrin and Pin [56] all present elements of automata theory, with different levels of mathematical sophistication.

Hartmanis and Stearns's paper [28] is often credited as the foundational paper of modern computational complexity, although other computer science pioneers introduced similar methods even earlier. Fortnow and Homer [22] give a concise history of the origins of computational complexity.

Alan Turing introduced the Turing machine in his revolutionary 1936 paper "On computable numbers, with an application to the *Entscheidungsproblem*" [78, 79], which gives a foundation to all modern computer science. Turing's original machine used a single tape for input, output, and scratch memory; proofs of equivalence between Turing's model and k-tape Turing machines can be found in basically every textbook mentioned at the beginning of these notes. Van Emde Boas [80] gives a comprehensive treatment of the equivalence of many models of computation.

The Church-Turing thesis is sometimes referred to using only Church's name, who first formulated it, but Robert Soare makes a convincing historical reconstruction [71, 73] showing that it was Turing who gave solid arguments for the thesis and showcased its importance; the name "Church-Turing thesis" seems therefore more historically accurate. Hofstadter [29] discusses the implications of the Church-Turing thesis in a somehow informal setting. It was recently suggested by authors such as Deutsch [19] that, since the Church-Turing thesis talks about computations that are physically implementable, it should be provable from the laws of physics; this line of thought highlighted interesting connections between physics and computational models [2, 11].

Hartmanis and Stearns's seminal paper [28] contains the first proof of the linear speed-up theorem. The asymptotic notation was first introduced in mathematics by Bachmann and popularized by Landau. Knuth established the notation in computer science in [38], where he also traces its historical origins; Graham et al. [26] provide a comprehensive mathematical treatment of the asymptotic notation.

The strong Church-Turing thesis gradually developed as "folk knowledge" in complexity theory; Slot and van Emde Boas give one of the earliest explicit statements of the thesis [70]. As with the standard Church-Turing thesis, the strong Church-Turing thesis should be interpretable as a consequence of the limits imposed by the laws of physics on (efficient) computational processes [1]. Under this view, Feynman first observed that deterministic or probabilistic Turing machines seem incapable of efficiently simulating certain quantum-mechanical phenomena [20]; and Shor presented an algorithm [34, 36, 49, 52, 65] that factors an n-bit integer in time $O(n^3)$ on a quantum computer, whereas no polynomial-time deterministic or even probabilistic algorithm for factoring is known. If large-scale computers based on quantum mechanics are implementable, the strong Church-Turing thesis will have to refer to *quantum Turing machines* [10] rather than ordinary Turing machines.

Strassen initiated the development of matrix multiplication algorithms asymptotically faster than the trivial $\Theta(n^3)$ algorithm. Strassen's algorithm [76] runs in time $O(n^{2.807})$ for $n \times n$ matrices; for many years, the asymptotically fastest algorithm was Coppersmith-Winograd's [15], which runs in time $O(n^{2.376})$. A recent breakthrough improved it to $O(n^{2.373})$ [75, 81]; related theoretical results suggest that there is still room for improvement [13, 58].

In a letter [21, 67] sent to von Neumann in 1956 and rediscovered in 1988, Gödel discussed the essence of the P/NP problem long before the birth of modern computational complexity. Cook [14] and Levin [42] independently introduced the notion of NP-completeness and produced the first NP-complete problems. Karp [33] showed how to prove that many other problems are NP-complete by polynomial-time reductions. There are several other types of reduction, with stricter bounds on the resources used, which are needed to study smaller classes than P and NP; any textbook on computational complexity discusses them at length. Garey and Johnson's survey [23] of NP-complete problems is still valuable but somehow outdated; Crescenzi and Kann maintain a more up-to-date list [17]. The paramount importance of the P/NP problem is testified by many facts, including its inclusion among the mathematical problems of the millennium by the Clay Mathematics Institute. Gasarch surveyed [24] the opinion of computer scientists and showed that a large majority of them believe that P and NP are not the same; Aaronson [3] adds an informal, sanguine, and persuasive argument for P$\neq$NP.

Von Neumann introduced the von Neumann architecture [83]; RAM and RASP machines [80] are modeled after it. There are many variants of abstract machines modeled after different computer architectures; for example, PRAM (Parallel RAM) is a family of machines to simulate parallel computation [80], and Knuth's MMIX [39] is a detailed model of RISC processors.

Markov introduced Markov chains and applied them to natural language processing [47]; a number of textbooks present Markov chains and their variants in detail [27, 61, 77]. Example 6.43 is adapted from Ajmone Marsan [6].

Motwani and Raghavan [51], Mitzenmacher and Upfal [50], Vazirani [82], and Ausiello et al. [9] discuss randomized (and approximate) algorithms in detail.

Solovay and Strassen gave the first randomized polynomial-time algorithm for primality testing [74]. Miller-Rabin [57] is another well-known randomized algorithm. Agrawal et al. [4] provide the first deterministic algorithm for primality testing.

References

1. Aaronson, S.: Limits on efficient computation in the physical world. Ph.D. thesis, University of California, Berkeley (2005)
2. Aaronson, S.: NP-complete problems and physical reality. SIGACT News **36**(1), 30–52 (2005). Complexity Theory Column
3. Aaronson, S.: Reasons to believe. http://www.scottaaronson.com/blog/?p=122 (2006)
4. Agrawal, M., Kayal, N., Saxena, N.: PRIMES in P. Ann. Math. **2**, 781–793 (2004)

5. Aho, A.V., Hopcroft, J.E., Ullman, J.D.: Data Structures and Algorithms. Addison-Wesley, Reading (1983)
6. Ajmone Marsan, M.: Stochastic petri nets: an elementary introduction. In: Rozenberg, G. (ed.) Advances in Petri Nets. Lecture Notes in Computer Science, No. 424, pp. 1–29. Springer, Berlin (1989)
7. Anderson, J.A.: Automata Theory with Modern Applications. Cambridge University Press, Cambridge (2006)
8. Arora, S., Barak, B.: Computational Complexity: A Modern Approach. Cambridge University Press, Cambridge (2009)
9. Ausiello, G., Crescenzi, P., Gambosi, G., Kann, V., Marchetti-Spaccamela, A., Protasi, M.: Complexity and Approximation. Combinatorial Optimization Problems and Their Approximability Properties. Springer, New York (1999)
10. Bernstein, E., Vazirani, U.: Quantum complexity theory. SIAM J. Comput. **26**, 1411–1437 (1997)
11. Blum, L., Cucker, F., Shub, M., Smale, S.: Complexity and Real Computation. Springer, New York (1997)
12. Bovet, D.P., Crescenzi, P.: Introduction to the Theory of Complexity. Prentice Hall, New York (1994)
13. Cohn, H., Kleinberg, R.D., Szegedy, B., Umans, C.: Group-theoretic algorithms for matrix multiplication. In: FOCS, pp. 379–388. IEEE Computer Society, Los Alamitos (2005)
14. Cook, S.A.: The complexity of theorem-proving procedures. In: STOC, pp. 151–158. ACM, New York (1971)
15. Coppersmith, D., Winograd, S.: Matrix multiplication via arithmetic progressions. J. Symbolic Comput. **9**(3), 251–280 (1990)
16. Cormen, T.H., Leiserson, C.E., Rivest, R.L., Stein, C.: Introduction to Algorithms, 3rd edn. MIT, Cambridge (2009)
17. Crescenzi, P., Kann, V.: A compendium of NP optimization problems. http://www.nada.kth.se/~viggo/problemlist/compendium.html (2005)
18. Cutland, N.: Computability: An Introduction to Recursive Function Theory. Cambridge University Press, Cambridge (1980)
19. Deutsch, D.: Quantum theory, the Church-Turing principle and the universal quantum computer. Proc. R. Soc. Lond. A **400**, 97–117 (1985)
20. Feynman, R.P.: Simulating physics with computers. Int. J. Theor. Phys. **21**(6/7), 467–488 (1982)
21. Fortnow, L.: Kurt Gödel (1906–1978). http://blog.computationalcomplexity.org/2006/04/kurt-gdel-1906-1978.html (2006)
22. Fortnow, L., Homer, S.: A short history of computational complexity. Bull. Eur. Assoc. Theor. Comput. Sci. **80**, 95–133 (2003)
23. Garey, M.R., Johnson, D.S.: Computers and Intractability: A Guide to the Theory of NP-Completeness. Freeman, San Francisco (1979)
24. Gasarch, W.I.: The P=?NP poll. SIGACT News **33**(2), 34–47 (2002)
25. Goldreich, O.: Computational Complexity: A Conceptual Perspective. Cambridge University Press, Cambridge (2008)
26. Graham, R.L., Knuth, D.E., Patashnik, O.: Concrete Mathematics: A foundation for computer science. Addison-Wesley, Reading (1994)
27. Häggström, O.: Finite Markov Chains and Algorithmic Applications. Cambridge University Press, Cambridge (2002)
28. Hartmanis, J., Stearns, R.E.: On the computational complexity of algorithms. Trans. Am. Math. Soc. **117**, 285–306 (1965)
29. Hofstadter, D.R.: Gödel, Escher, Bach: An Eternal Golden Braid. Basic Books, New York (1999)
30. Hopcroft, J.E., Motwani, R., Ullman, J.D.: Introduction to Automata Theory, Languages, and Computation, 3rd edn. Addison-Wesley, Boston (2006)

31. Ito, M.: Algebraic Theory of Automata and Languages. World Scientific Publishing Company, Singapore (2004)
32. Jones, N.D.: Computability and Complexity: From a Programming Perspective. MIT, Cambridge (1997)
33. Karp, R.M.: Reducibility among combinatorial problems. In: Thatcher, J.W., Miller, R.E. (eds.) Complexity of Computer Computations, pp. 85–103. Plenum Press, New York (1972)
34. Kaye, P., Laflamme, R., Mosca, M.: An Introduction to Quantum Computing. Oxford University Press, Oxford (2007)
35. Khoussainov, B., Nerode, A.: Automata Theory and Its Applications. Birkhäuser, Boston (2001)
36. Kitaev, A.Y., Shen, A.H., Vyalyi, M.N.: Classical and Quantum Computation. American Mathematical Society, Providence (2002)
37. Kleinberg, J., Tardos, É.: Algorithm Design. Addison-Wesley, Boston (2005)
38. Knuth, D.E.: Big Omicron and big Omega and big Theta. ACM SIGACT News **8**, 18–24 (1976)
39. Knuth, D.E.: The Art of Computer Programming (Volumes 1–4A). Addison-Wesley, Boston/London (2011)
40. Kozen, D.C.: Theory of Computation. Springer, London (2006)
41. Lawson, M.V.: Finite Automata. Chapman and Hall/CRC, Boca Raton (2003)
42. Levin, L.A.: Universal sorting problems. Probl. Inf. Transm. **9**, 265–266 (1973)
43. Lewis, H.R., Papadimitriou, C.H.: Elements of the Theory of Computation, 2nd edn. Prentice-Hall, Upper Saddle River (1997)
44. Linz, P.: An Introduction to Formal Languages and Automata, 4th edn. Jones and Bartlett, Sudbury (2006)
45. Mandrioli, D., Ghezzi, C.: Theoretical Foundations of Computer Sciences. Wiley, New York (1987)
46. Manna, Z.: Mathematical Theory of Computation. Dover, Mineola (2003)
47. Markov, A.A.: Extension of the limit theorems of probability theory to a sum of variables connected in a chain. In: Howard, R. (ed.) Dynamic Probabilistic Systems. Markov Chains (Appendix B), vol. 1. Wiley, New York (1971)
48. Mehlhorn, K., Sanders, P.: Algorithms and Data Structures: The Basic Toolbox. Springer, Berlin (2008)
49. Mermin, N.D.: Quantum Computer Science: An Introduction. Cambridge University Press, Cambridge (2007)
50. Mitzenmacher, M., Upfal, E.: Probability and Computing: Randomized Algorithms and Probabilistic Analysis. Cambridge University Press, Cambridge (2005)
51. Motwani, R., Raghavan, P.: Randomized Algorithms. Cambridge University Press, Cambridge/New York (1995)
52. Nielsen, M.A., Chuang, I.L.: Quantum Computation and Quantum Information. Cambridge University Press, Cambridge (2000)
53. Odifreddi, P.: Classical Recursion Theory. North Holland, Amsterdam (1999)
54. Papadimitriou, C.H.: Computational Complexity. Addison-Wesley, Reading (1993)
55. Papadimitriou, C.H., Steiglitz, K.: Combinatorial Optimization: Algorithms and Complexity. Dover, Mineola (1998)
56. Perrin, D., Pin, J.É.: Infinite Words. Pure and Applied Mathematics, vol. 141. Elsevier, Amsterdam (2004)
57. Rabin, M.O.: Probabilistic algorithm for testing primality. J. Number Theory **12**, 128–138 (1980)
58. Robinson, S.: Toward and optimal algorithm for matrix multiplication. SIAM News **38**(9), 354–356 (2005)
59. Rogers Jr., H.: Theory of Recursive Functions and Effective Computability. MIT, Cambridge (1987)
60. Rosenberg, A.L.: The Pillars of Computation Theory. Springer, Berlin (2009)
61. Ross, S.M.: Stochastic Processes. Wiley, New York (1995)
62. Sakarovitch, J.: Elements of Automata Theory. Cambridge University Press, Cambridge (2009)

63. Sedgewick, R., Wayne, K.: Algorithms, 4th edn. Addison-Wesley, Upper Saddle River (2011)
64. Shallit, J.: A Second Course in Formal Languages and Automata Theory. Cambridge University Press, Cambridge (2008)
65. Shor, P.W.: Polynomial-time algorithms for prime factorization and discrete logarithms on a quantum computer. SIAM J. Comput. **26**(5), 1484–1509 (1997)
66. Simon, M.: Automata Theory. World Scientific Publishing Company, Singapore (1999)
67. Sipser, M.: The history and status of the P versus NP question. In: STOC, pp. 603–618. ACM, New York (1992)
68. Sipser, M.: Introduction to the Theory of Computation, 2nd edn. Course Technology, Boston (2005)
69. Skiena, S.S.: The Algorithm Design Manual, 2nd edn. Springer, New York (2008)
70. Slot, C.F., van Emde Boas, P.: On tape versus core; an application of space efficient perfect hash functions to the invariance of space. In: STOC, pp. 391–400. ACM, New York (1984)
71. Soare, R.I.: Computability and recursion. Bull. Symb. Logic **2**, 284–321 (1996)
72. Soare, R.I.: Recursively Enumerable Sets and Degrees, 2nd edn. Springer, Berlin (1999)
73. Soare, R.I.: Turing oracle machines, online computing, and three displacements in computability theory. Ann. Pure Appl. Logic **160**, 368–399 (2009)
74. Solovay, R., Strassen, V.: Erratum: A fast Monte-Carlo test for primality. SIAM J. Comput. **7**(1), 118 (1978)
75. Stothers, A.J.: On the complexity of matrix multiplication. Ph.D. thesis, University of Edinburgh (2010)
76. Strassen, V.: Gaussian elimination is not optimal. Numer. Math. **14**(3), 354–356 (1969)
77. Stroock, D.W.: An Introduction to Markov Processes. Springer, Berlin (2005)
78. Turing, A.M.: On computable numbers, with an application to the Entscheidungsproblem. Proc. Lond. Math. Soc. **42**, 230–265 (1936). Corrections in [79]
79. Turing, A.M.: A correction. Proc. Lond. Math. Soc. **43**, 544–546 (1937)
80. van Emde Boas, P.: Machine models and simulations. In: van Leeuwen, J. (ed.) The Handbook of Theoretical Computer Science. Algorithms and Complexity, vol. I, pp. 1–61. MIT, Amsterdam (1990)
81. Vassilevska Williams, V.: Multiplying matrices faster than Coppersmith-Winograd. In: Karloff, H.J., Pitassi, T. (eds.) Proceedings of the 44th Symposium on Theory of Computing Conference, STOC 2012, New York, 19–22, May 2012, pp. 887–898. ACM, New York (2012)
82. Vazirani, V.V.: Approximation Algorithms. Springer, Berlin (2001)
83. von Neumann, J.: First draft of a report on the EDVAC. http://qss.stanford.edu/~godfrey/vonNeumann/vnedvac.pdf (1945)

Part II
Temporal Models in Modern Theory and Practice

The growing importance and pervasiveness of computer systems has required the introduction of new, richer, and more expressive temporal models, fostering their evolution from the basic "historical" models of Part I of this book. This evolution has inevitably modified the boundaries with the traditional ways of modeling time, often making them fuzzy. In particular, this happened with heterogeneous systems, which require the combination of different abstractions within the same model.

This second part of the book shows how the basic models have been refined and extended to meet more sophisticated and advanced specification needs, and in particular those typical of hybrid, critical, and real-time systems. As discussed in Chap. 3, these categories of systems are different and somehow complementary, but with significant areas of overlap.

The extension and combination of basic models, required by heterogeneous systems, carry on along the route followed in Part I of the book, which connected the "micro" and "macro" views of hardware, and then obtained the abstract "software" view. The path continues in Part II by extending from a *computer-centric* to a *system-centric* view. As the terms themselves suggest, the computer-centric view focuses on the computing devices – hardware and software – which may interact with the environment through input/output operations; the system-centric view, in contrast, encompasses a whole collection of heterogeneous components, of which computing devices are only a part. Most often, the system-centric view targets components whose dynamics range over widely different time scales and granularities (in particular, continuous and discrete components often coexist). In this sense, the path from the computer- to the system-centric view inverts the trend of Chap. 5, and reintroduces elements that had been abstracted away when moving to purely software models.

The *computer-centric* view considers systems where time is inherently discrete, and which can be described with state models. Concurrency models usually are strictly synchronous, with the system synchrony given by a global clock ticking. Nondeterminism often models concurrent computations at an abstract level. The computer-centric view emphasizes the extensive analyzability of the properties of interest; to this end, it is possible and preferred to abstract away from the detailed timed behavior in exchange for decidable formal models, amenable to automated verification.

An example of activity based on the computer-centric view is the design of field buses for process control: the focus is on discrete signals coming from several sensors and on their proper synchronization; the environment that generates the signals is hidden behind the interface provided by the sensors.

Conversely, the *system-centric* view targets the modeling, design, and analysis of whole systems, including processes to be controlled, sensors and actuators, network connections among the various elements, and the individual computing devices. Time may be continuous or discrete, depending on the application domain. Concurrency models often are asynchronous, whereas the evolution of elementary components is usually deterministic. For instance, a controlled chemical process would be described in terms of continuous time and interacting asynchronous deterministic processes; on the other hand a logistic process – such as a complex storage system – would be better described in terms of discrete time. The system-centric view puts particular emphasis on the input/output signals, the organization of components in modules, and the resulting information flow – similarly to some aspects of dynamical systems. The traditional division between hardware and software is blurred, and the systemic aspects emerge.

In practice, no model is usually taken to be totally computer-centric or system-centric; aspects of both views are often combined and tailored for the specific application needs. More generally, the computer- and system-centric views are unified within systems made of interconnected components that *react* with suitable actions (or computations) to stimuli coming from their *environment*. This notion of "*reactive* system" accommodates heterogeneous temporal domains and interaction mechanisms. For example, reactive systems interacting with the physical world are called *embedded* systems; their environment models may feature a continuous notion of time and differential equations such as those introduced in Chap. 4, whereas their computational components typically leverage models such as those of Chap. 6. The overall model of time is *hybrid*, part discrete (the computer) and part continuous (the physical environment). Another fundamental feature of reactive systems is that their computations are not meant to terminate and produce a final, correct result; rather, they are designed to go on *ad infinitum*, and their correctness is a property of the ongoing interactions of their components and the environment.

This second part of the book presents some broad classes of formal modeling languages for reactive systems in order to discuss what kinds of temporal models they introduce and what system features they are suitable for describing. Chapters 7 and 8 analyze operational formalisms, in a progression from those that are naturally characterized as "synchronous" to those that are "asynchronous". Chapter 9 discusses descriptive formalisms based on logic, in particular those representing recurring solutions to outstanding modeling issues and introducing influential modeling styles. Chapter 10 discusses algebraic formalisms and, in particular, process algebras and their extensions to accommodate timed behavior. Finally, Chap. 11 discusses "dual-language approaches", in which two languages of different natures (operational and descriptive) are integrated for modeling and analysis purposes.

Chapter 7
Synchronous Abstract Machines

This chapter analyzes state-based formalisms that extend the models of Chap. 6 with a notion of time. The discussion progresses from the simplest formalisms to richer and richer extensions that can express sophisticated timing properties.

Section 7.1 introduces the notion of transition systems, used throughout the chapter to provide the semantics of the formalisms discussed. Section 7.2 shows how the finite-state automata introduced in Chap. 6 can be used to represent temporal features of systems. Section 7.3 discusses two extensions of finite-state automata, "Timed Transition Models" and "Statecharts", which introduce specific mechanisms for modeling time. Section 7.4 presents formalisms that further extend finite-state automata to deal with a continuous notion of time. Section 7.5 shows how probabilities can be introduced in finite-state automata. Section 7.6 discusses some design and verification techniques and tools based on the formalisms presented in the previous sections.

7.1 Transition Systems

The notion of "transition system" provides a useful unifying framework for a wide range of operational formalisms, especially the state-based ones, such as those described in this chapter and in Chaps. 4–6.

A transition system represents behaviors as sequences of transitions between states. Every sequence starts from an initial state, and with each new transition the state of the system evolves. Hence, a sequence of transitions also corresponds to a sequence of states; time evolves as the sequence progresses, but different transition systems may represent the evolution of time in different ways, as will be discussed later in this chapter.

Transition systems capture the semantics of a large class of models; to demonstrate this, we show how some models introduced in Chaps. 4–6 can be interpreted as transition systems.

C.A. Furia et al., *Modeling Time in Computing*, Monographs in Theoretical Computer Science. An EATCS Series, DOI 10.1007/978-3-642-32332-4_7,
© Springer-Verlag Berlin Heidelberg 2012

Fig. 7.1 Transition system
TS_{sr} corresponding to the
finite-state automaton of
Fig. 5.13

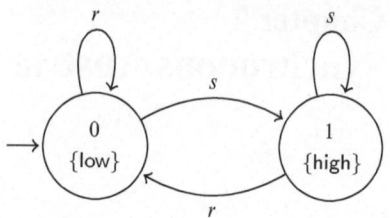

Definition 7.1 (Transition system). A transition system is a tuple $\langle S, S_0, I, \rightarrowtail,$
$AP, L\rangle$ where S is a set of states, $S_0 \subseteq S$ is a nonempty set of initial states, I
is a set of input symbols, AP is a set of atomic propositions used as state labels,
$L : S \rightarrow \wp(AP)$ is a labeling function that associates each state with a (possibly
empty) set of labels from the set AP, and $\rightarrowtail \subseteq S \times I \times S$ is a transition relation
between states and inputs of the system. In the most general definition, the sets of
states S and of input symbols I may be finite or infinite; the transition systems
themselves are correspondingly called "finite" or "infinite".   ∎

Example 7.2 (A finite-state automaton as transition system). Consider the finite-
state automaton of Fig. 5.13. If state "0" is its initial state, the correspond-
ing transition system TS_{sr} has $S = \{0, 1\}$, $S_0 = \{0\}$, $I = \{r, s\}$, and $\rightarrowtail \ =$
$\{\langle 0, s, 1\rangle, \langle 0, r, 0\rangle, \langle 1, s, 1\rangle, \langle 1, r, 0\rangle\}$. Labels represent atomic properties of interest
that are true in the states; for example, we define $AP = \{\mathsf{high}, \mathsf{low}\}$ and $L = \{0 \rightarrow$
$\{\mathsf{low}\}, 1 \rightarrow \{\mathsf{high}\}\}$.

Figure 7.1 shows a graphical representation of transition system TS_{sr} in which
the initial state is marked through an incoming arrow without a source state, and the
value of the labeling function for each state is written between curly braces.   ∎

Example 7.3 (A RAM program as transition system). Consider the RAM program
for primality testing introduced in Fig. 6.6 and repeated in Fig. 7.2 for reference.
The machine's state is given by the contents of its registers (the program counter
PC and the accumulator ACC), of the memory cells it uses ($M[1]$ and $M[2]$), and
of the output (with values 0, 1, or U, for "undefined" when no output is set). The
input symbols are the natural numbers $\mathbb{N}$, which represent the possible values read
at line 0, plus the symbol ε, denoting any instruction that does not require input (all
instructions except that on line 0).

In the initial state S_0, $PC = 0$, $ACC = 0$, $M[1] = M[2] = 0$, and the output
is U: the program is ready to execute instruction 0, and nothing has yet been stored
in memory or written on output. All states of the form

$$S_{1,n} = \langle PC = 1, ACC = 0, M[1] = n, M[2] = 0, U\rangle$$

are reachable with one transition with input symbol n from S_0, that is, are such that
$\langle S_0, n, S_{1,n}\rangle \in \rightarrowtail$. Then, for each $S_{1,n}$, $\langle S_{1,n}, \varepsilon, S_{2,n}\rangle \in \rightarrowtail$, with

$$S_{2,n} = \langle PC = 2, ACC = 1, M[1] = n, M[2] = 0, U\rangle,$$

Fig. 7.2 A RAM program
for primality testing

Label	Instruction	Comment
	READ 1	Store n into $M[1]$.
	LOAD= 1	If $n = 1$ then it is trivially prime
	SUB 1	and execution ends immediately
	JZ *yes*	with a positive answer.
	LOAD= 2	Initialize $M[2]$ to 2
	STORE 2	to start a loop of tests.
loop:	LOAD 1	When $M[2] = n$ exit the loop:
	SUB 2	all possible divisors have been tested
	JZ *yes*	hence the number is prime.
	LOAD 1	Assuming integer arithmetic, if
	DIV 2	$M[1]$ equals $(M[1] / M[2]) \times M[2]$ then
	MULT 2	$M[2]$ contains a divisor of n,
	SUB 1	hence n is not prime:
	JZ *no*	exit loop and report negative answer.
	LOAD 2	Increment $M[2]$ by 1
	ADD=1	to test the next divisor.
	STORE 2	
	JUMP *loop*	Repeat loop.
yes:	WRITE= 1	Output positive answer.
	HALT	
no:	WRITE= 0	Output negative answer.
	HALT	

instruction 1 stores 1 in the accumulator, and the PC points to instruction 2; and so on. All states such that $PC = 19$ correspond to the fact that the program reaches instruction 19; hence they have label "is_prime"; those such that $PC = 21$ have label "not_prime". ∎

Example 7.4 (A dynamical system as transition system). Consider the oscillator introduced in Example 4.3, described by the following equations:

$$\frac{d}{dt}i_L(t) = \frac{1}{L}V(t),$$

$$\frac{d}{dt}V(t) = \frac{1}{c}\left(i(t) - i_L(t) - \frac{V(t)}{R}\right).$$

The temporal evolution of the oscillator is captured by a transition system in which the state space $S = \mathbb{R} \times \mathbb{R}$ corresponds to the possible values of i_L and V, the initial state is $\langle 1, 0 \rangle$ (the values of i_L and V at time 0), the set of input symbols is $\mathbb{R}_{>0}$, corresponding to all possible time distances between states, and

$$\langle \langle i_{L,1}, V_1 \rangle, d, \langle i_{L,2}, V_2 \rangle \rangle \in \longmapsto$$

if and only if there is a t_1 such that

$$i_{L,1} = \cos\left(\frac{t_1}{\sqrt{LC}}\right) \cdot V_1 = -\sqrt{\frac{L}{C}} \sin\left(\frac{t_1}{\sqrt{LC}}\right),$$

and

$$i_{L,2} \quad = \quad \cos\left(\frac{t_1 + d}{\sqrt{LC}}\right) \cdot V_2 = -\sqrt{\frac{L}{C}} \sin\left(\frac{t_1 + d}{\sqrt{LC}}\right).$$

Then, state $\langle 1, 0 \rangle$ gets label "i_high", and state $\langle 0, 1 \rangle$ gets label "i_low". ∎

Exercise 7.5.

1. Define a suitable transition system for the Turing machine of Example 6.2.
2. Generalize the construction of item 1 to generic Turing machines as in Definition 6.1.
3. Define a suitable transition system for the cellular automaton "rule 110" described in Example 4.5. ∎

The transition systems of Examples 7.2–7.4 are *deterministic*: the initial state is unique and, for each state s, given an input symbol i, there is at most one state s' that is reachable from s with a transition labeled i, that is, such that $\langle s, i, s' \rangle \in \rightarrowtail$; hence, the choice of "next state" to be reached from s with input i is unique.

The transition systems of Examples 7.2–7.4 differ in the cardinality of their state spaces S and of their sets of input symbols I. In Example 7.2, both S and I are *finite* of size two; hence we say that the transition systems are themselves *finite*. On the other hand, in Example 7.3, both S and I are *infinite* but *countable*. Finally, in Example 7.4, S and I are uncountable (they have the cardinality of the real numbers). This shows how the notion of transition system can be adapted to a variety of cases, and in fact we will exploit it time and again in the rest of the book.

A transition system defines sequences of allowed transitions according to the following definition.

Definition 7.6 (Run of a transition system). Given a (possibly infinite) sequence $\sigma = i_1 i_2 i_3 \ldots$ of input symbols, a *run* r_σ for a transition system $\langle S, S_0, I, \rightarrowtail, AP, L \rangle$ is a sequence

$$s_0 \, i_1 \, s_1 \, i_2 \, s_2 \, i_3 \ldots ,$$

also written

$$s_0 \overset{i_1}{\rightarrowtail} s_1 \overset{i_2}{\rightarrowtail} s_2 \overset{i_3}{\rightarrowtail} \ldots ,$$

where $s_0 \in S_0$, and for all $k \geq 0$, $\langle s_k, i_{k+1}, s_{k+1} \rangle \in \rightarrowtail$. ∎

A state s' of a transition system TS is *reachable* if there exists a sequence of input symbols $\sigma = i_1 i_2 \ldots i_k$ and a corresponding finite run $r_\sigma = s_0 i_1 s_1 i_2 s_2 \ldots i_k s'$ of TS whose last element is s'.

Every sequence σ of input symbols that has a run r_σ in TS is a *trace* of TS. The sequence of labels $\pi = L(s_0) L(s_1) L(s_2) \ldots$ associated with the states traversed through run r_σ is also called a *trace* of the system. In the rest of the book, it will be clear from the context whether we are referring to the first or to the second type of trace.

Example 7.7 (Transition system runs). Let us consider the transition system of Example 7.2; given the sequence $\sigma = s\,r\,s\,r\,r\,s$, the corresponding run is

Fig. 7.3 Tree of runs with
common prefix $r s$

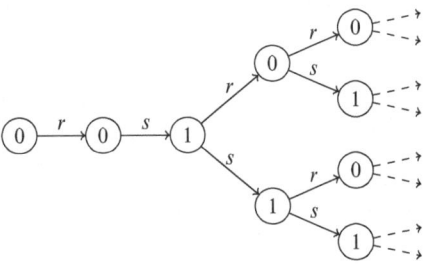

$r_\sigma = 0\,s\,1\,r\,0\,s\,1\,r\,0\,r\,0\,s\,1$. For the system of Example 7.3, a run corresponds to the computation of the RAM machine for the input provided at line 1. An example of fragment of computation is

$$\langle PC = 0, ACC = 0, M[1] = 0, M[2] = 0, U \rangle \, 5 \, \langle 1, 0, 5, 0, U \rangle \, \varepsilon$$

$$\langle 2, 1, 5, 0, U \rangle \, \varepsilon \, \langle 3, -4, 5, 0, U \rangle \, \varepsilon \, \langle 4, -4, 5, 0, U \rangle .$$

Finally, for the system of Example 7.4, assuming $L = C = 1$, an example of the first steps of a run is

$$\langle 1, 0 \rangle \, \frac{\pi}{6} \, \left\langle \frac{\sqrt{3}}{2}, \frac{1}{2} \right\rangle \, \frac{\pi}{3} \, \langle 0, 1 \rangle \, \frac{\pi}{4} \, \left\langle -\frac{\sqrt{2}}{2}, \frac{1}{2} \right\rangle ,$$

where the input symbols correspond to real-valued temporal distances between the states of the run. The concept of run for a transition system underlying a dynamical system is comparable to the notion of simulation of the continuous-time model (a sequence of states at variable temporal distances from each other); however, the run is an abstract concept, in which delays are not approximations of finite precisions, but actual real numbers (including irrationals).   ∎

Sets of runs that share a common prefix can be grouped in *trees*. Figure 7.3 shows part of a tree of the runs with prefix $r\,s$ for transition system TS_{sr} of Fig. 7.1.

As mentioned above, since in a transition system both states and transitions are labeled (the states through function L, the transitions with the input symbols associated with them), every run has two associated traces: the trace $\sigma = i_1\,i_2\,i_3 \ldots$ of input symbols, and the corresponding trace $\pi = L(s_0)\,L(s_1)\,L(s_2) \ldots$ of state labels. When one wants to state properties that involve both input symbols and state labels, the two sequences must be paired. This can be done either by associating input symbols to states, or, conversely, by labeling each transition outgoing from a state with the label of that state.

Let us consider Example 7.2 again. A property that holds for the transition system TS_{sr} is that after each s input, the signal is *high* (represented by proposition high being true). Let us transform TS_{sr} of Fig. 7.1 into a new transition system TS'_{sr} with only state labels in the following way (the general construction can be inferred

Fig. 7.4 Transition system TS'_{sr}

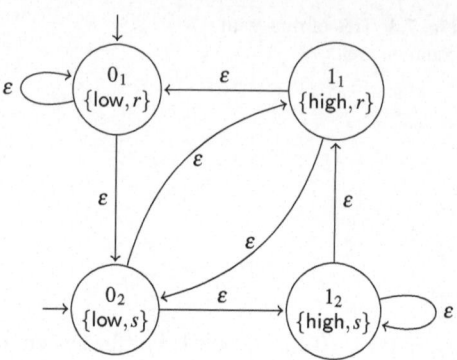

from this example). Each state s of TS_{sr} is replicated as many times as there are transitions outgoing from it – that is, twice for the states 0 and 1 in TS_{sr}. Let us call these new states s_1 and s_2 in TS'_{sr}. Each state s_k is labeled with both $L(s)$ and the input symbol i_k of the kth outgoing transition. Then, for each s' such that $\langle s, i_k, s' \rangle \in \rightarrowtail$ in TS_{sr}, if s'_1 and s'_2 are the replicas of state s' (from which two transitions also originate) in TS'_{sr}, we add the transitions $\langle s_k, \varepsilon, s'_1 \rangle$ and $\langle s_k, \varepsilon, s'_2 \rangle$ to the transition relation $\rightarrowtail'$ of TS'_{sr} (the symbol ε indicates absence of input).

Figure 7.4 depicts transition system TS'_{sr}; each replica of state 0, which is an initial state in TS_{sr}, is an initial state in TS'_{sr}. As mentioned in Sect. 6.2.8, multiple initial states are a form of nondeterminism (the state from which the computation starts is not uniquely defined), which seems to clash with the fact that the original transition system TS_{sr} of Fig. 7.1 is deterministic. In addition, all transitions in automaton TS'_{sr} have the empty label ε, which also seems to imply that when transforming TS_{sr} into TS'_{sr} one loses determinism (for example, state 0_1 has two outgoing transitions, both with the empty label). This apparent discrepancy is resolved by considering determinism as a property not of the transitions, but of the states (where all labels reside in the case of TS'_{sr}). Then, since each state of transition system TS'_{sr} has outgoing transitions that lead to states with different labels (for example, state 0_1 leads to either labeling $\{low, r\}$ or labeling $\{low, s\}$, which are distinct), TS'_{sr} is still deterministic. A possible run for TS'_{sr} is $0_1 \varepsilon 0_1 \varepsilon 0_2 \varepsilon 1_2 \varepsilon 1_1$, whose corresponding labeling π is $\{low, r\} \{low, r\} \{low, s\} \{high, s\} \{high, r\}$. This computation indeed satisfies the property that every s input is followed by the signal being high.

Dually, transition system TS_{sr} can be transformed into an equivalent one TS''_{sr} in which all labels are on the transitions. In this case, we augment the set I of input symbols with the set $\wp(AP)$ (that is, $I'' = I \times \wp(AP)$), then define that if $\langle s, i, s' \rangle \in \rightarrowtail$, then $\langle s, \langle i, L(s) \rangle, s' \rangle \in \rightarrowtail''$.

Figure 7.5 shows transition system TS''_{sr} corresponding to TS_{sr}. A possible run for TS''_{sr} is

$$0 \langle r, \{low\} \rangle \ 0 \ \langle r, \{low\} \rangle \ 0 \ \langle s, \{low\} \rangle \ 1 \ \langle s, \{high\} \rangle \ 1 \ \langle r, \{high\} \rangle,$$

Fig. 7.5 Transition system TS_{sr}''

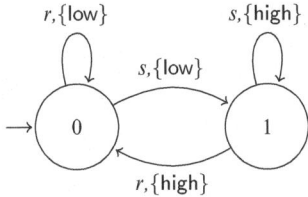

which has

$$\langle r, \{\text{low}\}\rangle \; \langle r, \{\text{low}\}\rangle \; \langle s, \{\text{low}\}\rangle \; \langle s, \{\text{high}\}\rangle \; \langle r, \{\text{high}\}\rangle$$

as a sequence of inputs σ. Again, σ shows that an input s is followed by a high signal.

7.1.1 Composition of Transition Systems

Transition systems can describe complex models by composing different modules. This section presents a few widely used composition mechanisms for transition systems; the literature mentioned in the bibliographic remarks describes other types of composition.

Given two transition systems

$$
\begin{aligned}
TS_1 &= \langle S_1, S_{0,1}, I_1, \rightarrowtail_1, AP_1, L_1 \rangle, \\
TS_2 &= \langle S_2, S_{0,2}, I_2, \rightarrowtail_2, AP_2, L_2 \rangle,
\end{aligned}
$$

let us denote as $TS_1 \bullet TS_2$ a generic composition operation.

The set of states of the composed system $TS_1 \bullet TS_2$ is the Cartesian product $S_1 \times S_2$ of the components' states for all notions of composition discussed in this chapter. Composition techniques differ in the ways in which the transition relation $\rightarrowtail_{1 \bullet 2}$ of the composed system is built from those of TS_1 and TS_2.

The *synchronous* composition rule requires that the composed system make a transition only if TS_1 and TS_2 agree on the input symbol. The composition is synchronous since a transition is taken in the composed system only if both TS_1 and TS_2 take a transition in lockstep. As a consequence, the runs of the composed system correspond to the input sequences that have a run in both TS_1 and TS_2, that is, that are in the intersection of the input sequences that can be processed by the two systems. For this reason, the synchronous composition of TS_1 and TS_2 is indicated as $TS_1 \cap TS_2 = TS_{1\cap2}$. Formally,

$$TS_{1\cap2} = \langle S_{1\cap2}, S_{0,1\cap2}, I_{1\cap2}, \rightarrowtail_{1\cap2}, AP_{1\cap2}, L_{1\cap2} \rangle,$$

Fig. 7.6 Transition system
TS_{s2r}

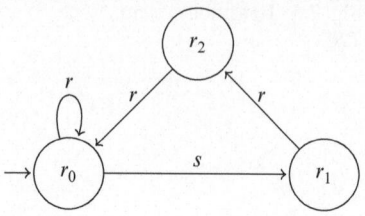

Fig. 7.7 Synchronous
composition $TS_{sr} \cap TS_{s2r}$ of
the systems of Figs. 7.1
and 7.6

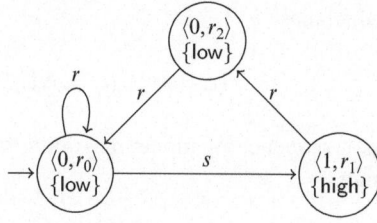

where

$$S_{1\cap2} = S_1 \times S_2,$$

$$S_{0,1\cap2} = S_{0,1} \times S_{0,2},$$

$$I_{1\cap2} = I_1 \cap I_2,$$

$$AP_{1\cap2} = AP_1 \cup AP_2,$$

$$L_{1\cap2}(\langle s_1, s_2 \rangle) = L_1(s_1) \cup L_2(s_2),$$

and

$$\langle \langle s_1, s_2 \rangle, i, \langle s_1', s_2' \rangle \rangle \in \longrightarrow_{1\cap2}$$

if and only if $\langle s_1, i, s_1' \rangle \in \longrightarrow_1$ and $\langle s_2, i, s_2' \rangle \in \longrightarrow_2$.

Consider, for example, the transition system TS_{s2r} of Fig. 7.6. It processes only input strings where between two successive s symbols there are at least two r symbols. Figure 7.7 shows the synchronous composition of TS_{s2r} with TS_{sr} of Fig. 7.1.

In this simple example, since any sequence of r and s can be processed by the transition system of Fig. 7.1 (i.e., it has an execution in the transition system), the effect of synchronizing the latter with TS_{s2r} is to keep only those executions that are also compatible with TS_{s2r}; hence the transition systems of Figs. 7.6 and 7.7 are structurally similar.

This is, of course, not true in general. Consider, in fact, the transition systems M_2 and M_3 in Fig. 7.8a and b, whose sets of input symbols are $\{0, 1\}$. M_2 and M_3 "count" the number of symbols 1 in the input: M_2 counts "modulo 2", and hence it is ready to input symbols 0 in its initial state A only after sequences with an even number of 1's; M_3 counts "modulo 3", and hence it is ready to input 0's only

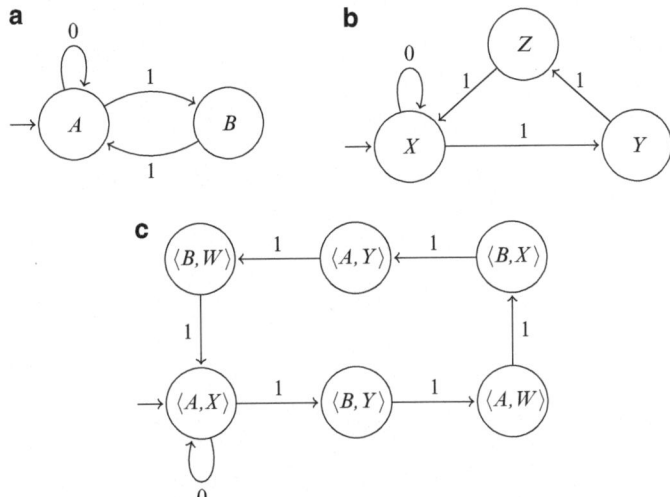

Fig. 7.8 Transition systems M_2 (**a**) counting "modulo 2", M_3 (**b**) counting modulo 3, and their synchronous composition M_6 (**c**) counting "modulo 6"

after sequences with a number of 1's that is a multiple of 3. Figure 7.8c shows the synchronous composition M_6 of the two modular counters; since it processes exactly the sequences processed by both composed transition systems, M_6 can read 0's precisely after input sequences with a number of 1's that is a multiple of both 2 and 3, hence a multiple of 6.

Another composition rule requires that TS_1 and TS_2 synchronize only on *some* transitions. This composition is called "parallel composition" and it is indicated by $TS_1 \parallel TS_2 = TS_{1\parallel2}$.[1] In parallel composition, it is customary to prescribe that TS_1 and TS_2 synchronize on the symbols in the intersection of their sets I_1 and I_2 of input symbols. For symbols that belong to only one of the two input sets, TS_1 and TS_2 move independently of each other. Formally,

$$TS_{1\parallel2} = \langle S_{1\parallel2}, S_{0,1\parallel2}, I_{1\parallel2}, \longrightarrow_{1\parallel2}, AP_{1\parallel2}, L_{1\parallel2}\rangle,$$

where

$$S_{1\parallel2} = S_1 \times S_2,$$
$$S_{0,1\parallel2} = S_{0,1} \times S_{0,2},$$

---

[1] Sometimes, "$\cap$" is referred to as "completely synchronous" composition, and "$\parallel$" as "synchronous" composition, to highlight the fact that in the first case the systems evolve in lockstep, while in the second they may synchronize but only sporadically.

Fig. 7.9 Transition system
TS_{psp}

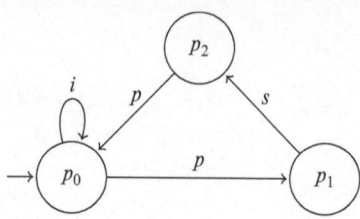

$$I_{1\|2} = I_1 \cup I_2,$$

$$AP_{1\|2} = AP_1 \cup AP_2,$$

$$L_{1\|2}(\langle s_1, s_2 \rangle) = L_1(s_1) \cup L_2(s_2),$$

and

$$\langle \langle s_1, s_2 \rangle, i, \langle s_1', s_2' \rangle \rangle \in \rightarrow_{1\|2}$$

if and only if:

(i) $i \in I_1 \cap I_2$, $\langle s_1, i, s_1' \rangle \in \rightarrow_1$ and $\langle s_2, i, s_2' \rangle \in \rightarrow_2$ (both TS_1 and TS_2 move, on a symbol that belongs to both input sets); or

(ii) $i \in I_1 \setminus I_2$, $\langle s_1, i, s_1' \rangle \in \rightarrow_1$, and $s_2 = s_2'$ (only TS_1 moves, on a symbol that belongs only to its input set, while TS_2 remains in the same state); or

(iii) $i \in I_2 \setminus I_1$, $\langle s_2, i, s_2' \rangle \in \rightarrow_2$, and $s_1 = s_1'$ (only TS_2 moves, on a symbol that belongs only to its input set, while TS_1 remains in the same state).

Consider, for example, the transition system TS_{psp} of Fig. 7.9, where every input symbol s is both preceded and followed by a p, and whose input symbols are i, p, and s. Sequences of $p\,s\,p$ fragments are separated by an arbitrary number of i symbols, which indicate the system being idle. The parallel composition $TS_{sr} \parallel TS_{psp}$ of the transition systems in Figs. 7.1 and 7.9 is represented in Fig. 7.10.

In $TS_{sr} \parallel TS_{psp}$, between any pair of p symbols there can only be one single s symbol; conversely, for any s symbol there is a pair of p symbols, one before and one after it. However, between an s symbol and the pair of p symbols enclosing it there can be an arbitrary number of r symbols. This is a consequence of the fact that symbol r is not shared by the two transition systems; hence TS_{sr} can evolve on that symbol independently of TS_{psp}. For example, the input trace $p\,r\,r\,s\,r\,p$ has a run in $TS_{sr} \parallel TS_{psp}$.

Parallel composition and synchronous composition are both commutative and associative: $TS_1 \parallel TS_2 = TS_2 \parallel TS_1$ and $TS_1 \parallel (TS_2 \parallel TS_3) = (TS_1 \parallel TS_2) \parallel TS_3$, and $TS_1 \cap TS_2 = TS_2 \cap TS_1$ and $TS_1 \cap (TS_2 \cap TS_3) = (TS_1 \cap TS_2) \cap TS_3$. Thus, the composition of more than two transition systems is uniquely defined, independently of the composition order.

Exercise 7.8. Show that parallel and synchronous compositions are commutative and associative. ∎

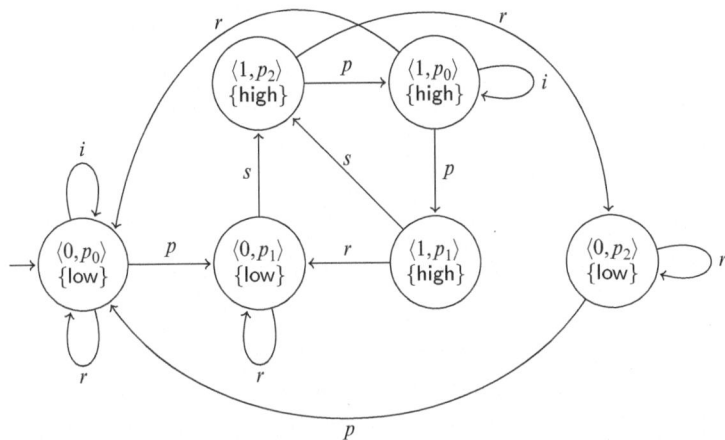

Fig. 7.10 Parallel composition $TS_{sr} \parallel TS_{psp}$ of the transition systems of Figs. 7.1 and 7.9

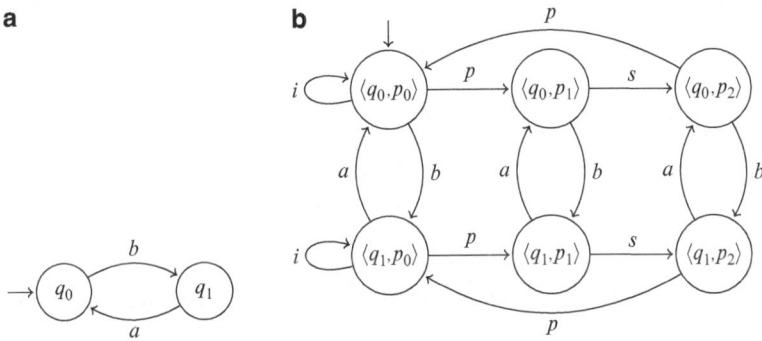

Fig. 7.11 Transition system TS_{ab} (**a**) and composition $TS_{ab} \parallel TS_{psp}$ (**b**)

Parallel and synchronous compositions coincide for transition systems with the same set of input symbols, because $I_1 = I_2 = I_1 \cap I_2$. On the other hand, the synchronous composition of transition systems with disjoint sets of input symbols is empty (it has no transitions), whereas their parallel composition processes all the *interleavings* of the sequences compatible with TS_1 and TS_2. Consider, for example, the transition system TS_{ab} of Fig. 7.11a, whose runs process sequences of alternating a's and b's. TS_{ab} does not have any symbols in common with the transition system T_{psp} of Fig. 7.9, so their parallel composition $T_{ab} \parallel T_{psp}$ – in Fig. 7.11b – processes sequences such as $a\,i\,p\,s\,b\,p\,i\,i\,a\,b\,a\,i\,p$, which are arbitrary interleavings of $a\,b\,a\,b\,a\,b\ldots$ and $i^{n_1}\,p\,s\,p\,i^{n_2}\,p\,s\,p\,i^{n_3}\,p\,s\,p\ldots$.

Parallel composition implicitly introduces synchronized transitions: those whose labels are shared by the composed systems. Synchronized transitions do not have a cause-effect relationship; namely, when two transition systems TS_1 and TS_2

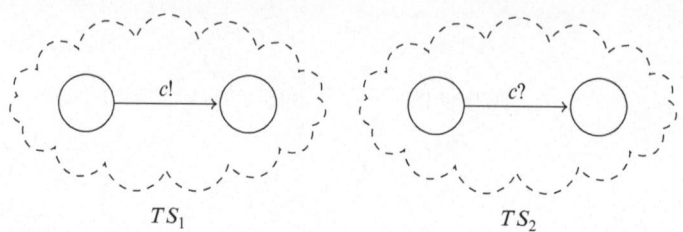

Fig. 7.12 Example of message passing between two transition systems TS_1 and TS_2

synchronize on a transition labeled i, it is not defined whether TS_1 causes TS_2 to make a transition with the same label, or vice versa. When including cause-effect relationship is required, we can use synchronization through **CHANNELS**. In this case, some transitions involve a message being passed from one system to the other through some *channel* c; a label "$c!$" represents the event of a message being sent through c, while "$c?$" represents the message being received, through that same channel c, in another component. Figure 7.12 shows the fragments of two transition systems TS_1 and TS_2 that include a message sent by TS_1 and received by TS_2 through a channel c.

Definition 7.9 (Channel composition). A transition system with channels is a tuple $\langle S, S_0, I, C, \rightarrowtail, AP, L \rangle$, where C is the set of channels through which messages can be exchanged. Transitions can be associated not only with symbols in I, but also with messages sent ($c!$) or received ($c?$) through some channel c in C; that is, we now have $\rightarrowtail \subseteq S \times (I \cup C! \cup C?) \times S$, where $C!$ is the set $\{c! \mid c \in C\}$ and $C?$ is $\{c? \mid c \in C\}$. The composition of transition systems with channels is quite naturally defined based on messages: given two transition systems (with set of channels C)

$$TS_1 = \langle S_1, S_{0,1}, I_1, C, \rightarrowtail_1, AP_1, L_1 \rangle,$$
$$TS_2 = \langle S_2, S_{0,2}, I_2, C, \rightarrowtail_2, AP_2, L_2 \rangle,$$

their composition $TS_1 \dashv\models TS_2$ is defined as a system

$$TS_{1\dashv\models2} = \langle S_{1\dashv\models2}, S_{0,1\dashv\models2}, I_{1\dashv\models2}, C, \rightarrowtail_{1\dashv\models2}, AP_{1\dashv\models2}, L_{1\dashv\models2} \rangle,$$

where

$$S_{1\dashv\models2} = S_1 \times S_2,$$
$$S_{0,1\dashv\models2} = S_{0,1} \times S_{0,2},$$
$$I_{1\dashv\models2} = I_1 \cup I_2,$$
$$AP_{1\dashv\models2} = AP_1 \cup AP_2,$$
$$L_{1\dashv\models2}(\langle s_1, s_2 \rangle) = L_1(s_1) \cup L_2(s_2).$$

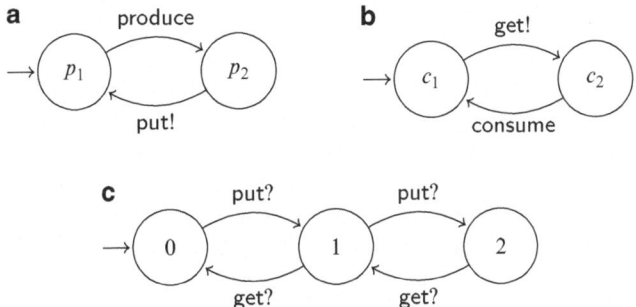

Fig. 7.13 Producer/consumer system using channel composition: producer (**a**), consumer (**b**), and buffer (**c**)

The transition relation $\rightarrowtail_{1 \dashv\vDash 2}$ is such that

$$\langle\langle s_1, s_2 \rangle, i, \langle s_1', s_2' \rangle\rangle \in \rightarrowtail_{1 \dashv\vDash 2}$$

if and only if:

(i) $i \in I_1$, $\langle s_1, i, s_1' \rangle \in \rightarrowtail_1$ and $s_2 = s_2'$; or
(ii) $i \in I_2$, $\langle s_2, i, s_2' \rangle \in \rightarrowtail_2$ and $s_1 = s_1'$; or
(iii) $\langle s_1, c!, s_1' \rangle \in \rightarrowtail_1$, $\langle s_2, c?, s_2' \rangle \in \rightarrowtail_2$, and $i = \varepsilon$ (absence of input); or
(iv) $\langle s_1, c?, s_1' \rangle \in \rightarrowtail_1$, $\langle s_2, c!, s_2' \rangle \in \rightarrowtail_2$, and $i = \varepsilon$. ■

Figure 7.13 shows three transition systems whose channel-based composition models a simple producer/consumer system with a two-element buffer.

Definition 7.9 entails that if TS_1 never sends or receives messages through a channel $\hat{c}$ of C (hence TS_1 does not have transitions labeled with channel $\hat{c}$), the composed system also does not have transitions labeled with channel $\hat{c}$, even if TS_2 does. As a consequence, the channel composition $\dashv\vDash$ of more than two systems is not associative and must be defined *globally* for all composed systems, not compositionally in pairs. More precisely, $TS_1 \dashv\vDash TS_2 \dashv\vDash \ldots \dashv\vDash TS_n$ has state space $S_1 \times S_2 \times \ldots S_n$ and, for *every* two systems i, j such that $\langle s_i, c!, s_i' \rangle \in \rightarrowtail_i$ and $\langle s_j, c?, s_j' \rangle \in \rightarrowtail_j$:

$$\langle\langle s_1, \ldots, s_i, \ldots, s_j, \ldots, s_n \rangle, \varepsilon, \langle s_1, \ldots, s_i', \ldots, s_j', \ldots, s_n \rangle\rangle \in \rightarrowtail_{\dashv\vDash} .$$

Despite the fact that the channel composition is global to several transition systems, message exchanges occur only between pairs of transition systems; if more than two systems are ready to exchange messages over the same channel $\hat{c}$, the choice of which pair will communicate is nondeterministic.

Exercise 7.10. Complete the definition of the composition $TS_1 \dashv\vDash \ldots \dashv\vDash TS_n$ of n systems through channels. ■

Exercise 7.11. Define three transition systems TS_1, TS_2, and TS_3 such that $(TS_1 \rightleftharpoons TS_2) \rightleftharpoons TS_3$ is different from $TS_1 \rightleftharpoons (TS_2 \rightleftharpoons TS_3)$. ∎

Messages exchanged through channels carry no information (such as data) other than the fact that the message is being exchanged over a certain channel. Message exchanges can only occur when one of the composing systems is ready to send and another to receive; that is, message passing in this case is a form of **HANDSHAKING**, and the systems *synchronize* on the exchange. In some notions of channel composition more sophisticated than the one presented in this chapter, channels include a buffer; hence message exchange becomes a form of *asynchronous coordination*. In these cases, messages typically carry also some data with them, which is enqueued in the channel buffer when the message is sent.

Exercise 7.12. The definition of multiple transition systems through channels entails that systems exchange messages *in pairs*. Define a different kind of composition through channels, in which messages can be *broadcast* to many receivers in the same transition. ∎

Channel-based compositions define *closed* systems, because messages disappear in the composed systems: by rules (iii) and (iv) the label on a transition in which a message is exchanged becomes ε (empty), and unused channels disappear from the composition, thus preventing $TS_1 \rightleftharpoons \ldots \rightleftharpoons TS_n$ from interacting with further components (as in the composite system of Fig. 7.13); the set C of channels is immaterial in composite closed systems. To have *open* composite systems, which can still be composed with other components, we have to revisit the above rules (i)–(iv) as follows: if $C_{\mathrm{env}} \subseteq C$ is the set of "open" channels used for interaction with the environment, rules (iii) and (iv) apply only to channels in $C \setminus C_{\mathrm{env}}$, while, for $c \in C_{\mathrm{env}}$, we apply the additional rules:

(i) $i = c!$ or $i = c?$, $\langle s_1, i, s_1' \rangle \in \longrightarrow_1$, and $s_2 = s_2'$;
(ii) $i = c!$ or $i = c?$, $\langle s_2, i, s_2' \rangle \in \longrightarrow_2$, and $s_1 = s_1'$.

Exercise 7.13. In the composition of open systems described above, channels in $C \setminus C_{\mathrm{env}}$ (used by just one system) disappear from the composition, as in the closed case. Define a composition in which channels used by only one system in the composition are still available for further compositions. Is your composition associative? ∎

Closed systems can show only one form of nondeterminism, which arises when, on a certain input from one state, the system can evolve into two different states; this behavior can be used to model transitions whose effects are unobservable outside the system. We call this *internal* nondeterminism, since it originates from internal choices made by the system (to go into one state or another on the same input). Open systems can show another form of nondeterminism, which we call *external* nondeterminism because it arises from the environment – which nondeterministically sends messages to the system. The difference between internal nondeterminism and external nondeterminism is similar to the difference between internal choice and external choice in some process algebras, presented in Chap. 10.

7.2 Automata Over Discrete Time

The formalism of *finite-state machines*, introduced in Chaps. 5 and 6, is very simple and intuitive, and is widely used to model a variety of systems. This section and the rest of this chapter first revisit them as special cases of transition systems, then progressively enrich them with features that increase their expressiveness and widen their applicability domains.

Finite-state machines (also called finite-state automata) are transition systems whose sets of states (S), input symbols (I), and atomic propositions (AP) are all *finite*. A common usage of finite-state machines is to distinguish between "good" and "bad" runs through the notion of *accepting states*.

Definition 7.14 (finite-state machine/finite-state automaton as acceptor of sequences). A finite-state machine (or automaton) A is a *finite* transition system TS in which a subset F of the state set S is the set of *final* (or *accepting*) states; that is, $A = \langle TS, F \rangle$ with $F \subseteq S$. A *finite* sequence $\sigma = i_1 i_2 i_3 \ldots i_n$ of input symbols is accepted by A if and only if there is a finite run r_σ of the automaton on σ such that the last state s_n of the run is accepting: $s_n \in F$. ∎

The simplest way to introduce the notion of time in finite-state machines is to adopt the convention that time advances whenever a transition is taken. This is the underlying assumption when determining the time complexity of a computation modeled through an automaton, discussed in Chap. 6. Under the assumption that every transition takes exactly one time unit, the automaton fragment in Fig. 7.14 defines that event e occurs three time units after event s, with no events in between (symbol τ represents the absence of events of interest).

By combining the convention that every transition takes one time unit with the notion of acceptance (Definition 7.14), we can precisely describe *finite* (but arbitrarily long) sequences of events with desired temporal properties. For example, the automaton in Fig. 7.15 accepts sequences in which every s event is followed, two time units later, by an e event (conversely, every e event must be preceded, three units earlier, by an s event). Since q_e is the only final state of the automaton, all accepted sequences must include at least one s event; in addition, they must terminate with either exactly one e event, or with a sequence of instants in which nothing of interest happens.

Exercise 7.15. Build a finite-state machine that accepts sequences of events a or b such that an a occurs exactly every three time units (the first occurrence being at time 3). b events may or may not occur at the other instants; however, at least one time unit must elapse between two successive occurrences of b. The machine can halt in an accepting state two time units after receiving a b. ∎

Under the convention that associates time advancement with transitions, time is **DISCRETE** and it advances **IMPLICITLY** whenever a transition is made. Without

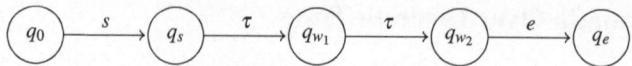

Fig. 7.14 A fragment of finite-state machine

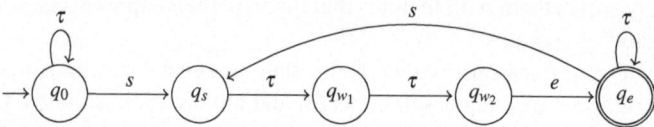

Fig. 7.15 A finite-state machine with "wait" τ transitions

further qualifications, there is no explicit notion of quantitative time. We have seen, however, that it is routine to introduce a simple **METRIC** by associating a time unit with the execution of a single transition. For example, in Fig. 7.15, we implicitly measure the three time units after which an s event occurs, by forcing the path from s to e to pass through two intermediate states via two "wait" transitions τ.

Time is, in general, unbounded in finite-state machine models, since sequences can have arbitrary length; according to Definition 7.14, however, only sequences of **FINITE** length are accepted, since they must be processed in their entirety before reaching an accepting state. In contrast, reactive systems typically engage in nonterminating computations. To model such infinite interactions, there are a number of acceptance conditions that can accommodate infinite sequences. Among them, the most widely used is the *Büchi acceptance condition*, which states that an *infinite* sequence σ of events is accepted by an automaton when, during the processing of σ, there is some accepting state that is visited an infinite number of times. Finite-state automata interpreted with the Büchi acceptance condition are called "Büchi automata".

Definition 7.16 (Büchi acceptance condition). Consider a finite-state machine $A = \langle TS, F \rangle$ and an *infinite* sequence (also called ω-sequence) $\sigma = i_1 i_2 i_3 \ldots$ of input symbols with run r_σ. Let $inf(r_\sigma)$ denote the set of states that are traversed *infinitely many times* during run r_σ. r_σ is *accepting* if and only if it traverses at least one accepting state infinitely many times: $inf(r_\sigma) \cap F \neq \emptyset$. Correspondingly, an infinite *input sequence* is accepted by A if and only if it has an accepting run.   ∎

Let us consider the automaton of Fig. 7.15 and interpret it using the Büchi acceptance condition; the sequences of events it accepts are now infinite ones such that there is at least one occurrence of s events and, either s and e occur infinitely often (that is, every e event is followed, sooner or later, by a new s event as the automaton moves from state q_e to state q_s) or, from some point on, there are no more s events (hence, no more e events either as the automaton stays forever in state q_e with an infinite sequence of τ events).

Fig. 7.16 Nondeterministic
Büchi automaton with no
equivalent deterministic
variant

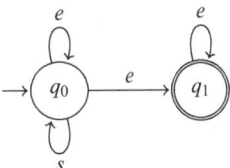

Exercise 7.17. What are the sequences of events accepted by the automaton of Fig. 7.15 assuming the Büchi acceptance condition if state q_s, instead of q_e, is the only final state? What if the only final states are q_{w_1} and q_{w_2}? ■

The expressiveness and other characteristics of Büchi automata differ from those of finite-state automata with the "finite" acceptance condition of Definition 7.14. For example, deterministic finite-state automata are as **EXPRESSIVE** as their nondeterministic variants under the acceptance condition of Definition 7.14, whereas nondeterministic Büchi automata can express a wider range of properties than their deterministic counterparts. This is exemplified by the automaton of Fig. 7.16, which accepts sequences of symbols such that there is a finite number of s symbols (that is, such that, from a certain point on, there are only e symbols). The automaton of Fig. 7.16 guesses, through a nondeterministic choice, one of the symbols e that is followed only by infinitely many e symbols, and triggers the transition from q_0 to q_1. No deterministic Büchi automaton can recognize the same sequences, since it is not possible to predict the point after which no more s events occur.

Exercise 7.18. Consider a system that inputs events of type a or b, or no event at all (the "wait" event τ); an input sequence is acceptable if there are at least two time units between any two a events, at least three time units between any two b events, and only a finite number of a events. Build a Büchi automaton that models such a system. Is your automaton deterministic? If it is nondeterministic, can you produce an equivalent deterministic one? ■

Example 7.19 (Resource allocator). As an example of use of Büchi automata, let us model a simplified resource allocator that works as follows. Clients can issue requests for the resource either with high priority (action hpr) or with low priority (action lpr). Whenever the resource is free and a high-priority request is raised, the resource is immediately granted and it becomes occupied. If it is free and a low-priority request is received, the resource is granted after two time units. Finally, if a high-priority request is received while the resource is occupied, it will be served as soon as the resource is released, whereas a low-priority request will be served two instants after the resource is released. Further requests received while the resource is occupied are ignored.

The deterministic automaton of Fig. 7.17 models the resource allocator: the requests and grant actions define the input alphabet and τ defines a "wait" transition. State **free** is the only accepting state; hence the resource must be free infinitely often. ■

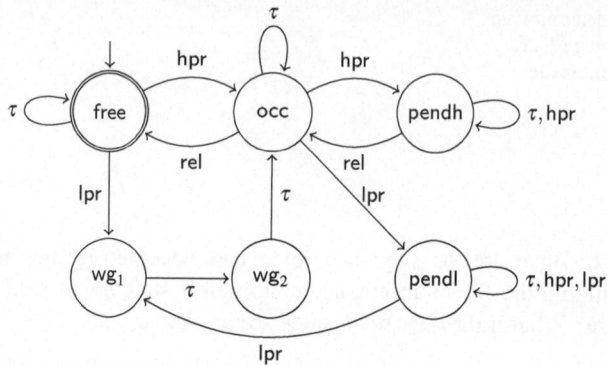

Fig. 7.17 Allocator system modeled through a Büchi automaton

Fig. 7.18 An automaton with variables

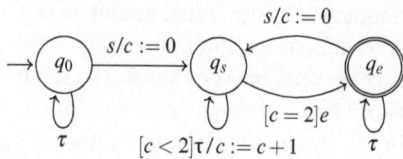

7.2.1   Extending Finite-State Machines with Variables

Extending states with *variables* makes automata representations more compact and potentially more powerful. Consider, for example, the automaton of Fig. 7.15; to count the number of instants between an *s* event and the next *e* event, we introduce a variable *c*, which is incremented every time an instant passes after the *s* event and before the *e* event. Then, event *e* must occur when the counter is 2, which means that two instants have passed after the *s* event.

For a set $X = \{x_1, x_2, \ldots, x_n\}$ of variables, where each variable x_i has a domain D_i, a *valuation* v is a function that associates with each variable a value in its domain, that is, $v : X \rightarrow D$, with D the union of all D_i's; an *extended state* is a pair $\langle s, v \rangle$, where $s \in S$, and v is a valuation. For example, $\langle q_s, v_1 \rangle$, with $v_1(c) = 1$, is an extended state of the automaton of Fig. 7.18. Transitions can include *guards* and *actions* that refer to variables. A guard is a Boolean predicate on the variables of the automaton (for example, $c < 2$ in the automaton of Fig. 7.18); an action is an assignment to one or more variables (such as $c := c + 1$). Then, a transition

$$\langle s, v \rangle \xrightarrow{\;[g]i/a\;} \langle s', v' \rangle \,,$$

where g is a guard and a is an action, is taken only if there is an arc with symbol i between states s and s' (with $s, s' \in S$), guard g holds for the values associated with variables by valuation v, and valuation v' is obtained from v by executing the assignments defined by action a. For example,

$$\langle q_s, v_0 \rangle \xrightarrow{[c<2]\tau/c:=c+1} \langle q_s, v_1 \rangle ,$$

where $v_0(c) = 0$ and $v_1(c) = 1$, since Boolean condition $c < 2$ holds in valuation v_0.

If variables only range over finite domains then the extended state also has finite cardinality $|S| \cdot |D_1| \cdot |D_2| \cdot \cdots \cdot |D_n|$; in this case, variables do not increase the expressiveness of finite-state automata. In contrast, infinite-domain variables do increase the expressive power.

Exercise 7.20. Solve again Exercise 7.18 using an automaton with variables. Compare the sets of states in the two solutions. ∎

Exercise 7.21. Define a Büchi automaton extended with one variable n – whose domain is the set $\mathbb{N}$ of natural numbers – that accepts infinite sequences of events a and b such that, infinitely often, the number of occurrences of a equals the number of occurrences of b. Is it possible to define a Büchi automaton using variables only on *finite* domains that accepts the same sequences? ∎

Exercise 7.22. Using automata with variables, define a variant of the resource allocator of Fig. 7.17 that buffers up to five pending low-priority requests and ten pending high-priority ones while the resource is occupied. Once the resource is released, one of the pending requests is granted the resource, starting with those with high priority. ∎

When the state is extended with variables, it is natural to allow different automata composed together to *share variables*: variables that appear in the guards and actions of two (or more) automata force the automata to *synchronize*, as they must agree on the values of these variables in order to proceed. Let us define, based on parallel composition, the *synchronization via variable sharing*.

Definition 7.23 (Synchronization via variable sharing). Let A_1 and A_2 be two automata with sets of variables X_1 and X_2, extended state spaces $S_1 \times V_1$ and $S_2 \times V_2$ (with V_1 and V_2, respectively, the sets of all valuations built on variables X_1 and X_2), and sets of input symbols I_1 and I_2. Their composition via variable sharing $A_1 \, |\!\!\lozenge\!\!| \, A_2$ has as extended state space $S_1 \times S_2 \times V$, with V the set of all valuations built on $X_1 \cup X_2$.

$$\langle \langle s_1, s_2, v \rangle, [g]i/a, \langle s_1', s_2', v' \rangle \rangle \in \mapsto_{1|\!\lozenge\!|2}$$

is one of its transitions if and only if one of the following holds:

(i) $i \in I_1 \setminus I_2$, $\langle \langle s_1, v(X_1) \rangle, [g]i/a, \langle s_1', v'(X_1) \rangle \rangle \in \mapsto_1$, $s_2 = s_2'$, and $v(X_2 \setminus X_1) = v'(X_2 \setminus X_1)$, where $v(X)$ is the restriction of valuation v to the variables of set X; or

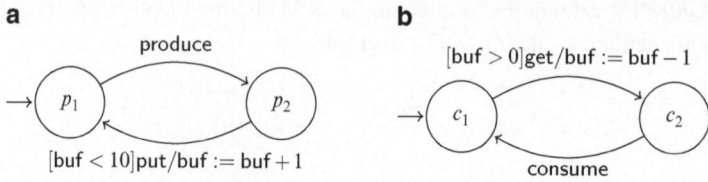

Fig. 7.19 A producer/consumer with buffer modeled through shared variables: producer (**a**), consumer (**b**)

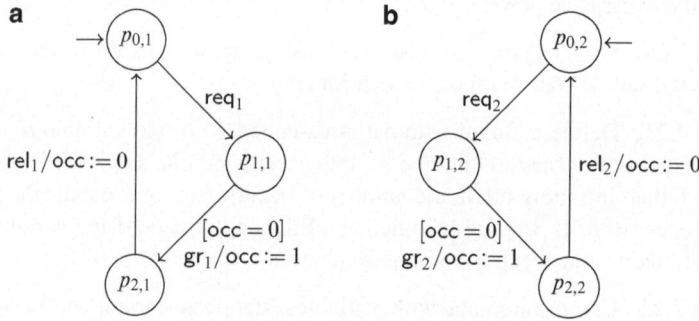

Fig. 7.20 Coordination of two processes through a lock

(ii) $i \in I_2 \setminus I_1$, $\langle\langle s_2, v(X_2)\rangle, [g]i/a, \langle s_2', v'(X_2)\rangle\rangle \in \rightarrowtail_2$, $s_1 = s_1'$, and $v(X_1 \setminus X_2) = v'(X_1 \setminus X_2)$; or

(iii) $i \in I_1 \cap I_2$, $\langle\langle s_1, v(X_1)\rangle, [g_1]i/a_1, \langle s_1', v(X_1)'\rangle\rangle \in \rightarrowtail_1$, $\langle\langle s_2, v(X_2)\rangle, [g_2]i/a_2, \langle s_2', v(X_2)'\rangle\rangle \in \rightarrowtail_2$, $g = g_1 \wedge g_2$, and $a = a_1 \cup a_2$. ∎

As before, when $I_1 = I_2$, we have the case of synchronous composition in the presence of shared variables, whereas if $I_1 \cap I_2 = \emptyset$ we have the interleaving of the two automata. Notice also that, for every $i \in I_1 \cap I_2$ and $x \in X_1 \cap X_2$, a synchronized transition in $\rightarrowtail_{1|\underline{X}|2}$ can occur only if $v_1(x) = v_2(x)$ and $v_1(x)' = v_2(x)'$ in the respective transitions in $\rightarrowtail_1$ and $\rightarrowtail_2$.

Example 7.24 (Producer/consumer with buffer). Figure 7.19 shows a variant of the producer/consumer example of Fig. 7.13, with a ten-element buffer represented through a shared variable buf counting the number of elements in the buffer. The producer and consumer automata do not share any input symbols; hence their composition $|\underline{X}|$ reduces to the interleaving of their transitions coordinated through the shared variable buf (for example, since initially we have buf $= 0$, the consumer cannot perform get before the producer has performed produce and then put at least once, after which guard buf > 0 holds). ∎

Example 7.25 (Mutual exclusion). Figure 7.20 shows a simple model of two processes, P_1 and P_2, which use a resource controlled through a locking mechanism, represented by a shared variable occ. After process P_i requests the resource (event req$_i$), it can acquire it only when shared variable occ is 0. Once the process acquires the resource, it sets occ to 1, to ensure it has exclusive access. After it has finished using the resource, it frees it by resetting variable occ to 0.   ■

Exercise 7.26 (♦). It is easy to extend the definitions concerning automata with extended states to include variables of type array of fixed, finite size. In fact, an array whose size is fixed and finite can be seen as a finite sequence of variables that can be accessed through an index. Using this intuition, define a two-automaton system modeling the two players in a tic-tac-toe game (with a classic 3×3 board).

■

Exercise 7.27. Faust and Margaret share usage of a car. Each usage session starts with picking up the car keys – which grant exclusive usage – followed by driving for an arbitrary amount of time, and ends with an optional refueling of the car before returning the keys. Driving consumes 1 L of gas per time unit. Since the gas tank has a volume of 60 L, no driving session can last longer than 60 time units. Faust and Margaret have, however, different refueling policies. Margaret always makes sure to return the car with five more liters of fuel than when she took it; if the tank already contains more than 55 L when she takes the car, she just returns a full tank. Faust is instead selfish, and never refuels after driving, unless the gas level is below 3 L, in which case he refuels to 7 L before returning the car. Faust's and Margaret's driving sessions can interleave in any way, but only one of them drives in any session. Model such a system using Büchi automata with variables.

(*Hint*: define the system as the composition of three automata, modeling Faust, Margaret, and the car).   ■

7.2.2 Finite-State Automata Features

According to the distinction introduced at the beginning of Part II, finite-state automata models are "computer-centric", focusing on simplicity and abstractness. In particular, time is modeled in a highly simplified way. Let us analyze their characteristics according to the dimensions of Chap. 3.

Independent of the acceptance condition, the notion of time is **DISCRETE**, **IMPLICIT**, and **METRIC**. If one adopts the Büchi acceptance condition instead of the one of Definition 7.14, however, system traces are **INFINITE** instead of finite.

The simplicity of both finite-state automata variants makes them amenable to automated **VERIFICATION**. Various techniques have been developed to analyze and verify finite-state automata, the most successful of which is probably model checking, discussed in Chap. 11.

The **NONDETERMINISTIC** versions of finite-state automata are particularly effective for characterizing multiple computation paths in the presence of the same inputs. **BRANCHING**-time models can characterize such nondeterministic behaviors, for example, through the notion of tree (see Sect. 3.3). The semantics of Büchi automata, however, typically relies on **LINEAR**-time models because linear time is usually considered more intuitive. For instance, the linear runs of the automaton model of Example 7.19 naturally represent the sequences of events that may take place in the resource allocator.

Models of complex systems can be built from the **COMPOSITION** of smaller automata using the mechanisms for transition systems presented in Sect. 7.1, with some modifications to take final states (and in particular the Büchi acceptance condition) into account.[2] Composed automata evolve **CONCURRENTLY**, with different semantics according to the composition rule applied. Time advancement in composite systems depends, in particular, on the composition semantics.

In the case of **SYNCHRONOUS** composition, the whole system takes a transition only when both components do, so there is a direct correspondence between the advancement of time in the components considered in isolation and in the overall system. Hence, if one considers the systems of Figs. 7.1 and 7.6 as Büchi automata (for example, by defining that all states are accepting) then TS_{s2r} prescribes that after an s signal there must be an r signal for at least two time units, and this property is preserved in the intersection of Fig. 7.7.

In the case of parallel composition, time advancement in the overall system may not reflect time advancement in the isolated components. Consider, for example, the systems of Figs. 7.1 and 7.9 and suppose all states are accepting. TS_{psp} is such that a symbol s is immediately preceded and followed by a signal p. This property, however, does not hold in the parallel composition of Fig. 7.10, since p is not a shared symbol; hence the two automata do not synchronize on it. In general, **METRIC** properties are not preserved by parallel composition under the assumption "transition = time step". In the next sections of this chapter we will see how relaxing the assumption makes it possible to preserve metric properties of individual components even without adopting strictly synchronous composition.

The simplicity of finite-state automata models is both a strength and a limitation. A strictly **SYNCHRONOUS** and discrete view of time is simple to understand and reason about, but it may hinder a "natural" modeling of continuous processes, where discretization may be too much an abstraction. In particular, some properties may not be preserved by discretization, such as periodicity for irrational periods, incommensurable with the duration of the step assumed in the discretization. Section 7.4.3 discusses a discretization technique that connects the semantics of finite-state automata over continuous and discrete time.

Other scenarios where discrete-time finite-state automata models are of limited applicability occur with heterogeneous systems whose components run at highly different speeds and time **GRANULARITIES**.

---

[2]The composition operations become exactly those presented in Sect. 7.1 when all states are final. See Sect. 11.1.1 and the bibliographic remarks for more details on this composition.

7.3 Decoupling Passing of Time and Transitions

As discussed at the end of Sect. 7.2, under the assumption that every transition implicitly advances time, the parallel composition of automata may alter the metric timing relationships between events. A possible way to avoid this problem is to decouple the passing of time from the taking of transitions, for example, by prescribing that time advances only when certain, suitably labeled, transitions are taken. This is the solution adopted, among others, in Timed Transition Models (discussed in Sect. 7.3.1) and in Statecharts (presented in Sect. 7.3.2).

7.3.1 Timed Transition Models

The core idea underlying the treatment of time in Ostroff's "Timed Transition Models" (TTMs) is to introduce a dedicated variable t which measures time. In addition, TTMs introduce a label, tick, that marks precisely the transitions where time advances: every tick transition increments t, and t never decreases (which corresponds to the fact that time is monotonic). For example, Fig. 7.21 shows an automaton with two tick transitions; since between event s and event e a tick occurs, e occurs in the next time instant as s; however, between p and s there is no tick transition, and hence these two events occur at the same time instant.

Now, consider the fragment of Fig. 7.22; it shares labels s and tick with the fragment of Fig. 7.21; hence, their parallel composition is compatible with only two sequences: tick p s tick e o, and tick p s tick o e; in both cases, events e and o occur exactly one instant after p and s, no matter the order in which they appear in the sequence, thus preserving metric properties in the composition – a solution to the problem highlighted at the end of Sect. 7.2.

TTMs use a model of time similar to that of Figs. 7.21 and 7.22, but without explicitly representing tick transitions or the t variable. First, an automaton such as the one depicted in Fig. 7.23, whose only role is to make time advance, is implicitly composed with every TTM.

Then, each TTM transition includes a guard, an action, and a pair $[l, u]$ which represents the lower bound and the upper bound for the transition to be taken after it is enabled. Consider a transition r originating from state s, with guard g and bound $[l, u]$; if T is the instant in which r becomes enabled (the system is in state s, and the guard g holds), then it *cannot* be taken before instant $T + l$ (until enough tick transitions are taken that variable t becomes $T + l$), and it *must* be taken before $T + u$, unless the transition is disabled beforehand. For example, the automaton of Fig. 7.18 corresponds to the TTM of Fig. 7.24 (in which all guards are simply **True**, and there are no actions). Note that the first s event need not occur exactly at the first instant, when the system starts, but it can have an arbitrary delay, as specified by the ∞ upper bound on the first transition.

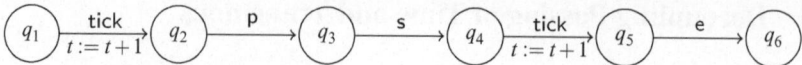

Fig. 7.21 Example of tick transitions

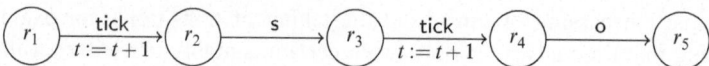

Fig. 7.22 A second example of tick transitions

Fig. 7.23 Global clock TTM

clock ⟲ tick/$t := t+1$

Fig. 7.24 TTM
corresponding to the
automaton of Fig. 7.18

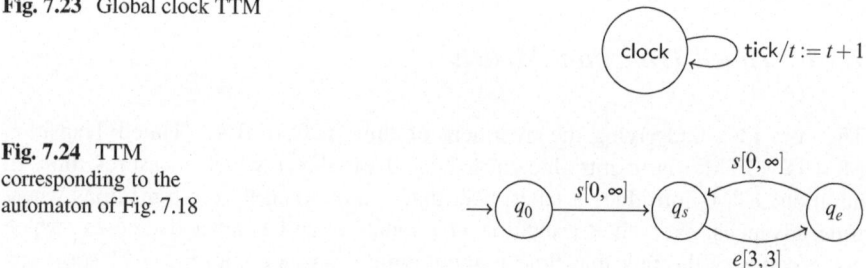

Example 7.28 (Resource allocator modeled with a TTM). In addition to the
IMPLICIT variable t counting the passing of time, TTMs can **EXPLICITLY**
include other variables, possibly with infinite domains. Figure 7.25 shows a TTM
representing a resource manager similar to the one presented in Example 7.19,
which manages high-priority (hpr) and low-priority (lpr) requests. Low-priority
requests can arrive only if there are no pending high-priority ones; hence the number
of pending high-priority requests must be counted. Each transition is annotated
with lower and upper bounds, a guard, and a variable update rule. For instance,
the transition rel_2 can be taken whenever the guard $occ > 1$ evaluates to true. The
model employs an integer-valued state variable occ to count the number of pending
high-priority requests. The effect of rel_2 is to update occ by decrementing it. Finally,
when rel_2 becomes enabled, it *must* be taken within a maximum of 100 clock ticks,
unless the state is left (and possibly reentered) by taking another (non-tick) enabled
transition (such as hpr_2, which is always enabled, since it has no guard).   ∎

In TTMs, time is modeled **EXPLICITLY** by means of lower and upper bounds
on variable t, which takes values in a **DISCRETE** time domain, and is updated
IMPLICITLY and **SYNCHRONOUSLY** by the occurrence of tick transitions. TTMs
have a **METRIC** notion of time, as they can express minimum and maximum delays
for transitions through lower and upper bounds.

The tick transition is a priori assumed to be fairly scheduled, that is, it must
occur infinitely often to prevent **ZENO** behaviors where time stops, as shown in the
following example.

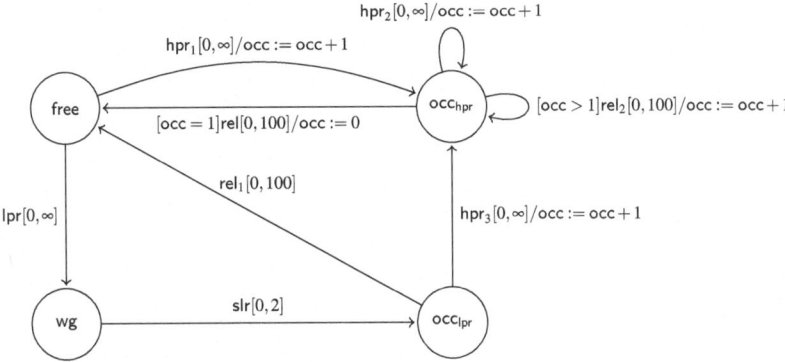

Fig. 7.25 A TTM representing a resource allocator

Fig. 7.26 A variation of the TTM of Fig. 7.24 that can exhibit Zeno behavior

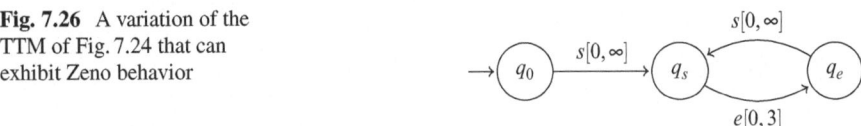

Example 7.29 (TTM with Zeno behavior). Figure 7.26 shows a TTM that is the same as the one of Fig. 7.24, except that the transition labeled e now has lower bound equal to zero. With this modification, event e must occur *no later than* three instants after event s occurs, but it can also occur in the same instant as s. Since also the transition from q_e to q_s has zero lower bound, e can be immediately followed by another s event, without the clock ticking in between, and so on. Then, an infinite sequence $s\,e\,s\,e\,\ldots$, in which no tick occurs (hence time does not advance), is compatible with the TTM of Fig. 7.26. Such a behavior corresponds to a system that executes an infinite number of operations in finite time. This kind of behavior has no physical meaning; hence the *semantics* of TTM rules it out a priori, by prescribing that tick transitions must occur infinitely often for an execution to be admissible. This is a kind of *fairness* constraint, which will be analyzed in greater detail in Chap. 9. ■

The TTM of Fig. 7.27 is another example of Zeno behavior; it requires that the first s event be immediately followed, in the same instant, by an e event, since the upper bound on the transition from q_s to q_e is zero. In addition, the e event must be followed, also in the same instant, by a new s event, and so on, *ad infinitum*. Then, the TTM of Fig. 7.27 *forces* an infinite sequence $s\,e\,s\,e\,\ldots$ to occur in zero time. As mentioned above, this is a behavior that has no physical meaning; hence the TTM of Fig. 7.27 is *syntactically* incorrect. More generally, any TTM in which there is a loop made of transitions with both lower and upper bounds equal to zero is *inadmissible*.

Fig. 7.27 A syntactically incorrect TTM

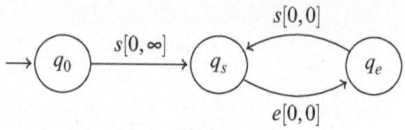

Fig. 7.28 Example of external events (e), internal events (i), and variables (a, b) in Statecharts

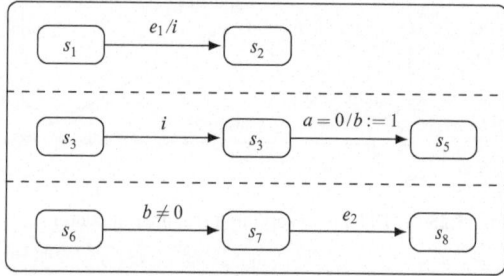

7.3.2 Statecharts

Statecharts are an automata-based formalism, invented by David Harel, where the taking of transitions is decoupled from the passing of time. Statecharts are supported by visual modeling tools that have gained popularity in many domains, especially in the software engineering community, where a version of Statecharts is part of the UML standard. Statecharts have a readable graphical syntax, which lends itself to a variety of interpretations, and over the years many different semantics (both formal and informal) have been defined for them.

Statecharts enrich finite-state automata with mechanisms for hierarchical abstraction, parallel composition, synchronization, and inter-component communication. They are an attempt to improve over bare finite-state automata, while retaining their advantages in terms of simplicity and natural graphical representation.

Statechart transitions include events of two kinds: *external* events E_e, and *internal* events E_i. The environment triggers external events, which constitute system input; internal events, in contrast, are used for communication between system components, and any Statechart transition can trigger them. Consider, for example, the fragment of Statechart shown in Fig. 7.28. It consists of one *AND node* made of three concurrent parts: the topmost one senses the external event e_1 and generates the internal event i when transitioning from state s_1 to state s_2, while the middle one senses the internal event i when transitioning from state s_3 to state s_4.

As Fig. 7.28 shows, Statecharts support variables: transitions may include guards on variables and actions with assignments. For example, the transitions from s_4 to s_5 and from s_6 to s_7 in Fig. 7.28 are guarded; the former also includes an assignment to variable b. Thus, synchronization in Statecharts can occur both through shared variables and through shared events.

Different semantics have been defined for Statecharts over the years; this section focuses on Harel and Naamad's for the STATEMATE tool, but it also hints at related variants when germane to the discussion.

Harel and Naamad's semantics supports two time advancement models. The first one, called "synchronous" by Harel and Naamad, prescribes that, at every instant, at most one step is taken in each automaton, in reaction to the events that occurred since the previous instant. According to this semantics, in the Statechart of Fig. 7.28, when the external event e_1 occurs, if the three parts are, respectively, in states s_1, s_3, and s_6, and both variables a and b are zero, the transition from s_1 to s_2 is taken in the first instant, which generates internal event i; then, at the next instant, transition $s_3 \to s_4$ is taken, as a reaction to event i generated at the previous instant; the instant after that, transition $s_4 \to s_5$ is taken, since its guard is enabled, which in turn enables the transition from s_6 to s_7, which is taken at the next instant. Hence, after four instants, the system will be in states $\langle s_2, s_5, s_7 \rangle$.

The second time advancement model in Harel and Naamad's semantics, called "asynchronous", prescribes that time advances only after *all* enabled transitions (whose guards evaluate to true, or that are triggered by internal events that have yet to be consumed) have been taken. The only exceptions happen with external events, which are sensed and acted upon only at the beginning of the instant in which they are generated by the environment. With the asynchronous semantics, in the Statechart of Fig. 7.28, when event e_1 occurs in configuration $\langle s_1, s_3, s_6 \rangle$, transition $s_1 \to s_2$ is taken, which generates event i; event i is *immediately* sensed by the middle part of the diagram, which takes transition $s_3 \to s_4$ before time advances; then, if variable a is equal to zero, the transition $s_4 \to s_5$ is also taken before time advances, which modifies the value of b, thus enabling the transition from s_6 to s_7; then, time advances only after transition $s_6 \to s_7$ is taken, thus bringing the system to configuration $\langle s_2, s_5, s_7 \rangle$, as there are no more enabled transitions. In this case, the system moves from configuration $\langle s_1, s_3, s_6 \rangle$ to configuration $\langle s_2, s_5, s_7 \rangle$ in one time instant, as the transition sequence $s_1 \to s_2, s_3 \to s_4, s_4 \to s_5, s_6 \to s_7$ occurs in zero time. These transition sequences occurring in zero time are called "supersteps" in Statecharts terminology. After this superstep, transition $s_7 \to s_8$ is taken only if event e_2 is generated by the environment at the beginning of the next superstep, because external events are sensed only at the beginning of a superstep, and not during the intervening zero-time steps. If, for example, the environment generates both e_1 and e_2 when the system is in configuration $\langle s_1, s_3, s_6 \rangle$, e_2 is discarded in that superstep, as no transition is enabled on e_2 from configuration $\langle s_1, s_3, s_6 \rangle$. Transition $s_7 \to s_8$ is taken only if, right at the beginning of the next superstep, e_2 is generated anew.

In Harel and Naamad's asynchronous Statechart semantics, time advances IMPLICITLY every time the system reaches a *stable* state without enabled transitions. When time advances, it usually advances by a single time unit; this implies that the notion of time is still DISCRETE. It is sometimes useful, for example, for

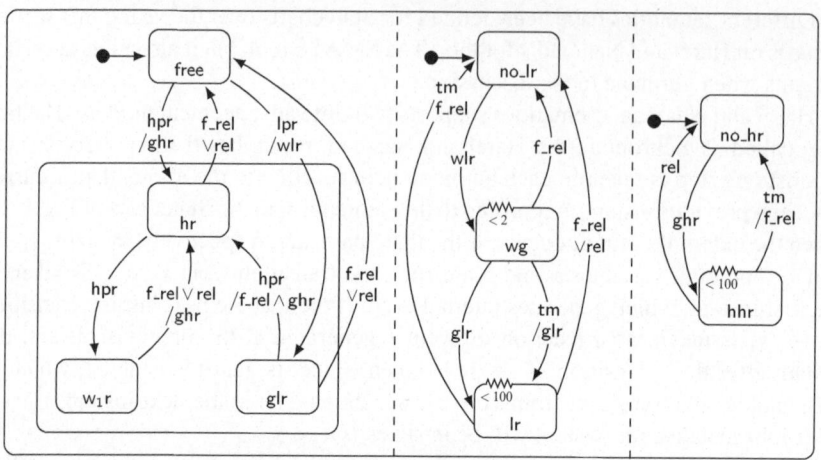

Fig. 7.29 A resource manager modeled through a Statechart

simulation purposes, to make time advance more than one time unit, to represent the fact that "nothing happens" for a certain time interval after some action was taken.

To illustrate other features of Statecharts, let us modify the resource manager from Example 7.28 so that: (i) (preemption) the arrival of a high-priority request while the resource is occupied by a low-priority one causes an immediate release of the resource to serve the high-priority request, and (ii) no more than one high-priority request can be pending while the resource is occupied. Figure 7.29 shows a Statechart of this version of the resource manager.

To represent the **METRIC** temporal constraint that, after any request has been granted, the resource must be released within 100 time units, we associate a *time-out* with some states, namely those represented with a short squiggle[3] on the boundary (such as hhr and wg in Fig. 7.29).

Thus, the transition that exits state hhr must be taken within 100 time units after hhr has been entered: if no external rel event has been generated by the environment within 100 time units, the time-out event tm is spontaneously generated exactly after 100 time units; the time-out event tm in turn produces the internal event f_rel, which is sensed by the other parts of the Statechart, forcing the release of the resource. If the environment produces event rel exactly 100 time units after entering state hhr, then both rel and tm events are available to the system, and the choice of which event is acted upon is nondeterministic. We use the same mechanism to model the maximum amount of time a low-priority request may have to wait for the resource to become available.

---

[3]The squiggle notation is the original one used to represent time-outs associated with states; more recent work on Statecharts have adopted a different syntax, which is, however, visually less clear.

Modeling time constraints using time-outs and exit events is consistent with the **IMPLICIT** representation of time advancement of the asynchronous semantics; in fact, a time-out of T from an event e is triggered when time advances T instants after event e occurred (after T supersteps from e, if each superstep increases time by one unit), no matter the actual transitions in the supersteps. More generally, since time-outs measure the passing of time without introducing references to explicit "tick" transitions, we can in principle seamlessly move from a discrete to a **CONTINUOUS** notion of time; for this, it is enough to posit that after each superstep time advances by an arbitrary amount – rational or irrational. This is the reason behind dubbing this semantics "asynchronous".

The example of Fig. 7.29 also illustrates the "AND (parallel) composition", another Statechart feature. The dashed lines separate three system components. The semantics of AND composition corresponds to the parallel composition of the sub-automata[4]; Statecharts' graphical representation, however, leaves the complete state space of the composition implicit, thus improving the readability of complex specifications. In the allocator example, we allow only one pending high-priority request to be "enqueued" while the resource is occupied; thus, the leftmost component is a finite-state automaton modeling whether the resource is free, serving a high-priority request with no other pending requests (state hr), or with one pending request (state w_1r), or serving a low-priority request (state glr).

Since in Statecharts all events – both internal and external – are broadcast over the whole system, labeling different transitions with the same event forces them to synchronize when they are simultaneously enabled. In Fig. 7.29, whenever the automaton is in the configuration $\langle w_1r, hhr, no\_lr \rangle$, a release event rel triggers the configuration to become $\langle hr, no\_hr, no\_lr \rangle$, and then cascade to $\langle hr, hhr, no\_lr \rangle$ because of the output event ghr triggered by the transition from w_1r to hr. Note that in the example above ghr and wlr are internal events in E_i, and they do not occur spontaneously in the environment but can only be generated internally for synchronization.

Statecharts feature three forms of **NONDETERMINISM**: first, the "usual" non-determinism of mutually exclusive transitions with the same input label (such as in Fig. 7.30a); second, states with time-outs are left nondeterministically *within* the prescribed bounds (Fig. 7.30b); third, "XOR composition", a different form of module composition which represents a nondeterministic choice from among different modules (Fig. 7.30c).

While overcoming some limitations of the basic finite-state automata models, Statecharts' rich syntax hides semantic subtleties that may let inconsistencies and errors go undetected. To illuminate potential sources of imprecision, many researchers have tried to define formally the most crucial aspects of the temporal semantics of Statecharts. The fact that different problems were unveiled

---

[4]There are some slight differences between the semantics of AND composition of sub-machines in Statecharts and the notion of parallel composition of finite-state machines. The interested readers will find references delving into this issue in the bibliographic remarks.

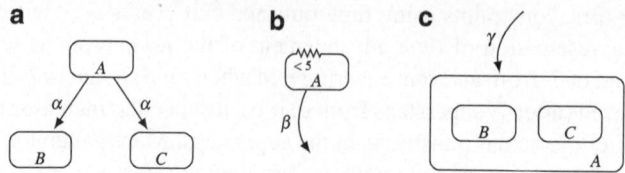

Fig. 7.30 Nondeterminism in Statecharts

only incrementally by different contributors is an indication of the difficulty of
finding a comprehensive, intuitive, non-ambiguous semantics to a language with
a deceivingly simple and plain syntax. We discuss here just a few examples; more
are referenced in the bibliographic remarks.

The apparently safe perfect **SYNCHRONY** assumption – the assumption that all
transition events in a superstep occur in zero time (despite the attribute "asynchronous" used in Statecharts parlance) – and the global "broadcast" availability
of all events – which are therefore non-local – generate some subtle difficulties in
obtaining a consistent semantics.

Consider the example of Fig. 7.29, and assume the system is in configuration
⟨glr, no_hr, lr⟩. If a high-priority request takes place, and thus an hpr external event
is generated by the environment, the system shifts to configuration ⟨hr, no_hr, lr⟩
in zero time. Simultaneously, the transition taken triggers the internal events f_rel
and ghr. Then, in zero time, event f_rel moves the system to ⟨hr, no_hr, no_lr⟩,
representing the low-priority request being forced to release the resource. Still
simultaneously, the latter ghr event triggers the transition from no_hr to hhr in the
middle sub-automaton. This conforms to our intuitive requirements; however, the
same f_rel generated event also triggers the first sub-automaton to the state free,
which is against the intuition that suggests that the event is only a message sent to
the other parts of the automaton.

Exercise 7.30. If we modify the leftmost component in Fig. 7.29 by replacing the
label hpr/f_rel ∧ ghr with the label hpr/f_rel in the transition from glr to hr, the
problem with shared events highlighted in the previous paragraph would not be
fixed. Can you explain why?   ∎

It is also not difficult to conceive scenarios in which the simultaneous occurrence
of some transitions causes an infinite sequence of states to be traversed during a
single superstep, thus causing a **ZENO** behavior.

How to properly disentangle such scenarios is not obvious. A possibility is, as
in Harel and Naamad's synchronous semantics, avoiding instantaneous transitions
altogether, attaching a nonzero time to transitions and forcing an ordering between
them or, symmetrically, disallowing zero-time residence in states. Other solutions
disallow loops of zero-time transitions, but accept a finite number of them; the
Esterel language, which is a "relative" of Statecharts', follows this approach.

Exercise 7.31. Define a Statechart in which an infinite number of transitions occur in a single superstep. ∎

7.4 Automata Over Continuous Time

Separating the passing of time from the taking of transitions facilitates moving from discrete to continuous time. Section 7.3.2 hinted at how this is possible with Statecharts using time-outs: time advances, after each superstep, by an arbitrary real amount; and a time-out expires when the cumulative elapsed time exceeds the value fixed by the time-out.

Separating passing of time and transitions also allows for a higher level of asynchrony among the parts composing a complex model: components need not synchronize on transitions for time to pass, and can evolve independently.

Various types of automata over continuous time feature such asynchronous behavior that separates time advancement and transitions. The rest of this section presents some of the most representative approaches in this field.

7.4.1 Timed Automata

"Timed automata" extend basic finite-state automata with real-valued *clock* variables. Although the term "timed automata" could generically denote automata including a description of time, here we specifically refer to the variant of the notation invented by Alur and Dill and used in the UPPAAL model checker.

In timed automata, the state has two parts: a finite discrete component called *location* (corresponding a state of finite-state automata), and a continuous component represented by a set of nonnegative real values each assigned to a finite number of variables called *clocks*. The state space is therefore *infinite*, since the clock components take value in the infinite set $\mathbb{R}_{\geq 0}$. Timed automata evolve in an alternation of instantaneous synchronous discrete "jumps" and continuous clock increases: when a timed automaton sits in some discrete state, each clock variable x increases as time elapses, that is, it evolves according to the differential equation $\dot{x} = 1$, thus effectively measuring time. External input events cause the location to switch; the corresponding transitions may reset the clocks to zero instantaneously. Both locations and transitions may include constraints on clocks; each constraint must be satisfied while sitting in a location and when taking a transition.

Definition 7.32 (Timed automaton). A timed automaton is a tuple $\langle L, A, C, E, l_0, I \rangle$, where

- L is a finite set of locations,
- A is a finite set of actions,
- C is a finite set of clocks,

Fig. 7.31 A resource manager modeled through a timed automaton

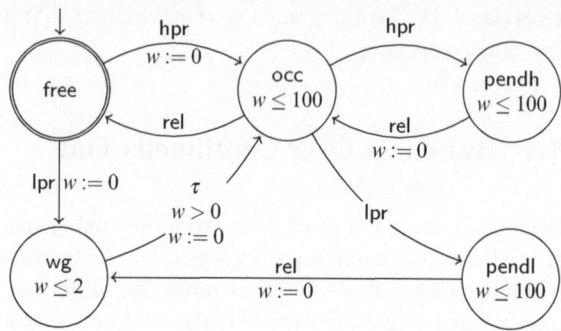

- $l_0 \in L$ is the initial location,
- E is a set of edges, with $E \subseteq L \times B(C) \times A \times \wp(C) \times L$, and
- I defines location invariants.

An edge of the form $\langle l, g, a, \hat{C}, l' \rangle$ in E denotes a transition from location l to location l' on action a that may occur only if the guard g holds, and that resets to zero the clocks in the set $\hat{C} \subseteq C$.

Guards are Boolean constraints from the set $B(C)$ of all conjunctions of comparisons between clocks and a constant – of the forms $x \sim n$ and $x - y \sim n$, where x and y are clocks in C, $n \in \mathbb{N}$, and $\sim \in \{<, \leq, =, \geq, >\}$.

The set of actions A corresponds to input symbols, and includes an "internal" action τ that does not read any input.

The function $I : L \to B(C)$ defines the location invariants: whenever the automaton sits in location l, $I(l)$ must hold. A timed automaton may also include a set F of final locations.   ∎

Example 7.33. Let us model, through a timed automaton, a variant of the resource manager of Fig. 7.29, where high-priority requests cannot preempt low-priority ones, and we allow a single low-priority request to be pending, waiting for the resource to become free. The resulting timed automaton – using a single clock w – is pictured in Fig. 7.31. The edge from location **free** to location **occ** has action **hpr** and resets clock w, but it has no guard (or, equivalently, has guard **True**). The edge from location **wg** to location **occ**, in contrast, has the internal action τ and a guard on clock w, which is also reset. All locations except **free** define an invariant on clock w.   ∎

We can use transition systems to define formally the semantics of timed automata. Since the state space of timed automata is infinite, we need infinite-state transition systems, whose states are pairs $\langle l, v \rangle$, where l is a location of the automaton, and v is a valuation of the clocks, that is, a function $C \to \mathbb{R}_{\geq 0}$ that assigns a nonnegative real value to each clock in the set C. Transitions are of two kinds: *actions* (namely, input symbols or the internal action τ), which may change location and reset clocks, and *delays*, which advance time without anything else changing. For example, the transition system for the automaton of Fig. 7.31 can reach the state $\langle \mathsf{wg}, v_{1.5} \rangle$, where v_x is such that $v_x(w) = x$. From there, it can make the sequence of transitions

$$\langle \mathsf{wg}, v_{1.5} \rangle \xrightarrow{\tau} \langle \mathsf{occ}, v_0 \rangle \quad \langle \mathsf{occ}, v_0 \rangle \xrightarrow{2.5} \langle \mathsf{occ}, v_{2.5} \rangle \quad \langle \mathsf{occ}, v_{2.5} \rangle \xrightarrow{\mathsf{lpr}} \langle \mathsf{pendl}, v_{2.5} \rangle \; ;$$

the second transition is a delay in location occ. The following definition formalizes the transition system semantics of timed automata.

Definition 7.34 (Semantics of timed automata). The semantics of a timed automaton $\langle L, A, C, E, l_0, I \rangle$ is given by a transition system $\langle S, S_0, J, \rightarrowtail \rangle$,[5] where $S = L \times V_C$ – with $V_C = \{ C \to R_{\geq 0} \}$ the set of all possible valuations of clocks C – $S_0 = \{ \langle l_0, v_0 \rangle \}$ – where v_0 is the valuation that assigns 0 to each clock in C – and $J = A \cup R_{\geq 0}$. Then, $\langle \langle l, v \rangle, e, \langle l', v' \rangle \rangle \in \rightarrowtail$ if and only if one of the following holds:

(i) $e \in A$ and there is an edge $\langle l, g, e, \hat{C}, l' \rangle$ such that v satisfies $I(l)$ (the values of the clocks in valuation v satisfy the invariant of location l), v' satisfies $I(l')$, v satisfies g, and v' is obtained from v by resetting all clocks in $\hat{C}$ while leaving all other clocks unchanged, written $v' = v[\hat{C} := 0]$;

or

(ii) $e \in R_{\geq 0}$, $l' = l$, and, for all $0 \leq d \leq e$, the valuation obtained by incrementing all clocks by d (written $v + d$) satisfies $I(l)$. ∎

In any run

$$\langle l_0, v_0 \rangle \xrightarrow{e_1} \langle l_1, v_1 \rangle \xrightarrow{e_2} \langle l_2, v_2 \rangle \ldots$$

of a transition system as in Definition 7.34, the sequence of e_i's consists of actions and delays in any order; however, we can always produce an equivalent sequence

$$\langle l_0, v_0 \rangle \xrightarrow{a_1, t_1} \langle l'_1, v'_1 \rangle \xrightarrow{a_2, t_2} \langle l'_2, v'_2 \rangle \ldots$$

where only actions appear, each paired with a corresponding *timestamp*, recording the absolute time at which the action occurs. The sequence $a_1 a_2 a_3 \ldots$ is the projection of the sequence $e_1 e_2 e_3 \ldots$ over the actions A, and if $a_i = e_j$ and $a_{i+1} = e_{j'}$, then $j' > j$ and, for all $j < k < j'$, $e_k \in R_{\geq 0}$. Then, each timestamp t_{i+1} is defined as

$$t_{i+1} = \begin{cases} t_i + \sum_{k=j+1}^{j'-1} e_k & i > 0, \\ 0 & i = 0 \text{ and } e_1 \in A, \\ \sum_{k=1}^{\bar{h}-1} e_k & i = 0, e_1 \in R_{\geq 0}, \text{ and} \\ & \bar{h} \text{ is the first action's index in the sequence } e_1 e_2 e_3 \ldots. \end{cases}$$

---

[5] The transition system does not include labels AP or labeling function L, which are not used in the semantics of timed automata.

Notice that the sequence of timestamps is monotonically nondecreasing: $t_{i+i} \geq t_i$ for all i. Such timed sequences

$$\langle a_1, t_1 \rangle \; \langle a_2, t_2 \rangle \; \langle a_3, t_3 \rangle \; \ldots$$

are called *timed words* and customarily taken to be of infinite length.

The automaton of Fig. 7.31 can process the timed word $\langle \mathsf{hpr}, 4.7 \rangle \; \langle \mathsf{lpr}, 53.9 \rangle \; \langle \mathsf{rel}, 64 \rangle \; \langle \tau, 65.1 \rangle \; \ldots$, which corresponds to the following sequence of states in the transition system:

$$\langle \mathsf{free}, 0 \rangle \xrightarrow{\mathsf{hpr}, 4.7} \langle \mathsf{occ}, 0 \rangle \xrightarrow{\mathsf{lpr}, 53.9} \langle \mathsf{pendl}, 49.2 \rangle \xrightarrow{\mathsf{rel}, 64} \langle \mathsf{wg}, 0 \rangle \xrightarrow{\tau, 65.1} \langle \mathsf{occ}, 0 \rangle \; \ldots .$$

In the sequence above, the timed automaton enters location occ at time 4.7 while instantaneously resetting clock w; thus, the new state becomes $\langle \mathsf{occ}, 0 \rangle$. Later, when the absolute time reaches 53.9 (or, equivalently, clock w reaches value 49.2), the automaton exits location occ and enters pendl without clock resets; this is compatible with the constraint $w < 100$ of location occ.

When a timed automaton includes a set of final locations F, a timed word is accepted if and only if it traverses at least one final location infinitely often.

Definition 7.35 (Büchi acceptance condition for timed automata).   Consider a timed automaton $TA = \langle L, A, C, E, l_0, I, F \rangle$, and let r be one if its (infinite) runs (corresponding to some timed word). Let $inf(r)$ denote the set of locations that are traversed *infinitely many times* during run r. r is *accepting* if and only if it traverses at least one accepting state infinitely many times: $inf(r) \cap F \neq \emptyset$. Correspondingly, a timed word $\sigma = \langle a_1, t_1 \rangle \; \langle a_2, t_2 \rangle \; \ldots$ is accepted if and only if: (i) the sequence of timestamps $t_1 \, t_2 \, \ldots$ diverges (that is, for all $T \in \mathbb{R}_{\geq 0}$ there is an $i \in \mathbb{N}$ such that $t_i > T$), and (ii) σ has a corresponding accepting run of TA.   ∎

The timed automata semantics introduces a **METRIC** treatment of time through timestamps. In some sense, timed words yield two different notions of time: a **DISCRETE** one, given by the position in the timed word, which defines a total ordering on events, and a **CONTINUOUS** and **METRIC** one, recorded by the timestamp values and controlled through the clocks. This decoupling of time is simple in principle, but it mars the natural perception of time as a unique flow.

Let us consider a few more dimensions of time modeling for timed automata.

- While timed automata are in general **NONDETERMINISTIC**, their semantics is usually defined through **LINEAR**-time models, such as the one outlined above based on runs and timed words.
- As for Büchi automata, deterministic timed automata are strictly less expressive than nondeterministic ones, but also more amenable to automated verification, so they may be preferred in some practical cases.
- Timed automata feature **IMPLICIT** references to absolute time, which is revealed only in the semantics based on timestamps. The *relative time* measured by clocks is in contrast **EXPLICITLY** referenced and set.

- Timed automata may exhibit **ZENO** *behaviors* when distances between transitions become increasingly smaller, so that the infinite sequence of timestamps $t_1 t_2 \dots t_i, \dots$ accumulates to a finite value $\bar{t}$. For instance, in the example of Fig. 7.31, if the two transitions hpr and rel are repeatedly taken at times $1, 1 + 2^{-1}, 1 + 2^{-1} + 2^{-2}, \dots, \sum_{k=0}^{n} 2^{-k}$, absolute time accumulates to $\sum_{k=0}^{\infty} 2^{-k} = 2$. Definition 7.35, however, rules these Zeno behaviors out of timed automata semantics a priori by requiring that timestamps diverge.

  The semantics presented in this section targets timed words with *weakly monotonic* timestamps ($t_{i+i} \geq t_i$ for all i), which can represent zero-time transitions. Other timed automata semantics – such as Alur and Dill's original semantics – use timed words with *strictly monotonic* timestamps ($t_{i+i} > t_i$ for all i), which implies that some positive time must elapse between consecutive transitions.

7.4.2   Hybrid Automata

In timed automata, time is real-valued, and clocks are real-valued variables used to keep track of the progress of time. As such, the time dynamics of clocks is very simple: when time advances by d units, every clock c advances by the same quantity – unless it is reset – obeying the differential equation $\dot{c} = 1$. "Hybrid automata" generalize timed automata by supporting real-valued variables whose timed behavior is more complicated than that of clocks. To define the dynamics of variables, hybrid automata locations include *flow conditions*, which constrain the derivatives of the real-valued variables, and *invariants*, which are sets of allowed values for the variables. Whenever a hybrid automaton sits in a location, its variables evolve over time according to the flow condition as long as their values are within the location's invariant. Upon reading input symbols, the automaton instantaneously *jumps* from one location to another, possibly setting the values of some variables according to additional transition constraints similar to timed automata resets.

Definition 7.36 (Hybrid automaton). A hybrid automaton is a tuple $\langle L, A, X, E, l_0, I, F, I_0 \rangle$, where L, A and $l_0 \in L$ are as for timed automata (see Definition 7.32).

- X is a finite set of real-valued variables.
- E is a set of edges, with $E \subseteq L \times B(X \cup X') \times A \times L$, where $B(X \cup X')$ is a set of Boolean constraints on variables in X, as well as on their primed counterparts in $X' = \{x' \mid x \in X\}$. The constraints relate variable values before the automaton jumps from the source location (unprimed variables) to variable values after the jump (primed variables); for this reason, the constraints associated with edges are called *jump constraints*.
- The function $I : L \rightarrow B(X)$ associates with each location an invariant on variables in X.
- $F : L \rightarrow B(X \cup \dot{X})$ represents *flow constraints*, which govern the evolution of the values of variables in X and of their time derivatives in $\dot{X} = \{\dot{x} \mid x \in X\}$.

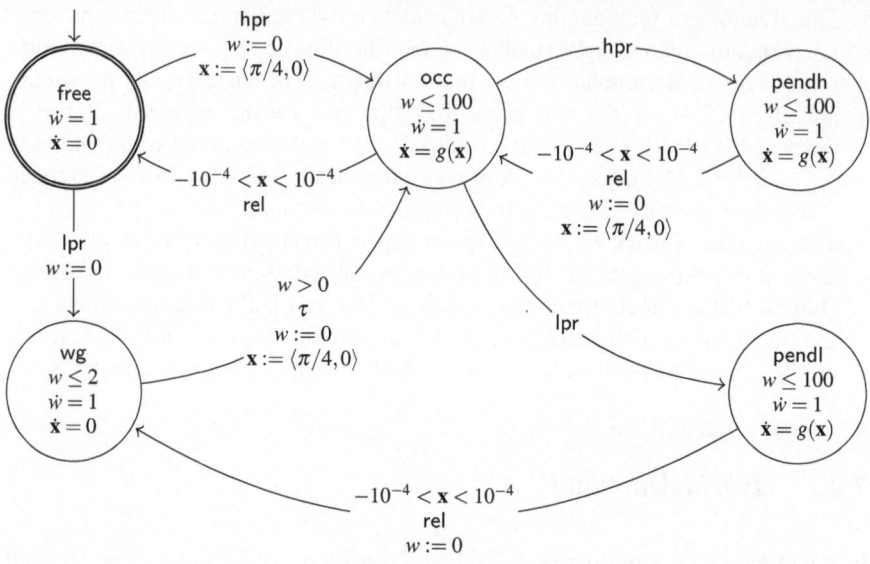

Fig. 7.32 A resource manager modeled through a hybrid automaton

- $I_0 \in B(X)$ is the *initial constraint*, which the variables in X must satisfy at the beginning of every execution. ∎

Example 7.37. Figure 7.32 shows an example of a hybrid automaton that is a variation of the timed automaton discussed in Example 7.33. w, x_1, and x_2 are the real-valued variables; for simplicity, consider x_1 and x_2 as components of the bidimensional vector $\mathbf{x}$. Each location contains the flow constraint $\dot{w} = 1$, which explicitly states that variable w grows at the same rate as time; hence w is a clock.

$\dot{\mathbf{x}} = g(\mathbf{x})$ is a nonlinear differential equation on x_1 and x_2, corresponding to the pair:

$$\dot{\mathbf{x}} = g(\mathbf{x}) = \begin{cases} \dot{x}_1 = x_2, \\ \dot{x}_2 = -\sin(x_1) - 0.4\, x_2. \end{cases}$$

When the resource is occupied (the automaton is in location occ, pendh, or pendl) x_1 oscillates around position 0 with ever decreasing arcs, as depicted in Fig. 7.33. When the oscillation range is small enough (as indicated by jump constraint $-10^{-4} < \mathbf{x} < 10^{-4}$), the automaton may switch location (and hence release the resource). The edge from location occ to location free has no variable assignments; hence we implicitly assume the jump constraint $w' = w \wedge x_1' = x_1 \wedge x_2' = x_2$: no variable changes its value when jumping. When the resource is not occupied, x_1 does not change due to flow constraint $\dot{x} = 0$. When the resource becomes occupied again, the oscillatory behavior restarts with $x_1 = \pi/4$ and $x_2 = 0$ ($\mathbf{x} := \langle \pi/4, 0 \rangle$ is shorthand for $x_1' = \pi/4 \wedge x_2' = 0$). ∎

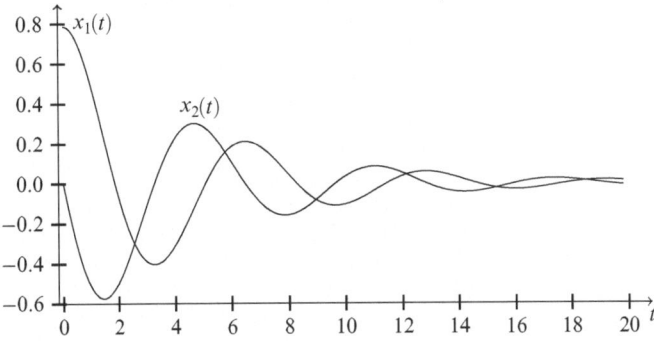

Fig. 7.33 Temporal behavior of $\dot{x}_1 = x_2$, $\dot{x}_2 = -\sin(x_1) - 0.4x_2$

As usual for the formalisms in this chapter, the semantics of hybrid automata is given in terms transition systems. The state space is $L \times V_X$, where V_X represents all possible valuations of variables in set X. For example, the automaton of Fig. 7.32 has $X = \{w, x_1, x_2\}$, and a possible valuation v is $v(w) = 0$, $v(x_1) = \pi/4$, $v(x_2) = 0$. As for timed automata (Definition 7.34), there are two kinds of transitions: actions, which cause instantaneous switches of location, and delays, which only advance time.

Definition 7.38 (Semantics of hybrid automata). The semantics of a hybrid automaton $\langle L, A, X, E, l_0, I, F, I_0 \rangle$ is given by a transition system $\langle S, S_0, J, \rightarrowtail \rangle$, where $S = L \times V_X$ – with $V_X = \{X \to \mathbb{R}\}$ the set of all possible valuations of variables X – $S_0 = \langle l_0, v_0 \rangle$ – where v_0 is a valuation that respects the initial constraint I_0 – and $J = A \cup R_{\geq 0}$. Then, $\langle \langle l, v \rangle, e, \langle l', v' \rangle \rangle \in \rightarrowtail$ if and only if one of the following holds:

(i) $e \in A$ and there is an edge $\langle l, j, e, l' \rangle$ such that v satisfies $I(l)$ (the values of the variables in valuation v satisfy the invariant of location l), v' satisfies $I(l')$, and the pair $\langle v, v' \rangle$ satisfies jump constraint j (which predicates on $X \cup X'$, and hence also on values *after* the jump);
   or
(ii) $e \in R_{\geq 0}$, $l' = l$, and there exists a differentiable function $f : [0, e] \to \mathbb{R}^{|X|}$ whose first-order derivative is $\dot{f} : (0, e) \to \mathbb{R}^{|X|}$, such that: $f(0) = v$; $f(e) = v'$; for all $0 \leq d \leq e$, $f(d)$ satisfies the invariant $I(l)$; and, for all $0 < d < e$, the pair $\langle f(d), \dot{f}(d) \rangle$ satisfies flow constraint $F(l)$. ■

Although in Definition 7.36 flow constraints are arbitrary as long as corresponding differentiable functions exist, the constraints chosen in practice are normally simpler and less general. A common choice is unions of *rectangular* constraints of the form $l \prec \mathscr{X} \prec u$, where $\mathscr{X}$ can be x, x', or $\dot{x}$, the lower bound l and the upper bound u are rational numbers or, respectively, $-\infty$ and ∞, and $\prec$ is either $<$ or $\leq$. For example, $0.5 < \dot{y} < 3$ is a rectangular constraint for a variable y (see Fig. 7.34 for a visual representation).

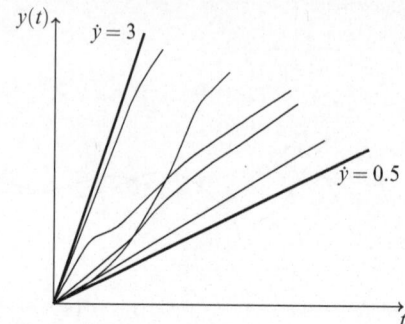

Fig. 7.34 Some behaviors compatible with the constraint $0.5 < \dot{y} < 3$

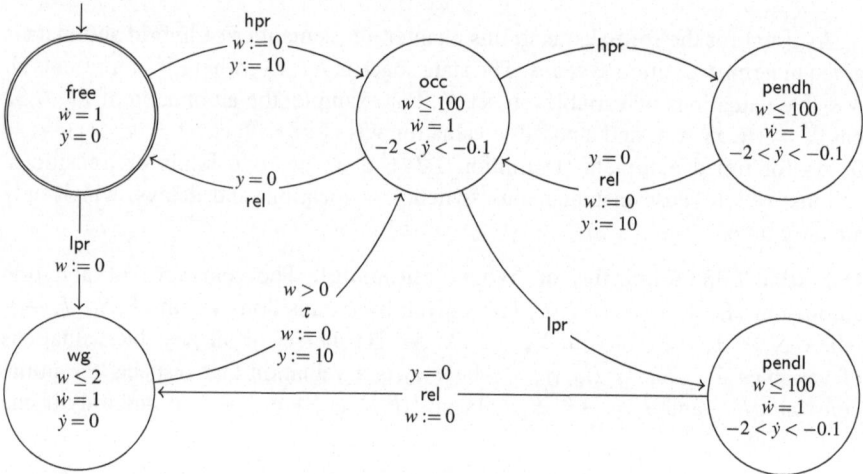

Fig. 7.35 A resource manager modeled through a rectangular automaton

Hybrid automata with only rectangular constraints are called "rectangular automata". Figure 7.35 shows a rectangular automaton with the same set of locations and variables as the hybrid automaton of Fig. 7.32.

Timed and hybrid automata support a **COMPOSITION** semantics similar to those of traditional automata and Statecharts, where concurrent components evolve in parallel, but synchronize on transitions in response to input symbols.

Part of the motivation for the development of timed and hybrid automata was the desire to extend and generalize to models with metric time the powerful automatic **VERIFICATION** techniques based on the combination of infinite-word finite-state automata and temporal logic (in particular, model checking, discussed in Chap. 11). The presence of real-valued variables renders, however, the verification problem much more difficult and, often, undecidable. In the case of hybrid automata, restrictions such as rectangular constraints trade some expressiveness for more tractable verification.

Exercise 7.39. Stopwatch automata are a simple extension of timed automata in which clocks can be stopped and restarted. Thus, each clock c can evolve with two possible derivatives: either $\dot{c} = 0$ or $\dot{c} = 1$.

1. Define the semantics of stopwatch automata based on transition systems.
2. Define a stopwatch automaton formalizing a scheduler that can manage two kinds of jobs: jobs of type A, which have higher priority, and which take up to five time units, and jobs of type B, which have lower priority, and which take up to 15 time units. The jobs are executed in slices of time of length 0.5 for those of type A, and of length 0.2 for those of type B. When both a job of type A and one of type B are under execution, they alternate (0.5 time units of job A are followed by 0.2 time units of job B, and so on, until the jobs are completed). In other words, the higher priority of type A jobs only means that they are preferred in the case of a choice and receive longer time slots, but no job can be preempted. The duration of a job of type A is not necessarily a multiple of 0.5 (similarly for jobs of type B). There cannot be more than one job of type A and one of type B executed at the same time; further requests during their execution are ignored. ∎

7.4.3 From Dense Back to Discrete Time: Digitization

The natural interpretation of timed automata uses dense-valued clocks; indeed, the use of dense time prompted the introduction of this formalism in the first place. If, however, we content ourselves with clocks that only record integer values – and correspondingly with discrete-valued timed words – timed automata essentially become a compact syntax for standard Büchi automata, with advantages in terms of reusability of simpler analysis techniques, algorithms, and tools.

Even when the application domain requires dense-time models, interpreting them over discrete-time may give approximations sufficiently precise for useful analysis and verification. The framework of *digitization* systematically addresses such problems of the correspondence between dense-timed words and discrete-timed approximations thereof. Given a timed word with dense timestamps, its ε-*digitization* is a timed word with discrete timestamps that round each dense-valued timestamp t to the previous integer $k = \lfloor t \rfloor$ if $t \leq k + \varepsilon$ and to the next integer $\lceil t \rceil$ if $t > k + \varepsilon$. For example, the ε-digitization of the following dense-timed word of the automaton of Fig. 7.31,

$$\langle \mathsf{hpr}, 4.7 \rangle \langle \mathsf{lpr}, 53.9 \rangle \langle \mathsf{rel}, 64 \rangle \langle \tau, 65.1 \rangle \ldots,$$

is the discrete-timed word

$$\langle \mathsf{hpr}, 5 \rangle \langle \mathsf{lpr}, 54 \rangle \langle \mathsf{rel}, 64 \rangle \langle \tau, 65 \rangle \ldots$$

for $\varepsilon = 0.5$, and it is the other discrete-timed word

$$\langle \mathsf{hpr}, 5 \rangle \langle \mathsf{lpr}, 54 \rangle \langle \mathsf{rel}, 64 \rangle \langle \tau, 66 \rangle \ldots$$

for $\varepsilon = 0.05$.

A *set* of timed words is *digitizable* if it contains all and only the ε-digitizations, for all $0 \leq \varepsilon < 1$, of its dense-timed words. Intuitively, the analysis of digitizable sets of words can focus on discrete-timed words only, because they represent the more complex dense-time behavior with arbitrarily precise approximation. Correspondingly, the properties of timed automata that define digitizable sets of words are essentially equivalent under the dense-time and the discrete-time semantics. While generic timed automata are not digitizable, some special classes of them, as well as other simpler operational formalisms, are; in particular, the timed-transition systems à la Manna and Pnueli – similar to the TTM of Sect. 7.3.1, but using dense time – are digitizable and correspond to a simple class of digitizable timed automata.

The equivalence between discrete- and dense-time semantics in digitizable models also allows for a form of *hybridness* (see Chap. 3), different from hybrid automata's, where heterogeneous models – conceived with discrete rather than dense time in mind – can describe different parts of a unique complex system. Chapter 9 will present a different notion of equivalence between discrete- and dense-time semantics based on the notion of *sampling*.

7.5 Probabilistic and Stochastic Automata

This section studies some of the most relevant formalisms that augment finite-state automata with a notion of probability; as usual, the focus is on time-related issues and models. Automata-based formalisms typically support stochastic behavior by replacing nondeterministic choice with probabilistic choice. A typical example is the probabilistic finite-state automata described in Chap. 6, which are finite-state automata where the next state is chosen according to a probability distribution that depends on the current state, rather nondeterministically. In the same vein, stochastic automata (discussed in Sect. 7.5.3) use clock resets randomly (colloquially, and somehow imprecisely, termed "random clocks") rather than nondeterministically as with the timed automata described in Sect. 7.4.1: the role of clock constraints, which restrain the nondeterministically chosen delays in timed automata, is assumed by distribution probabilities that shape the stochastic delays of clocks in stochastic automata.

Probabilistic and nondeterministic choices need not, however, be mutually exclusive, and in fact the two formalisms presented in this section combine nondeterministic and stochastic behaviors and provide two alternative probabilistic variants of timed automata. Namely, stochastic automata use clocks that are random variables, but choose the next location nondeterministically (as in timed automata); dually, Probabilistic Timed Automata (presented in Sect. 7.5.2) choose the next location according to probability distributions, but their clocks are reset nondeterministically (as in timed automata). As we did with the other models presented in this chapter, we first introduce (in the next Sect. 7.5.1) the notion of "stochastic transition system", which we then use to provide the semantics of both stochastic and probabilistic timed automata.

Notice that the attributes "probabilistic" and "stochastic" in the names of the formalisms presented in Sects. 7.5.2 and 7.5.3 stress the difference between the randomizations being applied to transitions, rather than time. As we discussed in Sect. 3.4.2, however, we prefer to use the two attributes as synonyms (except, of course, in the names of the notations themselves), while precisely characterizing the automata and their semantics in the text, without relying on any potentially ambiguous meanings of the terms.

7.5.1 Stochastic Transition Systems

"Stochastic transition systems" differ from the transition systems of Definition 7.1 in that their transition relations define *probability distributions* over target states, rather than just sets of target states, from a given source state on an input symbol. Given a set S of states, $Dist(S)$ denotes the set of all possible target probability distributions over S usable in definitions of stochastic transition systems.

A probability distribution μ over a *countable* set S is a function $\mu : S \rightarrow [0, 1]$ such that $\sum_{s \in S} \mu(s) = 1$; for every s in S, $\mu(s)$ is the probability of drawing s from μ. The *support* of μ is the set of elements in S that may be drawn from S, namely $support(\mu) = \{s \mid \mu(s) > 0\}$.

The definition of probability distributions over uncountable sets is more technical, because a function $\nu : U \rightarrow [0, 1]$ over an uncountable set U may be such that $\nu(u) = 0$ for all $u \in U$ – the probability of drawing each given value in U is 0 – but the cumulative probability of drawing any element in U is still 1. More precisely, a probability distribution over an uncountable set U is defined by the probabilities of drawing some element in $[u, u + \varepsilon]$ (with $u \in U$ and $\varepsilon > 0$) for every nonempty *interval* contained in U. In this case, a *probability density function* $\rho : U \rightarrow [0, 1]$ defines the probability distribution: the integral $\int_u^{u+\varepsilon} \rho(x)\mathrm{d}x$ gives the probability of drawing a value in $[u, u + \varepsilon]$. Alternatively, we can directly define a *cumulative distribution function* $R : U \rightarrow [0, 1]$ such that $R(u)$ is the probability of drawing a value less than or equal to u. The notion of support also requires generalization to handle distributions over uncountable sets U: the support is the smallest closed subset of U whose probability is cumulatively 1.

A probability distribution over any set S is then called *discrete* if either S is countable, or S is uncountable but its support is a countable set. The definitions of distribution and support used for countable sets are applicable to all discrete probability distributions (including those over uncountable sets). A probability distribution that is not discrete is called *continuous*.

Fig. 7.36 Discrete-time
Markov chain modeling an
allocator system

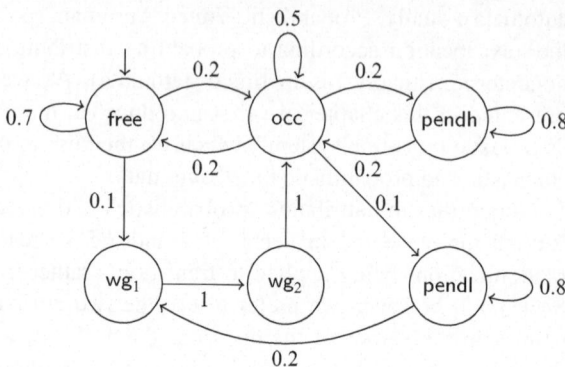

Exercise 7.40. The relation $\sum_{s \in S} \mu(s) = 1$ does not work in defining "normal-ization" in continuous probability distributions. How would you express the same condition for a continuous probability distribution with density function ρ? ∎

Definition 7.41 (Stochastic transition system). A stochastic transition system is a tuple $\langle S, s_0, I, \rightarrowtail, AP, L \rangle$ where S is a set of states, $s_0 \in S$ is the initial state, I is a set of input symbols, AP is a set of atomic propositions (labels), $L : S \rightarrow \wp(AP)$ is a labeling function that associates each state with a (possibly empty) set of labels from the set AP, and $\rightarrowtail \subseteq S \times I \times Dist(S)$ is a transition relation between the states of the system that associates a set of probability distributions over next states with every current state and input. ∎

Example 7.42 (Discrete-time Markov chain). The discrete-time Markov chain of Example 6.33 corresponds to a stochastic transition system in which $S = \{P, F\}$, there is only one input symbol $I = \{\tau\}$, and the transition relation contains only the two triples, $\langle P, \tau, \mu_P \rangle$ and $\langle F, \tau, \mu_F \rangle$, where $\mu_P(P) = 0.9$, $\mu_P(F) = 0.1$, $\mu_F(P) = 0.4$, $\mu_F(F) = 0.6$. Let us assume that P is the initial state to model the hypothesis that it is more likely to pass than to fail the first exam. ∎

Exercise 7.43. Modify the stochastic transition system of Example 7.42 to formal-ize more precisely the chances of passing the first exam. For example, how would you model the situation where there is an equal probability of passing or failing the first exam? ∎

Example 7.44 (Discrete-time Markov chain for an allocator system). Figure 7.36 shows a discrete-time Markov chain modeling a probabilistic variant of the allocator system used throughout this chapter. ∎

Exercise 7.45. Give a stochastic transition system version of the discrete-time Markov chain of Example 7.44. ∎

Example 7.46 (Markov decision process). Let us modify Examples 6.33 and 7.42 to associate different passing probabilities with different exams, and to have the probability of passing certain exams depend on which other exams have been

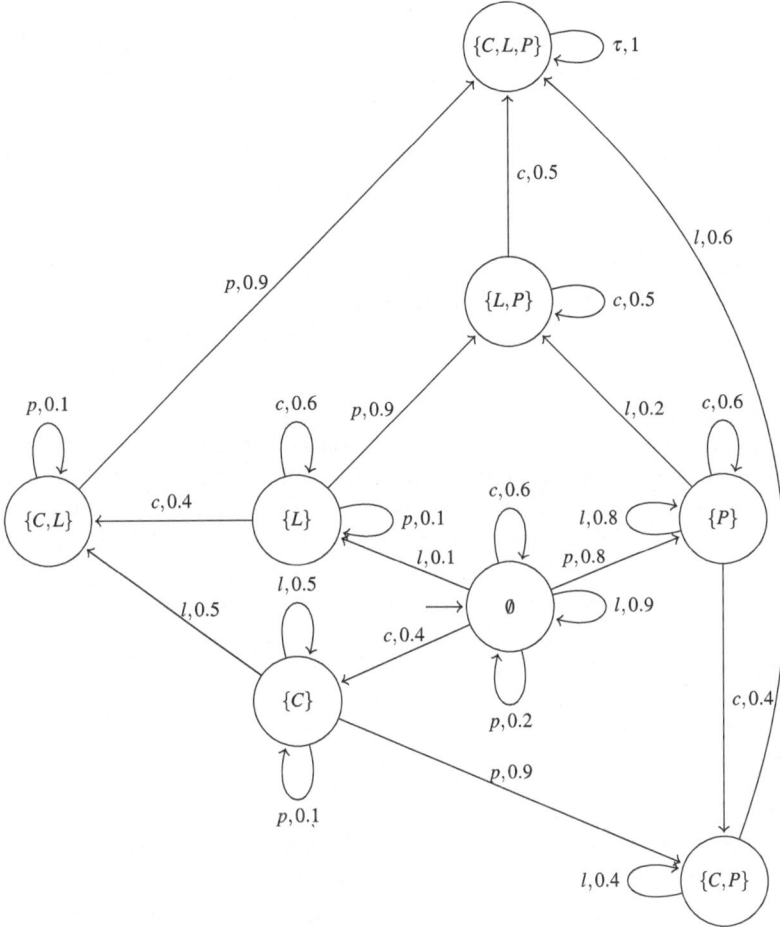

Fig. 7.37 Example of a Markov decision process as stochastic transition system

previously passed. For simplicity, let us consider only three courses: Calculus (C), Logic (L), and Programming (P). Each state corresponds to a possible combination of exams already passed; for example, state $\{P, C\}$ corresponds to having already passed the Programming and Calculus exams, but not Logic. The input symbols $\{c, l, p\}$ correspond to taking one of the exams (so a transition on input l from state $\{P, C\}$ corresponds to the Logic exam taken after having passed Programming and Calculus). Figure 7.37 shows the stochastic transition system describing some probabilities of passing/failing exams depending on the exams already passed. The stochastic transition system of Fig. 7.37 is an example of a *Markov decision process* (introduced in Sect. 6.4.1.2). ■

Exercise 7.47. With reference to Example 7.46, compute the probability of passing Logic within five attempts. Does this probability depend on the student's choices on which exams to try first? If so, what is the best strategy to maximize such a probability?   ■

Runs of stochastic transition systems are sequences of states and input symbols, where each next state is chosen from among those with nonzero probability under the chosen probability distribution.

Definition 7.48 (Run of stochastic transition systems). Given a sequence $\sigma = i_1 i_2 i_3 \ldots$ of input symbols, a *run* r_σ for a stochastic transition system $\langle S, s_0, I, \rightarrowtail,$ $AP, L\rangle$ is a sequence $s_0 i_1 s_1 i_2 s_2 i_3 \ldots$ such that, for all $k \geq 0$, there is a probability distribution μ_{k+1} such that $\langle s_k, i_{k+1}, \mu_{k+1}\rangle \in \rightarrowtail$ and $s_{k+1} \in support(\mu_{k+1})$. ■

Stochastic transition systems do not include a notion of time, though suitable ones can be encoded in them. For instance, Markov chains and Markov decision processes such as those of Examples 7.42, 7.44, and 7.46 are comparable to finite-state automata; hence, as discussed in Sect. 7.2, we can introduce a notion of time in these formalisms by simply assuming that time advances each time a transition is made. For example, under the correspondence "one transition made = one time unit = one week", the Markov decision process of Fig. 7.37 would model a student who takes one exam per week. Then, if the student adopts the strategy of taking the Logic exam each week until passed (i.e., the input sequence is $l\,l\,l\,l \ldots$ until the exam is passed), the probability of passing the Logic exam exactly after n weeks is $0.9^{n-1}\,0.1$. As mentioned in Sects. 6.4.1 and 7.2, the notion of time in this case is **DISCRETE**, **IMPLICIT**, and **METRIC**.

Exercise 7.49 (♦). (*For readers familiar with probability theory*). On the basis of the notion of **PARALLEL COMPOSITION** defined in Sect. 7.1, give a definition of parallel composition of stochastic transition systems.

(*Hint*: if p_1 and p_2 are the probabilities of events e_1 and e_2, the probability of e_1 immediately followed by e_2 is $p_1 \cdot p_2$).   ■

A number of **VERIFICATION** techniques and tools have been developed for discrete-time Markov chains and Markov decision processes. Some of these will be briefly described later in this chapter and in Chap. 11.

7.5.2 Probabilistic Timed Automata

"Probabilistic timed automata" combine the timing mechanisms of timed automata (presented in Sect. 7.4.1) with probabilistic choice: transitions to be executed are drawn at random from among those whose clock constraints are satisfied, according to a probability distribution that depends on the current state. The chosen transition determines the new location and the clocks to be reset.

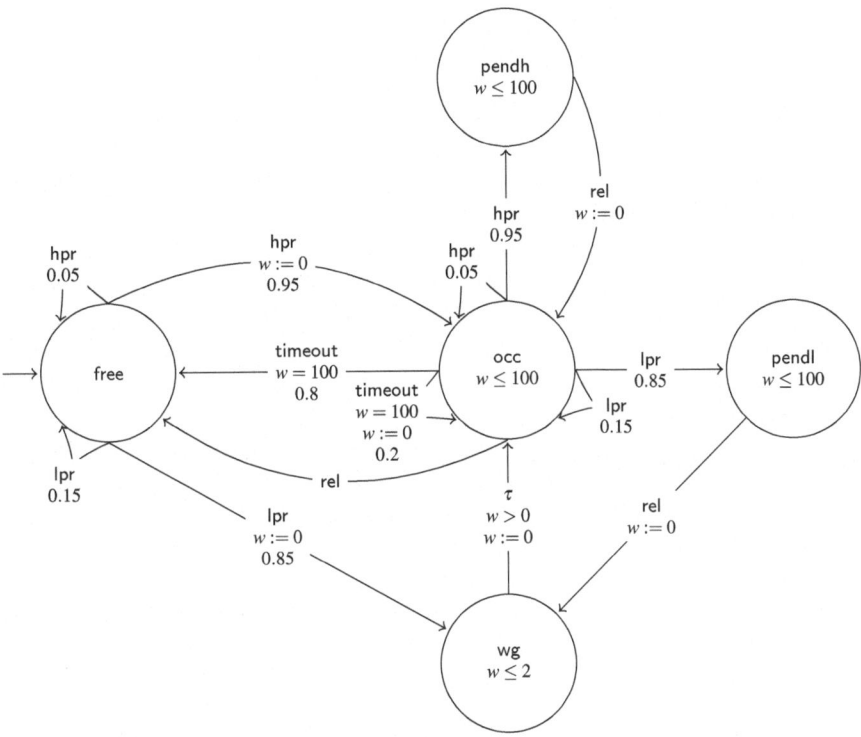

Fig. 7.38 A probabilistic timed automaton modeling a resource allocator with faulty requests

Definition 7.50 (Probabilistic timed automaton). A probabilistic timed automaton is a tuple $\langle L, A, C, E, l_0, I \rangle$, where L, A, C, l_0 and I are as in timed automata (Definition 7.32). E is a set of edges, with $E \subseteq L \times B(C) \times A \times Dist(\wp(C) \times L)$. An edge $\langle l, g, a, p \rangle$ in E denotes a *set* of transitions from location l, on action a, enabled when guard g holds: the *discrete* probability distribution p determines that a transition $\langle l, g, a, \hat{C}, l' \rangle$ to location l' which resets the clocks in $\hat{C}$ occurs with probability $p(\hat{C}, l')$. The interpretation of guards and clock resets is as in Definition 7.32. ∎

Example 7.51 (Resource allocator with probabilistic timed automata). Let us consider again the familiar resource allocator example (see, among others, Examples 7.33 and 7.37). This time, we want to model faulty communication of requests. Precisely, actions hpr or lpr may fail to produce a change of location, with higher probability for low-priority requests (action lpr) than for high-priority ones (action hpr). In addition, when the resource has been occupied for 100 time units by a high-priority request without requests pending (location occ), the allocator may grant the request another 100 time units with probability 0.2. Figure 7.38 shows a probabilistic timed automaton describing the faulty allocator. ∎

The semantics of probabilistic timed automata is given in terms of the stochastic transition systems introduced in Sect. 7.5.1. As with timed automata, there are two kinds of transitions: delays, and actions that change locations and may reset clocks. Delay transitions are as in timed automata (save with minor technicalities due to the presence of probabilistic distributions in the transition relation of stochastic transition systems); action transitions are stochastic events following the probability distribution of the edge taken.

Definition 7.52 (Semantics of probabilistic timed automata). The semantics of a probabilistic timed automaton $\langle L, A, C, E, l_0, I \rangle$ is given by a stochastic transition system $\langle S, s_0, J, \rightarrowtail \rangle$, where $S = L \times V_C$ – with $V_C = \{C \rightarrow \mathbb{R}_{\geq 0}\}$ the set of all possible valuations of clocks C – $s_0 = \langle l_0, v_0 \rangle$ – where v_0 is the valuation that assigns 0 to each clock in C – and $J = A \cup \mathbb{R}_{\geq 0}$. Then, $\langle \langle l, v \rangle, e, \mu \rangle \in \rightarrowtail$ if and only if one of the following holds:

(i) $e \in A$ and there is an edge $\langle l, g, e, p \rangle$ such that:

   (a) v satisfies the invariant $I(l)$ and the guard g;
   (b) for all pairs $\langle \hat{C}, l' \rangle$ such that $p(\hat{C}, l') > 0$, v' satisfies $I(l')$, where $v' = v[\hat{C} := 0]$ (as in Definition 7.34);
   (c) $\mu(l', v') = \sum_{\hat{C} \in Reset(v, v')} p(\hat{C}, l')$, where $Reset(v, v') = \{\hat{C} \mid v' = v[\hat{C} := 0]\}$;[6]

   or

(ii) $e \in \mathbb{R}_{\geq 0}$; for all $0 \leq d \leq e$ the valuation $v + d$ obtained by increasing all clock values in v by the quantity d satisfies $I(l)$, and $\mu = Dirac(\langle l, v + e \rangle)$, where $Dirac(q)$ is the distribution whose support is the singleton set $\{q\}$.

   This rule formalizes that a delay e is chosen nondeterministically and not stochastically; thus, after e is chosen nondeterministically, the corresponding delay transition occurs with probability 1.   ∎

Exercise 7.53. Define parallel composition of probabilistic timed automata.   ∎

Since the notion of time is the same in probabilistic timed automata as in timed automata, it retains the same characteristics discussed at the end of Sect. 7.4.1: **METRIC, CONTINUOUS, LINEAR**, and so on.

7.5.3 Stochastic Automata

In the probabilistic approaches studied in this chapter so far, probabilistic choices determine next states (in the case of Markov chains and Markov decision processes)

---

[6]$Reset(v, v')$ may contain more than one set of clocks when multiple disjoint clock resets lead to states with the same valuation v' and location l'. In particular, this may happen if valuation v already includes some clocks with value 0.

or locations (in the case of probabilistic timed automata). Time delays are in contrast nondeterministic: in Markov chains and decision processes time advances with each transition, whereas in probabilistic timed automata delays are chosen nondeterministically (within the bounds induced by guards and location invariants).

Stochastic automata, much like (probabilistic) timed automata, feature both discrete location changes and delays. When a discrete location change occurs, however, clocks are not reset to zero; instead, they are set to a value that is chosen randomly according to a distribution associated with each clock. Then, delay transitions *decrease* the values of the clocks, thus representing the fact that time is advancing. Correspondingly, stochastic automata edges do not include guards, but merely sets of clocks: a discrete transition is enabled only if all the clocks have *expired*, that is, have reached a negative or zero value. This explains why the probabilistic distributions associated with clocks play a role similar to guards in timed automata: they influence how much time must pass before a change of location is possible; the choice of "how much time must pass" is now stochastic.

Definition 7.54 (Stochastic automaton). A stochastic automaton is a tuple $\langle L, A, C, E, l_0, \kappa \rangle$, where L and A are, respectively, finite sets of locations and actions, and $l_0 \in L$ is the initial location. C is a finite set of clocks; each clock $c \in C$ is a random variable with probabilistic distribution $F_c : \mathbb{R} \to [0, 1]$, such that, for $d \in \mathbb{R}$, $F_c(d)$ is the probability of drawing a value less than or equal to d. $\kappa : L \to \wp(C)$ associates, with each location of the automaton, a set of clocks to be set (that is, for which a value must be drawn from their distributions) when the location is entered. E is a set of edges, with $E \subseteq L \times A \times \wp(C) \times L$. An edge $\langle l, a, \hat{C}, l' \rangle$ denotes a transition from location l to location l' on action a that can occur only if all the clocks in the set $\hat{C}$ have *expired* (that is, their values are less than or equal to 0). ∎

Example 7.55 (Stochastic resource allocator). Figure 7.39 shows a stochastic automaton representing yet another variant of the resource allocator example whose resource remains occupied for a random duration and whose requests arrive with random delays. The automaton has four clocks: h (or l) measures the delay between consecutive high-priority (or low-priority) requests; w_2 is the delay before a new low-priority request is granted; w_{100} represents how long the resource remains occupied. The random variables associated with clocks h and l are exponentially distributed, with rates $\lambda_h = 0.01$ and $\lambda_l = 0.001$. That is, $F_h(x) = 1 - \exp(-\lambda_h x)$, with mean value $1/\lambda_h = 100$ time units (recall that the mean value of an exponentially distributed random variable with rate λ is $1/\lambda$); similarly for clock l. Clocks w_2 and w_{100}, instead, are uniformly distributed over the intervals $[0, 2]$ and $[0, 100]$. That is, $F_{w_{100}}(x) = x/100$ if $0 \leq x \leq 100$, 0 if $x < 0$, and 1 if $x > 100$, with mean value 50; similarly for clock w_2. Notice that, in this example, the probability distributions are continuous. ∎

Definition 7.56 (Semantics of stochastic automata). The semantics of a stochastic automaton $\langle L, A, C, E, l_0, \kappa \rangle$ is given by a stochastic transition system

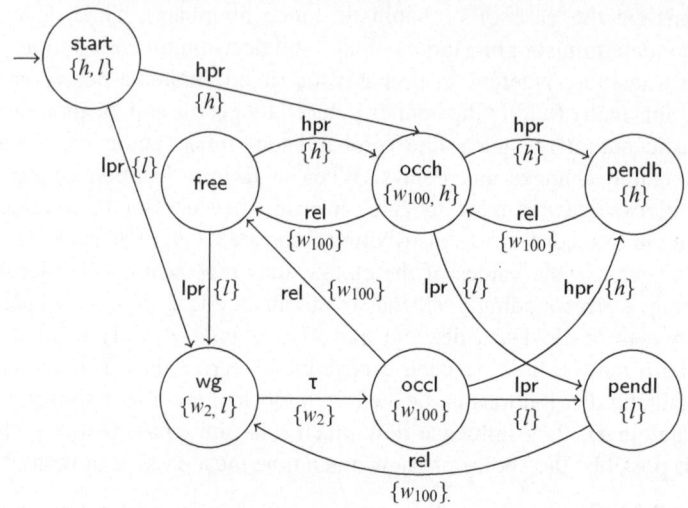

Fig. 7.39 Resource allocator with stochastic delays

$\langle S, s_0, J, \rightarrowtail \rangle$, where $S = L \cup \{l_{\text{init}}\} \times V_C$ – with $l_{\text{init}} \notin L$ and $V_C = \{C \to \mathbb{R}\}^7$ the set of all possible valuations of clocks C – and $J = A \cup \{\text{init}\} \cup \mathbb{R}_{\geq 0}$ – with init $\notin A$. The initial state $s_0 = \langle l_{\text{init}}, v_0 \rangle$ is such that v_0 is the valuation that assigns 0 to each clock in C. Then, $\langle \langle l, v \rangle, e, \mu \rangle \in \rightarrowtail$ if and only if one of the following holds:

(i) $e \in A$ and there is an edge $\langle l, e, \hat{C}, l' \rangle$ such that: (i.1) all clocks in set $\hat{C}$ are expired in valuation v; (i.2) $\mu(l'', v'') = 0$ for all $l'' \neq l'$; (i.3) $\mu(l', v') > 0$ only if v' is such that for all $c' \notin \kappa(l')$ the values of c' in valuations v and v' are the same ($v'(c') = v(c')$), and $\mu(l', v') = \prod_{c \in \kappa(l')} F_c(v'(c))$;
or

(ii) $e \in \mathbb{R}_{\geq 0}$, and (ii.1) there exists an edge $\langle l, a, \hat{C}, l' \rangle$ such that in valuation $v - e$, obtained by decreasing all clocks by e, all clocks in $\hat{C}$ have expired; (ii.2) there is no delay d with $0 \leq d < e$ and edge $\langle l, a', \hat{C}', l'' \rangle$ such that in valuation $v - d$ all clocks in $\hat{C}'$ have expired; (ii.3) $\mu = Dirac(\langle l, v - e \rangle)$;
or

(iii) $e = \text{init}$, $l = l_{\text{init}}$, $\mu(l'', v'') = 0$ for all $l'' \neq l_0$, and $\mu(l_0, v') > 0$ only if, for all $c' \notin \kappa(l_0)$, $v'(c') = 0$ and $\mu(l_0, v') = \prod_{c \in \kappa(l_0)} F_c(v'(c))$. ∎

Condition (ii.2) in Definition 7.56 implies that, as soon as an edge from the current location becomes enabled because its clocks have expired, some location transition *must* be made (if more than one edge becomes enabled, the choice among them is nondeterministic). This interpretation is called the "closed system

---

[7]Note that clock values in stochastic automata can be negative if time advances without resets.

view", because the environment cannot delay an action after the corresponding edge becomes enabled. An alternative is the "open system view", where the environment can introduce an additional delay before producing an action even after all clocks on the corresponding edge have expired; the delay represents the system waiting for the "right" input from the environment. Thus, in the "open system view", clocks count the random *minimum time* before an edge becomes enabled; after the random minimum time has passed, the choice of when to make the transition becomes nondeterministic.

We do not delve on this issue any further, and leave the formalization of the "open system view" as an exercise.

Example 7.57 (Continuous-time Markov chains as stochastic automata). Continuous-time Markov chains can be seen as a special case of stochastic automata with exponential distributions associated with clocks. In fact, as explained in Sect. 6.4.1.3, their *sojourn time* (the time spent in a state) follows a negative exponential whose parameter is the *rate* with which the state is exited. More precisely, if λ is the rate with which the system moves from state s to state s', $1 - \exp(-\lambda t)$ is the probability that the transition from s to s' occurs within t time units, provided there are no other competing transitions outgoing from s. If there are n transitions outgoing from state s, each one with rate λ_i ($i = 1, \ldots, n$), then the probability that state s is exited within t time units is $1 - \exp((\lambda_1 + \lambda_2 + \cdots + \lambda_n) t)$. This corresponds to a stochastic automaton (using the "closed system view") defined along the following lines:

1. The locations L are the same as the states of the continuous-time Markov chain;
2. Each transition from a location s to a location s' is associated with exactly one clock, whose distribution is a negative exponential with rate λ_i;
3. Whenever a location is entered, all clocks corresponding to the outgoing transitions are set according to their distributions.

Figure 7.40 shows a stochastic automaton corresponding to a continuous-time Markov chain modeling another variant of the resource allocator example, with minor differences in structure with respect to Example 7.55. In this case, the clocks (h, l, w, r) are associated with negative exponential probability distributions with rates $\lambda_h, \lambda_l, \lambda_w, \lambda_r$. ∎

Time in stochastic automata has features very similar to those found in timed automata. Under a standard timed-word semantics based on transition systems (see Sect. 7.4.1), there are two notions of time: a **DISCRETE** one given by the sequence of states in the run of the underlying stochastic transition system, and a **CONTINUOUS** one given by the timestamps. Also, the probability distributions associated with clocks introduce **METRIC** constraints on time (as in Example 7.55, where clock w_{100} cannot take values greater than 100; hence the resource must be released no later than 100 time units after it has become occupied). In the "**CLOSED** system view", time advances in a purely **RANDOM** manner, as determined by the clock dynamics; in the "**OPEN** system view", in contrast, the *minimum* time before the

Fig. 7.40 Stochastic automaton corresponding to a continuous-time Markov chain

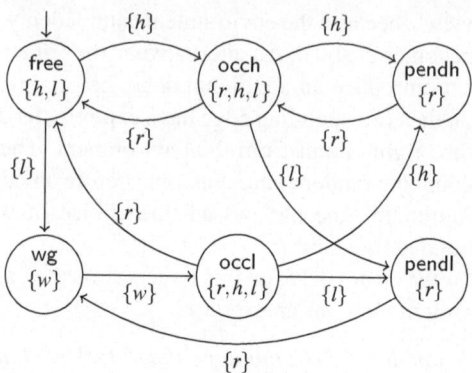

next discrete transition is random, but the overall delay until a discrete transition is made is **NONDETERMINISTIC**.

Stochastic automata can have **ZENO** runs. However, it can be shown that the probability of Zeno runs is non-null only in particular cases, such as when the probability distributions associated with the clocks have supports that are subsets of the nonpositive reals.

7.6   Methods and Tools Based on Synchronous Abstract Machines

Automata-based formalisms such as those presented in this chapter are at the basis of a wide range of tools for the design and analysis of software-based systems. Many of these tools are *model checkers*, which implement the verification techniques described in Sect. 11.1. Model checkers pursue a *dual-language* approach, in which an operational model is combined with a property expressed in a descriptive language (typically, a logic). Since model checking techniques rely on concepts introduced in Chap. 9, we leave a description of these tools to Chap. 11. Here, we briefly describe some tools that either are not, strictly speaking, model checkers, or are based on model checkers, but are used for purposes other than classic formal verification.

Statecharts have been a popular notation since their inception, even more so after their inclusion in the UML standard. As a consequence, all UML tools provide graphical editors to draw Statechart-like diagrams, but only a few support activities more complex than editing. In the last few years, especially with the advent of "model-driven approaches" to the development of complex systems, commercial tools have added capabilities such as automatic code generation and model-based testing (in some cases, also model checking). Examples of such tools are STATEMATE and Rhapsody. STATEMATE offers simulation capabilities based on the semantics described in Sect. 7.3.2, while Rhapsody additionally offers model checking and automatic test case generation capabilities. Artisan Studio is another tool that offers simulation and code generation capabilities from state diagram models.

The best known tool for building and analyzing timed automata models is UPPAAL. At its core, UPPAAL is a model checker that implements the technique described in Sect. 11.1.4. In addition, it offers simulation of timed automata, and several extensions for analysis other than pure verification. For example, UPPAAL-TRON (UPPAAL for Testing Real-time systems ONline) has the capability to drive (in real time) the testing of a system under development from timed automata-based models of its environment. The generation of the next input to be sent to the implementation under test (IUT) is done by exploring the timed automata-based models of the IUT itself and of its environment; the same models are used to check whether the responses provided by the IUT in reaction to the inputs sent are admissible or not. Another UPPAAL extension is UPPAAL-TIGA (UPPAAL for TImed GAmes), which implements algorithms to synthesize control strategies (also called "winning") from timed automata whose actions are partitioned into *controllable* and *uncontrollable*. Controllable actions are under system control; uncontrollable actions, on the other hand, are under the control of the environment. A control strategy is one such that the system achieves a chosen goal no matter what the decisions of the environment. Some examples of goals are *reachability* properties ("the system will eventually reach a desired state") or *safety* properties ("the system will never reach an undesired state"), which can be specified using temporal logics such as those introduced in Chap. 9.

Tools based on hybrid automata can typically solve reachability problems: "is a certain state reachable from an initial state?" For example, we could ask whether, in the automaton of Fig. 7.35, location **pendh** is reachable with the value of variable w equal to 90 (if it is not, then the upper bound of constraint $w \leq 100$ is never reached). The reachability problem is in general undecidable for hybrid automata; hence reachability algorithms may not terminate. Some special classes of hybrid automata are nevertheless decidable (the automaton of Fig. 7.35 belongs to one such class), and the tools are often able to provide answers even when the general case is undecidable. In addition, tools for hybrid automata are capable of parametric analysis, where a value in the model is left free, and the tool returns a value for the parameter, if this exists, such that the desired states are reached. Two well-known tools for hybrid automata analysis are HyTech (Hybrid Technology Tool) and the more recent PHAVer (Polyhedral Hybrid Automaton Verifier).

As mentioned above, tools dealing with Büchi, finite state, and probabilistic automata (such as SPIN, SMV/NuSMV, and PRISM) will be discussed, as model checkers, in Chap. 11.

7.7 Bibliographic Remarks

The concept of transition system is used to provide an underlying formal semantics to many different notations. Often, transition systems are used as the mathematical models on which model checking algorithms (see Chap. 11) are defined; Baier and Katoen [11] provide a comprehensive overview of different kinds of transition systems, and of their composition.

The Promela language [32], used in the SPIN model checker to describe concurrent processes, is a typical example of a language that supports coordination both through channels (which, in the case of Promela, can also carry data and have buffers) and through shared variables (in this case by interleaving access to common variables). Synchronous concurrency is pursued in several languages that constitute extensions and specializations of the basic infinite-word finite-state automaton, such as Esterel [14].

When coordination between components is fully synchronous, and there is little flexibility in the treatment of different speeds in the evolution of separate components of the system, it is rather difficult to achieve the technical requirement sometimes called compositionality [2, 5].

Timed Transition Models (TTMs) were introduced by Ostroff [40] as an extension of Manna and Pnueli's fair transition systems [37, 38]. The notation used in this book to represent them differs from the original one, for consistency with those of the other automata formalisms described, but the differences are limited to syntactic sugar.

Statecharts were introduced by Harel [22]. Their syntax has changed over the years, also due to their inclusion in the UML standard. The examples of Sect. 7.3.2 use Harel's original notation [22]. The UML variant of Statecharts, called "State Diagrams", is a standard of the Object Management Group [46]. The STATEMATE semantics is described by Harel and Naamad [23]. A number of articles deal with the semantic subtleties in the definitions of Statecharts [19,24,44,50]. Timing issues in Statecharts were investigated by Kesten and Pnueli [33] and by Peron [42].

In the literature, there are a number of automata-based formalisms augmented with a notion of time under the name of "timed automata" [7,8,36]. Those presented in this chapter were first proposed by Alur and Dill [1]. The original formalization of Alur and Dill permits constraints only on transitions; however, adding constraints to locations as well, as in the definition provided in this chapter, is a standard extension that does not impact on the salient features of the model (expressiveness, in particular). The definition of timed automata used in this book is that of the popular UPPAAL formal verification tool [13,49], which uses weakly monotonic sequences of timestamps, instead of the strongly monotonic ones of Alur and Dill's original definition. Other, different semantics of timed automata have been introduced and analyzed in the literature. Subtle differences may arise depending on which semantics is adopted; for instance, some interval-based "signal" semantics interpret timed automata over piecewise-constant functions of time, whose discontinuities trigger location changes [5, 9, 10].

Hybrid automata have been defined and studied by various authors [3, 26, 39]; later work has targeted decidability issues and verification techniques [4, 6, 29, 31]. In particular, it is known that the maximal class of hybrid automata for which the reachability problem is decidable is that of rectangular automata in which every time a flow condition on a variable changes due to a jump to a new location, the value of the variable is reset [31]. Stopwatch automata are a special class of hybrid automata that are strictly more expressive than timed automata [16, 31].

Henzinger, Manna, and Pnueli introduced the digitization framework [28], and showed that timed transition systems à la Manna and Pnueli [27] are digitizable. Ouaknine and Worrell [41] studied in detail the applicability of digitization techniques to timed automata.

The notion of "stochastic transition systems" presented in this chapter has been adapted from similar concepts defined by various authors [17, 18, 45]. Kwiatkowska et al. introduced probabilistic timed automata [34], and D'Argenio and Katoen defined stochastic automata [18]. For simplicity, the definition of stochastic automata introduced in this chapter restricts the state space, the set of actions, and the set of clocks in a stochastic automaton to be finite, though D'Argenio and Katoen's original definition [18] makes them countable. Baier et al. proved [12] that Zeno runs in continuous-time Markov chains have probability 0.

A few articles describe STATEMATE and other tools for Statecharts [15, 21, 25]. The UPPAAL model checker [35, 49] includes extensions for testing real-time systems [47] and for solving timed games [48]. Henzinger et al. describe the HyTech verification tool [30]. Frehse developed PHAVer [20, 43].

References

1. Alur, R., Dill, D.L.: A theory of timed automata. Theor. Comput. Sci. **126**(2), 183–235 (1994)
2. Alur, R., Henzinger, T.A.: Logics and models of real time: a survey. In: Huizing, C., de Bakker, J.W., Rozenberg, G., de Roever, W.-P. (eds.) Real Time: Theory in Practice. Lecture Notes in Computer Science, vol. 600, pp. 74–106. Springer, Heidelberg (1992)
3. Alur, R., Courcoubetis, C., Henzinger, T.A., Ho, P.H.: Hybrid automata: an algorithmic approach to the specification and verification of hybrid systems. In: Grossman, R.L., Nerode, A., Ravn, A.P., Rischel, H. (eds.) Hybrid Systems. Lecture Notes in Computer Science, vol. 736, pp. 209–229. Springer, Berlin/Heidelberg (1993)
4. Alur, R., Courcoubetis, C., Halbwachs, N., Henzinger, T.A., Ho, P.H., Nicollin, X., Olivero, A., Sifakis, J., Yovine, S.: The algorithmic analysis of hybrid systems. Theor. Comput. Sci. **138**, 3–34 (1995)
5. Alur, R., Feder, T., Henzinger, T.A.: The benefits of relaxing punctuality. J. ACM **43**(1), 116–146 (1996)
6. Alur, R., Madhusudan, P.: Decision problems for timed automata: a survey. In: Bernardo, M., Corradini, F. (eds.) Revised Lectures from the International School on Formal Methods for the Design of Computer, Communication and Software Systems: Formal Methods for the Design of Real-Time Systems (SFM-RT'04). Lecture Notes in Computer Science, vol. 3185, pp. 1–24. Springer, Berlin (2004)
7. Archer, M.: TAME: using PVS strategies for special-purpose theorem proving. Ann. Math. Artif. Intell. **29**(1–4), 139–181 (2000)
8. Archer, M., Heitmeyer, C.L.: Mechanical verification of timed automata: a case study. In: Proceedings of the IEEE Real Time Technology and Applications Symposium, Brookline, pp. 192–203 (1996)
9. Asarin, E.: Challenges in timed languages: from applied theory to basic theory. Bull. EATCS **83**, 106–120 (2004). (Concurrency Column)
10. Asarin, E., Caspi, P., Maler, O.: Timed regular expressions. J. ACM **49**(2), 172–206 (2002)

11. Baier, C., Katoen, J.P.: Principles of Model Checking. MIT, Cambridge (2008)
12. Baier, C., Haverkort, B., Hermanns, H., Katoen, J.P.: Model-checking algorithms for continuous-time Markov chains. Softw. Eng. IEEE Trans. **29**(6), 524–541 (2003)
13. Bengtsson, J., Yi, W.: Timed automata: semantics, algorithms and tools. In: Desel, J., Reisig, W., Rozenberg, G. (eds.) Lectures on Concurrency and Petri Nets, Advances in Petri Nets (from ACPN'03). Lecture Notes in Computer Science, vol. 3098, pp. 87–124. Springer, Berlin (2004)
14. Berry, G., Gonthier, G.: The esterel synchronous programming language: design, semantics, implementation. Sci. Comput. Program. **19**(2), 87–152 (1992)
15. Bienmüller, T., Damm, W., Wittke, H.: The STATEMATE verification environment—making it real. In: Proceedings of the 12th International Conference on Computer Aided Verification (CAV'00). Lecture Notes in Computer Science, vol. 1855, pp. 561–567. Springer, Berlin (2000)
16. Cassez, F., Larsen, K.G.: The impressive power of stopwatches. In: Proceedings of the 11th International Conference on Concurrency Theory, CONCUR '00, University Park, pp. 138–152 (2000)
17. Cattani, S., Segala, R., Kwiatkowska, M., Norman, G.: Stochastic transition systems for continuous state spaces and non-determinism. In: Sassone, V. (ed.) Foundations of Software Science and Computational Structures. Lecture Notes in Computer Science, vol. 3441, pp. 125–139. Springer, Berlin/Heidelberg (2005)
18. D'Argenio, P.R., Katoen, J.P.: A theory of stochastic systems (parts I and II). Inf. Comput. **203**(1), 1–74 (2005)
19. Eshuis, R.: Reconciling Statechart semantics. Sci. Comput. Program. **74**, 65–99 (2009)
20. Frehse, G.: PHAVer: algorithmic verification of hybrid systems past HyTech. In: Proceedings 8th International Workshop, Hybrid Systems: Computation and Control, HSCC 2005, Zurich, March 9–11, 2005. Lecture Notes in Computer Science, vol. 3414, pp. 258–273 (2005)
21. Giese, H., Tichy, M., Burmester, S., Flake, S.: Towards the compositional verification of real-time UML designs. In: Proceedings of ESEC/SIGSOFT FSE 2003, Helsinki, pp. 38–47 (2003)
22. Harel, D.: Statecharts: a visual formalism for complex systems. Sci. Comput. Program. **8**(3), 231–274 (1987)
23. Harel, D., Naamad, A.: The STATEMATE semantics of Statecharts. ACM Trans. Softw. Eng. Methodol. **5**(4), 293–333 (1996)
24. Harel, D., Pnueli, A., Schmidt, J.P., Sherman, R.: On the formal semantics of Statecharts. In: Proceedings of the 2nd IEEE Symposium on Logic in Computer Science (LICS'87), Ithaca, pp. 54–64 (1987)
25. Harel, D., Lachover, H., Naamad, A., Pnueli, A., Politi, M., Sherman, R., Shtull-Trauring, A., Trakhtenbrot, M.B.: STATEMATE: a working environment for the development of complex reactive systems. IEEE Trans. Softw. Eng. **16**(4), 403–414 (1990)
26. Henzinger, T.A.: The theory of hybrid automata. In: Proceedings of the 11th Annual Symposium on Logic in Computer Science (LICS), pp. 278–292. IEEE Computer Society Press, Los Alamitos (1996)
27. Henzinger, T., Manna, Z., Pnueli, A.: Timed transition systems. In: de Bakker, J., Huizing, C., de Roever, W., Rozenberg, G. (eds.) Real-Time: Theory in Practice. Lecture Notes in Computer Science, vol. 600, pp. 226–251. Springer, Berlin/New York (1992)
28. Henzinger, T.A., Manna, Z., Pnueli, A.: What good are digital clocks? In: Kuich, W. (ed.) Proceedings of the 19th International Colloquium on Automata, Languages and Programming (ICALP'92). Lecture Notes in Computer Science, vol. 623, pp. 545–558. Springer, Berlin/Heidelberg (1992)
29. Henzinger, T.A., Nicollin, X., Sifakis, J., Yovine, S.: Symbolic model checking for real-time systems. Inf. Comput. **111**(2), 193–244 (1994)
30. Henzinger, T.A., Ho, P.H., Wong-Toi, H.: HYTECH: a model checker for hybrid systems. Int. J. Softw. Tools Technol. Transf. **1**(1–2), 110–122 (1997)
31. Henzinger, T.A., Kopke, P.W., Puri, A., Varaiya, P.: What's decidable about hybrid automata? J. Comput. Syst. Sci. **57**(1), 94–124 (1998)
32. Holzmann, G.J.: The SPIN Model Checker: Primer and Reference Manual. Addison-Wesley, Reading/Harlow (2003)

33. Kesten, Y., Pnueli, A.: Timed and hybrid Statecharts and their textual representation. In: Vytopil, J. (ed.) Proceedings of the 2nd International Symposium on Formal Techniques in Real-Time and Fault-Tolerant Systems (FTRTFT'92). Lecture Notes in Computer Science, vol. 571, pp. 591–620. Springer, Nijmegen (1992)

34. Kwiatkowska, M., Norman, G., Parker, D., Sproston, J.: Verification of real-time probabilistic systems. In: Merz, S. Navet, N. (eds.) Modeling and Verification of Real-Time Systems: Formalisms and Software Tools, pp. 249–288. Wiley, Hoboken (2008)

35. Larsen, K.G., Pettersson, P., Yi, W.: UPPAAL in a nutshell. Int. J. Softw. Tools Technol. Transf. **1**(1–2), 134–152 (1997)

36. Lynch, N., Vaandrager, F.W.: Forward and backward simulations–part II: timing-based systems. Inf. Comput. **128**(1), 1–25 (1996)

37. Manna, Z., Pnueli, A.: How to cook a temporal proof system for your pet language. In: Conference Record of the 10th Annual ACM Symposium on Principles of Programming Languages, Austin, pp. 141–154 (1983)

38. Manna, Z., Pnueli, A.: The Temporal Logic of Reactive and Concurrent Systems: Specification. Springer, New York (1992)

39. Nicollin, X., Olivero, A., Sifakis, J., Yovine, S.: Hybrid automata: an algorithmic approach to the specification and verification of hybrid systems. In: Grossman, R.L., Nerode, A., Ravn, A.P., Rischel, H. (eds.) Hybrid Systems. Lecture Notes in Computer Science, vol. 736, pp. 149–178. Springer, Berlin/Heidelberg (1993)

40. Ostroff, J.S.: Deciding properties of timed transition models. IEEE Trans. Parallel Distrib. Syst. **1**(2), 170–183 (1990)

41. Ouaknine, J., Worrell, J.: Revisiting digitization, robustness, and decidability for timed automata. In: Proceedings of the 18th IEEE Symposium on Logic in Computer Science (LICS 2003), Ottawa, 22–25 June 2003, pp. 198–207. IEEE Computer Society (2003)

42. Peron, A.: Synchronous and asynchronous models for Statecharts. Tech. Rep. TD-21/93, Dipartimento di Informatica, Università di Pisa (1993)

43. PHAVer: polyhedral hybrid automaton verifier. http://www-verimag.imag.fr/~frehse/phaver_web/

44. Pnueli, A., Shalev, M.: What is in a step: on the semantics of Statecharts. In: Ito, T., Meyer, A.R. (eds.) Proceedings of the International Conference on Theoretical Aspects of Computer Software (TACS'91). Lecture Notes in Computer Science, vol. 526, pp. 244–264. Springer, Berlin/Heidelberg (1991)

45. Segala, R.: Modeling and verification of randomized distributed real-time systems. Ph.D. thesis, MIT (1995)

46. UML 2.0. http://www.omg.org/spec/UML/2.0/ (2005)

47. UPPAAL for testing real-time systems online. http://people.cs.aau.dk/~marius/tron/

48. UPPAAL for timed games. http://people.cs.aau.dk/~adavid/tiga/

49. The UPPAAL model checker. http://www.uppaal.org

50. von der Beeck, M.: A comparison of Statecharts variants. In: Proceedings of the 3rd International Symposium on Formal Techniques in Real-Time and Fault-Tolerant Systems, Lübeck. Lecture Notes in Computer Science, vol. 863, pp. 128–148. Springer (1994)

Chapter 8
Asynchronous Abstract Machines: Petri Nets

This chapter presents Petri nets as one of the most significant examples of asynchronous abstract machines.

Petri nets are named after their inventor, Carl Adam Petri. Since their introduction, they have become rather popular both in the academic and, to some extent, in the industrial world as a fairly intuitive graphical notation for modeling concurrent systems. In particular, they inspired the activity diagrams of the UML standard.

There are several variants of Petri nets and their semantics. This chapter first introduces the basic model, which is the most widely used one in the literature and in applications. Then, it presents a few of its significant extensions, with the usual focus on modeling of timing aspects.

8.1 Basic Petri Nets

We introduce Petri nets with an example of the typical problem of managing a resource shared by multiple processes. With minor differences with respect to the many examples of this type considered in Chap. 7, we consider a system composed of two processes P_1 and P_2, (with equal priority) competing for the use of a shared resource, which is managed by an allocator A. We first model the system by means of the operational formalisms discussed in Chap. 7; a critical analysis of the result will motivate the introduction of Petri nets.

Precisely, we model process P_1 by the simple finite-state automaton in Fig. 8.1a, with states id_1, wg_1, and hl_1 representing, respectively, the situation where P_1 is idle, is waiting for the resource, and is holding it; and with transitions labeled req_1, gr_1, and rel_1 representing the actions of requesting the resource, of being granted it by the allocator, and of releasing it. Process P_2 is modeled similarly, as shown in Fig. 8.1c. The allocator A (Fig. 8.1b) has states free, to_1, and to_2 representing the resource being available, begin assigned to P_1, and being assigned to P_2, and transitions labeled gr_1, gr_2, rel_1, and rel_2, representing the same actions as those of P_1's and P_2's transitions.

C.A. Furia et al., *Modeling Time in Computing*, Monographs in Theoretical Computer Science. An EATCS Series, DOI 10.1007/978-3-642-32332-4_8,
© Springer-Verlag Berlin Heidelberg 2012

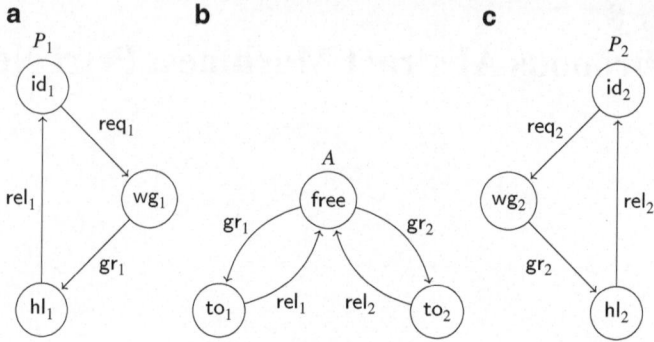

Fig. 8.1 Two processes P_1 (**a**) and P_2 (**c**) competing for a resource managed by allocator A (**b**)

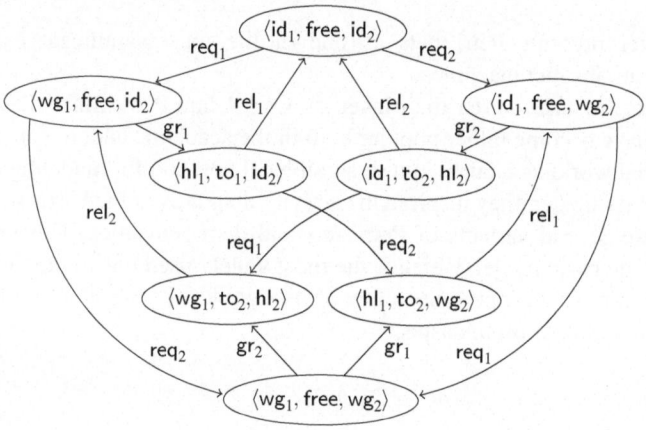

Fig. 8.2 Parallel composition of the automata in Fig. 8.1

A shortcoming of this system description by means of three separated automata is that it does not show explicitly the overall configurations of the composed system and the interactions occurring among the system components. The customary way to compose finite state automata that model interacting subsystems is by constructing the *parallel composition* of such automata: we skip a few intermediate steps of the construction – which are now obvious from Chap. 7 – and show the result in Fig. 8.2.

The model in Fig. 8.2 contains, in a single automaton, all possible overall system states and the possible transitions between them. Still, it does not explicitly convey the fundamental notion that the two processes P_1 and P_2 can evolve in parallel and synchronize, through the allocator, at suitable time instants and when the proper conditions occur, to get or release the resource.

Exercise 8.1. The automata in Fig. 8.1 and their composition in Fig. 8.2 are simple but enough to motivate the introduction of a different model based on Petri nets.

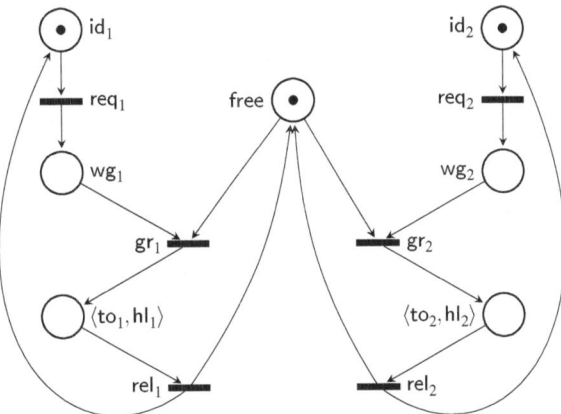

Fig. 8.3 A simple Petri net modeling the processes in Fig. 8.1

However, still using the automata-based notations introduced in Chap. 7, we could aim at a more sophisticated system description, for instance, by explicitly modeling that the allocator *receive* requests req_1 and req_2 from P_1 and P_2 and *grant* the resource to either one of them, so that req_1 is an output of P_1 and an input to A, and conversely the grant action gr_1 is an output of A and an input to P_1, and so on.

1. Model such a more detailed description by means of Statecharts.
2. Argue that the gist of the criticism about Figs. 8.1 and 8.2 still applies to the more sophisticated Statechart model (or to any similar model).   ∎

Petri nets support the notion of distributed state combined with synchronization through shared transitions; this makes for compact and elegant models of **DISTRIBUTED, ASYNCHRONOUS** systems. The allocator system mentioned above can be described by the Petri net in Fig. 8.3.

In the Petri net, the presence of a "token" (a small black disk) in a state (called "place" in Petri net terminology) indicates that the component process is currently in that state, and hence ready to perform the action represented by the "transition" connected to the state (explicitly depicted by a thick bar). When an action is shared (see the gr_1 and gr_2 transitions), it can be carried out only if all participating processes are ready to execute it, that is, they are in the state connected to the transition (by means of a directed arrow from the corresponding place to the transition).

The Petri net in Fig. 8.3 adequately models the system composed of the two processes P_1 and P_2 plus the allocator A for managing a single resource. The same Petri net can model a variety of different, possibly more complex, situations just by varying the initial placement of tokens. In the following we introduce more formally the Petri net notation by referring to a net with the same topology (in terms of places, transitions and connections among them) as that of Fig. 8.3, but with generic names and no initial token placement to make the discussion independent of the particular system considered in the above introductory example. Correspondingly, we will refer to Figs. 8.4 and 8.6 instead of Fig. 8.3.

Fig. 8.4 A basic Petri net

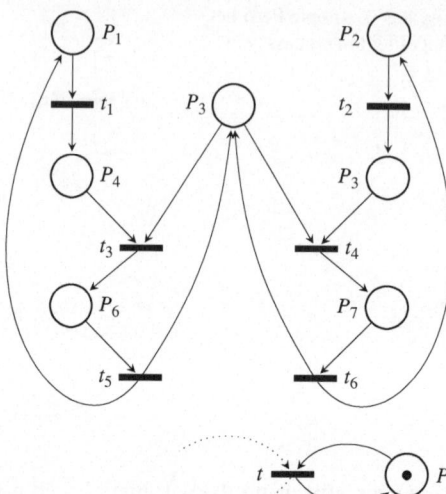

Fig. 8.5 A Petri net fragment

Definition 8.2 (Basic Petri net). A *Petri net* is a triple $\langle P, T, A \rangle$ where P is a finite set of *places*, T a finite set of *transitions*, and A a set of *arcs* connecting places to transitions and transitions to places. The customary graphic representation of a Petri net represents places as circles, transitions as thick line segments, and arcs as directed arrows.

The *preset* of a transition t (or of a place p) is the set of places (or transitions) directly connected to t (or p) by an arc. The *postsets* of transitions and places are defined similarly.   ■

Example 8.3. Figure 8.4 shows an example of Petri net with six transitions, $t_1, t_2, \ldots, t_6$, seven places $P_1, P_2, \ldots, P_7$, and sixteen arcs. The preset of transition t_4 is $\{P_3, P_5\}$; the postset of place P_7 is $\{t_6\}$.   ■

The behavior of a Petri net is defined by the location changes of *tokens* in the places, occurring as a consequence of the firing of transitions. Places can store one or more *tokens*; a place is *marked* when it contains at least one token. The number of tokens stored in each place defines the state of the net, also called *marking*.

When all the places in the preset of a transition are marked, the transition is *enabled*. An enabled transition can *fire*; the firing removes a token from each place in the transition's preset and deposits one into each place in the transition's postset. If more than one transition is enabled in a given marking, the choice of which transition will fire is **NONDETERMINISTIC**. Notice that a place may be both in the preset and in the postset of the same transition t, such as place P in the net fragment of Fig. 8.5. When a transition such as t fires, a token in P is consumed and replaced by a new one immediately.

Example 8.4 (Sequence of markings). Figure 8.6 shows a marking of the net of Fig. 8.4 where both t_1 and t_2 are enabled and hence can fire. Suppose t_1 fires: P_1 becomes empty and P_4 becomes marked; hence both t_3 and t_2 are enabled after the

Fig. 8.6 A marking of the
Petri net of Fig. 8.4

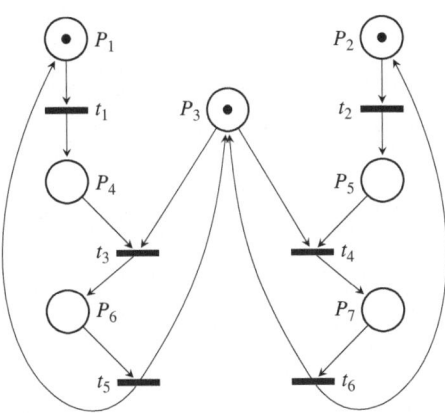

firing of t_1. Let t_2 fire next, leaving exactly one token in each of P_3, P_4, and P_5 and enabling t_3 and t_4. If t_3 fires, P_3 and P_4 become empty and P_6 becomes marked. The removal of the token in P_3 also disables t_4, which was enabled before t_3 fired. ∎

Example 8.5 (Concurrent processes with Petri nets). Petri nets are a natural model of concurrent activities: the markings of the places represent the availability of the resources; transitions model actions that can take place only if certain resources are available, and whose effect may make resources available and thus enable transitions. For example, by generalizing the introductory remarks that led to Fig. 8.3, the net in Fig. 8.4 may model two generic concurrent interacting processes. Unlike the actions corresponding to the other transitions t_1, t_2, t_5, and t_6, which can proceed **CONCURRENTLY**, if t_3 and t_4 happen to be enabled at the same time, the shared resource corresponding to P_3 allows only one process to proceed while the other has to wait for the resource to become available again; in this case, we say that the two transitions are "in conflict". ∎

A *firing sequence* originating in a marking m_0 is a sequence of transitions $t_{i_1} t_{i_2} \ldots t_{i_n}$ such that: m_0 enables t_{i_1}, the firing of t_{i_1} produces a marking m_1 that enables t_{i_2}, and so on. A marking m is *reachable* from m_0 if there exists a firing sequence originating in m_0 which produces the marking m.

Exercise 8.6.

- Describe all firing sequences of length 5 originating in the marking of Fig. 8.6.
- Describe all firing sequences of any length originating in the marking of the net in Fig. 8.4 where P_1 stores n tokens and all other places are empty, for $n > 0$.
- Find a marking m of the net in Fig. 8.4 such that exactly one infinite and periodic firing sequence originates in m. ∎

The firing of a transition with a preset larger than the postset (such as t_3 in Fig. 8.6) decreases the total number of tokens in the net, whereas the firing of a transition with a postset larger than the preset (such as t_5 in Fig. 8.6) increases it.

These simple observations entail that the total number of tokens in a Petri net may increase, possibly in an unbounded manner, during the evolution of the net. The definition of *bounded* Petri net is given accordingly.

Definition 8.7 (Bounded Petri net).   A Petri net with a given initial marking m_0 is:

- *k-bounded* if no place stores more than k tokens in any marking reachable from m_0.
- *Bounded* if it is k-bounded for *some* k. It is *unbounded* if it is not k-bounded for any k.   ■

Exercise 8.8.

- Show that the Petri net with marking in Fig. 8.6 is 1-bounded.
- Can you find an initial marking that makes the net unbounded? If not, why not?
- Modify the net fragment in Fig. 8.5 to make it unbounded (for the given initial marking).   ■

Let us now evaluate the essential features of the Petri net formalism with the help of the taxonomy introduced in Chap. 3.

Implicit time model.   Petri nets support no **EXPLICIT** notion of time, but a time model can implicitly be associated with the semantics of the net. More precisely, there are two major approaches to formalizing the semantics of Petri nets.

The simpler semantics is based on *interleaving*, where the behavior of a net consists of all its firing sequences. The interleaving semantics introduces a **TOTAL ORDERING** of the events modeled by the firing of transitions that may fail to adequately capture the asynchronous nature of the notation. For example, consider the transitions t_1 and t_2 in Fig. 8.6. The structure of the net is such that the firing of t_1 and t_2 can occur asynchronously or simultaneously. Any firing sequence, however, introduces a forced ordering between the firings of the two transitions that is immaterial; consider, for example, the two sequences $t_1\, t_2\, t_3\, t_5$ and $t_2\, t_1\, t_3\, t_5$ which interleave t_1 and t_2.

Given the limitations of interleaving semantics, a *true concurrency* (fully **ASYN-CHRONOUS**) approach is often preferred to formalizing the semantics of Petri nets. This approach models time as a **PARTIAL ORDER**, as opposed to the total order of the events modeled by transition firings. For example, the two firing sequences $t_1\, t_2\, t_3\, t_5$ and $t_2\, t_1\, t_3\, t_5$ are coalesced into a true concurrency description $\{t_1, t_2\}\, t_3\, t_5$, where the first item $\{t_1, t_2\}$ denotes that t_1 and t_2 are not ordered with respect to each other.

Nondeterminism.   We already remarked that Petri nets are **NONDETERMINIS-TIC** operational models. There are two sources of nondeterminism in Petri nets. The first is the nondeterministic choice from among concurrent but independent transitions, such as t_1 and t_2, both enabled in Fig. 8.6. The true concurrency semantics focuses on capturing such a source of nondeterminism. The second form of nondeterministic choice is responsible for selecting from among transitions that are in conflict because they share enabling places, such as t_3 and t_4, which share P_3 in Fig. 8.4.

Metric.   The time model of basic Petri nets has no **METRICS**; hence behaviors express only (possibly partial) order among firing events. The convention of associating one time unit with each firing event – often adopted to introduce a metric in synchronous operational models, as thoroughly exploited in Chap. 7 for automata-based formalisms – is unsuitable for Petri nets, as it contrasts sharply with the asynchronous nature of the notation.

Composition.   Achieving a workable notion of **COMPOSITION** is usually more involved for Petri nets than for most synchronous operational formalisms – automata in particular. The inherently asynchronous and nondeterministic nature of Petri nets introduces a nonlocality in the behavior that multiplies the possible interactions between composed nets. This combinatorial explosion – inherent in the composition of all operational formalisms but especially prominent in Petri nets – renders reasoning about the global behavior often unintuitive, difficult, and expensive.

Assume, for example, that two nets synchronize on a transition t: t fires in the composed net only when it would fire simultaneously on both component nets. This notion of composition is similar to that of the synchronous composition of automata and Statecharts discussed in Chap. 7. When composing automata, however, the current state of each sub-automaton determines precisely the transitions enabled. In Petri nets, in contrast, the state of each component is defined by the tokens "scattered" among all the places; hence, a transition firing asynchronously anywhere in the net may move tokens around so as to enable or disable potentially any synchronized transition.

Several notions of **COMPOSITION** of Petri nets have been proposed to mitigate these problems and to facilitate the modeling of structured systems. Section 8.6 discusses a few significant approaches along these lines.

Verification and complexity.   Basic Petri nets do not achieve the full computational power of Turing machines. Still, the algorithmic analysis of many interesting properties is intractable or even undecidable. For example, *reachability* – the problem of determining whether a given marking is reachable from an initial marking – is decidable, but even the best algorithms require exponential space, because the problem is EXPSPACE-complete (see Sect. 6.2). Another decidable problem is *boundedness*: determining whether a Petri net is bounded; it also requires exponential space. On the other hand, it is undecidable whether two Petri nets are equivalent, and whether a generic temporal logic property holds for the behavior of a net (the *model checking* problem; see Chap. 11).

The feature of the Petri net model responsible for the high complexity of analysis is the possibility for places to store an unbounded number of tokens in the general case. Correspondingly, the complexity is significantly reduced if we consider only *bounded* nets. In this case, Petri nets are essentially equivalent to succinct descriptions of finite state automata; Exercise 8.14 asks you to prove this fact. For bounded nets, the problems of determining whether a marking is reachable and whether a temporal logic property holds for the behavior of the net are PSPACE-complete. Equivalence is also decidable, with different complexities depending on which notion of equivalence is considered; most notions of equivalence determine EXPTIME-complete or even EXPSPACE-complete problems.

Exercise 8.9. Consider the net of Fig. 8.6 with an initial marking where P_3 stores two tokens instead of one. Does the set of possible firing sequences change? Describe the change in terms of both the interleaving semantics and the true concurrency model. On the basis of your analysis comment on how the change affects the properties, discussed in Example 8.5, of the system modeled by the net.
■

Exercise 8.10 (♣). Compare the various notions of composition of synchronous machines discussed in Chap. 7 with the notion of process composition underlying Petri nets, with interleaving and with true-concurrency semantics.   ■

Example 8.11 (The dining philosophers). The *dining philosophers* problem is a famous abstract model of some recurring problems in the description and analysis of concurrent processes. The problem was originally introduced by Edsger Dijkstra in 1965 as a synchronization problem for five computers accessing shared peripherals. The problem was soon retold by Tony Hoare with the colorful metaphor of the dining philosophers, the version in which it has been best known ever since.

Imagine five philosophers sitting at a round table as in Fig. 8.7.[1] The activity of each philosopher consists of "thinking" or "eating" (or trying to eat) from a dish containing spaghetti. A philosopher needs two forks to eat.[2] There are only five forks on the table, each positioned between two adjacent philosophers; thus, when a philosopher wants to eat, he must first grasp the two forks at his sides. This is possible only if the forks are not held by a neighbor philosopher and are available.

Let us illustrate how the philosophers problem can describe, in a simple and intuitive way, several fundamental problems and recurring features of concurrent systems. Assume, for instance, that each philosopher adopts the following strategy to acquire the forks: first he acquires the fork to his right; then he tries to get the fork to his left without releasing the other fork; if a fork is not on the table because another philosopher holds it, he waits (and thinks). With this strategy if, at some point, all philosophers have acquired the forks on their right, none of them will ever be able to proceed and the system will be blocked in a *deadlock* situation.

Another scenario is the following: Plato and Socrates first get their forks 1, 2, 3, and 4 and eat for a while; then they release the forks and think for another while. While Plato and Socrates are thinking, Confucius and Voltaire do the same sequence of actions: they acquire their forks 2, 3, 4, and 5, eat, release the forks, and think. After Voltaire and Confucius are done eating, the sequence of actions starts

---

[1]The philosophers' pictures are courtesy of Benjamin D. Esham/Wikimedia Commons.

[2]As has been repeatedly observed, the setup does not perfectly fit normal eating habits, at least Italian ones, where spaghetti are eaten using only one fork. A more realistic version may assume bowls of rice eaten with chopsticks. However, we liked it better to present the "historical" version of the problem.

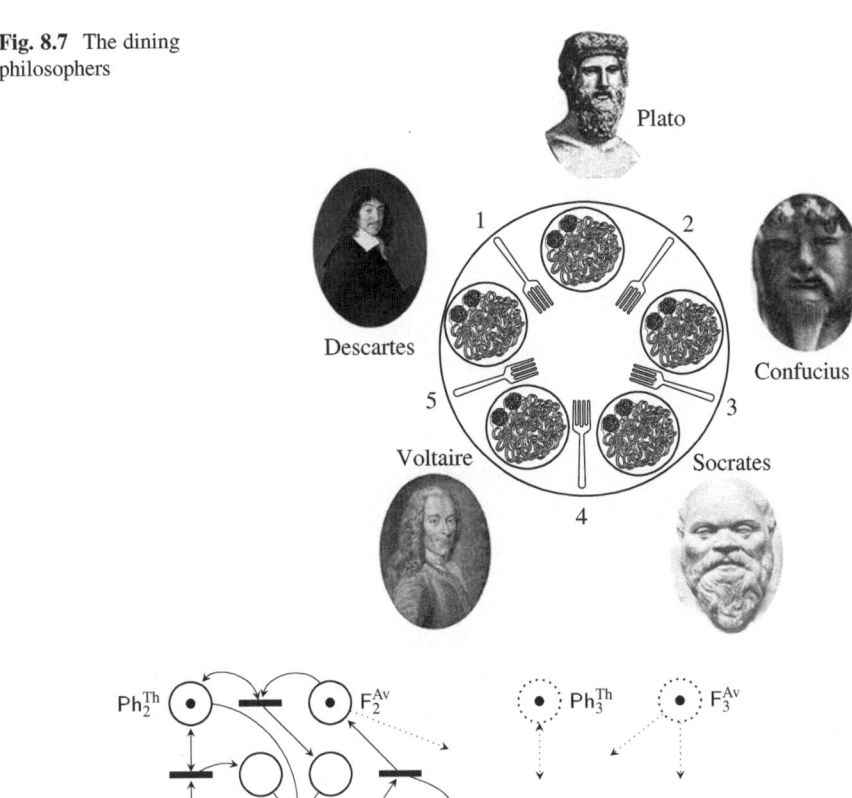

Fig. 8.7 The dining philosophers

Fig. 8.8 A Petri net fragment modeling the dining philosophers

over with Plato and Socrates who preempt fork 1 before Descartes has a chance to take it. If this loop is repeated indefinitely, Descartes will never be able to eat and will die of *starvation*. Indeed, the usage of this term to denote a process which is always prevented from executing, for instance, because the scheduler always grants the resource to higher priority processes, originated from this famous example of Dijkstra's.

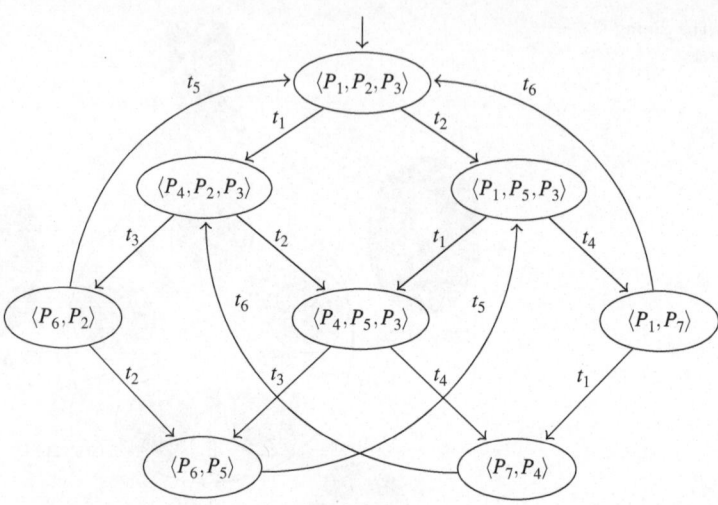

Fig. 8.9 A finite-state automaton equivalent to the Petri net of Fig. 8.6

Petri nets offer a simple and natural way to formalize the dining philosophers problem. Figure 8.8 sketches a possible formalization, where a token in place $\mathsf{Ph}_k^{\mathrm{Th}}$ denotes that philosopher k is thinking, in place $\mathsf{Ph}_k^{\mathrm{Ea}}$ denotes that philosopher k is eating, and in place $\mathsf{F}_k^{\mathrm{Av}}$ denotes that fork k is available. ∎

Exercise 8.12.

- Complete the model of the dining philosophers in Fig. 8.8.
- Build an initial marking and a firing sequence corresponding to a system going into a deadlock.
- Build an initial marking and a firing sequence corresponding to a philosopher starving. ∎

Exercise 8.13. Modify the Petri net formalization of the dining philosophers in Fig. 8.8 to constrain each philosopher to grabbing both his forks only simultaneously and if they are both available. Can the system still deadlock with these additional constraints? Can starvation occur?

Exercise 8.14. Figure 8.9 shows a finite state automaton that is equivalent to the Petri net of Fig. 8.6 in the following sense: its states represent all possible markings of the Petri net with initial marking as in Fig. 8.6, and the set of all possible firing sequences of the net from the given initial marking coincides with the set of all possible sequences of transitions of the automaton from its initial state $\langle P_1, P_2, P_3 \rangle$. The following questions ask you to build a proof that bounded Petri nets and finite state automata are equivalent in expressive power, and to suggest new angles to compare the two formalisms.

- Describe how to generalize the construction to transform any given *k-bounded* Petri net with initial marking m_0 into an equivalent finite state automaton. Provide an upper bound on the number of states of the automaton as a function of the bound k on the number of tokens and of the number of places and transitions of the Petri net.
- Conversely, show how to build a bounded Petri net equivalent to a given finite state automaton. Can you always make the equivalent net 1-bounded? How many places and transitions does the net have, as a function of the number of states of the automaton?
- (♣) As a corollary of the previous two items, k-bounded Petri nets are "equivalent" to 1-bounded ones for some notion of equivalence. Provide a suitable definition of equivalence between Petri nets.
- (♣) Previously in this section, we claimed that the Petri net of Fig. 8.6 is equivalent to the automaton of Fig. 8.9. However, the Petri net is clearly nondeterministic, whereas the "equivalent" automaton is deterministic. Argue about the reasons for this discrepancy. Can you translate the same Petri net into an equivalent *nondeterministic* finite-state automaton N? If you transform N into its equivalent deterministic version D, how does D compare with the automaton in Fig. 8.9?
- (♣) Chapter 7 presented various notions of synchronization among concurrent automata, typically based on events synchronizing transitions or on shared variables. Discuss similarities and differences with respect to the way synchronization occurs in Petri nets. ■

8.2 Variations and Extensions of Basic Petri Nets

One remarkable feature of the Petri net model is its amenability to syntactic and semantic changes and enrichments, which can significantly affect the expressiveness of a notation and the complexity of its analysis.

A first, straightforward extension of basic Petri nets allows multiple arcs to connect the same pair of place/transition or, equivalently, associates a weight with each arc to denote its multiplicity. Correspondingly, a transition is enabled when each place of its preset stores at least as many tokens as the multiplicity of the arc connecting the place to the transition, and the firing of a transition consumes (or produces) for each place in the preset (or postset) as many tokens as the multiplicity of the connecting arc.

It is easy to verify that such an extension is just a matter of convenience since it does not augment the expressive power of the formalism: it suffices to add auxiliary places and transitions so that the effect of, say, a double arc is split into two subsequent firings as intuitively suggested in Fig. 8.10. Then, the behavior of the modified net is identical to the original one's as long as we abstract away the intermediate markings of the auxiliary places and the intermediate firings of

Fig. 8.10 The elimination
of multiple arcs in Petri nets

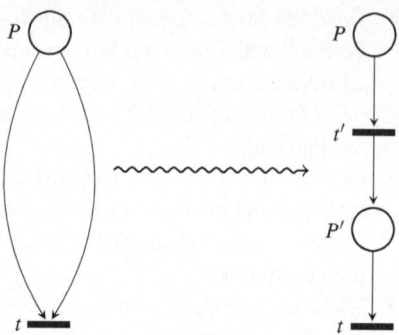

the auxiliary transitions. We leave the details of the complete construction and the related proof of equivalence as an exercise (in connection with the notion of Petri net equivalence suggested in Exercise 8.14).

Among the many other variations of the basic Petri net model that have been proposed in the literature, let us now discuss the addition of *inhibitor arcs*. An inhibitor arc is a special kind of arc connecting a place to a transition. A transition reached by an inhibitor arc is enabled only if the place connected to it through the inhibitor arc is *empty*. Graphically, an inhibitor arc is terminated by a small circle in place of an arrowhead.

Inhibitor arcs are useful, for example, for modeling the behavior of biological or chemical systems, where the presence of certain chemicals or proteins prevents some reactions from occurring. Example 8.15 shows another natural application of inhibitor arcs, to model the priority of requests, in a new variation of the resource allocator discussed in Chap. 7.

Example 8.15. Let us consider a new variant of the example of the resource manager used throughout Chap. 7 with the following characteristics. First, since we are presently considering Petri nets without a notion of time, we assume no metric time constraint. Second, we disallow low-priority requests while the resource is occupied, or high-priority requests while there is a pending low-priority request. Conversely, we introduce a mechanism to keep track of the number of consecutive high-priority requests that occur while the resource is occupied. The manager makes sure that all of them are served (consecutively) before the resource becomes free again. This behavior is modeled by the Petri net in Fig. 8.11, with places free, occ, pendh, wr, and wg, and transitions hpr_1, hpr_2, hpr_3, lpr, rel_1, rel_2, and rel_3. The Petri net is unbounded; more precisely, the only place where tokens can accumulate is pendh, with each token representing a pending high-priority request. Finally, an *inhibitor arc*, from place pendh to transition rel_2, guarantees that transition rel_2 is enabled only if place pendh stores no tokens, that, is the resource is released only when all pending high-priority requests have been served.   ■

Analysis and Verification. Inhibitor arcs increase the expressive power of basic Petri nets and make them as powerful as Turing machines. This entails that

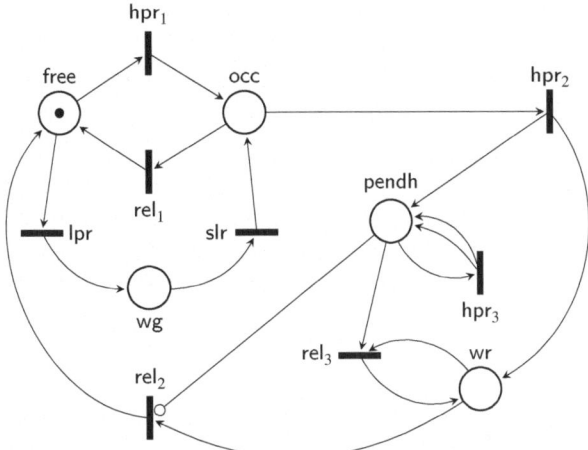

Fig. 8.11 A resource manager modeled though a Petri net with inhibitor arcs

practically all interesting analysis problems, such as determining the reachability of a marking or deciding whether the net is bounded, are undecidable for Petri nets with inhibitor arcs.

Introducing other restrictions while preserving the semantics of inhibitor arcs can restore the decidability of interesting properties. For example, bounded Petri nets with inhibitor arcs are still equivalent in expressive power to finite state automata (see Exercise 8.16).

Primitive nets constitute a larger subclass of Petri nets with inhibitor arcs where properties such as reachability and boundedness are decidable. A Petri net with inhibitor arcs is *primitive* if there exists a positive integer p for every place P attached to an inhibitor arc such that either P never stores more than p tokens or P stores more than p tokens at some point and it never empty afterward; hence the transitions connected to P via an inhibitor arc are permanently disabled. Thus, primitive Petri nets generalize bounded Petri nets.

Exercise 8.16. Exercise 8.14 showed that *bounded* basic Petri nets and finite state automata are equivalent. Show that the same holds for *bounded Petri nets with inhibitor arcs* with the following approach.

- Show how to transform any k-bounded Petri net with inhibitor arcs into an equivalent net (still with inhibitor arcs) that is 1-bounded.
- Show how to transform any 1-bounded Petri net with inhibitor arcs into an equivalent net *without* inhibitor arcs.   ∎

Exercise 8.17. Show that Petri nets with inhibitor arcs are equivalent to Turing machines.

(*Hint*: Start with an appropriate notion of equivalence between Turing machines and Petri nets, such as by describing both as acceptors of event sequences, or as computers of integer functions. Then, devise a suitable encoding of Turing machine configurations into Petri net markings. Finally, use inhibitor arcs to implement conditional branching ("**if-then-else**"), and use that to fill up the remaining details of the construction).   ∎

Exercise 8.18 (Prioritized Petri nets).   Prioritized Petri nets extend basic Petri nets by associating a positive integer with each transition, representing its priority. Priorities constrain the firing of transitions as follows: an enabled transition can fire only if no transition with higher priority is concurrently enabled.

Show that prioritized Petri nets have the same expressive power as Petri nets with inhibitor arcs.   ∎

8.3   Timed Petri Nets

This section shows how to introduce a notion of metric time in the Petri net notation, and how to use this feature to model hard real-time systems. The approach presented here mainly refers to Merlin and Farber's proposal, which is one of the earliest, arguably the most intuitive, and the best-known timed extension of Petri nets.

Definition 8.19 (Timed Petri net).   A timed Petri net is a Petri net where a *minimum* and a *maximum* firing time – also called *lower* and *upper bound* – are attached to each transition. Both firing times are nonnegative values; they can both be 0 and the maximum can be ∞.

A transition enabled by the presence of tokens in its preset will fire within the relative time interval defined by the minimum and maximum firing times associated with it. If a place in the preset becomes unmarked before the transition fires, the transition is disabled until all places in the preset are marked again. Every time a transition becomes enabled, the "stopwatch" that measures the firing time is reset.
                                                                                    ∎

Example 8.20 (Dining philosophers, continued from Example 8.11).   The dining philosophers problem and its formalization suggested in Example 8.11 lend themselves to illustrating the use of timed Petri nets. Assume that the thinking session of every philosopher lasts at least th_m and at most th_M time units, after which the philosopher is hungry again and tries to grab the forks in any order, as soon as they are available. An eating session starts when he has both forks and lasts at least ea_m and at most ea_M time units. This timed version of the dining philosophers is formalized by the timed Petri net of Fig. 8.12 (for simplicity, the picture shows only the fragment for one philosopher, the others being identical).

A suitable choice of the thinking and eating times of each philosopher can make the system free of deadlocks and prevent starvation. A straightforward way to meet such requirements is the following: let $th_m = th_M = 4 \cdot ea_m = 4 \cdot ea_M$ and modify the net so that the philosophers start their first eating session in a sequential fashion, each following the previous one exactly after ea_m time units. These choices,

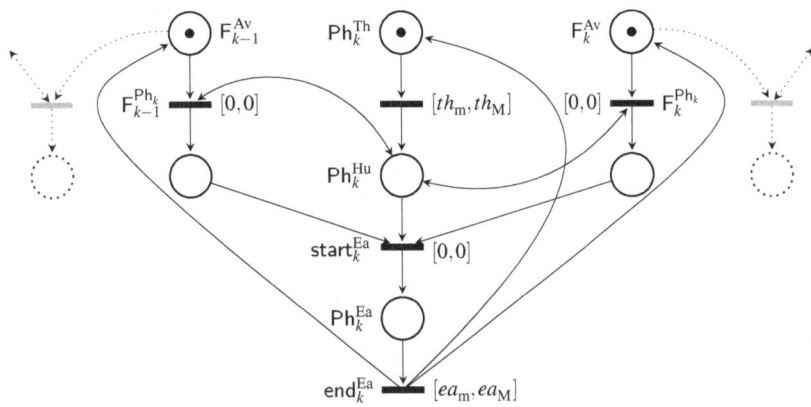

Fig. 8.12 A timed Petri net formalizing a "timed" dining philosopher

however, make the system fully deterministic, and may be too restrictive for a specification, as it permits a unique implementation that does not make a very efficient usage of the resources. The next exercise asks for less constraining choices for the time parameters. ∎

Exercise 8.21. Consider again the timed version of the dining philosophers, introduced in Example 8.12.

- Suggest a set of constraints on the firing times th_m, th_M, ea_m, and ea_M different from those in Example 8.20 that allows for some parallelism during the eating sessions but still prevents deadlock and starvation.
- Consider now the constraint that each philosopher grabs both his forks only simultaneously and if they are both available (Exercise 8.13 introduced this constraint in the untimed version of the problem). Modify the timed model of Fig. 8.12 to include this constraint and suggest a choice of firing times th_m, th_M, ea_m, and ea_M that prevents deadlock and starvation. Compare the new set of constraints with the one given in the previous point.
- Consider the change in the fork acquisition policy that introduces a maximum time during which a philosopher can hold a single fork; if he cannot acquire the other fork before this time elapses, he releases the fork he is holding. Modify the Petri net model to reflect this change and analyze possible deadlock and starvation behaviors. ∎

8.3.1 Issues in the Semantics of Timed Petri Nets

The semantics of timed Petri nets, introduced informally in Definition 8.19, is deceptively intuitive: every attempt to make it rigorous and formal has to face subtle intricacies. The present section illustrates the main ones and hints at how they can be unraveled and clarified.

Weak vs. Strong Time Semantics. Is it possible that a transition is enabled continuously up to and including its maximum firing time without actually firing? Is the transition *forced* to fire or not? What happens if the transition does not have to fire (for example, until the firing of some other transition resets the places in its preset)?

There are arguments in favor of both choices – forced and optional firing. The most common assumption is the former: a transition enabled continuously for its maximum firing time *must* fire within its time bounds. This is usually called *strong time semantics*, and was implicit in the timed dining philosophers of Example 8.20. There are cases, however, where the other assumption – called *weak time semantics* – is preferred because it is considered more consistent with the traditional Petri nets semantics, where a transition is never forced to fire.

Imagine, for instance, a student who has to deliver her thesis: once she has a printed copy and the written approval of her advisor, there is a fixed time window for officially turning the thesis in to the administration. The student is not forced to actually deliver the thesis; if she forfeits the delivery, the delivery session expires and the two prerequisites (tokens in a Petri net model) of having a printed copy and the advisor's approval are not available anymore. Perhaps they are still valid in a successive session or maybe a renewal of the advisor's approval is requested. Such scenarios can be naturally modeled by means of a timed Petri net with a weak time semantics.

Remark 8.22. The timed automata described in Sect. 7.4.1 could be considered to have a *weak time semantics*, because their transitions are *not forced* to be taken when the upper bound of some constraint is reached; rather, all that is prescribed is that *when* (and *if*) a transition is taken by a timed automaton, its corresponding constraint (and those of the source and target locations) *must* be satisfied.

Consider, for instance, two transitions that exit from a state s, one with guard $c = 10$ and one with guard $c = 20$ on the same clock c, where both transitions reset the clock. In a strong time semantics, the second transition would never be taken, because the first one would be forced to fire earlier. According to the semantics of Definition 7.32, however, both transitions may be taken and the clock would be reset nondeterministically either at time 10 or at time 20.   ■

A potentially undesirable feature of timed Petri nets under the weak time semantics is the generation of "dead" tokens that cannot be used to fire any transition because all the firing times have expired. Such dead tokens can accumulate in a place, spoiling the clarity of the graphical representation of a net's sequence of markings – just like the obsolete versions of documents scattered among folders on hard drives.

Exercise 8.23. Given each of the fragments of timed Petri nets in Fig. 8.13, for which the weak time semantics is assumed, provide equivalent Petri net fragments having the strong time semantics.

Fig. 8.13 Two timed Petri net fragments with weak time semantics

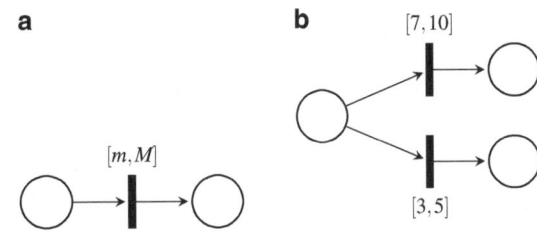

Transitions as Events vs. Transitions as Actions. The firing of transitions is normally considered instantaneous, consistently with the interpretation of firings as occurrences of events. There are cases, however, where we may want transitions to represent *actions* with an appropriate duration. Nevertheless, the normal assumption of instantaneous transition firing does not affect generality, since an activity with non-null duration can be modeled by a pair of transitions with a place in between them: the first transition models the beginning of the activity and the second one models its end.

Zero-Time Transitions. If the minimum firing time associated with a transition is 0, the transition can fire immediately once enabled: a case of zero-time transition or, in more precise terminology, *zero-time firing* of a transition. A few such transitions are present in Fig. 8.12 providing the timed model of the dining philosophers. As discussed in Chap. 3, transitions taking no time can be a useful abstraction whenever the duration of the event modeled by the transition can be neglected with respect to the time taken by other activities in the overall system.

On the other hand, a sequence of transitions with zero firing time can produce multiple firings that are consecutive *and* occur at the same time. Even worse, a loop of transitions with zero firing time can generate **ZENO** behaviors where an infinite number of firings occur without the time ever advancing. To rule out such unfeasible behaviors **A PRIORI**, loops of transitions with zero minimum firing time are usually syntactically forbidden in the construction of timed Petri nets. For instance, if we take $ea_m = th_m = 0$ in the fragment of Fig. 8.12 (representing a dining philosopher), we have such a loop of "zero-time transitions": the model would include a behavior (physically unfeasible) where a philosopher gets the forks, eats, and thinks infinitely many times in zero time.

Exercise 8.24. Timed Transition Models (TTMs) define similar constraints to prevent Zeno sequences of zero-time transitions a priori (see Sect. 7.3.1, and Figs. 7.26 and 7.27 in particular). Discuss similarities and differences between the constraints in TTMs and those discussed in the previous paragraph for Petri nets.

Recharge Time. Consider the net fragment with marking in Fig. 8.14, and assume a strong time semantics. If the transitions t_{i_1} and t_{i_2} fire simultaneously at absolute time τ, then t_o is enabled and must (according to the strong time semantics) fire at time $\tau + 3$.

Fig. 8.14 A timed Petri net
fragment illustrating
simultaneous firings

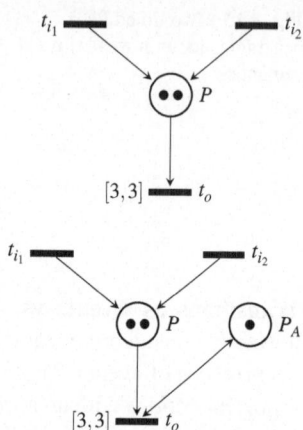

Fig. 8.15 A timed Petri net
fragment illustrating the
recharge time semantics

   The firing of t_o consumes either one of the two tokens that have been deposited
in P at time τ. Hence, the question is: when will t_o fire again? If the other token
can be used immediately, then t_o fires twice at the same time $\tau + 3$. Otherwise, we
should wait for another three time units, and t_o will fire again at time $\tau + 6$ using
the second token.

   As with the strong vs. weak time semantics, there are cases where either of the
two interpretations is the most appropriate. For example, if the transition models
the access to a replicated resource, such as a memory bank or a processor, there
is no reason to prevent a double simultaneous firing, given that multiple available
resources can be accessed in parallel. On the other hand, if tokens in the net represent
the input data to be processed by a sequential algorithm, the two firings must be put
in sequence to represent properly the access to the unique resource.

   The semantics where multiple simultaneous firings are not allowed is sometimes
referred to as the "recharge time semantics". The recharge time semantics is actually
a special case of the semantics allowing simultaneous firings, as every transition
can be given the recharge time semantics with the addition of an auxiliary place, as
shown in Fig. 8.15 with reference to the net fragment of Fig. 8.14: whenever t_o fires,
the token in P_A is consumed and restored immediately; this forces exactly three
more time units to elapse before t_o can fire again, taking the second token in P.

   As a "graphical embellishment", the auxiliary place P_A could be hidden in the
graphical notation and a special symbol, say a thin rectangular box, could replace
the normal transition representation to denote its recharge time behavior.

Infinite Upper Bounds. If the maximum firing time M_v of a transition v is equal
to ∞, then behaviors in which v never fires, and tokens entering v's preset stay there
forever, is admissible, regardless of the firing semantics. This is consistent with the
traditional assumption that timed Petri nets reduce to classical Petri nets when, for
every v, the minimum and maximum firing times are, respectively, $m_v = 0$ and
$M_v = \infty$.

The time interval $[m_v, \infty]$ also carries the intuitive interpretation of "the limit for $M_v \to \infty$ of $[m_v, M_v]$". Since in the strong time semantics this means that v *must* fire in time t, such that $m_v = t = M_v$ unless previously disabled, we could deduce that the meaning of $[m_v, \infty]$ is that v *must* fire in some time $t < \infty$, which is different from saying that the semantics of the net is the same as in untimed Petri nets, where an enabled transition may *never* fire. Equivalently, we could assume *fairness* for the firing of enabled transitions; a transition cannot remain enabled for infinite time without firing. In some sense, such a new, "fair" interpretation is more consistent with the strong time semantics, whereas the traditional one is more consistent with the weak time semantics. Since, however, in some cases both interpretations may be applicable, we will assume the original "unfair" semantics for the interval $[m_v, \infty]$. We will use the different symbol "$\diamond$" to denote an unbounded but finite value, that, is the fact that the transition cannot remain enabled for infinite time without firing.

Remark 8.25. From the point of view of the theory of computation, the "unfair" semantics described by intervals of the form $[m, \infty]$ can be perfectly simulated by a nondeterministic Turing machine, whereas the "fair" semantics of $[m, \diamond]$ is not exactly computable. Intuitively, a Turing machine can use nondeterminism to increment an integer counter an arbitrary number of times before deciding to output its value and halt; such a behavior also includes "unfair" nonterminating computations where the counter is forever incremented and never output: it is impossible to define a Turing machine that excludes precisely only those nonterminating computations, and hence one that is *guaranteed* to stop at *any* finite time.

Intervals of the form $[m, \diamond]$ are still perfectly appropriate, and useful, in formal *specifications*, where they constrain implementations to guarantee fairness without unnecessary details about a specific upper bound. For instance, an implementation that guarantees the firing of a transition within ten seconds satisfies, but is not equivalent to, a timing specification $[0, \diamond]$. Section 9.1.1 will show that an "eventually" fundamental operator of temporal logic has the same semantics (and uses the same symbol) as "$\diamond$". ∎

8.3.2 Formalizing the Semantics of Timed Petri Nets

Once the issues outlined in the previous Sect. 8.3.1 have been clarified, the behavior of timed Petri nets can be formalized through several approaches:

- Attaching timestamps to transition firings;
- Modeling nets as dynamical systems;
- Extending the time domain with infinitesimal numbers.

The following subsections describe these approaches. Another approach based on temporal logic will be presented in Sect. 11.3.

Note that, no matter what the semantics adopted, time is modeled in timed Petri nets in an **EXPLICIT** and **METRIC** way – whether **DISCRETE** or **CONTINUOUS**.

Timestamps. The timing information can be introduced in a semantics based on firing sequences by attaching a timestamp to each element of a sequence. The timestamp denotes the absolute time at which the firing occurred, similarly to the timed words introduced in Sect. 7.4.1.

For example, the sequence

$$\langle F_1^{Ph_1}, 2\rangle \quad \langle F_5^{Ph_1}, 2\rangle \quad \langle start_1^{Ea}, 2\rangle \quad \langle end_1^{Ea}, 2 + ea_m\rangle \quad \cdots$$

models a feasible behavior of the timed Petri net in Fig. 8.12: at time 2, the philosopher performs in zero time the sequence of actions: "grab fork 1", "grab fork 5", "start eating"; then, after ea_m time units, the eating ends. Timestamps represent the values of an additional variable in the system's state; more precisely, of *many* additional variables, one for each token.

As remarked in Chap. 7, two different time models coexist in models based on timestamps: the time implicitly subsumed by the order of the firing sequence and the time modeled by the timestamps. A sound model must guarantee consistency of the two time models. For instance, switching the order in which the two forks are acquired in the firing sequence described above yields a different but still feasible behavior:

$$\langle F_5^{Ph_1}, 2\rangle \quad \langle F_1^{Ph_1}, 2\rangle \quad \langle start_1^{Ea}, 2\rangle \quad \langle end_1^{Ea}, 2 + ea_m\rangle \quad \cdots ;$$

in contrast, the transition "start eating" cannot precede "grab fork 5" even if they both occur at the same time; hence

$$\langle F_1^{Ph_1}, 2\rangle \quad \langle start_1^{Ea}, 2\rangle \quad \langle F_5^{Ph_1}, 2\rangle \quad \langle end_1^{Ea}, 2 + ea_m\rangle \quad \cdots$$

is not a valid behavior. Similarly, the firing of nonconflicting transitions must still respect the ordering of the timestamps, so the following is not a feasible behavior:

$$\langle F_1^{Ph_1}, 2\rangle \quad \langle F_3^{Ph_3}, 4\rangle \quad \langle F_5^{Ph_1}, 2\rangle \quad \langle start_1^{Ea}, 2\rangle \quad \langle end_1^{Ea}, 2 + ea_m\rangle \quad \cdots .$$

Dynamical Systems. Dynamical systems, described in Chap. 4, can formalize the semantics of a timed Petri net. The marking of the net defines the system state, which evolves as a function of time according to the net's semantics. To make this construction rigorous, a few technical subtleties must be handled properly.

The first subtlety concerns the "identity" of tokens. In Petri nets without time, tokens in the same place are indistinguishable and only their number matters. In timed Petri nets, tokens in the same place may have different "ages", hence the formalization must introduce some sort of timestamp to record this additional necessary piece of information.

Consider, for instance, the net fragment of Fig. 8.16 and suppose that one token is produced by transition t_{i_1} at time 3, and another token is produced by t_{i_2} at time 4. According to the strong-time semantics without recharge time discussed in Sect. 8.3.1, the output transition t_o will fire at time $6 = 3 + 3$ and then again

Fig. 8.16 An example Petri net fragment

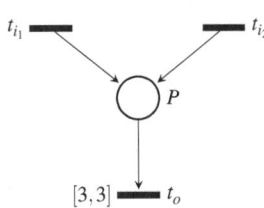

at time $7 = 4 + 3$. A state description that does not distinguish between multiple tokens would only record the fact that there are two tokens in P at time 4; this is insufficient for determining correctly the behavior of the net.

Other subtleties in the formalization of timed Petri nets as dynamical systems stem from the presence of zero-time transitions. A transition taking no time associates two states (the markings before and after the transition firing) with the same time; hence, strictly speaking, the system state cannot be formalized as a function of the independent variable "time".

Consider, for example, the firing sequence

$$\langle F_1^{Ph_1}, 2 \rangle \quad \langle F_5^{Ph_1}, 2 \rangle \quad \langle start_1^{Ea}, 2 \rangle \quad \cdots$$

for the net of Fig. 8.12. The ordered firing of three transitions at the same absolute time produces three different markings, all at the same time 2, which is clearly inconsistent with the notion of "system state".

Behaviors originating from zero-time transitions can still be rendered adequately with a dynamical system model, as long as loops of zero-time transitions are forbidden. Then, every sequence of zero-time firings is necessarily finite and can be modeled by considering only the marking reached at the end of the sequence, assuming that tokens flow instantaneously through intermediate places without actually changing the marking.

Infinitesimal Numbers. A general approach to dealing uniformly with zero-time firings extends the time domain with *infinitesimal* numbers used, for example, in non-standard analysis. The model describes zero-time firings as consuming a non-null but *infinitesimal* amount of time. This preserves the intuition associated with the abstraction of zero-time transitions, because any infinitesimal number is smaller than any "regular" positive number but still larger than zero.

The mathematical formalization and analysis of behavior become simpler and perhaps more elegant than in more conventional approaches. For example, the markings before and after any zero-time transition will actually occur at two distinct time instants that differ only by an infinitesimal amount; hence this kind of Petri nets can be formalized as a dynamical system without incurring any mathematical inconsistency.

Example 8.26 (Railroad crossing problem (KRC)). This example studies a railroad crossing guarded by an automatic gate. The gate must be controlled so that cars

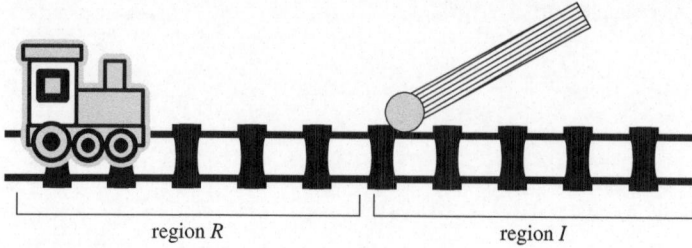

Fig. 8.17 Railroad crossing

cannot cross the tracks when a train is incoming. The system, consisting of the trains, the cars, and the gate, is real time because the timing of the gate movement must be accurately designed, taking into account the speed of trains and the topology of the crossing, so as to guarantee that the gate will be closed whenever a train is about to cross.

Several variants of the railroad crossing problem have been often used as a benchmark to demonstrate the flexibility and analyzability of formal notations for real-time systems. In this example we consider a highly simplified version of the problem, known as "kernel railroad crossing" (KRC), and we model and analyze it with timed Petri nets.

Consider a system with a single track and a single train moving from left to right as in Fig. 8.17. The behavior of the system is the following:

- The speed of the train, in meters per second, varies between a minimum $V_{\min}$ and a maximum $V_{\max}$.
- The gate guards a region called "critical region" denoted by I. The critical region extends over l_I meters.
- The critical region is adjacent to a "monitoring region" denoted by R. R precedes I and extends over l_R meters.
- A sensor positioned at the beginning of R signals the approaching of the train; another sensor at the end of I signals the exit of the train from that region.
- The gate takes exactly γ seconds to go from open to close (and vice versa).
- Whenever the sensor at the entry of R is activated by a train, the control issues a command to close the gate; conversely, when the train exits I, the corresponding sensor issues a command to open it.

The timed Petri net of Fig. 8.18, where

$$R_{\mathrm{m}} = \frac{l_R}{V_{\max}}, \qquad R_{\mathrm{M}} = \frac{l_R}{V_{\min}}, \qquad I_{\mathrm{m}} = \frac{l_I}{V_{\max}}, \quad \text{and} \quad I_{\mathrm{M}} = \frac{l_I}{V_{\min}},$$

formalizes the informal description above.

The control subsystem responsible for activating the gate works correctly if two properties are satisfied:

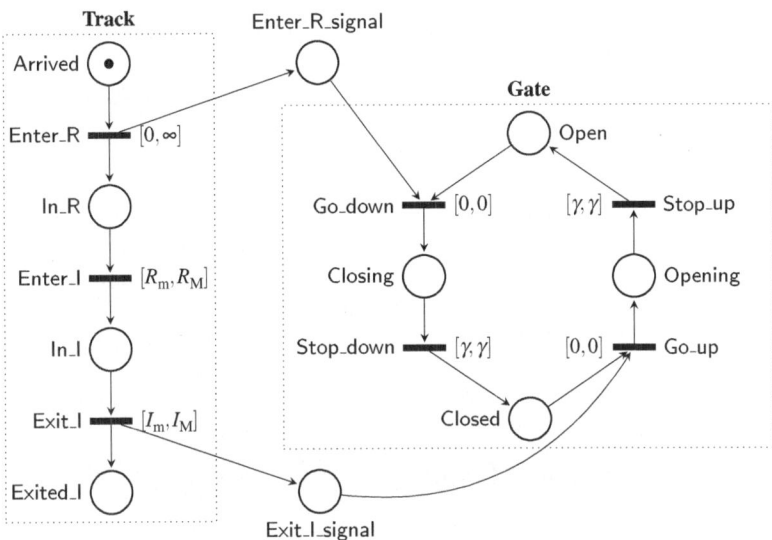

Fig. 8.18 A timed Petri net formalizing the KRC system

Safety property: whenever the train is in the critical region the gate is closed.
Utility property: the gate is not closed unless necessary. (A system where the gate
   is always closed trivially satisfies the safety property but is not very useful).

The behavior of the system model in Fig. 8.18 can be analyzed to determine
conditions under which the safety and utility properties hold. We see that the safety
property holds if $R_m = \gamma$ or, equivalently, $l_R = V_{max} \cdot g$. The constraint is tight:
if $R_m < \gamma$, it is possible that a train enters the critical region before the gate is
completely closed. On the other hand, if R_m is significantly larger than γ, the gate
may be closed long before the train enters the critical region; hence the system
would fail to achieve utility.

Another aspect that should be clear from this example is the need for a notation
supporting metric time to model and analyze the problem appropriately. In the
untimed version of the model, it would be impossible to guarantee that whenever
there is a token in place In_I there is also a token in Closed.   ∎

Exercise 8.27. Generalize the KRC problem of Example 8.26 by allowing several
trains to access the same track. Assume that the arrival times of trains are at least D
seconds apart.

• What is a reasonable value for D, with respect to the other parameters of the
   problem?
• Assuming this reasonable constraint on D, modify the Petri net of Fig. 8.18 to
   support multiple trains.
• Does the new system satisfy safety and utility? Under which conditions?   ∎

Fig. 8.19 A Petri net with
time constraints associated
with places

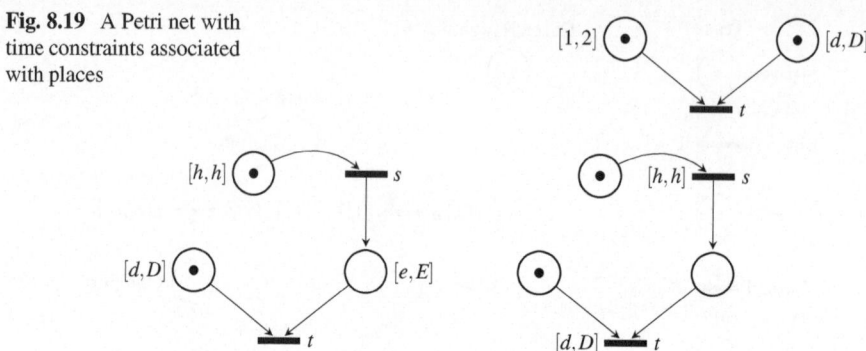

Fig. 8.20 Two timed extensions of Petri nets

8.3.3 Timed Petri Nets: Complexity and Expressiveness

Real-time constraints on transitions increase the expressive power of Petri nets to
match that of Turing machines. An immediate consequence is that problems such
as reachability, boundedness, and model checking are undecidable for timed Petri
nets.

As with basic Petri nets (without time), bounding the number of tokens stored in
every place reduces the expressiveness of timed Petri nets and yields a class where
reachability and model checking of some real-time properties are fully decidable.
In fact, bounded timed Petri nets and timed automata (Chap. 7) have the same
expressiveness with respect to the timestamp semantics (discussed in Sect. 8.3.2);
hence the same basic analysis techniques can be used for both notations.

Exercise 8.28 (Place-timed Petri nets). A different timed extension of Petri nets
associates a time interval to *places* rather than transitions. The semantics prescribes
that a place P with interval $[m, M]$ contribute to enabling the transitions in its
postset only if P has been storing a token for at least m and at most M time units.
Any transition can fire when enabled simultaneously by *all* places in its preset.

Describe all possible firing times of transition t in the timed Petri net in Fig. 8.19,
with a strong time semantics and with a weak time semantics, when the marking is
initialized at time 0 and:

- $d = 0$ and $D = 8$;
- $d = 3$ and $D = 4$.

Consider now the two Petri nets in Fig. 8.20, one associating a time interval with
transitions and the other with places. Assuming a strong time semantics and that the
marking is initialized at time 0, determine constraints over the parameters d, D, e,
E, and h such that the two nets generate exactly the same timed firing sequences.

(*Hint*: first consider what happens when $d = D$).   ■

Fig. 8.21 A fragment of a
timed Petri net with inhibitor
arcs

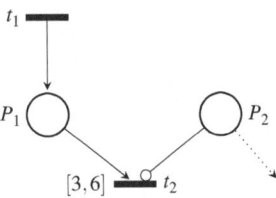

8.4 Timed Petri Nets with Inhibitor Arcs

Inhibitor arcs and real-time constraints on the firing of transitions are complementary ways, among many others, of increasing the expressive power of basic Petri nets. For certain modeling purposes, it may be useful to combine both extensions to avail of their complementary features. Once again, the seemingly intuitive semantics of the resulting model hides subtleties and pitfalls that require clarifications.

Consider the net fragment of Fig. 8.21 and suppose that a token is produced in P_1 at time 1, and a token is in P_2 at time 0 and is consumed at time 8. When is t_2 enabled for firing?

A reasonable answer is that t_2 can fire at any time between 11 and 14. This corresponds to assuming that a transition becomes enabled when all places in its preset are not marked if they connect to inhibitor arcs and are marked otherwise. Once the transition becomes enabled, it can (or must, under the strong time semantics) fire within the given time bounds. We call this firing semantics "timed inhibition semantics" (TI semantics).

Another reasonable semantics for the same scenario separately considers the marking of places connected to inhibitor arcs and those connected to "positive" (non-inhibitor) arcs. Namely, a transition becomes enabled when all places in its preset connected to positive arcs are marked, but it can fire within the given time bounds only when the places connected to inhibitor arcs are empty, regardless of when they became empty. We call this alternative firing semantics "untimed inhibition semantics" (UI semantics).

According to the UI semantics, t_2 would not fire because it would be inhibited by the token in P_2 throughout the time interval [4, 7]. Thus, the token in P_1 would become "dead", similarly to what can happen under the weak time semantics even without inhibitor arcs. If, instead, the token in P_2 is consumed before time 7, t_2 could (or must) fire within time 7.

Both interpretations (TI and UI) are reasonable in different contexts. For instance, if the presence of a token in a place inhibiting a transition represents an obstacle to performing the action modeled by the transition, it may be natural to assume that the action is performed as soon as the obstacle is removed, as in the UI semantics. On the other hand, TI treats the presence and absence of tokens in places enabling or inhibiting a transition more symmetrically than UI, which may be preferable in other situations.

Fig. 8.22 A net fragment
with the TI semantics (**a**) and
one with the UI semantics (**b**)
that implements the TI
semantics of (**a**)

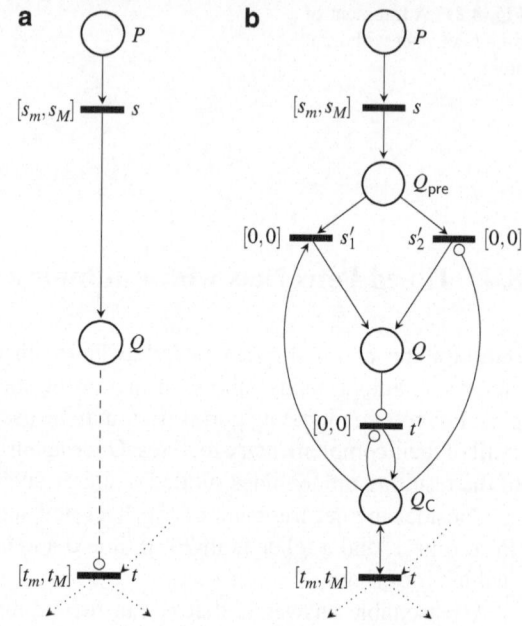

UI is at least as expressive as TI, in that the behaviors prescribed by TI can
always be implemented in a net with the UI semantics. Figure 8.22 illustrates the
transformation that achieves this if the net is 1-bounded (the transformation in the
general case of unbounded nets is left as an exercise). In Fig. 8.22a, the dashed
inhibitor arcs follow the firing semantics TI. Figure 8.22b uses in contrast only
inhibitor arcs with the UI semantics, but it produces the same timed sequence of
transitions s and t as the net in Fig. 8.22a does. This requires, in particular, the
introduction of an auxiliary place Q_C to represent the "complement" of place Q:
Q_C is empty if and only if Q is marked.

The transformation is also possible in the other direction: every Petri net with TI
semantics can be transformed in an equivalent one with UI semantics. Hence, the TI
and UI semantics have exactly the same expressive power.

Example 8.29. Figure 8.23 augments the untimed Petri net with inhibitor arcs of
Fig. 8.11 with real-time constraints on transitions. These timing constraints ensure
that each use of the resource takes no longer than 100 time units since the last request
occurred, and that a low-priority request is served within two time units.

For instance, place wg may receive a token as the result of a low-priority request
(lpr) issued. The token in wg enables the transition slr, which can fire at any
time between zero and two time units; this expresses the fact that the request is
granted and the resource becomes occupied *within* two time units. As in the untimed
version of Example 8.15, additional high-priority requests accumulate tokens in
pendh. Then, the resource is not released (transition rel$_2$) until all such pending
high-priority requests have been served and pendh becomes empty. Each granted

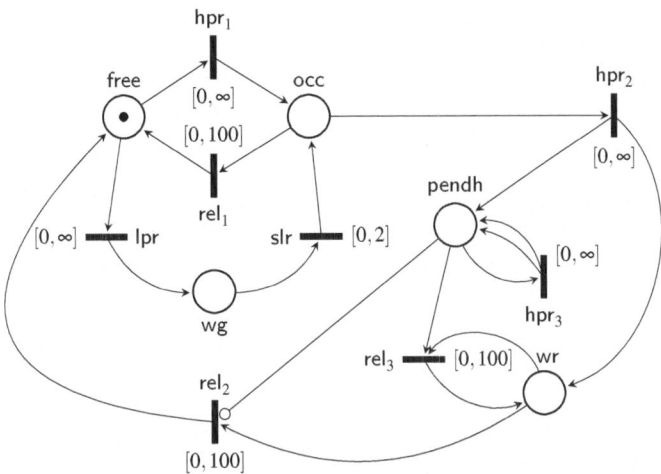

Fig. 8.23 A resource manager modeled through a timed Petri net with inhibitor arcs

high-priority request must release the resource at a time between 0 and 100 time units.

Notice that the behavior of the inhibitor arc in the net of Fig. 8.23 is the same for both semantics TI and UI, since **pendh** only becomes empty simultaneously with the deposit of a token in **wr** (remember that, when a transition is both in the preset and in the postset of a place, a token is consumed and a new one is produced immediately).   ∎

Exercise 8.30. Consider the Petri net with inhibitor arcs of Fig. 8.23. It can be shown that, if a new high-priority request is received while the resource is already occupied, then it can in fact remain occupied for more than 100 time units.

- Find a sequence of markings that has such a behavior.
- Modify the net in such a way that the resource is never occupied continuously for more than 100 time units (remember that the resource is occupied not only when the place **occ** is marked).
- An alternative semantics is the "age memory" semantics, which works as follows: the "stopwatch" of a transition, introduced in Definition 8.19, is not reset when a transition is disabled (even if it does not fire), but rather when it fires; in addition, when a transition is disabled the stopwatch is stopped, and it restarts when the transition is enabled anew.
  Discuss how interpreting the net of Fig. 8.23 under the age memory semantics impacts the constraint on the resource not being occupied for more than 100 time units.   ∎

Exercise 8.31 (♣). Exercise 8.28 introduced "place-timed Petri nets", a variant of timed Petri nets where time intervals are associated with places rather than with transitions. Consider now extending place-timed Petri nets with inhibitor arcs.

- Discuss how the two semantics TI and UI can be adapted to this new model. Is one of the two preferable?
- Show how to implement the UI semantics under TI (this is the converse of what was done in this section for timed Petri nets with real-time constraints on transitions). ∎

Exercise 8.32. Define the semantics of some versions of Petri nets using transition systems (defined in Sect. 7.1). You can consider basic Petri nets, timed Petri nets, and timed Petri nets with inhibitor arcs.

8.5   Stochastic Petri Nets

An alternative approach to introducing a notion of time in Petri nets associates *random delays* with transitions; this approach is pursued by an extension of untimed Petri nets known as *stochastic Petri nets*. Stochastic Petri nets provide a notation for defining continuous-time Markov chains (introduced in Sect. 6.4.1.3 and further discussed in Example 7.57) in a concise and easy to understand way; as such, they are based on a **CONTINUOUS** notion of time.

In a stochastic Petri net, every transition t has an associated probability density function characterizing the random delay with which t fires after being enabled. To guarantee the equivalence of stochastic Petri nets with continuous-time Markov chains and facilitate their analysis, the random delay of each transition t has an exponential distribution with real parameter ρ_t called *firing rate*. Intuitively, the firing rate determines the frequency with which t fires: the higher its firing rate, the more likely it is that a transition fires soon after being enabled.

The semantics of stochastic Petri nets combines the usual behavior of the underlying untimed Petri net with randomized firing times. To determine which transition fires from among those enabled in the current marking, we sample the probability distributions associated with every enabled transition. The transition which samples the smallest value d will fire next, after a delay of exactly d time units. The firing determines a new marking, where the process can start over. In the literature, this policy for choosing the next firing transition is called *race policy with resampling*, as the probability distributions are resampled with every new marking.

The behavior of stochastic Petri nets is best understood in terms of the average of all possible outcomes. Consider the stochastic Petri net in Fig. 8.24 (which could represent, for example, the behavior of a lamp randomly switching on and off), and assume that t_1 and t_2 have firing rates $\rho_1 = 6$ and $\rho_2 = 3$, respectively. The firing rates determine the *steady-state* or *nominal behavior* (see Sect. 6.4.1.3) of the net, where t_1 and t_2 respectively fire *exactly* $1/6$ and $1/3$ time units after they are enabled; hence, at steady state, the lamp will alternately stay on for $1/6$ time instants, and off for $1/3$ time instants. Correspondingly, the steady-state probability of the lamp being on is 33 %, while the probability of the lamp being off is 67 %; in fact, over the course of a time unit, at steady state the lamp will be

Fig. 8.24 A stochastic Petri
net fragment

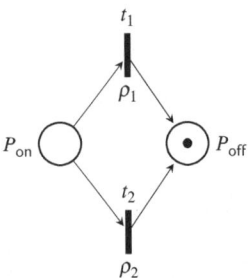

on for $1/6 + 1/6 = 1/3$ time units, and off for $1/3 + 1/3 = 2/3$ time units.
Notice, however, that the probability distribution involved makes all possible delays
admissible (there could be executions in which, at some point, the lamp stays on for,
say, 100 time units).

Since the negative exponential probability function is a *memoryless* drawing
process (as pointed out in Sect. 6.4.1.3), for a given marking the sampled values
for each transition are independent of the previously drawn values. This has several
noticeable consequences on the semantics of the resulting model.

- Since every enabled transition has nonzero probability of firing, the set of
  reachable markings of a stochastic Petri net is independent of the firing rates
  and corresponds to that of the underlying untimed Petri net. Correspondingly,
  the decidability of properties such as reachability and boundedness (and hence
  finite-stateness) for stochastic Petri nets follows from the decidability of the same
  properties for basic Petri nets and can be analyzed by means of the very same
  algorithms (and tools).
- As mentioned above, every stochastic Petri net is equivalent to a continuous-time
  Markov chain (and vice versa). Unbounded stochastic Petri nets generate infinite-
  state continuous-time Markov chains; similarly, bounded stochastic Petri nets
  generate finite-state continuous-time Markov chains, called probabilistic finite-
  state automata in Chap. 6.

The memoryless property also suggests the following alternative characterization
of the semantics of a stochastic Petri net: whenever a new marking enables some
transition t that has not been enabled since its last firing, draw some delay d from
the probability distribution associated with t and reset a countdown timer to the
value d. The timer will measure the time during which t is enabled in the future,
possibly discontinuously: if a new marking disables t, pause the timer and resume
it as soon as t becomes enabled again. When the countdown reaches zero, t fires.
In the literature, the policy of choosing the next transition to fire according to this
countdown mechanism is called *race policy with age memory*. This policy for the
interpretation of stochastic Petri nets attaches a notion of "activity" to transitions:
when a transition t is enabled, the corresponding activity starts, and it completes
when t fires; the activity can be interrupted (if t is disabled), and resumed when t is
newly enabled.

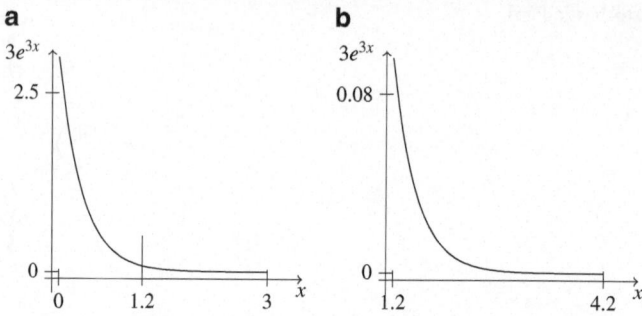

Fig. 8.25 Shape of the negative exponential $3e^{3x}$ in the ranges $[0, 3]$ (**a**) and $[1.2, 4.2]$ (**b**), normalized to the same height

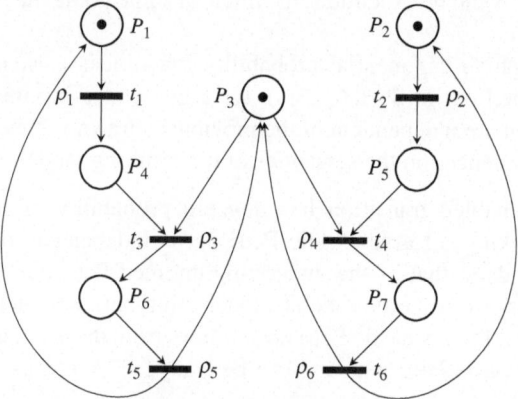

Fig. 8.26 A stochastic version of the Petri net of Fig. 8.6

The memoryless property of the negative exponential distribution function associated with transitions makes the two interpretations, with resampling and with age memory, equivalent. In fact, the distribution function of the random variable corresponding to the time *remaining* before a transition t fires after it has been enabled for some time d is also a negative exponential with mean $1/\rho_t$, independent of d. Hence, it does not matter whether the timer is resampled when a new marking is entered or not. This feature can be graphically seen by looking at the normalized shape of the negative exponential depicted in Fig. 8.25: if we consider only the values of $\rho_t \exp(-\rho_t x)$ for $x \geq d$ and re-normalize, the new function has exactly the same shape as that of the former.

Let us show through an example how a continuous-time Markov chain is derived from a stochastic Petri net. Consider the net of Fig. 8.26, which is a stochastic version of the net of Fig. 8.6 obtained by associating with each transition t_i a firing rate ρ_i.

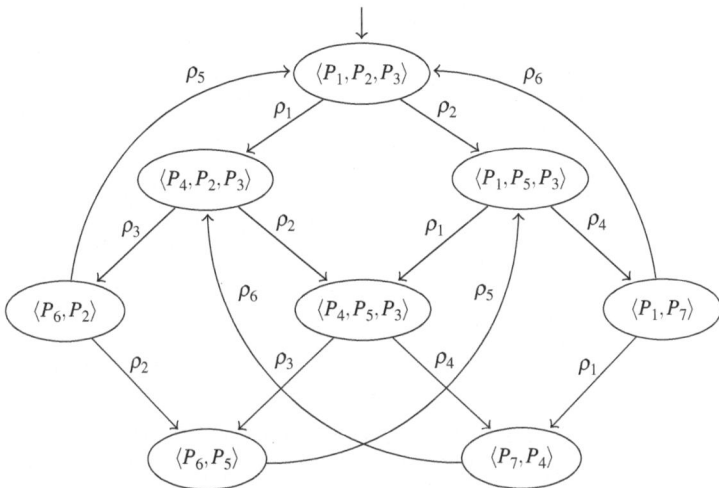

Fig. 8.27 Continuous-time Markov chain corresponding to the stochastic Petri net of Fig. 8.26

The net is 1-bounded; hence the corresponding continuous-time Markov chain will be finite-state. In fact, the states of the continuous-time Markov chain correspond to the markings of the underlying untimed Petri net, as depicted in Fig. 8.27. The transition rate from a marking m_i to a marking m_j in the Markov chain is the sum of the firing rates of the transitions t_k enabled in m_i such that, if t_k fires in m_i, it produces marking m_j. In the case of the net of Fig. 8.26, for each marking, the firing of different transitions produces different markings; hence the transition rate from m_i to M_j is simply the firing rate of the unique transition that produces m_j from m_i, as depicted in Fig. 8.27.

A simple generalization of the stochastic Petri net model allows firing rates to depend on the marking of the net. This does not change any basic property of the model because every sampling of the probability distribution function determined by the firing rate is independent of the others.

Stochastic Petri nets have been further extended into "generalized stochastic Petri nets", which introduce also the possibility of *immediate* transitions (that is, transitions that fire as soon as they are enabled). This supports the description of nondeterministic choices between alternative execution paths (that is, choices that are typically made instantaneously and that do not have a duration, or have negligible duration with respect to the dynamics of the system), which may be useful for modeling non-purely stochastic complex systems.

Exercise 8.33. In stochastic Petri nets, the set of reachable markings does *not* depend on the firing rates. Build a timed Petri net whose set of reachable markings *does* depend on the firing delay intervals. ∎

Exercise 8.34. This exercise compares three interpretations of the Petri net in Fig. 8.28:

Fig. 8.28 A stochastic Petri net

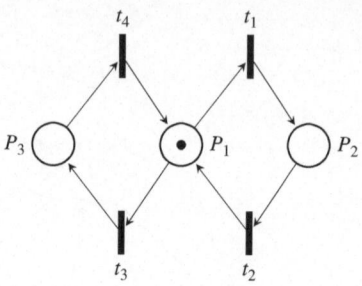

S: a stochastic Petri net where the firing rates of t_1 and t_2 are both ρ, and the firing rates of t_3 and t_4 are both λ.

T: a timed Petri net where t_1 and t_2 both have $[d, D]$ as firing delay interval, and t_3 and t_4 have $[h, H]$, assuming a strong time semantics.

T^w: as in **T**, but with a weak time semantics.

Let $d = D = 1/\rho$ and $h = H = 1/\lambda$; that is, each transition in the timed net has a firing delay equal to the average delay in the stochastic net.

- Which transitions fire in every behavior of **S**, **T**, or **T^w**?
- Which transitions fire in no behavior of **S**, **T**, or **T^w**?
- Compare the sets of all firing sequences of **S**, **T**, and **T^w**. Which is the largest? Which is the smallest?   ∎

Analysis and Verification. The decidability of fundamental properties such as reachability and boundedness is essentially equivalent in stochastic Petri nets and basic Petri nets. In addition, the theory of continuous-time Markov chains can be leveraged to calculate some quantitative properties of a stochastic Petri net, such as the *probability* that a certain marking is reached or the *average number* of tokens in a place. Such figures are typically computed through numerical analysis algorithms; the complexity of these algorithms often depends on the number of reachable markings of the Petri net.

8.6 Languages, Methods, and Tools Based on Petri Nets

Several practical specification languages are based on Petri nets. Among them, UML activity diagrams feature a syntax clearly inspired by Petri nets, though their semantics is not formalized completely. Figure 8.29 displays a simple activity diagram for the concurrent processes described in Example 8.5. The informal meaning of the diagram is the following: when the system starts, processes P_1 and P_2 start activities in which they simply idle away, while the resource is being prepared to be assigned; then, process P_1 (or P_2) starts an activity in which it is getting ready to acquire the resource (which typically involves requesting it); after the resource is ready to be assigned and one of the two processes is chosen, it gets

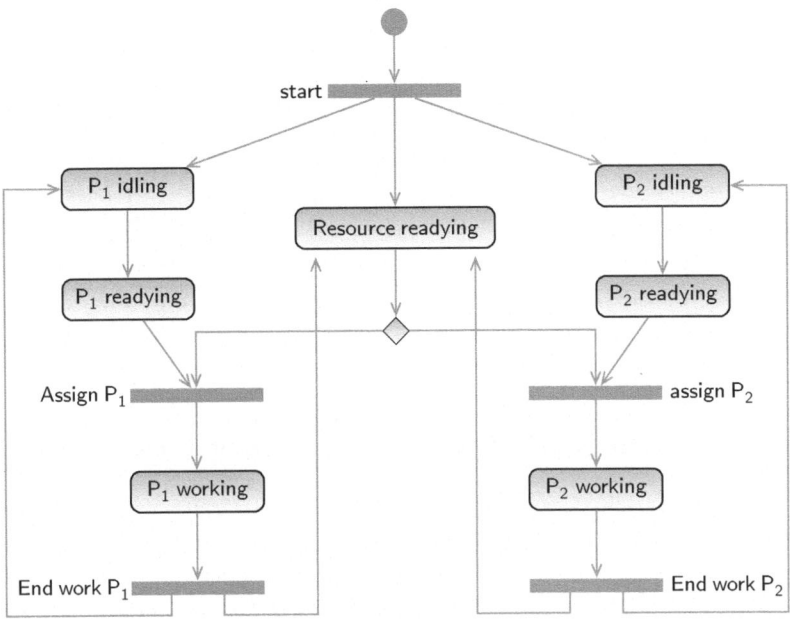

Fig. 8.29 An example of UML activity diagram

the resource, and then releases it, thus starting anew the readying activity of the resource (and its own idling activity).

8.6.1 CO-OPN2 and the Railroad Crossing Problem

CO-OPN is a significant example of a comprehensive specification environment for concurrent and real-time systems, built on top of "vanilla" Petri nets. This section presents the essential features of CO-OPN and demonstrates them on the railroad crossing problem (introduced in Example 8.26).

CO-OPN combines Petri nets with abstract data types and object-oriented constructs to facilitate the integration of heterogeneous components. CO-OPN2 is a version of CO-OPN that supports real-time modeling through features inspired by timed Petri nets with inhibitor arcs (as introduced in the present chapter).

The fundamental composition mechanism of CO-OPN2 is the synchronization between objects, illustrated by the *synchronized real-time Petri net* in Fig. 8.30.

The two *objects* O_1 and O_2 essentially behave as asynchronous timed Petri nets. The *interface transitions* **Move** and **Put**, called *methods* in the object-oriented terminology customary in CO-OPN2, are connected by a dashed line representing a *synchronization* mechanism: neglecting some inessential semantic details, once enabled, they must fire simultaneously within a time in the intersection of their

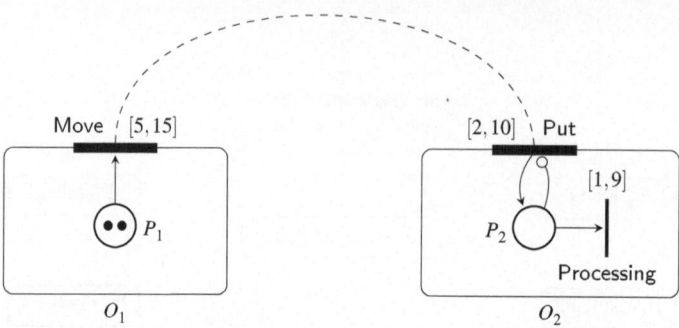

Fig. 8.30 A synchronized real-time Petri net

respective time intervals ([5, 10] in Fig. 8.30, assuming they are both enabled at time 0).

Let us now consider the "generalized railroad crossing" (*GRC*), the version of the railroad crossing problem (introduced in Example 8.26) with a parametric number of tracks, where any number of trains can be in the regions R and I at the same time. Figure 8.31 proposes a possible formalization of the GRC problem by means of the synchronized real-time Petri nets of CO-OPN2. The two components model a train and a gate and correspond to CO-OPN2 *objects*. The synchronized transitions define the *interface* of the two components, with Enter_R_signal synchronized with Enter_R, and Exit_I_signal synchronized with Exit_I. The object Train keeps track of the arrival and exit of trains. Place P is used to prevent the simultaneous entry (Enter_R) of multiple trains at the same time (assuming $t_1' > 0$).

Place Counter stores a number of tokens equal to the number of trains occupying the regions R and I. Hence, the inhibitor arc attached to Counter prevents the opening of the gate as long as R and I are occupied by some train. In fact, the GRC problem requires that the command to close the gate be issued exactly when the first train enters region R.

Exercise 8.35. In the GRC model of Fig. 8.31, which relations among the various time parameters guarantee the safety of the system (that is, the gate is fully closed when a train is in I)?   ■

Exercise 8.36. In the GRC model of Fig. 8.31, which semantics for inhibitor arcs (TI or UI) is more appropriate for describing consistently the behavior of the system?   ■

8.6.2 Tools for Petri Nets

The rest of this chapter discusses the main features of Petri net models with respect to decidability, analyzability, and complexity. As we have seen, most properties are

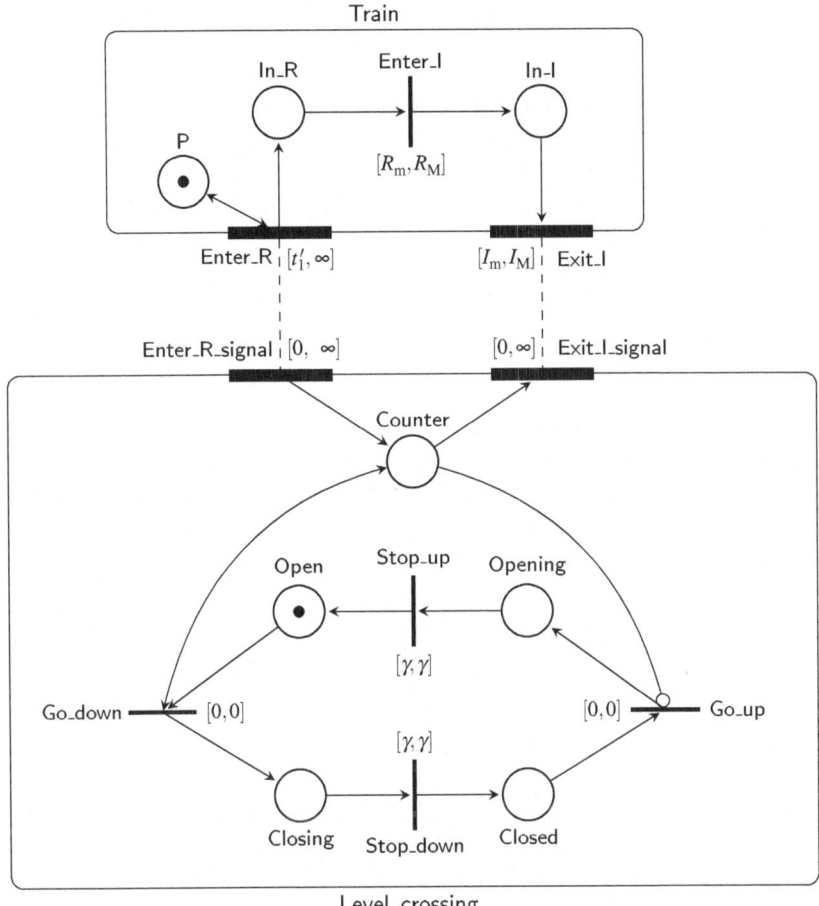

Fig. 8.31 A synchronized real-time Petri net specifying the GRC system

undecidable or intractable in the most general Petri net models, which rules out the availability of efficient analysis algorithms. Even the simpler goal of building interpreters for Petri nets that analyze their properties through simulation faces problems of combinatorial explosion due to the intrinsic nondeterminism of the model.

Nevertheless, some of these difficulties have been shown to be practically surmountable, and interesting tools for the analysis of both untimed and timed Petri nets are available. Among them we mention TINA, based on Berthomieu and Diaz's algorithm for the reachability problem of timed Petri nets. This algorithm, which assumes that the time bounds associated with transitions are rational numbers, has pioneered algorithms for similar models. The adoption of rational values (which always admit a least common multiple) for the transitions' time bounds allows for the partitioning of the Petri net's state space into a finite set of equivalence classes, and hence for the definition of exhaustive enumeration techniques for its analysis. Another interesting tool is ROMEO, which exploits more recent approaches based on the translation of timed Petri nets into timed automata. Several tools are also available for the analysis of stochastic Petri nets, or variations thereof. Most of them rely on algorithms that exploit the theory of stochastic processes and queuing theory and allow for the simulation of the modeled system. Let us mention, among them, GreatSPN and TimeNET.

8.7 Bibliographic Remarks

The literature on Petri nets, especially of the timed versions, is quite vast and diverse. When this model, intuitive and appealing in its basic version, is enriched with a notion of time, it lends itself to numerous variations that are apt to satisfy modeling and specification needs in a great variety of applicative situations.

Many monographs give a comprehensive presentation of Petri nets and their variants [13, 17, 19, 27, 29, 38, 41].

Carl Adam Petri introduced the first net model in 1963 [39]. His original model slightly differs from the basic Petri nets of Sect. 8.1, in that a transition is enabled only if the places in its postset are empty.

Esparza and Nielsen survey the decidability and complexity of basic (and bounded) Petri nets [21, 22]. Busi extends the analysis to Petri nets with inhibitor arcs [12].

The timed Petri net model introduced in Sect. 8.3 follows Merlin and Farber [32]. Merlin and Farber called their model "time Petri nets", whereas the name "*timed* Petri nets" usually identifies a different model introduced independently by Ramchandani [40]. This chapter ignores this, slightly confusing, terminology, and it is systematically based on Merlin and Farber. Wang and Penczek and Polrola [37, 45] gave extensive presentations of the various time (and timed) Petri nets.

The large number of variations in the definition and semantics of timed extensions of Petri nets calls for a systematic comparison of the expressiveness of the several variants. We refer to works pursuing this task [6, 9, 10, 14–16, 26, 30]. In particular, Ghezzi et al. [26] introduced the distinction between weak and strong time semantics, and Cerone and Maggiolo-Schettini [16] performed a detailed comparison of the expressive power of timed Petri nets under these two semantics.

The subtleties of the timed Petri net semantics are analyzed in great detail in the authors' work [23, 26]. Gargantini et al. [25] suggested the use of infinitesimal numbers to represent zero-time transitions of timed Petri nets.

Stochastic Petri nets were introduced by Natkin [36] and, independently, by Molloy [34]. Generalized stochastic Petri nets were defined by Ajmone Marsan et al. [2, 3], who also provided [1] an introduction to the model, and comprehensive references [4]. Generalized stochastic Petri nets can be analyzed through the GreatSPN [5] and TimeNET [48] tools.

Activity diagrams are part of the UML standard [44]. Synchronized real-time Petri nets in the context of the CO-OPN2 formalism are also described by several authors [11, 18].

The dining philosopher problem was introduced as a synchronization problem by Dijkstra [20], who credits Tony Hoare for suggesting the metaphor of the philosophers [33]. It can be found, in many variants and presentations, in many books and articles.

Heitmeyer and Mandrioli's book [28] presents formalizations of the railroad crossing problem with various real-time notations.

Petri nets have been applied to the modeling and analysis of biological systems [31], for which the tool Snoopy [42] has been developed.

Another field in which Petri nets (untimed, timed, and stochastic) have been widely applied is the modeling and control of flexible manufacturing systems [35, 46, 47].

Berthomieu and Diaz's reachability algorithm for timed Petri nets [7] is implemented in the Tina tool [8]. The Romeo tool is in contrast based on the translation [24] of timed Petri nets into timed automata.

The "Petri Nets World" Web site [43] maintains a comprehensive list of tools based on Petri nets.

References

1. Ajmone Marsan, M.: Stochastic Petri nets: an elementary introduction. In: Rozenberg, G. (ed.) Advances in Petri Nets. Lecture Notes in Computer Science, vol. 424, pp. 1–29. Springer, Berlin/New York (1989)
2. Ajmone Marsan, M., Balbo, G., Chiola, G., Conte, G.: Generalized stochastic Petri nets revisited: random switches and priorities. In: Proceedings of the International Workshop on Petri Nets and Performance Models, pp. 44–53. IEEE-CS Press, Los Alamitos (1987)
3. Ajmone Marsan, M., Balbo, G., Conte, G.: A class of generalized stochastic Petri nets for the performance analysis of multiprocessor systems. ACM Trans. Comput. Syst. 2(1), 93–122 (1984)
4. Ajmone Marsan, M., Balbo, G., Conte, G., Donatelli, S., Franceschinis, G.: Modelling with Generalized Stochastic Petri Nets. Wiley (1995). Available online at http://www.di.unito.it/~greatspn/GSPN-Wiley/
5. Baarir, S., Beccuti, M., Cerotti, D., De Pierro, M., Donatelli, S., Franceschinis, G.: The GreatSPN tool: recent enhancements. SIGMETRICS Perform. Eval. Rev. 36(4), 4–9 (2009)
6. Bérard, B., Cassez, F., Haddad, S., Lime, D., Roux, O.H.: Comparison of the expressiveness of timed automata and time Petri nets. In: Pettersson, P., Yi, W. (eds.) FORMATS. Lecture Notes in Computer Science, vol. 3829, pp. 211–225. Springer, Heidelberg (2005)
7. Berthomieu, B., Diaz, M.: Modeling and verification of time dependent systems using time Petri nets. IEEE Trans. Softw. Eng. 17(3), 259–273 (1991)
8. Berthomieu, B., Vernadat, F.: Time Petri nets analysis with TINA. In: QEST, pp. 123–124. IEEE Computer Society, Los Alamitos (2006)
9. Bouyer, P., Haddad, S., Reynier, P.A.: Timed Petri nets and timed automata: On the discriminating power of Zeno sequences. Inf. Comput. 206(1), 73–107 (2008)
10. Boyer, M., Roux, O.H.: On the compared expressiveness of arc, place and transition time Petri nets. Fundam. Inform. 88(3), 225–249 (2008)

11. Buchs, D., Guelfi, N.: A formal specification framework for object-oriented distributed systems. IEEE Trans. Softw. Eng. **26**(7), 635–652 (2000)
12. Busi, N.: Analysis issues in Petri nets with inhibitor arcs. Theor. Comput. Sci. **275**(1–2), 127–177 (2002)
13. Cassandras, C.G., Lafortune, S.: Introduction to Discrete Event Systems. Springer, Boston (2007)
14. Cassez, F., Roux, O.H.: Structural translation from time Petri nets to timed automata. J. Syst. Softw. **79**(10), 1456–1468 (2006)
15. Cerone, A.: A net-based approach for specifying real-time systems. Ph.D. thesis, Università degli Studi di Pisa, Dipartimento di Informatica (1993). TD-16/93
16. Cerone, A., Maggiolo-Schettini, A.: Time-based expressivity of time Petri nets for system specification. Theor. Comput. Sci. **216**(1–2), 1–53 (1999)
17. David, R., Alla, H.: Discrete, Continuous, and Hybrid Petri Nets. Springer, Berlin/London (2010)
18. Di Marzo Serugendo, G., Mandrioli, D., Buchs, D., Guelfi, N.: Real-time synchronised Petri nets. In: Proceedings of the 23rd International Conference on Application and Theory of Petri Nets (ICATPN'02), Adelaide. Lecture Notes in Computer Science, vol. 2360, pp. 142–162. Springer, Berlin (2002)
19. Diaz, M.: Petri Nets: Fundamental Models, Verification and Applications. Wiley, Hoboken (2009)
20. Dijkstra, E.W.: Hierarchical ordering of sequential processes. Acta Inf. **1**(2), 115–138 (1971)
21. Esparza, J.: Decidability and complexity of Petri net problems–an introduction. In: Rozenberg, G., Reisig, W. (eds.) Advances in Petri Nets. Lecture Notes in Computer Science, vol. 1491, pp. 374–428. Springer, Berlin (1998)
22. Esparza, J., Nielsen, M.: Decidability issues for Petri nets–a survey. Bull. Eur. Assoc. Theor. Comput. Sci. **52**, 245–262 (1994)
23. Felder, M., Mandrioli, D., Morzenti, A.: Proving properties of real-time systems through logical specifications and Petri net models. IEEE Trans. Softw. Eng. **20**(2), 127–141 (1994)
24. Gardey, G., Lime, D., Magnin, M., Roux, O.H.: Romeo: a tool for analyzing time Petri nets. In: Etessami, K., Rajamani, S.K. (eds.) CAV. Lecture Notes in Computer Science, vol. 3576, pp. 418–423. Springer, Heidelberg (2005)
25. Gargantini, A., Mandrioli, D., Morzenti, A.: Dealing with zero-time transitions in axiom systems. Inf. Comput. **150**(2), 119–131 (1999)
26. Ghezzi, C., Mandrioli, D., Morasca, S., Pezzè, M.: A unified high-level Petri net formalism for time-critical systems. IEEE Trans. Softw. Eng. **17**(2), 160–172 (1991)
27. Haas, P.J.: Stochastic Petri Nets: Modelling, Stability, Simulation. Springer, New York (2010)
28. Heitmeyer, C.L., Mandrioli, D. (eds.): Formal Methods for Real-Time Computing. Wiley, Chichester/New York (1996)
29. Jensen, K., Kristensen, L.M.: Coloured Petri Nets: Modelling and Validation of Concurrent Systems. Springer, Dordrecht/New York (2009)
30. Khansa, W., Denat, J.P., Collart-Dutilleul, S.: P-time Petri nets for manufacturing systems. In: Proceedings of the International Workshop on Discrete Event Systems (WODES'96), Edinburgh, pp. 94–102 (1996)
31. Krepska, E., Bonzanni, N., Feenstra, K.A., Fokkink, W., Kielmann, T., Bal, H.E., Heringa, J.: Design issues for qualitative modelling of biological cells with Petri nets. In: Proceedings of Formal Methods in System Biology (FMSB'08), Cambridge, pp. 48–62 (2008)
32. Merlin, P.M., Farber, D.J.: Recoverability and communication protocols: implications of a theoretical study. IEEE Trans. Commun. **24**(9), 1036–1043 (1976)
33. Misa, T.J.: An interview with Edsger W. Dijkstra. Commun. ACM **53**(1), 41–47 (2010)
34. Molloy, M.K.: On the integration of delay and throughput measures in distributed processing models. Ph.D. thesis, University of California, Los Angeles (1981)
35. Moody, J., Antsaklis, P.J.: Supervisory Control of Discrete Event Systems Using Petri Nets. Kluwer, Boston (1998)

36. Natkin, S.: Les reseaux de Petri stochastiques et leur application a l'evaluation des systèmes informatiques. Thèse de Docteur Ingénieur, CNAM, Paris, France (1980)
37. Penczek, W., Polrola, A.: Advances in Verification of Time Petri Nets and Timed Automata: A Temporal Logic Approach. Springer, Berlin/New York (2006)
38. Peterson, J.L.: Petri Net Theory and the Modelling of Systems. Prentice-Hall, Englewood Cliffs (1981)
39. Petri, C.A.: Fundamentals of a theory of asynchronous information flow. In: Proceedings of IFIP Congress, pp. 386–390. North Holland Publishing Company, Amsterdam (1963)
40. Ramchandani, C.: Analysis of asynchronous concurrent systems by timed Petri nets. Ph.D. thesis, Massachussets Institute of Technology (1974)
41. Reisig, W.: Petri Nets: An Introduction. EATCS Monographs on Theoretical Computer Science. Springer, Berlin/New York (1985)
42. Rohr, C., Marwan, W., Heiner, M.: Snoopy–a unifying Petri net framework to investigate biomolecular networks. Bioinformatics **26**(7), 974–975 (2010)
43. Petri nets world. http://www.informatik.uni-hamburg.de/TGI/PetriNets/
44. UML 2.0 superstructure specification. Tech. Rep. formal/05-07-04, Object Management Group (2005)
45. Wang, J.: Timed Petri Nets, Theory and Application. Kluwer, Boston (1998)
46. Wang, J., Deng, Y.: Incremental modeling and verification of flexible manufacturing systems. J. Intell. Manuf. **10**(6), 485–502 (1999)
47. Zhou, M., Venkatesh, K.: Modeling, Simulation, and Control of Flexible Manufacturing Systems: A Petri Net Approach. World Scientific, Singapore/River Edge (1999)
48. Zimmermann, A., Knoke, M.: TimeNET 4.0 user manual. Tech. Rep. 2007-13, Technische Universität Berlin, Faculty of EE&CS (2007)

Chapter 9
Logic-Based Formalisms

The modeling notations presented in Chaps. 7 and 8 are operational in nature; this chapter deals with the use of *logics* for modeling time features. Logic is a descriptive notation: it can express *properties* of the modeled systems, rather than their states, transitions, and actions.

A straightforward approach to applying logics to the description of temporal features, very much in line with approaches reported in Chaps. 2 and 4, is to introduce – in the predicates and functions that constitute the alphabet of the logic – an additional argument representing absolute time.

The modeling of parallel, reactive, time-critical systems is, however, most often done using *temporal logics*, which are a type of *modal logic* where modalities are given an interpretation in terms of time. Many temporal logics have been introduced in the last few decades, characterized by features such as the underlying time model (linear or branching) and their ability to deal with discrete or dense time or to express relevant quantitative properties such as durations, delays, and time-outs.

The widespread adoption of temporal logics is due to their ability to express properties in a concise, intuitive way, and to the fact that the numerous existing variations of this notation can provide the expressive power needed in many application fields and are often amenable to effective analysis and verification. As a consequence, most of the dimensions of the time modeling problem introduced in Chap. 3 will naturally arise and will be discussed in connection with the features of the various temporal logics analyzed in the present chapter.

Sections 9.1–9.4 present several types of temporal logics. We start in Sect. 9.1, with an illustration and multifaceted comparison of the simplest ones: linear- and branching-time temporal logics with only future-time operators and a discrete time model, without metric operators.

After presenting these basic temporal logics, we introduce features that can increase their modeling ability, such as the presence of past-time operators in addition to the future ones (Sect. 9.2), the inclusion of metric operators (Sect. 9.3), and the adoption of dense time (Sect. 9.4).

C.A. Furia et al., *Modeling Time in Computing*, Monographs in Theoretical Computer Science. An EATCS Series, DOI 10.1007/978-3-642-32332-4_9, © Springer-Verlag Berlin Heidelberg 2012

Section 9.5 presents temporal logics that adopt time *intervals* as the fundamental entities of time modeling. Section 9.6 discusses issues arising from the need to combine in a unique logic-based model different time granularities. Section 9.7 discusses some logic-based formalisms which adopt an explicit notion of time and are therefore not based on the modal operators of classical temporal logic. Section 9.8 introduces probabilistic logic-based formalisms, where events are assigned probabilities, so that we can express requirements on the probability of certain system evolutions occurring. Section 9.9 provides an overview of the main methods and tools for analyzing logic-based temporal models, possibly in connection with operational ones.

9.1 Linear- and Branching-Time Temporal Logics

This section starts with the simple and intuitive "linear temporal logic" (LTL), and then proceeds to the richer and more complex branching-time temporal logic, which includes all the operators of LTL plus additional operators for quantifying over execution paths of transition systems.

9.1.1 Linear Temporal Logic

As a first, very simple example of temporal logic, let us consider propositional "LINEAR temporal logic" (LTL) over discrete time. LTL formulae are built from atomic propositions with the usual Boolean connectives and the temporal operators $\bigcirc$ ("next", also "X"), $\Diamond$ ("eventually in the future", also "F"), $\Box$ ("always in the future" or "globally", also "G"), and U ("until"). These have rather natural and intuitive meanings, as LTL formulae are interpreted over linear sequences of states, each state corresponding to a time instant. LTL formulae implicitly refer to a "current time instant", and temporal connectives assert that some propositions hold in time instants different from the current one: formula $\bigcirc\phi$ means that proposition ϕ holds at the state that immediately follows the one where the formula is interpreted, $\Diamond\phi$ means that ϕ will hold at some state following, or including, the current one, $\Box\phi$ means that ϕ will hold at all future and present states, and ϕ U ψ means that there is some future state in which proposition ψ will hold and ϕ holds from now until then. Propositions appearing in LTL formulae label the elementary facts that may hold in each state.

Let us consider the resource allocator first introduced in Chap. 7; the following formula specifies that, if a high-priority request is issued (proposition hpr is true) at a time when the resource is free (modeled by proposition free), then the resource will be granted and become occupied (proposition occ) at the second successive state in the sequence:

$$\Box((\text{hpr} \wedge \text{free}) \Rightarrow \bigcirc\bigcirc\text{occ}) .$$

LTL is well suited to specifying **QUALITATIVE** time relations, such as eventuality, invariance, and ordering among events. For instance, the following LTL formula, still concerning the resource allocator, describes a possible assumption about incoming resource requests, that no low-priority request can follow a high-priority request until a release of the resource occurs:

$$\Box\,(\mathsf{hpr} \Rightarrow \bigcirc((\neg\mathsf{lpr})\ \mathsf{U}\ \mathsf{rel}))\,.$$

Though LTL is not expressly equipped with temporal operators to express metric properties on time, we can use the next operator $\bigcirc$ for this purpose: for instance, $\bigcirc^3 p$ (i.e., $\bigcirc\bigcirc\bigcirc p$) indicates that proposition p holds three time units in the future. The use of $\bigcirc^k$ (i.e., $\bigcirc$ repeated k times) to denote the time instant at k time units in the future is only possible, however, under the condition that there is a one-to-one correspondence between the states of the sequences over which formulae are interpreted and the time points of the temporal domain; we will see in the following sections that this is not the case for some variants of temporal logic. Formula $\bigcirc^k \phi$, strictly speaking, is not compliant with the syntax of LTL: it is a shorthand for the well-formed formula $\overbrace{\bigcirc\bigcirc\cdots\bigcirc}^{k\ \text{times}}\phi$, whose size, measured in terms of nested operator occurrences, grows linearly with the *value* of the distance k of the referenced state from the current implicit time; this fact may influence the performance of analysis and verification algorithms, which, as we will see in Sect. 9.9, typically depends on the size of the formulae encoding the analyzed property.

Definition 9.1 (LTL syntax and semantics). Given a set AP of atomic propositions, LTL formulae are defined as follows, with $p \in AP$:

$$\phi \quad ::= \quad p \mid \neg\phi \mid \phi_1 \wedge \phi_2 \mid \bigcirc\phi \mid \Diamond\phi \mid \Box\phi \mid \phi_1\ \mathsf{U}\ \phi_2\,.$$

The semantics of LTL interprets formulae over linear sequences of states $\sigma = s_0 s_1 s_2 s_3 \ldots$ that are infinite in the future direction only. Each state s is an assignment of values to all atomic propositions in AP, that is, an element of $\wp(AP)$. If we denote as σ^i the suffix sequence starting from state s_i, that is, $\sigma^i = s_i s_{i+1} s_{i+2} \ldots$, the satisfaction relation $\models$ between sequences and LTL formulae is defined as follows (for $p \in AP$):

$\sigma \models p$	if and only if	$p \in s_0$;
$\sigma \models \neg\phi$	if and only if	it is not the case that $\sigma \models \phi$;
$\sigma \models \phi_1 \wedge \phi_2$	if and only if	$\sigma \models \phi_1$ and $\sigma \models \phi_2$;
$\sigma \models \bigcirc\phi$	if and only if	$\sigma^1 \models \phi$;
$\sigma \models \Diamond\phi$	if and only if	there exists i such that $i \geq 0$ and $\sigma^i \models \phi$;
$\sigma \models \Box\phi$	if and only if	for all i such that $i \geq 0$, $\sigma^i \models \phi$;
$\sigma \models \phi_1\ \mathsf{U}\ \phi_2$	if and only if	there exists $j \geq 0$ such that $\sigma^j \models \phi_2$ and, for all i such that $0 \leq i < j$, $\sigma^i \models \phi_1$.

The semantics of an LTL formula ϕ is then defined as the set $\text{Words}(\phi)$ of infinite words (state sequences) over which the formula is satisfied:

$$\text{Words}(\phi) \quad = \quad \{\sigma \in \wp(AP)^{\omega} \mid \sigma \models \phi\}. \qquad \blacksquare$$

From the above definition of $\sigma \models \phi$, it is apparent that ϕ is conventionally interpreted with reference to the first state of the sequence, that is, the first state of σ serves as the implicit "current" time for the overall formula ϕ.

Remark 9.2. From the LTL semantics in Definition 9.1, it follows that the operators $\diamond$ and $\square$ are not primitive, as they can be derived from the until operator U as

$$\diamond\phi = \textbf{True } \mathsf{U} \ \phi$$

and

$$\square\phi = \neg\diamond\neg\phi.$$

This is not surprising, because the formula $\phi_1 \ \mathsf{U} \ \phi_2$ includes an existential quantification on the occurrence of ϕ_2, as it states that *there exists a future time point* where ϕ_2 holds. Therefore the definition $\diamond\phi = \textbf{True } \mathsf{U} \ \phi$ simply removes any requirement on the first argument of the until operator, leaving the future occurrence of ϕ as the only remaining one, while $\square\phi = \neg\diamond\neg\phi \equiv \neg(\textbf{True } \mathsf{U} \ \neg\phi)$ is a typical example of a definition of a universal quantification ("property ϕ will hold *for every future time* instant") in terms of the negation of an existential one ("there is no future time instant where ϕ does not hold").

Based on these relations, in the following semantic definitions, operators $\diamond$ and $\square$ will not be introduced directly, and operators $\bigcirc$ and U will be considered the only two fundamental time operators of LTL. $\qquad \blacksquare$

Example 9.3. Still with reference to the resource allocator example, consider the set of atomic propositions $AP = \{\text{free}, \text{hpr}, \text{lpr}, \text{occ}, \text{rel}\}$, the LTL formulae

$$\phi_1 = \square((\text{free} \wedge \text{hpr}) \Rightarrow \bigcirc\bigcirc\text{occ}),$$

$$\phi_2 = \square(\text{hpr} \Rightarrow \bigcirc(\neg\text{lpr } \mathsf{U} \text{ rel})), \text{ and}$$

$$\phi_3 = \square\diamond\text{hpr}.$$

We described ϕ_1 and ϕ_2 in the previous section; ϕ_3 states that hpr occurs infinitely often in the future. Let us now consider two state sequences:

$\sigma_1 = \{\text{free}\}\,\{\text{free}, \text{hpr}\}\,\{\text{free}\}\,\{\text{occ}\}\,\{\text{rel}\}\,\{\text{free}\}\,\{\text{free}, \text{hpr}\}\,\{\text{free}\}\,\{\text{free}\}\,\{\text{occ}\}$

$\{\text{occ}\}\,\{\text{rel}\}\,(\{\text{free}\})^{\omega}$ and

$\sigma_2 = \{\text{free}\}\,\{\text{free}, \text{hpr}\}\,\{\text{free}\}\,\{\text{occ}\}\,\{\text{rel}\}\,\{\text{free}\}, \{\text{free}, \text{hpr}\}\,\{\text{free}, \text{lpr}\}\,\{\text{occ}\}$

$\{\text{rel}\}\,(\{\text{free}, \text{hpr}\}\,\{\text{free}\}, \{\text{occ}\}\,\{\text{rel}\}\,\{\text{free}\})^{\omega}.$

We have

- $\sigma_1 \not\models \phi_1$, because the property $(\mathsf{free} \wedge \mathsf{hpr}) \Rightarrow \bigcirc\bigcirc\bigcirc\mathsf{occ}$ does not hold at time 6, namely, $\sigma_1^6 \not\models (\mathsf{free} \wedge \mathsf{hpr}) \Rightarrow \bigcirc\bigcirc\bigcirc\mathsf{occ}$: in fact, the state $\{\mathsf{free}, \mathsf{hpr}\}$ is not followed, two instants later, by a state in which occ holds; notice that, instead, the subformula $(\mathsf{free} \wedge \mathsf{hpr}) \Rightarrow \bigcirc\bigcirc\bigcirc\mathsf{occ}$ holds at the previous instants; in particular, we have $\sigma_1^1 \models (\mathsf{free} \wedge \mathsf{hpr}) \Rightarrow \bigcirc\bigcirc\bigcirc\mathsf{occ}$, because state $\{\mathsf{free}, \mathsf{hpr}\}$ at time 1 is followed, at time 3, by state $\{\mathsf{occ}\}$;
- $\sigma_1 \models \phi_2$, because every hpr is eventually followed by a rel in σ_1, and lpr does not hold in any state; hence subformula $\bigcirc(\neg\mathsf{lpr}~\mathsf{U}~\mathsf{rel})$ holds in every state in which hpr holds;
- $\sigma_1 \not\models \phi_3$, because σ_1 contains only a finite number of occurrences of proposition hpr; hence, for $i \geq 7$, we have $\sigma_1^i \not\models \Diamond\mathsf{hpr}$;
- $\sigma_2 \models \phi_1$, because every occurrence of the hpr proposition is followed in σ_2, two states later, by an occurrence of the occ proposition;
- $\sigma_2 \not\models \phi_2$, because $\sigma_2^6 \not\models \mathsf{hpr} \Rightarrow \bigcirc(\neg\mathsf{lpr}~\mathsf{U}~\mathsf{rel})$, as the occurrence of hpr at time 6 is followed by an occurrence of lpr at time 7, which precedes the occurrence of rel at time 9; notice that $\sigma_2^1 \models \mathsf{hpr} \Rightarrow \bigcirc(\neg\mathsf{lpr}~\mathsf{U}~\mathsf{rel})$, because the first occurrence of hpr is not followed by any lpr in the states following it up to time 4, where proposition rel holds;
- $\sigma_2 \models \phi_3$, because the sequence of states $\{\mathsf{free}, \mathsf{hpr}\} \{\mathsf{free}\} \{\mathsf{occ}\} \{\mathsf{rel}\} \{\mathsf{free}\}$ included in the ω operator is repeated an infinite number of times, so that we have $\sigma_2^i \models \Diamond\mathsf{hpr}$ for every i. ∎

It is also possible to interpret LTL formulae with reference to transition systems, introduced in Sect. 7.1. In this case, the alphabet of LTL formulae is typically $AP \cup I$, including both state labels AP and transition labels I. For instance, with reference to the resource allocator modeled as a transition system in Example 7.19, the LTL formula $\Box((\mathsf{free} \wedge \mathsf{hpr}) \Rightarrow \bigcirc\mathsf{occ})$ states the property that "whenever the resource is free and a high-priority request occurs, at the next time instant the resource is occupied", which holds for the transition system model of the resource allocator depicted in Fig. 7.17. The example also shows that, when interpreting an LTL formula with reference to transition systems, it is customary to evaluate the items of the input alphabet labeling a transition at the same time instant as the propositional letters labeling the state from which the transition exits. The interpretation of LTL formulae with respect to transition systems can be formalized by introducing the notion of input-inclusive states and input-inclusive state sequences, and by interpreting LTL formulae, with respect to input-inclusive state sequences, with exactly the same semantics as in Definition 9.1. An input-inclusive state is an assignment of values

$$\underline{s} \in \{s \cup \{i\} \mid s \in \wp(AP) \text{ and } i \in I\},$$

and an input-inclusive trace is a sequence of input-inclusive states. For each run of a transition system $s_0~i_1~s_1~i_2~\ldots~s_k~i_{k+1}~\ldots$ there exists a corresponding

input-inclusive state sequence $\underline{s}_0\,\underline{s}_1\,\underline{s}_2\,\cdots\,\underline{s}_k\,\cdots$ where, for each k, $\underline{s}_k = s_k \cup \{i_{k+1}\}$.
Then a transition system TS satisfies an LTL formula ϕ, denoted by $TS \models \phi$, if
and only if *all* input-inclusive traces of TS starting from an initial state satisfy ϕ.
If Words (TS) denotes the set of all possible input-inclusive traces of TS starting
from an initial state, then

$$TS \models \phi \quad \text{if and only if} \quad \text{Words}\,(TS) \subseteq \text{Words}\,(\phi)\,.$$

In the rest of the book, when using LTL or other temporal logics to describe
properties of transition systems, we will freely use both *state* and *transition* labels
as propositions.

From the above definition it is also clear that the set Words (TS) is in general
a *subset* of all possible state sequences over which ϕ is interpreted as true: there
might exist state sequences that satisfy ϕ, but are not generated by any execution
of the transition system. This circumstance is often described by saying that the
transition system provides a possible *implementation* of the specification ϕ, and any
other transition system TS', such that Words $(TS') \subseteq$ Words (ϕ), also constitutes
an admissible, alternative implementation.

9.1.1.1 Expressing Properties in Linear Temporal Logic

We recall some important classes of properties of parallel reactive systems, namely
safety, liveness, fairness, and *reachability,* and illustrate the extent to which they can
be specified in LTL. In fact, typically in the design of such systems, an operational
model of the system under development is provided by means of a transition system,
and its expected properties are stated in temporal logic.

Intuitively, a *safety* property states that "something good occurs continually"
or, equivalently, "nothing bad ever happens". A safety property can be disproved
by exhibiting a *finite* execution where "good" does not hold at a certain point in
time (or "bad" eventually occurs). Somewhat symmetrically, a *liveness* property
states that some desired circumstance eventually occurs from any instant (we are
considering *unbounded* liveness, where the length of time to the occurrence of the
desired event is not limited). A *liveness* property can be disproved by exhibiting an
infinite computation in which the desired event never occurs from a certain point
on. A special case of liveness is *responsiveness*: such a property typically states
that whenever a service is requested it will eventually be provided. Both safety and
liveness are easily expressible in LTL: for instance, if proposition ϕ represents an
undesired event, the safety property that it will never take place is expressed by
the LTL formula $\square(\neg\phi)$; if proposition ϕ represents a service request and ψ the
granting of this request then formula $\square(\phi \Rightarrow \diamond\psi)$ specifies responsiveness to the
request.

The condition of *fairness* is often adopted as a necessary assumption for proving
liveness properties in parallel systems competing for shared resources, for instance,
in the scheduling of processes. We illustrate three versions of fairness, using as

reference a set of processes competing for a shared processor. Formula ψ holds whenever a process is enabled, and ϕ when the process gets its turn to be executed.

Unconditional fairness   asserts that the process is executed infinitely often: $\Box\Diamond\phi$;
Strong fairness   states that every process enabled infinitely often will get its turn infinitely often: $\Box\Diamond\psi \Rightarrow \Box\Diamond\phi$;
Weak fairness   specifies that if a process is continually enabled from some time on, then it will get its turn infinitely often: $\Diamond\Box\psi \Rightarrow \Box\Diamond\phi$.

Fairness is sometimes referred to as *freedom from starvation*: for instance, in the dining philosophers Example 8.11, the property that philosopher i gets the opportunity to eat infinitely often corresponds to the LTL formula $\Box\Diamond\mathsf{Ph}_i^{\mathrm{Ea}}$.

The reachability of certain states for transition systems can be easily expressed in LTL if it is intended in a strong, *universal* meaning, that is, if the desired property refers to all possible system traces: in this case, it simply amounts to a form of liveness. For instance, with reference to the transition system of Fig. 8.2, the property that all executions reach the state where propositions id_1, to_2, and hl_2 hold simultaneously is stated by the formula $\Diamond(\mathsf{id}_1 \wedge \mathsf{to}_2 \wedge \mathsf{hl}_2)$; however, $\Diamond(\mathsf{id}_1 \wedge \mathsf{to}_2 \wedge \mathsf{hl}_2)$ does not hold for the considered transition system, as shown by the execution trace $(\langle \mathsf{id}_1, \mathsf{free}, \mathsf{id}_2 \rangle \langle w_1, \mathsf{free}, \mathsf{id}_2 \rangle \langle \mathsf{hl}_1, \mathsf{to}_1, \mathsf{id}_2 \rangle)^\omega$. In some cases, we may be interested in expressing a different property, called *existential* reachability, asserting that a configuration where $\mathsf{id}_1 \wedge \mathsf{to}_2 \wedge \mathsf{hl}_2$ holds is reached by *some* (not necessarily all) computations of the transition system. To establish that a state characterized by $\mathsf{id}_1 \wedge \mathsf{to}_2 \wedge \mathsf{hl}_2$ is reachable in some traces, we can build formula $\Box\neg(\mathsf{id}_1 \wedge \mathsf{to}_2 \wedge \mathsf{hl}_2)$, which asserts that the desired configuration is never reached in all possible executions, and disprove it (by means of some of the methods and tools discussed in Sect. 9.9): if it is not the case that for all paths a state where $(\mathsf{id}_1 \wedge \mathsf{to}_2 \wedge \mathsf{hl}_2)$ holds is never reached, then there is at least one path σ, starting from the initial state, such that $\sigma \models \Diamond(\mathsf{id}_1 \wedge \mathsf{to}_2 \wedge \mathsf{hl}_2)$. Thus, existential reachability from the initial state can be expressed, though somewhat indirectly. In contrast, existential reachability from any state is not expressible in LTL. For instance, if we want to state that, starting from a state where $\mathsf{id}_1 \wedge \mathsf{to}_2 \wedge \mathsf{hl}_2$, there is at least one computation leading to a configuration where $\mathsf{hl}_1 \wedge \mathsf{to}_1 \wedge w_2$, the formula $\Box((\mathsf{id}_1 \wedge \mathsf{to}_2 \wedge \mathsf{hl}_2) \Rightarrow (\mathsf{hl}_1 \wedge \mathsf{to}_1 \wedge w_2))$ is not appropriate because it asserts that *every* computation that, at a certain time, will reach a state satisfying $\mathsf{id}_1 \wedge \mathsf{to}_2 \wedge \mathsf{hl}_2$, will subsequently reach a state where $\mathsf{hl}_1 \wedge \mathsf{to}_1 \wedge w_2$, which is not what we mean. Other attempts to express the desired property in LTL are similarly bound to fail.

Exercise 9.4. With reference to the above discussion on existential reachability, explain the meaning of the formulae

$$\Box((\mathsf{id}_1 \wedge \mathsf{to}_2 \wedge \mathsf{hl}_2) \Rightarrow \Diamond(\mathsf{hl}_1 \wedge \mathsf{to}_1 \wedge w_2)),$$

$$\Diamond(\mathsf{id}_1 \wedge \mathsf{to}_2 \wedge \mathsf{hl}_2) \Rightarrow \Diamond(\mathsf{hl}_1 \wedge \mathsf{to}_1 \wedge w_2), \text{ and}$$

$$\Diamond((\mathsf{id}_1 \wedge \mathsf{to}_2 \wedge \mathsf{hl}_2) \Rightarrow \Diamond(\mathsf{hl}_1 \wedge \mathsf{to}_1 \wedge w_2)),$$

Fig. 9.1 Nondeterministic
Büchi automaton for a traffic
light

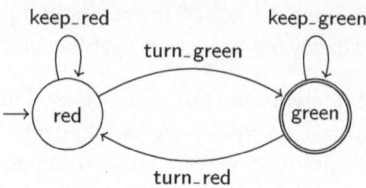

and argue why none of them expresses the property that, starting from a state where $id_1 \wedge to_2 \wedge hl_2$ holds, there is at least one computation leading to a configuration satisfying $hl_1 \wedge to_1 \wedge w_2$. ∎

9.1.1.2 Linear Temporal Logic and Nondeterministic Büchi Automata

It is interesting to compare the expressive power of LTL, a typical descriptive modeling notation, with that of operational formalisms such as automata. In this regard, since states essentially are assignments of Boolean values to a finite set of atomic propositions, the state space, over which LTL subformulae are evaluated at every time instant, is *finite*. We might therefore conjecture that some of the finite-state operational formalisms discussed in the previous chapters might be as expressive as LTL. Finite-state automata on finite words, presented in Definition 7.14, are not suitable for a comparison with LTL, because the languages defined by finite-state automata consist of finite sequences, whereas LTL formulae specify infinite behaviors, satisfied only by state sequences that effectively include unbounded series of events. For instance, consider the LTL formula $\Box\Diamond p$, for "p is repeated infinitely often": for any sequence $\sigma = s_0 \, s_1 \, s_2 \, s_3 \, \ldots$ such that $\sigma \models \Box\Diamond p$ we have that $\forall i > 0 \exists j > i : p \in s_j$. The operational formalism of Büchi automata, introduced in Sect. 7.2, is the best candidate for comparison with LTL. In fact, the following property holds:

> For every LTL formula ϕ there exists a nondeterministic Büchi automaton that accepts exactly the sequences satisfying ϕ.

Example 9.5. For a traffic light with only green and red lights, an essential property is that the green light appears infinitely often, expressed by the LTL formula $\Box\Diamond\text{green}$. This is modeled by the nondeterministic Büchi automaton with $AP = \{\text{green}, \text{red}\}$ and $I = \{\text{turn\_green}, \text{turn\_red}, \text{keep\_green}, \text{keep\_red}\}$ in Fig. 9.1. ∎

The correspondence between LTL and nondeterministic Büchi automata constitutes the basis of automata-based LTL model checking (presented in Sect. 11.1.1), a well-established verification technique which, given a system modeled by a Büchi automaton and a property stated as an LTL formula, establishes whether all possible system behaviors (i.e., its execution sequences) satisfy the stated property.

Remark 9.6 (Equivalence of LTL and Büchi automata). Since for any LTL formula ϕ it is possible to construct a nondeterministic Büchi automaton that accepts exactly the state sequences that satisfy ϕ, the question arises of whether LTL and nondeterministic Büchi automata are equivalent. It can be proved that this is not the case: nondeterministic Büchi automata are strictly more expressive than LTL, as suggested in Exercise 9.8. ∎

Exercise 9.7. Consider the variation of the traffic light specification of Example 9.5, where it is required that green and red lights alternate, with no particular requirement on the duration of each color.

- Specify the required property on color alternation by a suitable LTL formula.
- Define a nondeterministic Büchi automaton equivalent to the above-defined LTL formula. ∎

Exercise 9.8. Consider the following set of sequences of states over the alphabet $AP = \{p\}$:

$$S = \{s_0 \, s_1 \, s_2 \ldots \mid p \in s_i \text{ for all even } i\}.$$

Notice that, in the sequences of S, the states s_j, when j is an odd number, may or may not include a.

- Define a nondeterministic Büchi automaton which accepts exactly the sequences in S.
- Argue that you cannot characterize the set S by means of an LTL formula over the same singleton alphabet AP. ∎

9.1.2 Branching-Time Temporal Logic

The interpretation of LTL formulae is defined with reference to state sequences, according to the notion that, in linear time, every state has no more than one successor. When LTL formulae are used to describe system properties, it is usually intended that *all* system executions (each one represented as a state sequence on which the formula can be interpreted) satisfy it.

In a more formal setting, if the system is modeled as a transition system TS, then the semantics of an LTL formula ϕ with reference to that model is that $TS \models \phi$ if and only if *all* its possible executions satisfy ϕ. There are circumstances, however, in which we are interested in formalizing, through temporal logic, properties that hold in *some*, but not necessary all, computations of the modeled system. An example of such a property is existential reachability, discussed in Sect. 9.1.1. **BRANCHING** time temporal logic satisfies this specification need: it supports the specification of properties that hold in some, but not necessarily all, executions of a transition system.

Correspondingly, branching-time temporal logic supports two kinds of modalities: linear operators and path quantifiers. *Linear operators*, such as $\bigcirc$, U, $\square$, and $\Diamond$, come from linear temporal logic and refer to successive states in a given sequence. Path quantifiers are, in contrast, a distinctive feature of branching-time logic suggested by the fact that, in transition systems, a state might have more than one successor if the transition relation is not functional. Therefore, we can, in general, consider the *set* of successor states of a given state, and hence the set of executions departing from it. Path quantifiers denote *all* paths or *some* path departing from a given state by *universal* ($\forall$) or *existential* ($\exists$) quantification. The various existing branching-time temporal logics differ in the way in which the two kinds of modalities – linear operators and path quantifiers – are composed. In the following, we introduce a first branching-time temporal logic, called "computation tree logic" (CTL), in which linear operators and path quantifiers strictly alternate. CTL is, as a result, quite constrained in its expressive power and, despite the generalization due to the presence of path quantifiers in addition to linear operators, is incomparable with LTL: there are properties that can be expressed in LTL but not in CTL, and vice versa. A second branching-time temporal logic, CTL*, does not impose any constraint on the way in which the two kinds of operators can be composed; therefore, CTL* subsumes both LTL and CTL with regard to expressiveness.

9.1.2.1 Computation Tree Logic

Based on the presence, in CTL, of two modalities, we distinguish between state and path formulae.

CTL *state* formulae over a set AP of atomic propositions use standard logic connectives and path quantifiers; they are defined as follows:

$$\Phi ::= p \mid \neg\Phi \mid \Phi_1 \wedge \Phi_2 \mid \exists\phi \mid \forall\phi,$$

where $p \in AP$ and ϕ is a path formula. *Path* formulae introduce the linear operators, and are defined as

$$\phi ::= \bigcirc\Phi \mid \Diamond\Phi \mid \square\Phi \mid \Phi_1 \cup \Phi_2,$$

where Φ, Φ_1, and Φ_2 are state formulae. We will always indicate state formulae by *capital* Greek letters, and path formulae by lowercase ones. According to the above definitions, every instance of a linear operator ($\bigcirc$, $\Diamond$, $\square$, or $\cup$) is immediately preceded by a path quantifier ($\exists$ or $\forall$), and every path quantifier is immediately followed by a linear operator, so that, in fact, CTL formulae can include only pairs "path quantifier/linear operator" such as $\forall\bigcirc$ and $\exists\square$. Every state formula is interpreted with reference to a transition system state (which, as is customary in temporal logic, is left implicit): the universal path quantifier "$\forall$" means "for all paths starting from the current state", whereas the existential path quantifier "$\exists$" means "for some (at least one) of the paths starting from the current state". Conversely, path formulae are interpreted over sequences of states of a transition system, and the meaning of the linear operators ($\bigcirc$, $\Diamond$, $\square$, $\cup$) is exactly the same as in LTL.

To show some typical and intuitive properties expressible in CTL, consider a system where processes can request a resource or service (proposition req) and the request for the resource or service can be granted by a server process (proposition gr). The following CTL formula asserts that in every execution a request will be eventually satisfied in some of the sequences departing from the state of the request:

$$\forall \Box (\mathsf{req} \Rightarrow \exists \Diamond \mathsf{gr}) \, .$$

The following formula asserts, in contrast, that there exists a computation where all requests get eventually satisfied in every possible successive evolution:

$$\exists \Box (\mathsf{req} \Rightarrow \forall \Diamond \mathsf{gr}) \, . \tag{9.1}$$

In formula (9.1), because of the $\Box$ linear operator, the set of possible requests that are certainly granted is potentially infinite (a single execution path can contain an unbounded number of requests); however, because of the outermost $\exists$ path quantifier, there might still be an unbounded number of requests granted in no subsequent system executions: these consist of all requests that are outside the single sequence whose existence is required by the path quantifier $\exists$.

Yet another CTL formula specifies that, in every circumstance, requests will be eventually satisfied:

$$\forall \Box (\mathsf{req} \Rightarrow \forall \Diamond \mathsf{gr}) \, .$$

Even these very simple examples show how the path quantifiers and linear operators of branching-time temporal logic may interact in quite subtle ways.

Similarly as with LTL, a satisfaction relation "$\models$" between transition systems and formulae defines the semantics of CTL. Since CTL formulae are partitioned into the two categories of state and path formulae, the satisfaction relation must be defined accordingly, in two different ways, referring to states and sequences of states. The satisfaction relation $\models$ between states and state formulae is then defined as follows (for any $p \in AP$):

$s \models p$	if and only if	$p \in s$;
$s \models \neg \Phi$	if and only if	it is not the case that $s \models \Phi$;
$s \models \Phi_1 \wedge \Phi_2$	if and only if	$s \models \Phi_1$ and $s \models \Phi_2$;
$s \models \exists \phi$	if and only if	$\sigma \models \phi$ for some path σ starting from state s;
$s \models \forall \phi$	if and only if	$\sigma \models \phi$ for all paths σ starting from state s.

The satisfaction relation between sequences and path formulae is defined as follows (only the $\bigcirc$ and U linear operators are covered, since the other ones derive from them with the same definitions as in LTL):

$\sigma \models \bigcirc \Phi$	if and only if	$s_1 \models \Phi$;
$\sigma \models \Phi_1 \, \mathsf{U} \, \Phi_2$	if and only if	if there exists $j \geq 0$ such that $s_j \models \Phi_2$ and, for all i such that $0 \leq i < j$, $s_i \models \Phi_1$.

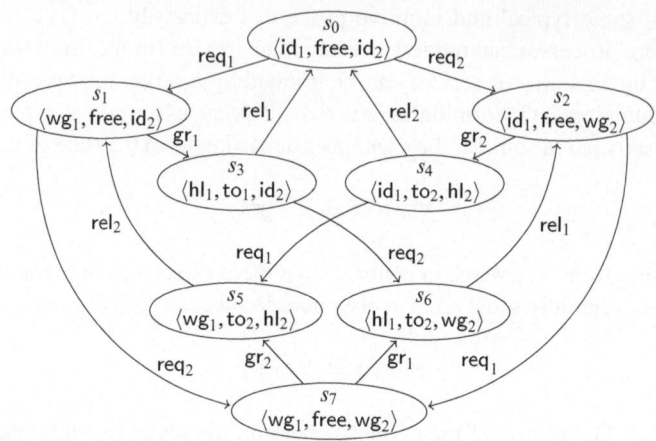

Fig. 9.2 Figure 8.2 repeated

Ultimately, the semantics of CTL is defined by means of the relation between *state* formulae and transition systems:

$$TS \models \Phi \quad \text{if and only if} \quad s_0 \models \Phi \text{ for every initial state } s_0 \text{ of } TS ;$$

hence, unless otherwise specified, the term CTL formula is intended to denote a CTL state formula.

Example 9.9. We illustrate the interpretation of CTL formulae with reference to the transition system TS of Fig. 8.2, duplicated in Fig. 9.2, where, for ease of reference, identifiers $s_0, \ldots, s_7$ are attached to the states. The transition system of Fig. 9.2 has labels on both states and transitions, while the semantics of CTL references only state labels. Section 7.1, however, showed that every transition system where both states and transitions are labeled can be transformed into an equivalent transition system with labels on states only. This transformation could be applied to the transition system in Fig. 9.2, but, for simplicity, we keep referring to the original representation with labels on both states and transitions, leaving the transformation implicit.

It is instructive to depict a finite portion of the infinite tree of executions of a transition system in order to reason about sequences of states and the satisfaction of CTL formulae. In the tree of Fig. 9.3, we terminate a branch when it reaches a state already drawn.

Consider the following CTL formulae.

$\forall \Box (\mathsf{wg}_1 \Rightarrow (\forall \Diamond \mathsf{hl}_1))$: whenever process P_1 is waiting for the resource, in every successive execution the resource will eventually be assigned it; this formula does not hold for the transition system, as witnessed by the counterexample $s_0 (s_1 s_7 s_5)^\omega$, where process P_1 remains forever waiting for the resource without

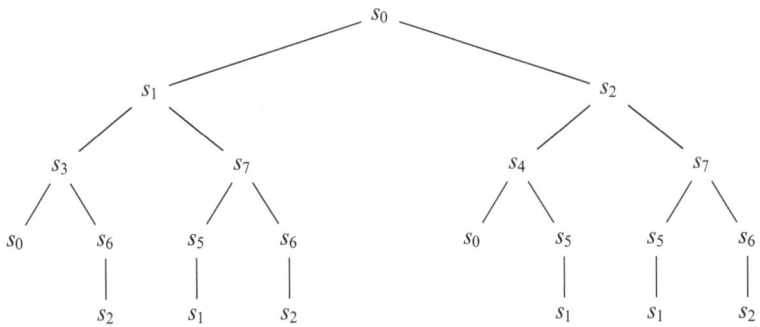

Fig. 9.3 Execution tree for a transition system

ever being assigned it; as a side remark, the transition system is not "fair" for process P_1 (the notion of fairness will be further discussed in Sect. 9.1.2.2).

$\forall\square(wg_1 \Rightarrow (\exists\lozenge hl_1))$: whenever process P_1 is waiting for the resource, it will eventually be assigned it in some of the executions departing from the waiting state; this formula holds for the transition system: from every state where wg_1 holds (i.e., states s_1, s_5, and s_7) a state is reachable where hl_1 holds (indeed, states s_3 and s_6, where hl_1 holds, are both reachable from any of the states s_1, s_5, and s_7).

$\forall\square(wg_1 \Rightarrow \neg\exists(\neg hl_1 \ U \ id_1))$: in every system execution when process P_1 is waiting for the resource it can never become idle before being assigned the resource; this formula is also satisfied by the transition system: in all state sequences departing from states s_1, s_5, and s_7, either no state where hl_1 holds is ever reached or a state where hl_1 holds is reached without going through a state where id_1 holds; indeed, reaching any of the states s_0, s_2, s_4 where id_1 holds from any one of the states s_1, s_5, s_7 where wg_1 holds requires to traverse one of the states s_3, s_6 where hl_1 holds. ∎

Exercise 9.10. Consider again the resource allocator of Example 7.19; specify the following properties by means of suitable CTL formulae and discuss whether they hold or not.

- For all paths, an hpr is eventually issued.
- For all paths, there always exists a path for which no hpr is issued.
- There exists a path in which every lpr may not be followed, in any of the subsequently reached states, by an hpr. ∎

9.1.2.2 Expressiveness of LTL and CTL

It is interesting to relate the semantics of LTL and CTL, possibly to compare their expressive power. To this end, we adopt the convention of evaluating LTL formulae with reference to transitions systems, as we also do for CTL formulae. Then, an

LTL formula ϕ and a CTL formula Φ are equivalent if, for every transition system TS, $TS \models \phi$ if and only if $TS \models \Phi$. The satisfaction relation symbol "$\models$" obviously refers to LTL semantics in the case of formula ϕ and to CTL semantics for formula Φ.

Since CTL includes the path quantifiers in addition to the linear operators of LTL, we might conjecture that CTL is at least as expressive as LTL – for every LTL formula there is an equivalent CTL formula – and possibly that CTL is even more expressive than LTL – there are CTL formulae for which no equivalent LTL formula exists.

It turns out that neither of these properties holds; that is, LTL and CTL are incomparable: for each of the two, we can find a formula that does not have an equivalent counterpart in the other logic.

In fact, the strict alternation between linear operators and path quantifiers in the syntax of CTL imposes quite strong a restriction on the classes of behaviors and models that can be characterized by its formulae. We do not report here a formal proof of the fact the CTL and LTL are incomparable (see the bibliographic remarks); rather, we provide a set of illustrative examples to foster intuition on the subtleties involved in the interpretation of LTL and, especially, CTL formulae.

Recall that an LTL formula holds with reference to a transition system if and only if it is satisfied by *all* state sequences that start from any initial state. Then, consider a CTL formula $\mathcal{U}$ containing only universal state modal operators. If we drop all universal path quantifiers in $\mathcal{U}$, is the obtained LTL formula equivalent to $\mathcal{U}$? We will see that this is not necessarily the case. However, the following theorem, which we state without proof, provides a very useful result concerning the way in which one can obtain an LTL formula equivalent to a given CTL formula, *if there exists one*.

Theorem 9.11. *Given a CTL formula Φ, if there is an LTL formula ϕ equivalent to it, then ϕ is obtained from Φ by removing all path quantifiers (both existential $\exists$ and universal $\forall$).* ∎

As a consequence of Theorem 9.11, if the LTL formula ϕ obtained from a CTL formula Φ by dropping all path quantifiers is not equivalent to Φ, then there does not exist any LTL formula equivalent to Φ.

The following example presents a case where Theorem 9.11 holds non-vacuously: starting from a CTL formula, an equivalent LTL formula is obtained by eliminating from it all path quantifiers.

Example 9.12. Consider the CTL formula $\forall \Box \forall \Diamond p$ and the LTL formula $\Box \Diamond p$. It is rather easy to show, for any transition system TS, that $TS \models \forall \Box \forall \Diamond p$ if and only if $TS \models \Box \Diamond p$.

In fact, if $TS \models \forall \Box \forall \Diamond p$ then, for every state sequence σ of TS and for every state s_i in σ, we have $s_i \models \Diamond p$; but since i is arbitrary, $\sigma \models \Box \Diamond p$; hence, since σ is also arbitrary, $TS \models \Box \Diamond p$.

On the other hand, if $TS \models \Box \Diamond p$, consider any reachable state s_i of TS and any path starting from it. Then, since the path starting from s_i is a suffix of some

Fig. 9.4 A transition system
TS such that
$TS \models \Diamond(p \wedge \bigcirc p)$ but
$TS \not\models \forall\Diamond(p \wedge \forall\bigcirc p)$

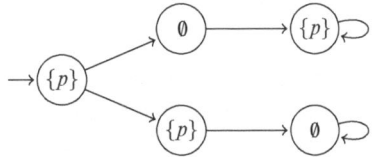

path starting from an initial state of TS, there is a state s_j, with $j \geq i$, on this path such that $s_j \models p$, and hence $s_i \models \Diamond p$; thus (since the path from s_i is arbitrary), $s_i \models \forall\Diamond p$ and, since s_i is also arbitrary, $TS \models \forall\Box\forall\Diamond p$. ∎

The following example illustrates the case in which the LTL formula obtained by dropping path quantifiers from a given CTL formula is not equivalent to it; hence the CTL formula does not have an equivalent LTL counterpart.

Example 9.13. The CTL formula $\forall\Diamond(p \wedge \forall\bigcirc p)$ and the LTL formula $\Diamond(p \wedge \bigcirc p)$ are not equivalent. This can be shown by considering the transition system TS of Fig. 9.4.

$TS \models \Diamond(p \wedge \bigcirc p)$: in both of its sequences ($\sigma_1 = \{p\}\{p\}\emptyset\emptyset\emptyset \ldots$ and $\sigma_2 = \{p\}\emptyset\{p\}\{p\}\{p\} \ldots$) there are, at some point, two consecutive p's. On the other hand, it is not the case that $TS \models \forall\Diamond(p \wedge \forall\bigcirc p)$, because in σ_1 there is no state s such that $s \models p \wedge \forall\bigcirc p$ (notice that the initial state has one successor state where p does not hold).

A general, intuitive explanation for the nonequivalence of $\forall\Diamond(p \wedge \forall\bigcirc p)$ and $\Diamond(p \wedge \bigcirc p)$ derives from the observation that the CTL formula $\forall\Diamond(p \wedge \forall\bigcirc p)$, because of the nested universal path quantification, imposes a sort of synchronization on the occurrences of p, matching the first p in the formula and the one required to follow immediately: *all* the successor states of the state where the first p occurs must also satisfy p. The LTL formula $\Diamond(p \wedge \bigcirc p)$, in contrast, admits the possibility that the two required consecutive occurrences of p may have different positions in different execution sequences. ∎

Example 9.13 also illustrates a fundamental feature of the CTL logic that differentiates it from LTL: nested path quantifiers (which, because of the alternation of the two kinds of operators imposed by the CTL syntax, are always separated by some linear operator) may refer to distinct sets of paths. This facilitates the expression of some kinds of properties, but also hinders the formalization of other kinds.

Consider, for instance, existential reachability, which Sect. 9.1.1.1 showed as not being expressible in LTL; in CTL, it can be expressed quite naturally by using two nested path quantifiers. Referring again to the transition system of Fig. 9.2, the property that, starting from a state satisfying $\mathsf{id}_1 \wedge \mathsf{to}_2 \wedge \mathsf{hl}_2$, there is at least one computation leading to a state where $\mathsf{hl}_1 \wedge \mathsf{to}_1 \wedge \mathsf{wg}_2$ holds is expressed – in a simple, direct way – by the CTL formula $\forall\Box((\mathsf{id}_1 \wedge \mathsf{to}_2 \wedge \mathsf{hl}_2) \Rightarrow \exists\Diamond(\mathsf{hl}_1 \wedge \mathsf{to}_1 \wedge \mathsf{wg}_2))$. The fact that the set of sequences referred to by the inner path quantification "$\exists$" (which is the set of sequences departing from a state where the subformula $\mathsf{hl}_1 \wedge \mathsf{to}_1 \wedge \mathsf{wg}_2$

holds) is not the same as the set of sequences quantified by the outermost "∀" quantifier (the set of all sequences departing from an initial state of the transition system) makes it possible for the formula to express the intended property. Let us recall, in contrast, that the LTL formula $\Box((id_1 \wedge to_2 \wedge hl_2) \Rightarrow \Diamond(hl_1 \wedge to_1 \wedge wg_2))$ does not express the same reachability property because it holds for a transition system only if the whole formula holds for *all* of its sequences.

Existential reachability of a state from any given state might have a significant applicative meaning: for instance, it can mean that the described system is always able to recover from an error.

The very same features of CTL that make it suitable for expressing reachability properties prevent it from expressing another fundamental class of properties, namely *fairness*, which we have shown to be easily expressible in LTL. Consider, for instance, the strong fairness property, which is formalized in LTL as $\Box\Diamond\psi \Rightarrow \Box\Diamond\phi$. This fairness assumption can seamlessly be composed in LTL formulae to express that some desired property π holds under the hypothesis of fairness, through the following LTL formula: $(\Box\Diamond\psi \Rightarrow \Box\Diamond\phi) \Rightarrow \pi$. On the other hand, CTL is unfit to express the strong fairness assumption, and consequently to express properties to be analyzed under the fairness hypothesis. The reason for this is that fairness requires the property $\Box\Diamond\psi \Rightarrow \Box\Diamond\phi$ to hold for *every computation* of the modeled system; hence one should write, in CTL style, a formula like $\forall(\Box\Diamond\psi \Rightarrow \Box\Diamond\phi)$, which, however, is not a legal CTL formula, since $\Box\Diamond\psi \Rightarrow \Box\Diamond\phi$ is not a path formula, as it violates the strict alternation between the two kinds of operators.

As a further example, consider the system of three processes in Fig. 8.1, and the transition system of Fig. 8.2, which globally models all possible executions of this system. A strong fairness assumption concerning process P1 can be stated in LTL by the following formula: $\Box\Diamond wg_1 \Rightarrow \Box\Diamond hl_1$; if P_1 is waiting for the resource infinitely often then it will hold it infinitely often. This fairness property does not hold for the transition system of Fig. 8.2, as witnessed by the path $\langle id_1, free, id_2 \rangle (\langle wg_1, free, id_2 \rangle \langle wg_1, free, wg_2 \rangle \langle wg_1, to_2, hl_2 \rangle \langle wg_1, to_2, hl_2 \rangle)^\omega$, that does not satisfy it. The fairness property cannot be expressed in CTL: for instance, the CTL formula $\forall\Box\forall\Diamond wg_1 \Rightarrow \forall\Box\forall\Diamond hl_1$ does not express fairness; it holds in the transition system, as the antecedent $\forall\Box\forall\Diamond wg_1$ of the implication does not hold (it is not true that for all possible executions process P1 will be waiting for the resource infinitely often); another possible candidate CTL formula for expressing fairness, $\forall\Box\exists\Diamond wg_1 \Rightarrow \forall\Box\exists\Diamond hl_1$, is also inadequate: it is satisfied by the transition system, as the consequent $\forall\Box\exists\Diamond hl_1$ of the implication holds (a state where P_1 holds the resource is reachable from every state of the transition system).

Such limitations of CTL regarding fairness concern its expressiveness, and a practical escape may consist of incorporating a priori a fairness hypothesis in the definition of CTL semantics and in the implementation of automatic analysis tools: unfair sequences are ruled out by defining the semantics of CTL in terms of a satisfaction relation which considers only fair sequences, and implementing analysis tools that consider only executions that satisfy appropriate fairness conditions.

Fig. 9.5 Strict inclusion of
both LTL and CTL by CTL*

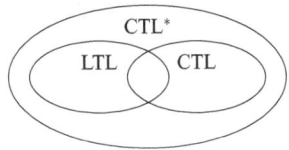

9.1.2.3 The Branching Logic CTL*

CTL* extends CTL by supporting every possible combination of the two kinds of modalities: linear operators over state sequences ($\Diamond$, U, etc.) and path quantifiers ($\forall$ and $\exists$).

The syntax of CTL* adopts, similarly to CTL, the notion of state and path formulae, with a few essential differences. CTL* *state* formulae over a set AP of atomic propositions are defined as

$$\Phi ::= p \mid \Phi \mid \Phi_1 \wedge \Phi_2 \mid \exists \phi \mid \forall \phi \, ,$$

where $p \in AP$ and ϕ is a path formula. CTL* *path* formulae are defined as

$$\phi ::= \Phi \mid \bigcirc \phi \mid \Diamond \phi \mid \Box \phi \mid \phi_1 \, U \, \phi_2 \, ,$$

where Φ is a state formula, and ϕ, ϕ_1, and ϕ_2 are path formulae. Hence, the only syntactic difference with CTL is in the definition of path formulae: in CTL*, every state formula is also a path formula, and the $\bigcirc$, $\Diamond$, $\Box$, and U operators are applicable to path formulae (and hence also to state formulae). This relaxes the strict alternation between linear operators and path quantifiers that characterizes CTL syntax and limits its expressiveness. Furthermore, since formulae without path quantifiers are valid CTL* formulae, all LTL formulae are also well-formed CTL* formulae. This shows that CTL* subsumes both LTL and CTL. Since LTL and CTL are incomparable and CTL* can combine any LTL with any CTL formula, it follows that CTL* strictly includes not only LTL and CTL, but also their union, as shown in Fig. 9.5.

For example, we can write the following CTL* formula that includes both a fairness assumption (not expressible in CTL) and a property of possible recovery from any error (not expressible in LTL):

$$\forall (\Box \Diamond \mathsf{req} \Rightarrow \Box \Diamond \mathsf{gr}) \wedge \forall \Box (\mathsf{error} \Rightarrow \exists \Diamond \mathsf{recover}) \, .$$

9.2 Future and Past Temporal Logics

Linear Temporal Logic was originally proposed for modeling correctness properties of computer programs. In this context, operators referring to the future – like $\bigcirc$, $\Diamond$, $\Box$, U – suffice for expressing basic program properties such as termination,

Future LTL operator	Past PLTL operator	Meaning
$\bigcirc \phi$	$\ominus \phi$	ϕ held at the previous instant ("yesterday"; also "Y")
$\Diamond \phi$	$\diamondsuit \phi$	ϕ held at some previous instant ("previously"; also "P")
$\Box \phi$	$\boxminus \phi$	ϕ held at all previous instants ("historically"; also "H")
$\phi_1 \cup \phi_2$	$\phi_1 S \phi_2$	ϕ_2 occurred at some past time and ϕ_1 has held since that time until the present ("since")

Fig. 9.6 Past operators for linear temporal logic

invariance, and computation of a correct output from a given input. As the applications of temporal logic extended to the specification, design and verification of complex properties of reactive systems, researchers considered the inclusion of modalities corresponding to past tenses, with the purpose of making descriptions more terse and intuitive, and possibly to increase the expressiveness of the notation.

"Linear Temporal Logic with Past" (PLTL) extends LTL with a set of linear operators that are the analogues, in the past, of the future operators presented in the previous section. Figure 9.6 associates every future operator of LTL with a past operator of PLTL and provides a brief, intuitive explanation of its meaning.

As in LTL, the semantics of PLTL is traditionally provided over ω-words of states, representing mono-infinite computation paths, because temporal logic was introduced to study properties of programs or computational systems having an initial state. We next provide a formal semantics for PLTL; for brevity we only provide the clauses for the basic temporal operators "next" $\bigcirc$, "yesterday" $\ominus$, "until" $\cup$, and "since" S (the operators "previously" $\diamondsuit$ and "historically" $\boxminus$ are derivable from the "since" operator S). Since PLTL formulae can refer to time instants preceding the current implicit one, suffix sequences (used in Sect. 9.1) are not suitable for defining the semantics. We therefore introduce a variant of the semantic relation "$\models$" having the current time instant as an additional argument.

Definition 9.14. For a PLTL formula ϕ, a state sequence $\sigma \in \wp(AP)^\omega$, and a time instant $k \in \mathbb{N}$, "$\sigma, k \models \phi$" means that formula ϕ is satisfied by sequence σ at time k, and it is defined as follows.

$\sigma, k \models \bigcirc \phi$	if and only if	$\sigma, k+1 \models \phi$;
$\sigma, k \models \phi_1 \cup \phi_2$	if and only if	there exists $j \geq 0$ such that $\sigma, k+j \models \phi_2$ and, for all i such that $0 \leq i < j$, $\sigma, k+i \models \phi_1$;
$\sigma, k \models \ominus \phi$	if and only if	$k > 0$ and $\sigma, k-1 \models \phi$;
$\sigma, k \models \phi_1 S \phi_2$	if and only if	there exists $j \geq 0$ such that $k - j \geq 0$ and $\sigma, k-j \models \phi_2$ and, for all i such that $0 \leq i < j$, $\sigma, k-i \models \phi_1$.

As customary, PLTL formulae are evaluated at time 0: PLTL formula ϕ is satisfied by a state sequence σ if and only if $\sigma, 0 \models \phi$.   ∎

Do the past operators introduced in PLTL increase the expressiveness of the logic with respect to LTL? Are they actually useful in practice, that is, are there significant classes of properties that can be described more concisely by using also past operators rather than by using future operators only?

Concerning the question of expressiveness, the addition of past operators does not add expressive power to future-only LTL. Moreover, a constructive "separation theorem" allows for the elimination of past operators, obtaining equivalent LTL formulae. The elimination of past operators may, however, in general introduce a nonelementary blowup in the alternation depth of future and past operators of the original formula. Furthermore, an exponential lower bound on the blowup in formula length when eliminating past operators also holds: there exists some PLTL formula ϕ with past operators whose every equivalent LTL formula has size $\Omega(2^{|\phi|})$.

On the other hand, it is widely recognized that the extension of LTL with past operators makes for specifications that are easier to write, shorter and more easily understandable. As a very simple, but also very illustrative example, consider the property "Every alarm is due to a fault", which may be found in the specification of fault-tolerant systems. A PLTL formula can use the "previously" operator to formalize the property simply as

$$\Box(\text{alarm} \Rightarrow \Diamond\!\!\!\!-\, \text{fault}),$$

whereas the following is arguably one of the simplest LTL versions of the same specification:

$$\neg(\neg\text{fault } U \text{ (alarm} \wedge \neg\text{fault)}).$$

Clearly, the latter version is much harder to read and understand.

The semantics of PLTL is defined with reference to state sequences which are infinite only in the future. This gives rise to a series of technical intricacies due to the fact that past operators must be assigned a conventional (true or false) value when they refer to time points before the initial state, and hence outside the temporal domain or, stated differently, to non-existing system states. We call this kind of semantic complication "border effect problem", and we illustrate it in the following example.

Example 9.15 (Transmission line with unit delay). Consider a transmission line which receives a signal at one end (modeled by proposition in) and delivers it at the opposite end (modeled by proposition out) with a fixed, unit delay. An LTL specification of this system is

$$\Box(\text{in} \Longleftrightarrow \bigcirc\text{out}). \tag{9.2}$$

The following, seemingly equivalent specification uses the $\ominus$ operator, rather than $\bigcirc$, expressing the property that every received message is delivered and no spurious message is generated: in other words, every out predicate is preceded by a corresponding in, and if there is no out then there was no previous in:

$$\Box((\text{out} \Rightarrow \ominus\text{in}) \wedge (\neg\text{out} \Rightarrow \ominus\neg\text{in})). \tag{9.3}$$

The evaluation of the past operators in formula (9.3) is problematic at the very first time instant, because the semantics of the yesterday operator assigns it a conventional false value when it references a time point outside the temporal domain. The definition

$$\sigma, k \models \ominus\phi \qquad \text{if and only if} \qquad k > 0 \ \text{and} \ \sigma, k - 1 \models \phi$$

implies that $\sigma, 0 \not\models \phi$ for any ϕ. Hence, it turns out that not only is (9.3) not equivalent to (9.2), but it is even unsatisfiable, because, at instant 0, both $\ominus\neg$in and $\ominus$in are false. Therefore, at instant 0, if out holds then formula out $\Rightarrow \ominus$in is false; otherwise, if out does not hold at 0 then $\neg$out $\Rightarrow \ominus\neg$in is false. But if we write $\neg\ominus$in instead of $\ominus\neg$in in (9.3), the formula becomes satisfiable. This is because $\neg\ominus$in may have a different value than $\ominus\neg$in.   ∎

A solution to the border effects problems of the kind highlighted by Example 9.15 consists of providing, for each operator that may refer to time points outside the temporal domain, two dual versions, which are interpreted differently, and which can be employed, with care and attention, to obtain the desired effect in every specific situation.

For instance, the dual version $\oplus$ of the yesterday operator $\ominus$ is equivalent to $\ominus$ at every time point but 0, where it is conventionally evaluated to true, regardless of its argument:

$$\sigma, k \models \oplus\phi \qquad \text{if and only if} \qquad k > 0 \ \text{implies} \ \sigma, k - 1 \models \phi$$

where "$k > 0$ implies $\sigma, k - 1 \models \phi$" is conventionally true when $k = 0$, because the antecedent of the implication is false. Using both versions of the yesterday operator, the incorrect (9.3) can be fixed as follows:

$$\Box((\text{out} \Rightarrow \ominus\text{in}) \wedge (\neg\text{out} \Rightarrow \oplus\neg\text{in})) . \tag{9.4}$$

Exercise 9.16. Consider the following variation of formula (9.4):

$$\Box((\text{out} \Rightarrow \oplus\text{in}) \wedge (\neg\text{out} \Rightarrow \oplus\neg\text{in})) . \tag{9.5}$$

Is (9.5) a valid correction of (9.3) too? If it is not, can you further modify it in a way that makes it a valid correction of (9.3) different from (9.4)?   ∎

The introduction of additional, dual operators, however, makes the logic more complex and less intuitive. A simpler and more effective solution to the same border effect problem allows for interpretations of temporal logic with past operators over *bi-infinite* (still discrete) state sequences – that is, sequences whose states can be mapped over the set $\mathbb{Z}$ of integer numbers – instead of the mono-infinite state sequences (used for LTL in Definition 9.14). Formally, a bi-infinite state sequence may be defined as a function $\sigma : \mathbb{Z} \to \wp(AP)$. The semantics of PLTL over bi-infinite state sequences is then a straightforward generalization of that of LTL over mono-infinite ones: the semantic definitions of PLTL are unchanged for the future operators $\bigcirc$ and U when σ is a bi-infinite state sequence and k is an index ranging on $\mathbb{Z}$ instead of $\mathbb{N}$, while the semantic clauses of the past operators "yesterday" $\ominus$ and "since" S become exactly symmetrical to those of their future counterparts "next" $\bigcirc$ and "until" U.

$\sigma, k \models \bigcirc\phi$ if and only if $\sigma, k+1 \models \phi$;

$\sigma, k \models \phi_1 \, U \, \phi_2$ if and only if   there exists $j \geq 0$ such that $\sigma, k+j \models \phi_2$
and, for all i such that $0 \leq i < j$, $\sigma, k+i \models \phi_1$;

$\sigma, k \models \ominus\phi$ if and only if $\sigma, k-1 \models \phi$;

$\sigma, k \models \phi_1 \, S \, \phi_2$ if and only if   there exists $j \geq 0$ such that $\sigma, k-j \models \phi_2$
and, for all i such that $0 \leq i < j$, $\sigma, k-i \models \phi_1$.

When interpreting PLTL formulae over bi-infinite state sequences, two new derived operators "sometimes" "Som" and "always" "Alw" may be introduced as suitable generalizations of the existential temporal operators $\Diamond$ and $\diamondsuit$ and, respectively, of the universal temporal operators $\Box$ and $\boxminus$:

$$\text{Alw}\phi := \Box\phi \wedge \boxminus\phi,$$

$$\text{Som}\phi := \Diamond\phi \vee \diamondsuit\phi.$$

Formulae of the kind "Alwϕ" and "Somϕ" are frequently used in system specification: they characterize time-invariant properties, since the outermost "Alw" or "Som" operator performs a sort of universal or existential temporal quantification: if such formulae hold at some time point, they hold at every time instant. Therefore, the time instant 0, which constitutes a privileged reference in the evaluation of LTL and PLTL formulae over mono-infinite state sequences, is immaterial for the interpretation of PLTL formulae "Alwϕ" and "Somϕ".

Exercise 9.17. Consider the digital controller of an air-conditioning system. The controller takes as input two sensor signals low_temp and high_temp, respectively meaning that the temperature is below or above the acceptable range, and issues the control commands cool (to activate the refrigerator), heat (to activate the heater), and no_action if the temperature is acceptable and no device need be active. The following tentative PLTL specification of the controller defines the conditions under which every command must be issued, and states that, at any time, exactly one of the signals cool, heat, or no_action, must hold:

$$\Box \begin{pmatrix} (\text{heat} \Longleftrightarrow \ominus\text{low\_temp}) \wedge \\ (\text{cool} \Longleftrightarrow \ominus\text{high\_temp}) \wedge \\ (\text{no\_action} \Longleftrightarrow \ominus(\neg\text{high\_temp} \wedge \neg\text{low\_temp})) \wedge \\ (\text{heat} \vee \text{cool} \vee \text{no\_action}) \wedge \\ \neg(\text{heat} \wedge \text{cool}) \wedge \neg(\text{heat} \wedge \text{no\_action}) \wedge \neg(\text{cool} \wedge \text{no\_action}) \end{pmatrix}. \quad (9.6)$$

Discuss the satisfiability of this specification, and of the specification obtained from it when the $\ominus$ operator is substituted by its dual $\oplus$, with reference to a mono-infinite state sequence. What changes if the specification (9.6) is interpreted with reference to a bi-infinite state sequence.   ∎

Exercise 9.18. A set S of bi-infinite sequences is called "shift invariant" if, for every bi-infinite state sequence $\sigma \in S$, S also includes every bi-infinite state sequence σ' which is composed of the same states as those appearing in sequence σ and in the same order, but with indices shifted by some given integer constant k: $\forall i \in \mathbb{Z}$ $(\sigma'(i) = \sigma(i + k))$. Argue that every set of bi-infinite state sequences satisfying a PLTL formula of the kind "Alwϕ" or "Somϕ" is shift invariant.   ∎

Exercise 9.19 (♦). With reference to a discrete bi-infinite time domain, consider a semantics of linear temporal logic based on "global satisfiability": a formula of the logic holds in a state sequence if and only if it evaluates to true for every time instant of the sequence. Show that PLTL is strictly more expressive than LTL under this notion of global satisfiability.   ∎

The adoption of a bi-infinite domain for the interpretation of temporal logics with past, besides simplifying their semantics, also has significant advantages from the practical viewpoint. Analogously to the mono-infinite case where termination may be ignored, interpreting a PLTL specification on bi-infinite time is convenient for dealing with system models where initialization may also be ignored. This increases the level of abstraction, since we can write specifications that are simpler and more easily understandable, because they do not include the description of the operations (such as configuration or installation) typically performed at system deployment time. This is particularly useful for reactive systems embedded into devices that continuously monitor or control some process, whose initialization is often irrelevant, by focusing only on regime behavior.

Consider for instance a railroad crossing like the one modeled using Petri nets in Example 8.26. The control system of the railroad crossing, consisting of sensors, actuators, and controllers, is installed and initialized at a certain time, and will be shut down some time (possibly several decades) afterwards. A PLTL specification may abstract away system installation (just as it usually ignores system shutdown), and provide a model that ignores the start of the system (where presumably one should describe restrictions on train circulation and a special initial system configuration that holds just once, at installation time), as if it always operated on regime behavior; for example, designers may state and prove a property asserting that if the bar is in the up position, either no train ever circulated in the past or the bar was raised after the last train exited the critical region.

9.3 Temporal Logics with Metric on Time

Section 9.1 showed how LTL is well suited for describing qualitative time properties, such as eventuality, invariance, and precedence. For instance, we have already discussed that the formula

$$\Box(\mathsf{req} \Rightarrow \Diamond \mathsf{gr}) \tag{9.7}$$

states that any request will eventually be satisfied, without, however, imposing any finite time bound within which this must occur. There exist, however, many applications where computer-based systems, in charge of monitoring or controlling an unpredictable environment, must react to incoming stimuli in a *timely fashion*: the system output must not only be correct in its value, but it must also be provided at some time that satisfies quantitative constraints. These are the "real-time systems", discussed in Sect. 3.2: real-time requirements are quite frequent in critical applications such as industrial plant monitoring and controlling, and the engineering of medical, weapon, transportation, and avionics systems.

LTL has very limited ability to specify quantitative time properties, due both to its simplicity – its limited repertoire of operators – and to the restrictive assumptions on the underlying time domain. For instance, the qualitative property (9.7) of "eventual" satisfaction of a request can be strengthened into a quantitative one, stating that the grant event occurs exactly five time units after the request:

$$\Box(\mathsf{req} \Rightarrow \bigcirc\bigcirc\bigcirc\bigcirc\bigcirc\mathsf{gr}) .$$

The notation soon becomes lengthy and cumbersome, for example, if the property to be formalized is even slightly relaxed: if the request must be satisfied *within* five time units the formula expressing the requirement is the wordy

$$\Box\,(\mathsf{req} \Rightarrow (\bigcirc\mathsf{gr} \vee \bigcirc\bigcirc\mathsf{gr} \vee \bigcirc\bigcirc\bigcirc\mathsf{gr} \vee \bigcirc\bigcirc\bigcirc\bigcirc\mathsf{gr} \vee \bigcirc\bigcirc\bigcirc\bigcirc\bigcirc\mathsf{gr})) .$$

Another major limitation of LTL in expressing metric properties derives from the presence of the next time operator $\bigcirc$, which implies that the underlying time domain is discrete: this does not, per se, prevent the description of quantitative time properties, but it is too restrictive in many applications, especially in distributed, asynchronous systems where components evolve independently without a shared, synchronized clock.

The same criticism applies to CTL, as well as to every logic using the same set of linear-time operators. We now discuss how temporal logic can be adequately endowed with a QUANTITATIVE notion of time: this requires us to modify and extend the set of time operators to accommodate numerical constants that represent time measures, such as the length of a time interval over which a property must continually hold, or the maximal delay within which a system must provide a response to a stimulus coming from the environment. We will illustrate two possible ways of incorporating metric time into linear temporal logic, exemplified by the temporal logics MTL and TRIO.

9.3.1 Metric Temporal Logic

"Metric temporal logic" (MTL) extends LTL and its underlying time framework. The time domain $\mathbb{T}$ (also called set of instants, moments, or time points) is still

MTL operator	Meaning
$\Diamond_{=d}\phi$	ϕ will occur at a time point exactly d time units in the future (with respect to the implicit current time)
$\Diamond_{>d}\phi$	ϕ will occur at some time point which is more than d time units in the future
$\Box_{<d}\phi$	ϕ will hold at all future time points whose distance from current time is less than d
$\phi_1 \, S_{\leq d} \, \phi_2$	ϕ_2 occurred in the past no sooner than d time units ago, and ϕ_1 has held since

Fig. 9.7 A few significant examples of MTL quantitative time operators

assumed to be linear: a totally ordered structure $\langle \mathbb{T}, < \rangle$. On top of that, the MTL time framework includes a function $d \,:\, \mathbb{T} \times \mathbb{T} \,\rightarrow\, \Delta$ which assigns a *distance* $d(t, t') \in \Delta$ to each pair of time points $t, t' \in \mathbb{T}$. Δ is a structure equipped with addition operation "$+$" and null value "0". In most practical cases, the sets $\mathbb{T}$ and Δ will be related or even coincide: for instance, they could both be equal to the set $\mathbb{N}$ of natural numbers, or to the set of nonnegative real or rational numbers. MTL's overall framework is, however, quite general and can accommodate a great variety of more complex metric temporal structures, which we do not discuss here.

MTL constitutes a very natural extension of LTL: it eliminates the next-time $\bigcirc$ operator (which becomes redundant as a means for referring to "the time point at distance 1"), it does not require a discrete time structure, and it generalizes the linear-time operators by adding a quantitative time parameter, which represents a distance in time $d \in \Delta$, qualified with a relational symbol from among "$=$", "$<$", "$\leq$", "$>$", and "$\geq$" to represent an exact, maximal, or minimal value for the distance. Figure 9.7 lists a few significant examples of MTL quantitative time operators, with an informal explanation of their meanings.

For instance, the aforementioned typical real-time property that a request is always satisfied at a distance of exactly five time units is stated by the following MTL formula

$$\Box(\mathsf{req} \Rightarrow \Diamond_{=5} \mathsf{gr}),$$

while the property that the request is granted a response *within* five time units is a simple variation of that:

$$\Box(\mathsf{req} \Rightarrow \Diamond_{\leq 5} \mathsf{gr}).$$

As another significant example, the following MTL formula specifies that an event represented by the atomic proposition p eventually takes place at some time and from that point on it repeats itself with period d

$$\Diamond(p \wedge \Box(p \Rightarrow \neg p \, \widetilde{\mathsf{U}}_{=d} \, p)),$$

where we use the strict variant "$\widetilde{\mathsf{U}}$" of the until operator which does not require its first argument to hold at the current time.

A variation of this property in which the repetition of p is not required to last indefinitely, but only over a time interval of length k, is formalized as follows

$$\Diamond(p \wedge \Box_{\leq k}(p \Rightarrow \neg p \ \widetilde{\mathsf{U}}_{=d} \ p)).$$

The original definition of MTL allows for explicit quantification of variables ranging over the set Δ, thus giving considerable expressive power to the logic. In particular, all qualitative LTL linear-time operators can be expressed in terms of MTL quantitative operators and first-order quantification over variables representing time distance explicitly transliterating the semantics of the LTL operators. As an example, the formula $\forall x(x \geq 0 \Rightarrow \Diamond_{=x} \phi)$ is obviously equivalent to $\Box\phi$, while $\exists y(y \geq 0 \wedge \Diamond_{=y} \phi_2 \wedge \forall x(0 \leq x < y \Rightarrow \Diamond_{=x} \phi_1))$ is equivalent to $\phi_1 \ \mathsf{U} \ \phi_2$. Most applications of MTL, however, only used its more constrained, propositional version.

9.3.2 TRIO

The expressive power given by first-order quantification over time is fully exploited in the TRIO temporal logic, introduced by the authors. TRIO assumes an underlying **LINEAR** temporal structure like LTL and MTL, and it features a **QUANTITATIVE** description of time through a single basic modal operator, called "Dist". TRIO supports different models of time, but the current chapter focuses on bi-infinite dense time such as the real numbers. The formula $\text{Dist}(p, d)$ means that proposition p holds at a time instant exactly d time units from the current one. This formula refers to the future if $d > 0$, to the past if $d < 0$, or to the present if $d = 0$. Many other TRIO *derived operators* are defined by means of first-order quantification over the time parameter of the basic operator Dist. We list the most significant ones in Fig. 9.8, together with definitions in terms of the fundamental Dist operator and informal explanations of their meanings.

Clearly, all operators of LTL and MTL are derivable from the Dist operator by means of propositional connectives and first-order quantification; for instance, $\phi \vee$ SomF(ϕ) is equivalent to the LTL formula $\Diamond\phi$, $\phi \wedge$ AlwF(ϕ) to $\Box\phi$, $\phi \wedge$ Lasts(ϕ, d) to $\Box_{<d}\phi$, and $\phi \vee$ WithinF(ϕ, d) to $\Diamond_{<d}\phi$. Existential quantifications over time variables yield requirements that some property holds or an *event* occurs in one time point of a specified set, while universal quantifications state that a property holds over a set of time points, often corresponding to a time interval; properties that hold over time intervals are often interpreted as *states* of the described system.

Still with reference to the example of the request/grant mechanism, the following TRIO formula asserts that every request will be satisfied within five time units:

$$\text{AlwF}(\mathsf{req} \Rightarrow \text{WithinF}(\mathsf{gr}, 5)).$$

TRIO operator	Definition	Meaning
$\text{Futr}(\phi,d)$	$d > 0 \wedge \text{Dist}(\phi,d)$	ϕ occurs sometimes in the future
$\text{Past}(\phi,d)$	$d > 0 \wedge \text{Dist}(\phi,-d)$	ϕ occurred sometimes in the past
$\text{Lasts}(\phi,d)$	$\forall t(0 < t < d \Rightarrow \text{Dist}(\phi,t))$	ϕ holds for the next d time units
$\text{Lasted}(\phi,d)$	$\forall t(0 < t < d \Rightarrow \text{Dist}(\phi,-t)))$	ϕ held for the last d time units
$\text{SomF}(\phi)$	$\exists t(t > 0 \wedge \text{Dist}(\phi,t))$	ϕ occurs sometimes in the future
$\text{SomP}(\phi)$	$\exists t(t > 0 \wedge \text{Dist}(\phi,-t))$	ϕ occurred sometimes in the past
$\text{Som}(\phi)$	$\exists t(\text{Dist}(\phi,t))$	ϕ occurs sometimes
$\text{AlwF}(\phi)$	$\forall t(t > 0 \Rightarrow \text{Dist}(\phi,t))$	ϕ holds always in the future
$\text{AlwP}(\phi)$	$\forall t(t > 0 \Rightarrow \text{Dist}(\phi,-t))$	ϕ held always in the past
$\text{Alw}(\phi)$	$\forall t(\text{Dist}(\phi,t))$	ϕ always holds
$\text{WithinF}(\phi,d)$	$\exists t(0 < t < d \wedge \text{Dist}(\phi,t))$	ϕ will occur within d time units
$\text{WithinP}(\phi,d)$	$\exists t(0 < t < d \wedge \text{Dist}(\phi,-t))$	ϕ occurred within the last d time units
$\text{Until}(\phi,\psi)$	$\exists t(\text{Futr}(\psi,t) \wedge \text{Lasts}(\phi,t))$	ψ will occur in the future and ϕ will hold until then
$\text{Until}^{\text{w}}(\phi,\psi)$	$\text{Until}(\phi,\psi) \vee \text{AlwF}(\phi)$	Weak until: ψ may never occur in the future
$\text{Since}(\phi,\psi)$	$\exists t(\text{Past}(\psi,t) \wedge \text{Lasted}(\phi,t))$	ψ occurred in the past and ϕ held since then
$\text{Since}^{\text{w}}(\phi,\psi)$	$\text{Since}(\phi,\psi) \vee \text{AlwP}(\phi)$	Weak since: ψ may have never occurred in the past
$\text{NowOn}(\phi)$	$\exists t(t > 0 \wedge \text{Lasts}(\phi,t))$	there exists a future nonempty interval starting from now in which ϕ holds
$\text{UpToNow}(\phi)$	$\exists t(t > 0 \wedge \text{Lasted}(\phi,t))$	there exists a past nonempty interval immediately preceding now in which ϕ holds

Fig. 9.8 Some typical TRIO derived operators

The following one states that any two requests must be at least 50 time units apart:

$$\text{Alw}(\text{req} \Rightarrow \text{Lasts}(\neg\text{req}, 50))\,.$$

Remark 9.20 (Including or excluding interval endpoints). For the operators that incorporate a quantification over a time interval (for example, Lasts), the definitions provided in Fig. 9.8 exclude the endpoints of the specified time interval, i.e., intervals are open. Variants of the operators including either one or both of the endpoints can be easily defined; for notational convenience, to indicate respectively inclusion or exclusion of the interval's endpoints, the subscripts "i" or "e" are appended to the operator's name. Figure 9.9 shows a few examples regarding the operators Lasts, Since, Until, AlwP, and SomP. ■

9.3.3   Specifying Properties in First-Order Metric Temporal Logics

First-order temporal logics include in their alphabet predicates and functions. Functions with arity 0 (argumentless functions), are usually called "constants", as

TRIO operator	Definition
$\text{Lasts}_{ie}(\phi, d)$	$\forall t(0 \leq t < d \Rightarrow \text{Dist}(\phi, t))$
$\text{Lasted}_{ie}(\phi, d)$	$\forall t(0 < t \leq d \Rightarrow \text{Dist}(\phi, -t))$
$\text{WithinF}_{ei}(\phi, d)$	$\exists t(0 < t \leq d \wedge \text{Dist}(\phi, t))$
$\text{WithinP}_{ei}(\phi, d)$	$\exists t(0 \leq t < d \wedge \text{Dist}(\phi, -t))$
$\text{WithinP}_{ie}(\phi, d)$	$\exists t(0 < t \leq d \wedge \text{Dist}(\phi, -t))$
$\text{Until}_{ie}(\phi_1, \phi_2)$	$\exists t(t > 0 \wedge \text{Futr}(\phi_2, t) \wedge \text{Lasts}_{ie}(\phi_1, t))$
$\text{Until}^w_{ei}(\phi_1, \phi_2)$	$\exists t(t > 0 \wedge \text{Futr}(\phi_2, t) \wedge \text{Lasts}_{ei}(\phi_1, t) \vee \text{AlwF}(\phi_1))$
$\text{Since}_{ii}(\phi_1, \phi_2)$	$\exists t(t > 0 \wedge \text{Past}(\phi_2, t) \wedge \text{Lasted}_{ii}(\phi_1, t))$
$\text{Since}_{ie}(\phi_1, \phi_2)$	$\exists t(t > 0 \wedge \text{Past}(\phi_2, t) \wedge \text{Lasted}_{ie}(\phi_1, t))$
$\text{AlwF}_i(\phi)$	$\forall t(t \geq 0 \Rightarrow \text{Dist}(\phi, t))$
$\text{SomP}_i(\phi)$	$\exists t(t \geq 0 \wedge \text{Dist}(\phi, -t))$

Fig. 9.9 Examples of TRIO derived operators with explicit indication of interval endpoint inclusion. The first subscript refers to the left endpoint of the interval and the second one refers to the right endpoint

their value does not depend on any parameter but is assigned in a unique way by the interpretation structure. In a first-order temporal logic, however, the value assigned to a "constant" function may change with time, just as the Boolean value assigned to atomic propositions in propositional LTL may be different at different time instants. In first-order temporal logic, we must therefore clearly distinguish between the notion of a "variable" as a nullary function whose value can be different at every time instant, and that of a logic variable used as a placeholder in first-order quantifications. We call *"time-dependent* variables" the former type of variables, and "first-order, *rigid, global,* or *logic* variables" – or simply "variables" without qualifications – the latter (the attributes *rigid* and *global* emphasize the fact that their variable values do not depend on the time instant of evaluation and are not affected by the temporal operators).

The notion of time-dependent versus time-independent entity also applies to predicates: a "time-dependent predicate" is associated with a possibly distinct relation at every time instant; therefore, time-dependent predicates are first-order generalizations of the atomic propositions in LTL, whose Boolean values may change with time.

As a simple example of the combined use of time-dependent variables and first-order variables, consider the specification of an air-conditioning system where a relevant property is that if the temperature is above 25°, then it will be decreased by more than 5° within D time units. This is specified in TRIO by the following formula, where the temperature is represented by the time-dependent variable temp, while the first-order variable *tempVal* denotes any temperature, that is, any possible *value* assumed by the time-dependent variable temp:

$$\text{Alw}\left(\forall tempVal\left(\begin{pmatrix} \text{temp} = tempVal \\ \wedge \ tempVal > 25 \end{pmatrix} \Rightarrow \text{WithinF}(\text{temp} < tempVal - 5, D)\right)\right). \tag{9.8}$$

The two occurrences of the time-dependent variable temp in (9.8) denote the value of the temperature at distinct time instants, and therefore, in general, have different values, whereas in every interpretation all occurrences of the first-order variable *tempVal* have the same value at every instant of time.

The inclusion of the primitive quantitative operator Dist and of first-order quantification in the TRIO logic provides the full expressiveness of predicate logic – where, as illustrated in Chap. 2, functions and predicates include a time argument to represent the change in time of the modeled values and properties – but with terser and more readable formulae. Consider, for instance, Example 2.10, which provided a first-order logic model of a safe. In TRIO, we can specify the same system through the atomic propositions Open, Closed, and Correct_code_entered, which respectively correspond to the homonymous functions of one time variable that were included in that example:

$$\text{Alw} \left(\begin{array}{l} (\text{Open} \iff \neg\text{Closed}) \quad \wedge \\ (\text{Open} \iff \text{WithinP}_{\text{ie}}(\text{Correct\_code\_entered}, 3)) \end{array} \right). \tag{9.9}$$

A close comparison with Example 2.10 also shows that the two quantifications over time variables in formula (2.6) are substituted in (9.9) by the two operators Alw and WithinP. Thanks to the implicit notion of current time embedded in the linear-time operators, the TRIO formalization closely corresponds to the description of the system features in natural language (e.g., "it is always the case that the safe is open iff the safe is not closed, and at any time the safe is open iff a correct code was entered within the previous three time units"), which makes it rather intuitive and easily understandable.

Unsurprisingly, the presence of metric features in MTL and TRIO enables them to specify quantitative timing properties more concisely than in LTL equipped with next time "○" as the only metric operator.

Consider again a transmission line where every message entering it (represented by the atomic proposition in) at the input end will exit the output end exactly k time units later (represented by the atomic proposition out). This is specified in LTL by the formula

$$\Box(\text{in} \iff \overbrace{\bigcirc\bigcirc\cdots\bigcirc}^{k\ \text{times}} \text{out}). \tag{9.10}$$

Notice that (9.10) requires the transmission line not to lose any message (this is ensured by the "⟹" implication) nor to generate any spurious message (as guaranteed by the "⟸" implication). As we have already noticed, the example shows that an LTL formula that denotes a time point at distance k from the current implicit time must use k nested occurrences of the ○ operator. The same property can be specified in TRIO by the following formula

$$\text{Alw}(\text{in} \iff \text{Futr}(\text{out}, k)), \tag{9.11}$$

where the operator Futr uses the numerical constant k as a second parameter. This makes the metric temporal logic formula exponentially shorter (assuming the constant k is encoded in binary in TRIO) than every equivalent LTL formula.

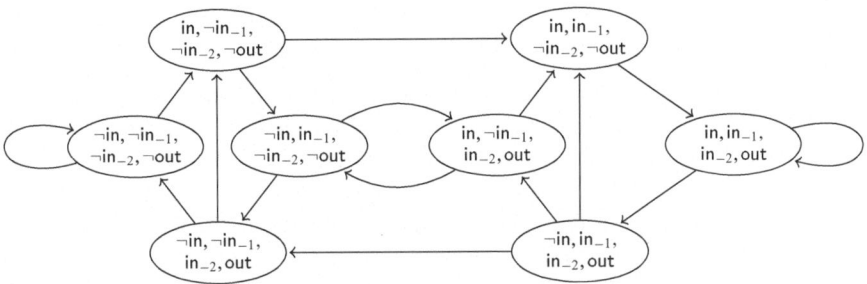

Fig. 9.10 A transition system modeling a transmission line with delay $k = 2$

It is also particularly interesting how the same property of the transmission line would be modeled by operational notations, say by a transition system. For the sake of simplicity, let $k = 2$; Fig. 9.10 shows a transition system that models this property. It is based on the set of atomic propositions $AP = \{\text{in}, \text{in}_{-1}, \text{in}_{-2}, \text{out}\}$: propositions in and out are true in a state if and only if a message is input to and output from the line when the modeled system is in that state; propositions in_{-1} and in_{-2} have a similar meaning with reference to the previous two time instants (they hold in a given state if and only if proposition in was true, respectively, in the previous state and in the state preceding it).

In the transition system, when going from one state to the next, the value of in is taken by proposition in_{-1}, and the value of in_{-1} is taken by proposition in_{-2}: in other words, in_{-1} and in_{-2} store the last two previous values of in in every state. The use of states to store the memory of previous occurrences of some significant event or the previous values of some variable is a typical feature of operational models: since, in transition systems, the next state is only determined by the current state and by the input, the state must incorporate, by means of suitable labels, all the information, from the execution of the model up to the current time, which can determine the system transition to the next state. This justifies the eight states of the transition system of Fig. 9.10: the system must "remember", by means of propositions in_{-1} and in_{-2}, the last two values of the in proposition, and, moreover, it must take into account the current value of in, for a total of $2 \cdot 2 \cdot 2 = 8$ possible combinations of Boolean values (proposition out is included in AP only for the sake of readability, as it is always identical to in_{-2}).

An easy generalization leads us to consider that $k+1$ state variables are necessary to model a transmission line with a delay k, so that the transition system has, in that case, 2^{k+1} states. Therefore, if we compare a specification written as a TRIO formula (9.11) (whose size is $\Theta(\log k)$ if k is encoded in binary) with a transition system used to specify the same property (whose number of states is $\Theta(2^k)$), we conclude that the number of states of the transition system is larger than the size of the TRIO formula by a double exponential factor. This is a simple example of how descriptive notations can provide very concise, implicit characterizations of system behavior, while operational ones can be more explicit, sometimes easier to understand, but also often more verbose.

Example 9.21 (Real-time allocator). Consider a real-time allocator which serves a set of client processes competing for a shared resource. Each client process p may request the resource by issuing the message $\mathsf{req}(p, \delta)$, which registers it to the allocator with a time-out δ for the request to be serviced. If the allocator is able to satisfy p's request within the indicated time-out, then it grants the resource to p by a $\mathsf{gr}(p)$ signal; otherwise the request is completely ignored. Once the process has used the resource granted to it by the allocator, it frees the resource with a **free** signal.

The allocator follows a FIFO policy in assigning the resource to processes: at any time, it will grant the free resource to the least recent pending request. The following is a first, partially formal version of the specification:

$$\text{Alw} \left(\text{grant the resource to process } a \Longleftrightarrow \left(\begin{array}{l} \text{the resource is free } \land \\ \text{there is a valid pending request by } a \quad \land \\ \text{there is no other process } a' \text{ with an earlier valid pending request} \end{array} \right) \right).$$

Its refinement in terms of a TRIO formula is based on the use of predicates and propositions representing the signals exchanged between the processes and the allocator, and variables a, a' and b, ranging over the set of processes, δ and δ', ranging over the set of possible values of the request time-out, and t and t', representing time distances and lengths of time intervals in the Past and Lasted operators:

$$\text{Alw} \left(\forall a \left(\mathsf{gr}(a) \Longleftrightarrow \exists t \exists \delta \left(\begin{array}{l} \text{Since}^w(\neg \exists b\, \mathsf{gr}(b), \mathsf{free}) \land \\ \mathrm{Past}(\mathsf{req}(a, \delta), t) \land t \leq \delta \land \\ \mathrm{Lasted}(\neg\mathsf{gr}(a), t) \land \\ \neg \exists a' \exists \delta' \exists t' \left(\begin{array}{l} \mathrm{Past}(\mathsf{req}(a', \delta'), t') \land \\ a' \neq a \quad \land \\ t' < \delta' \quad \land \\ t' > t \quad \land \\ \mathrm{Lasted}(\neg\mathsf{gr}(a'), t') \end{array} \right) \end{array} \right) \right) \right). \quad \blacksquare$$

Exercise 9.22. With reference to the real-time allocator of Example 9.21, provide a TRIO specification of the following properties.

Allocator fairness: if a process that does not obtain the resource always requests it again immediately after the request is expired, then if the process requests the resource it will eventually obtain it.

(*Hint*: to give a formally precise meaning to the adverb "immediately", remember that the time domain is currently discrete).

Process precedence: the allocator never grants the resource to a process a that has requested it later than another process b, unless it has already granted b the resource.

Does the system specification in Example 9.21 satisfy the two properties above?

Requests rotation: a process requires the resource only after all other processes have requested and obtained it. Is your specification formula satisfiable over $\mathbb{Z}$? Is it over $\mathbb{N}$? If not, how could you modify it in such a way that it is satisfiable both over $\mathbb{Z}$ and over $\mathbb{N}$?
(*Hint*: watch out for border effects).   ∎

Exercise 9.23 (Timer reset lamp).   Consider the following timed device, called the "timer reset lamp". The lamp has two buttons, on and off: when the on button is pressed the lamp turns on and it remains on, if no other event occurs, for Δ time units, after which it spontaneously turns off. The lamp can be turned off before the Δ time units expire, by the pushing of the off button, or it can stay on for another Δ time units by the pushing again of the on button.

- Write an MTL specification of the timer reset lamp.
  (*Hint*: use on and off to denote the pressing of the buttons and L to hold when the lamp is on).
- Write the same specification as a transition system, in the same style as in the transition system modeling the transmission line in Fig. 9.10, for $\Delta = 3$.
- Compare the two formalizations of the timer reset lamp, paying special attention to how the two specifications change depending on the actual value of the constant Δ; for instance, what happens if the value of Δ is doubled from 3 to 6?   ∎

9.4   Discrete and Dense Time Domains

Temporal logics such as LTL and CTL, as originally devised to model executions of digital computers, are unable to describe continuous phenomena: their assumptions of discrete time and stepwise execution of transition systems lead to models based on finite or denumerable state sequences. Discrete state sequences adequately model program execution but, as we argued in Sect. 3.1.1, are unsuitable for describing fully asynchronous systems, where time distances between related events do not have a lower bound, or physical processes whose models must include variables that change in time in a continuous way (such as temperature, voltage, or the position of a mechanical actuator).

   In this section, we discuss the consequences of adopting a DENSE set, such as the set of real numbers, as the time domain over which formulae in a METRIC temporal logic with past operators are interpreted. We introduce the notions of *point-based predicate* (a predicate that holds at isolated time points) and *interval-based predicate* (a predicate that holds over nonempty temporal intervals). The adoption of a dense, possibly continuous, time domain allows us to model ASYNCHRONOUS systems where distinct events may occur at arbitrarily close time instants. As a consequence, ZENO behaviors, where, for instance, an unbounded number of events

Fig. 9.11 Visualization of point-based (**a**) and interval-based (**b**) predicates

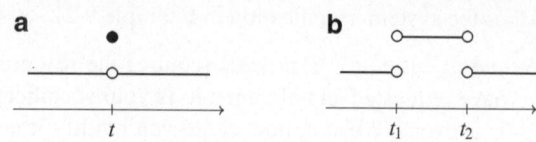

take place in a finite time interval, become possible and must be explicitly ruled out through axioms or ad hoc semantic assumptions. Finally, we generalize the definitions of point-based and interval-based predicates to time-dependent variables ranging over denumerable or uncountable domains. All the notions presented are formally defined in terms of TRIO formulae that are intended to be always valid, and hence implicitly enclosed in the outermost Alw operator.

9.4.1   Dense-Time Modeling

"Point-based" (or simply "point") predicates are time-dependent predicates that hold only at isolated time points, and are false elsewhere. Their behavior is pictured in Fig. 9.11a. Their formal definition in terms of TRIO is as follows.

Definition 9.24 (Point-based predicate). A time-dependent predicate E is "point-based" if and only if

$$E \Rightarrow \mathrm{UpToNow}(\neg E) \wedge \mathrm{NowOn}(\neg E) .$$ ∎

A point-based predicate has therefore a null duration. This definition is clearly an abstraction, since no physical activity has truly null duration. This kind of abstraction is very common among formalisms for modeling real-time systems, as null duration predicates are suitable for formalizing events whose duration is small with respect to the reaction times of the system components.

Example 9.25. In the "kernel railroad crossing" described in Example 8.26, point predicates can represent commands given to the bar: Go_up and Go_down. When the bar receives a Go_up command and it is closed, it starts opening. If it is open and it receives a Go_down command, it starts closing. These commands can be thought as control pulses having a very small duration, whose precise value is negligible.

∎

In contrast to point-based predicates, an "interval-based" (or simply "interval") predicate may keep its value for nonempty time intervals. Interval predicates hold true or false for intervals with non-null duration; thus they are never true or false at isolated time points. A diagrammatic representation of interval behavior is pictured in Fig. 9.11b.

Fig. 9.12 Left- (**a**) and right-continuous (**b**) interval predicates

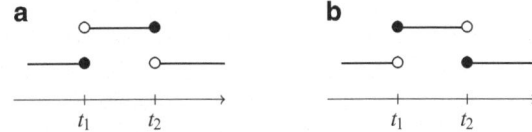

Definition 9.26 (Interval-based predicate). A time-dependent predicate I is "interval-based" if and only if

$$(I \Rightarrow (\text{UpToNow}(I) \vee \text{NowOn}(I))) \wedge (\neg I \Rightarrow (\text{UpToNow}(\neg I) \vee \text{NowOn}(\neg I))).$$

∎

Since an interval predicate I cannot keep its value at isolated time points, if I is true (or false) at a time point then there is an interval, following or preceding it, in which I is true (false). Notice that this definition does not constrain the value of I at the precise instants when it changes value from true to false or vice versa (instants t_1 and t_2 in Fig. 9.11b). Thus, the time intervals in which a predicate keeps its value may be open or closed (they may or may not contain their endpoints). Several choices can be made about the predicate value at such instants, but there are two common possible assumptions (shown in Fig. 9.12): a predicate is "left continuous" if it has the value it has had in its left neighborhood; a predicate is "right continuous" if it has the value that it will have in its right neighborhood. Left- and right-continuous predicates are formalized in TRIO as follows.

Definition 9.27 (Left- and right-continuous predicate). A predicate I is a "left-continuous interval-based predicate" if and only if

$$(I \Rightarrow \text{UpToNow}(I)) \wedge (\neg I \Rightarrow \text{UpToNow}(\neg I));$$

it is a "right-continuous interval-based predicate" if and only if

$$(I \Rightarrow \text{NowOn}(I)) \wedge (\neg I \Rightarrow \text{NowOn}(\neg I)).$$

∎

Remark 9.28 (On the physical meaning of left and right continuity). At the very instant a right-continuous predicate switches value, it already holds the value it will have in its right neighborhood: when it becomes false, it is already false, and when it becomes true it is already true. This characterization might be questionable if we consider its physical meaning: it looks like right-continuous predicates can somehow foresee their changes. If we consider, for instance, the case of an event causing a change in the value of the state (this situation is common in reactive systems), with right-continuity the cause (the event) and the effect (the state taking the new value) are simultaneous, against the common intuition that effects should strictly follow their causes.

Left continuity is not affected by such objection, because at the very instant a right-continuous predicate switches value, it still has its previous value before the change: when it becomes false it is still true, and it is false only afterwards, strictly after the change occurs. Left-continuous models can, however, be criticized too, because when the change occurs the value of the predicate has not changed yet. If we consider, for instance, the case of an event signaling the change in value of the state, with left-continuity the cause (in this case, the change of state value) and the effect (the change event occurring) are again simultaneous.

Therefore, both modeling assumptions are subject to symmetrical criticisms and support arguments. Probably, the choice most respectful of physical causality is to leave predicate values undefined in the changing instants. However, undefined values in formal models may lead to other problems (such as when the result of operations on undefined values is not defined either) and may complicate the semantics of the adopted formal notation. In all, it is preferable to use entities (predicates and variables) with a value defined at every instant, therefore considering them as total functions of time.   ∎

The closure properties of interval-based predicates are other important features in the choice of which kinds of entities to adopt. We can see that by composing predicates with the same type of continuity through logical connectives, we get derived predicates with the same features; for example, the conjunction of two left-continuous interval-based predicates is a left-continuous interval-based predicate, whereas the conjunction of a left-continuous predicate with a right-continuous one may not even be a generic interval-based predicate. These properties make it more convenient, when modeling a system by means of a continuous-time temporal logic, to use consistently one of the two conventions (left or right continuity), which prevents inconsistencies and simplifies specification and analysis.

Exercise 9.29. Let L_1 and L_2 (or R_1 and R_2) be two generic left-continuous (or right-continuous) interval-based predicates.

- Prove that $L_1 \wedge L_2$, $L_1 \vee L_2$, and $\neg L_1$ are all left-continuous interval predicates.
- Prove that $R_1 \wedge R_2$, $R_1 \vee R_2$, and $\neg R_1$ are all right-continuous interval predicates.
- Show an example of behavior for L_1 and R_1 such that $L_1 \wedge R_1$ is a point-based predicate.   ∎

Exercise 9.30 (♣). Solve Exercise 9.22 again, assuming the two continuous-time domains $\mathbb{R}$ and $\mathbb{R}_{\geq 0}$. Discuss if and how you had to change the original solution, and how the interpretation and satisfiability of the formulae changed when moving from discrete to continuous time.   ∎

9.4.1.1   Non-Zeno Requirement for Time-Dependent Predicates

Consider the following formula (9.12), which specifies the occurrences of an event E in the proximity of the current instant *now*, where t and n are logic variables of type real and natural (unlike most other formulae of this section, (9.12) is not implicitly quantified by the operator Alw but evaluated only at *now*):

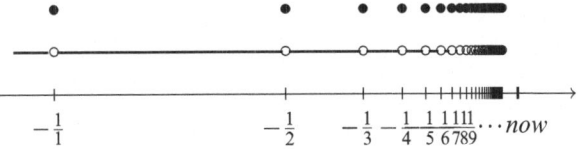

Fig. 9.13 A Zeno event sequence

$$\forall t \left(\mathrm{Past}(E,t) \iff \exists n \left(t = \frac{1}{n} \right) \right). \tag{9.12}$$

As shown in Fig. 9.13, E occurs precisely at distances of $1, 1/2, 1/3, 1/4, \ldots$ time units in the past with respect to *now* (where the formula is interpreted), but it is false exactly at *now*.

E has Zeno behavior, as it occurs infinitely many times near the current instant *now*. Predicate E does not satisfy the formula

$$\neg E \Rightarrow \exists \varepsilon \, \mathrm{Lasted}(\neg E, \varepsilon),$$

meaning that, if E is false now then there is a left neighborhood of now where E is false; in fact, E occurs an infinite number of times immediately before *now*.

The axiomatic definition of non-Zenoness is straightforward for first-order metric temporal logics like MTL or TRIO when it is applied to simple entities like predicates or variables ranging over finite domains. It can be more complex in the most general case of real-valued time-dependent variables.

Informally, a predicate is non-Zeno if it has finite variability, that is, its truth value changes a finite number of times over every finite interval. Correspondingly, a predicate P is non-Zeno if we require that there exist a time interval before or after every time instant in which P is constantly true or constantly false. This constraint is expressed by the following TRIO formula:

$$(\mathrm{UpToNow}(P) \lor \mathrm{UpToNow}(\neg P)) \land (\mathrm{NowOn}(P) \lor \mathrm{NowOn}(\neg P)). \tag{9.13}$$

The first conjunct (with UpToNow) guarantees that there is no accumulation point of changing instants before the current instant, whereas the second one (with NowOn) guarantees the same in the future. For a non-Zeno predicate P, expressions such as "the value of P immediately before (or after) the current time" are well defined. The predicate E in the formula (9.12) does not satisfy the non-Zeno requirement (9.13): at the current instant neither $\mathrm{UpToNow}(E)$ nor $\mathrm{UpToNow}(\neg E)$ hold, because there is no $\varepsilon > 0$ such that $\mathrm{Lasted}(E, \varepsilon)$ or $\mathrm{Lasted}(\neg E, \varepsilon)$.

Remark 9.31 (On the distance between events). As we remarked in Sect. 3.6, the non-Zeno requirement does not imply that there is a minimum time distance between any two changes in value of a predicate that is defined a priori for every possible system behavior. Non-Zenoness does, however, require that the notion

Fig. 9.14 Arbitrarily close
occurrence of independent
events

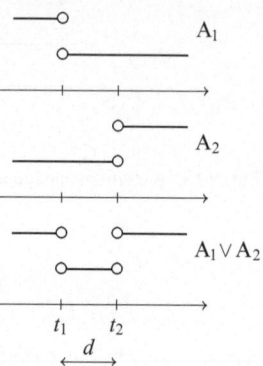

of "next occurrence" of any event E be well-defined and unique at every instant (except when E never happens in the future). Consider, for example, two non-Zeno predicates A_1 and A_2 and the derived predicate $A_1 \vee A_2$, whose behavior is depicted in Fig. 9.14. $A_1 \vee A_2$, as well as any other derived predicate, is still guaranteed to be non-Zeno, provided the basic predicates are; however, if A_1 and A_2 are independent (for example, two inputs coming from separate channels in an asynchronous system), the distance d between a change in A_1 and a change in A_2 may be arbitrarily small.

In the case of asynchronous systems, in general, d has no positive lower bound; hence the distance between two consecutive changes of $A_1 \vee A_2$ can be arbitrarily small. In other words, non-Zenoness is fully compatible with asynchronous behavior, because it does not require *bounded* variability but only *finite* (yet arbitrarily fast) variability.

In the case of synchronous, clock-based systems, in contrast, every input arrives at time instants that are multiples of the clock period (or some other fixed quantity); hence is not only $A_1 \vee A_2$, as well as any other derived predicate, guaranteed to be non-Zeno, but the distance d also has a positive lower bound. Some formal languages mirror such synchronous behavior and assume a fixed minimum delay between actions.   ∎

9.4.1.2 Non-Zeno Requirement for Time-Dependent Variables

Applying the requirement of finite variability to time-dependent variables ranging over countable domains (such as the natural numbers or subsets thereof) is straightforward. In this case, formula (9.13), which refers to the two possible values (true and false) that a predicate can take, naturally generalizes through quantification over a countable domain to state that variables have a constant value both in an immediately preceding and in an immediately following nonempty interval.

Definition 9.32 (Non-Zeno time-dependent variable over countable domain).
A time-dependent variable v over a countable domain D is non-Zeno if and only if

$$\exists a \in D(\text{UpToNow}(v = a)) \wedge \exists b \in D(\text{NowOn}(v = b)).$$

We do not impose any requirement at isolated time points, where the variable may have any value. ∎

Next, we consider time-dependent variables on uncountable domains, for instance, the reals or intervals of the reals. This is the most general case, and also quite frequent in practice: real-time systems are often hybrid systems involving both real-valued physical variables and digital components. The definition previously provided for countable domains, which requires a variable to be piecewise constant, cannot be applied to uncountable domains, because real-valued quantities can change continuously, spanning an infinite number of values in a finite time interval: consider, for instance, a sinusoid or a ramp. We also cannot accept every possible behavior for real-valued time-dependent variables because they typically represent real-world entities subject to physical laws. In particular, we expect real-valued variables satisfying some non-Zenoness requirement to give rise to non-Zeno predicates when composed via arithmetic and relational operators. For instance, the time-dependent variable b, previously illustrated in Sect. 3.6, whose value changes in time according to the law

$$b(t) \quad = \quad \begin{cases} \exp\left(-\frac{1}{(t-t_0)^2}\right) \sin\left(\frac{1}{t-t_0}\right) & t \neq t_0, \\ 0 & t = t_0 \end{cases}$$

is continuous with continuous derivatives of all orders (denoted by $b \in C^\infty$), but its value changes sign an infinite number of times in any interval containing t_0, so that the formula $b = 0$ is a Zeno point-based predicate.

Informally, a non-Zeno time-dependent variable v ranging over an uncountable domain D (representing, for instance, the current temperature in a thermostat application) is piecewise analytic when considered as a function of time.[1]

Formally, a variable v with values in a domain D is non-Zeno if, at every time, there exist two functions $f, g : \mathbb{R} \rightarrow D$ that are analytic at 0, and such that v is equal to f in a nonempty right interval and to g in a nonempty left interval of the current time; if $C^\omega(0)$ denotes the set of functions that are analytic at 0, the non-Zeno requirement can be formalized as follows.

Definition 9.33 (Non-Zeno time-dependent variable over uncountable domain).
A time-dependent variable v over an uncountable domain is non-Zeno if and only if

---

[1]A function is analytic at a given point if it possesses derivatives of all orders and agrees with its Taylor series about that point. It is piecewise analytic if it is analytic over finitely many contiguous (open) intervals.

Fig. 9.15 Examples of Zeno and non-Zeno behaviors (for readability, all plots are scaled, and (**b**) and (**c**) are also slightly vertically shifted)

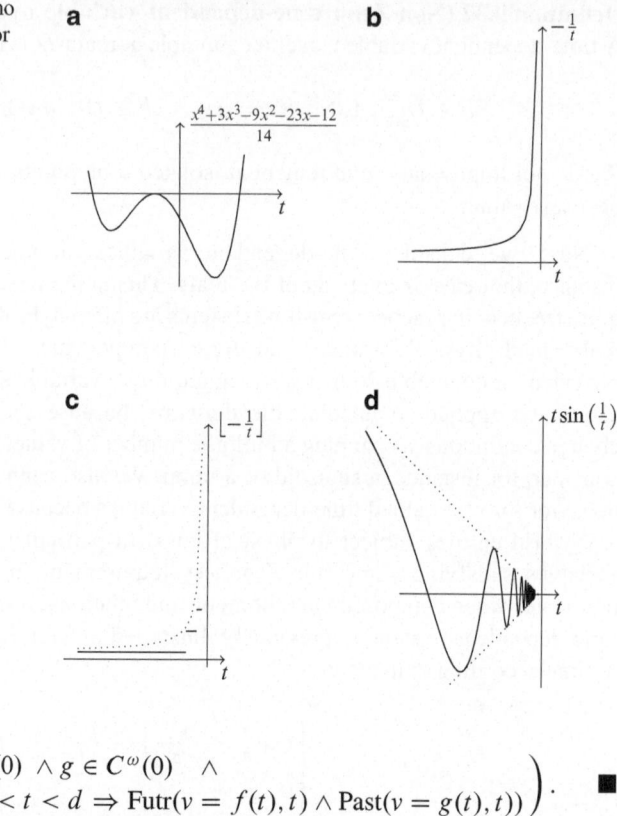

$$\exists f \exists g \left(\begin{array}{l} f \in C^{\omega}(0) \;\wedge\; g \in C^{\omega}(0) \quad \wedge \\ \exists d \,\forall t \,(0 < t < d \Rightarrow \mathrm{Futr}(v = f(t), t) \wedge \mathrm{Past}(v = g(t), t)) \end{array} \right). \quad \blacksquare$$

Analyticity is quite a strong smoothness requirement on functions; it guarantees that any constant line is intercepted only finitely many times over any finite interval. Hence, any formula of the kind $v = K$, where K is a value in D, is guaranteed to be a non-Zeno predicate.

According to Definition 9.33, constants are non-Zeno, and so are time-dependent variables whose graph over time is polynomial (as in Fig. 9.15a), sinusoidal (sin and cos), or exponential. Time-dependent variables that are piecewise constant, linear, or harmonic are non-Zeno. Sums, differences, and products of non-Zeno variables are all non-Zeno; division is non-Zeno if the denominator is always different from zero. Variables with constant or bounded derivatives are non-Zeno.

Instances of *Zeno* variables (expressed as functions of time) are $-1/t$ (Fig. 9.15b), $\lfloor -1/t \rfloor$ (Fig. 9.15c), and $t \cdot \sin(1/t)$ (Fig. 9.15d).

The examples in Fig. 9.15 show that the weaker requirement of continuity is insufficient to guarantee non-Zenoness, because it does not guarantee that predicates built on variables are non-Zeno. For example, if v is a variable that behaves as the continuous function in Fig. 9.15d, the formula $v = 0$ is Zeno because there is an accumulation point of isolated zeros in a neighborhood of the origin. Indeed, neither UpToNow($v = 0$) nor UpToNow($v \neq 0$) holds at the origin; v takes and leaves the value 0 infinitely often. An even more compelling example is the function $b(t)$ presented above; b has continuous derivatives of every order (it belongs to the class C^∞) but it is not analytic, and indeed it does not satisfy our intuitive requirements for non-Zenoness, because it changes sign an infinite number of times in every interval that includes t_0.

In all, piecewise analytic variables do not generate Zeno behaviors, and are sufficiently general to include common timed behaviors, such as sinusoids or ramps, routinely used to model real-word quantities.

Example 9.34. In the "kernel railroad crossing" (KRC) problem (Example 8.26), an example of a non-Zeno time dependent-variable over an uncountable domain is the angle Gate_angle of the bar, with the domain being the interval of real numbers between 0 and 90.   ■

Exercise 9.35 (♦). Using the notions of point- and interval-based predicates, define a translation of timed automata (as described in Sect. 7.4.1) into TRIO formulae.   ■

Example 9.36 (Kernel railroad crossing in TRIO). We provide a formalization in TRIO of the "kernel railroad crossing" (KRC) discussed in Example 8.26 with reference to timed Petri nets.

Train movements inside the crossing region are modeled by the following predicates. Enter_R, Enter_I, and Exit_I are point-based predicates respectively representing a train entering region R, entering region I, and exiting region I. In_R and In_I are interval-based predicates representing a train being in region R (the train has entered R but has not exited it yet) and in region I. Since the KRC problem considers only one train, each event Enter_R, Enter_I, and Exit_I occurs at most once.

(K1) Enter_R $\Rightarrow$ AlwP($\neg$Enter_R) $\wedge$ AlwF($\neg$Enter_R)
(K2) Enter_I $\Rightarrow$ AlwP($\neg$Enter_I) $\wedge$ AlwF($\neg$Enter_I)
(K3) Exit_I $\Rightarrow$ AlwP($\neg$Exit_I) $\wedge$ AlwF($\neg$Exit_I)

Formulae (K4)–(K7) model the train movement in the various regions, and refer to the minimum and maximum times necessary to cross regions R and I called, respectively, R_m, R_M, I_m, I_M. The train takes time between R_m and R_M to go from region R to region I.

(K4) Enter_R $\Rightarrow$ $\exists t(R_m \leq t \leq R_M \wedge$ Futr(Enter_I, t))

The train takes time between I_m and I_M to exit region I once it has entered it.

(K5) Enter_I $\Rightarrow$ $\exists t(I_m \leq t \leq I_M \wedge$ Futr(Exit_I, t))

Every exit event occurs only after the corresponding enter event.

(K6) Enter_I $\Rightarrow$ $\exists t(R_m \leq t \leq R_M \wedge$ Past(Enter_R, t))
(K7) Exit_I $\Rightarrow$ $\exists t(I_m \leq t \leq I_M \wedge$ Past(Enter_I, t))

The time bounds on train movements satisfy the following inequalities:

(K8) $0 < R_m \leq R_M \wedge 0 < I_m \leq I_M$

A train is inside region R or I if it has not exited that region since it entered it.

(K9) $\mathsf{In\_R} \iff \mathsf{Since_{ie}}(\neg\mathsf{Enter\_I}, \mathsf{Enter\_R})$
(K10) $\mathsf{In\_I} \iff \mathsf{Since_{ie}}(\neg\mathsf{Exit\_I}, \mathsf{Enter\_I})$

To complete the specification, we describe bar movements. The "up" and "down" commands are formalized by the mutually exclusive $\mathsf{Go\_up}$ and $\mathsf{Go\_down}$ point-based predicates, while the bar state is modeled by four mutually exclusive interval-based predicates Open, Closed, $\mathsf{Opening}$, and $\mathsf{Closing}$, with obvious meanings. After a $\mathsf{Go\_down}$ or $\mathsf{Go\_up}$ command, the bar reaches the final position in γ time units (with $\gamma < R_m$) with the same speed of movement in both directions. When the bar is moving, any command is ignored: the bar continues its movement until it reaches the final position. Only then will the bar be ready to accept a new command. The bar movements are modeled by the following formulae.

When the bar is closed and receives a $\mathsf{Go\_up}$ command, it moves upwards for γ time units and then it stays open until the other command, $\mathsf{Go\_down}$, is issued. A symmetrical formula holds for the opposite position and direction of movement.

(M1) $\mathsf{UpToNow(Closed)} \wedge \mathsf{Go\_up} \Rightarrow \left(\begin{array}{l} \mathsf{Lasts_{ei}(Opening}, \gamma) \wedge \\ \mathsf{Futr}(\mathsf{Until_{ei}^w(Open, Go\_down)}, \gamma) \end{array} \right)$

(M2) $\mathsf{UpToNow(Open)} \wedge \mathsf{Go\_down} \Rightarrow \left(\begin{array}{l} \mathsf{Lasts_{ei}(Closing}, \gamma) \wedge \\ \mathsf{Futr}(\mathsf{Until_{ei}^w(Closed, Go\_up)}, \gamma) \end{array} \right)$

Initially, before any operation takes place, the bar is open (we assume that the bar is installed before trains arrive).

(M3) $\mathsf{AlwP}(\neg\mathsf{Go\_down}) \Rightarrow \mathsf{Open}$

Operations on the bar are specified by the following formulae, which constitute a *specification* for the design of the device that issues commands to the bar.

(C1) $\mathsf{Go\_down} \iff \mathsf{Past}(\mathsf{Enter\_R}, R_m - \gamma)$
(C2) $\mathsf{Go\_up} \iff \mathsf{Exit\_I}$

The constant $R_m - \gamma$ in formula 9.36 takes into account the time γ necessary to reach the closed position once the bar movement has started, and conservatively considers the minimum time R_m necessary for any train to reach region I from region R. On the other hand, the bar is immediately raised when a train exits region I.

These commands are carefully defined to guarantee the two fundamental properties of *safety* and *utility*: the bar is closed when a train is inside region I, but at the same time it is not closed longer than necessary. Safety and utility are formalized in TRIO as follows.

(R1) *Safety*: $\mathsf{In\_I} \Rightarrow \mathsf{Closed}$
(R2) *Utility*: $\mathsf{Lasted_{ii}}(\neg\mathsf{In\_I}, \gamma) \vee \mathsf{Lasts_{ii}}(\neg\mathsf{In\_I}, R_M - R_m + \gamma) \Rightarrow \mathsf{Open}$

The two constants γ and $R_M - R_m + \gamma$ in the property 9.36 represent the time γ taken by the bar to open up and the conservative lower bound $R_M - R_m + \gamma$ on the time before the bar is lowered upon train arrival – a consequence of the uncertainty about the precise train position. ∎

Exercise 9.37 (♦). Based on the model of Example 9.36, provide a high-level proof (for instance, a sequence of lemmas) of the safety property 9.36.

(*Hint*: you can find some inspiration in Sect. 11.3, where a TRIO proof of the safety property is sketched based on a timed Petri net model of the KRC). ∎

Exercise 9.38. In the KRC Example 9.36, formalize the behavior of a more sophisticated kind of bar, whose movement in the up direction can be interrupted by a Go_down command. ∎

Exercise 9.39 (♦). Formalize a "*generalized* railroad crossing", where multiple trains are traveling on various rail tracks.

(*Hint*: assume that there is a maximum number of trains that can enter regions R and I per time unit, and that each train is identified by an integer). ∎

9.4.2 Sampling as a Means to Reconcile Discrete- and Dense-Time Semantics

Modeling different components of heterogeneous systems may require the combination of time domains with contrasting characteristics. In particular, hybrid systems, often mentioned in previous chapters, combine elements operating over discrete time (such as digital computers) with other components evolving over dense domains (such as physical processes). Temporal logics such MTL and TRIO can model elements having different characteristics, because their formulae are applicable to both discrete and dense time domains (even if the present chapter has focused on their dense-time semantics). The problem is that the discrete-time and dense-time semantics are quite different in general; hence the "global meaning" of system models that encompass formulae referring to both semantics can be inconsistent, or diverge considerably from intuitive expectations.

As a first simple example, consider LTL. Section 9.3 showed how LTL can express simple metric properties with the next operator over discrete time domains. A possible semantics of LTL formulae with next over dense time could state that the next operator refers to a time point at a fixed, predefined distance, such as one time unit. A formula such as $\Box(\mathsf{req} \Rightarrow \bigcirc\bigcirc\mathsf{gr})$ would then mean "every request is followed by a grant after two time units" also under a dense-time interpretation.

Even with such an assumption about the duration of consecutive discrete steps, changing the time domain can turn discrete-time satisfiable formulae into dense-time unsatisfiable ones, and vice versa. For example, the formula

$$\Box(\mathsf{req} \Rightarrow \bigcirc\mathsf{gr} \vee \bigcirc\bigcirc\mathsf{gr}) \tag{9.14}$$

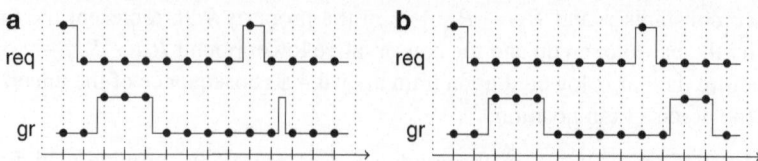

Fig. 9.16 Sampling a dense-time behavior with (**a**) and without (**b**) information loss

says that every request is followed by a grant after one or two time units. The meaning of (9.14) over discrete time is clear, but it becomes fuzzy over dense time. What if a request is followed by a grant exactly after 1.5 time units? This makes (9.14) false over dense time, but it is questionable whether this is the "right" interpretation of the formula in this context.

A first consequence of such ambiguities is the need for more expressive notations, such as MTL and TRIO, to model metric properties over dense time; Sect. 9.4.1 commented on the subtleties of dense-time modeling with these metric temporal logics.

Even if we take temporal logics with metric operators, the related problem of uniformly reconciling discrete-time and dense-time semantics in hybrid system models is still open. The framework of "sampling" addresses this problem; as the name suggests, the framework adapts some notions of classical sampling theory to the domain of metric temporal logics. Consider a generic MTL formula α interpreted over a dense time domain, say $\mathbb{R}$. α determines the set $dense_\mathbb{R}(\alpha)$ of all behaviors over $\mathbb{R}$ that satisfy it. In a hybrid system where α describes a component A operating over dense time (for example, the evolution of temperature in an air-conditioned room), other units B operate over discrete time (for example, the air-conditioning controller) and interact with A through a sampler, such as pictured in Fig. 9.16.

If A's dynamics over dense time is too fast with respect to the sampling period, module B will miss some events that occur over dense time. Figure 9.16a, for example, shows a behavior where every request is followed by a grant within two time units *in dense time*, but the second grant holds for too short and it is missed by the sampler; hence the second request is not followed by a grant within two time units *in discrete time*. The signals in Fig. 9.16b, in contrast, hold their values for at least one sampling period; hence the bounded response property of requests and grants holds in dense time and in its discrete-time sampling.

Similarly to the classical theory of sampling, which guarantees no information loss only for signals with limited frequency, the sampling framework assumes dense-time behaviors that are non-Berkeley (see Sect. 3.6): the time that elapses between two consecutive state changes is not less than that of the sampling period. The behaviors in Fig. 9.16b satisfy this requirement, unlike those in Fig. 9.16a. Under the assumption of non-Berkeleyness, the sampling framework defines how to transform a generic formula α interpreted over dense time into another formula β to be interpreted over a discrete time domain, say $\mathbb{Z}$, such that the set of discrete-time behaviors $discrete_{\mathbb{Z}}(\beta)$ that satisfy β correspond to the events module B reads in discrete time from the sampler. β is a simple syntactic transformation of α; hence β can be regarded as the "best" discrete-time approximation of α. For example, the MTL formula (to be interpreted over dense time)

$$\Box(\text{req} \Rightarrow \Diamond_{<2} \text{gr}) \tag{9.15}$$

corresponds to the formula (to be interpreted over discrete time)

$$\Box(\text{req} \Rightarrow \Diamond_{\leq2} \text{gr}), \tag{9.16}$$

which relaxes the grant time from "strictly less than 2" to "less than or equal to 2"; sampling at every integer instant any non-Berkeley dense-time behavior that satisfies (9.15) gives a discrete-time behavior that satisfies (9.16). The relaxation is indeed necessary: in a behavior where a request occurs at a sampling instant t and the corresponding grant at $t + 1.99$, the grant is recorded only at the integer sampling instant $t + 2$, which does not satisfy the constraint "< 2" of (9.15). The sampling framework also defines suitable transformations of discrete-time MTL formulae into dense-time formulae with the "same" semantics.

Chapter 7 presented the notion of "digitization", which provides an alternative framework to reconcile discrete-time and dense-time semantics. For metric temporal logics, the two frameworks of sampling and digitization are complementary, in that there exist digitizable MTL formulae that are not transformable with the rules of sampling, and conversely some MTL formulae amenable to sampling are not digitizable. However, while all flat MTL formulae – which do not nest temporal operators – are amenable to sampling techniques, finding ample syntactically definable sets of digitizable MTL formulae is an open problem. For this reason, digitization is more commonly applied to operational formalisms such as transition systems and process algebras.

9.5 Interval-Based Temporal Logics

All temporal logics considered so far adopt time points (instants) as fundamental entities: every state is associated with a time instant, and formulae are interpreted at some time instants. The so-called "*interval*(-based) temporal logics" adopt instead time intervals as fundamental temporal entities, and time points are derived from intervals as secondary entities, if not completely ignored. In principle, choosing intervals rather than points as elementary time notions may be a matter of subjective preference once we acknowledge that intervals are sets of points, and, conversely, points are special cases of intervals with a single element. In formal logic, however, seemingly limited variations in the alphabet of symbols and operators may considerably impact expressiveness, decidability, and the complexity of analysis and verification.

Over the years, interval temporal logics have been developed into a prolific research field, which has produced many related analysis and verification procedures and tools. This chapter presents "duration calculus", a representative example of interval temporal logic that has influenced several derived approaches. The bibliographic remarks mention other examples of temporal logics based on intervals.

Duration calculus formulae are interpreted with reference to a time interval, typically left implicit, much in the way that the current time is left implicit in formulae of linear temporal logic.

Duration calculus is a **LINEAR**-time logic, normally defined on the **DENSE** time domain of the nonnegative reals:

$$\mathbb{T} := \mathbb{R}_{\geq 0} \,.$$

Systems are modeled through *observable entities*: in the terminology introduced in Sect. 9.3.3, these include time-dependent variables, functions, and predicates. In the interpretation structures of the logic, observable entities correspond to functions of the time domain (and possibly of other domains). I denotes the semantic interpretation function that associates with every observable entity X a function

$$I(X) : \mathbb{T} \to domain(X) \,,$$

where "$domain(X)$" denotes the domain associated with entity X.

A hypothesis of finite variability (as discussed in Sect. 9.4.1) is assumed for every observable entity, so **ZENO** behaviors are excluded a priori. Duration calculus is a first-order language: formulae can include global variables (in the sense of Sect. 9.3.3), i.e., variables that maintain the same value for each of their occurrences in the formula and do not change with time, and are used in existential or universal quantifications.

In duration calculus, the notion of time interval is not really primitive, in that system states are initially associated with time points; however, the syntax of the language lifts the level of the description to intervals, and consequently duration calculus describes system features in terms of intervals where given properties hold, and possibly of their length, which makes duration calculus a **METRIC** temporal logic. Correspondingly, the syntax of duration calculus distinguishes between *state assertions*, concerning properties of states at individual time points, *duration terms*, which denote the length of time intervals, and *duration formulae*, which express properties by predicating on interval lengths. The rest of the section presents the main features of duration calculus; the presentation omits some inessential details and occasionally is informal, focusing on the syntactic and semantic features that characterize interval-based logics. The basic building blocks of duration calculus formulae are *state assertions*, defined by the following syntax:

$$\pi := 0 \mid 1 \mid P \mid X = d \mid \neg\pi \mid \pi_1 \wedge \pi_2 \,,$$

where 0 and 1 stand for **False** and **True**, P is a Boolean observable entity (a predicate or propositional letter), X is a non-Boolean observable entity and d is a value in X's domain. Any state assertion π is interpreted by the semantic function I at points of the temporal domain, that is, the semantics of π is a function $I[\pi] : \mathbb{T} \to \{0, 1\}$, defined as follows.

$$I[0](t) := 0$$

$$I[1](t) := 1$$

$$I[P](t) := I(P)(t)$$

$$I[X = d](t) := \begin{cases} 1 & I(X)(t) = d \\ 0 & \text{otherwise} \end{cases}$$

$$I[\neg\pi](t) := 1 - I[\pi](t)$$

$$I[\pi_1 \wedge \pi_2](t) := I[\pi_1](t) \cdot I[\pi_2](t)$$

Example 9.40 (State assertions in duration calculus). With reference to the kernel railroad crossing Example 9.36, the following are state assertions:

$$\text{In\_R}, \qquad \text{Gate\_angle} = 90, \qquad \text{In\_I} \Rightarrow \text{Closed},$$

where In_I, In_R, and Closed are predicate letters meaning that the train is currently in region I, that the train is currently in region R, and that the gate is closed, while Gate_angle is a real-valued observable entity denoting the current position of the gate. ∎

Building atop state assertions, duration calculus shifts the level of system description from points to intervals by means of *duration terms*, which express the cumulative duration of a state assertion over an interval using the special integral operator: a duration term θ is defined as

$$\theta := x \mid \ell \mid \int \pi,$$

where x is a global variable, and ℓ is a special symbol that represents the length of the current (implicit) time interval; for simplicity, we do not present functions, which are a straightforward extension.

The semantic interpretation function I assigns values to duration terms using a valuation function V for global variables and an interval $[b, e] \subseteq \mathbb{T}$, according to the following clauses:

$$I[x](V, [b, e]) := V(x)$$

$$I[\ell]([b, e]) := e - b$$

$$I[\int \pi]([b, e]) := \int_b^e I[\pi](t)\mathrm{d}t$$

Notice that, since the integral operator semantics is actually given in terms of an integral operation, its value is insensitive to the values of observable entities at isolated time points.

Full-fledged duration calculus formulae coincide with the syntactic category D of *duration formulae*, defined by

$$D := \mathbf{True} \mid p(\theta_1, \ldots, \theta_n) \mid \neg D \mid D_1 \wedge D_2 \mid \exists x\, D \mid D_1 \; ; \; D_2 \,,$$

where the θ_i's are duration terms, p is an n-ary predicate symbol (p does not denote an observable entity: its interpretation is a *global* relation, which does not change with time), x is a global variable, and " ; " is the chop operator presented next. The constant **False**, universal quantification, and other Boolean connectives are derived as usual.

Example 9.41 (Predicates involving duration terms). Still with reference to the kernel railroad crossing Example 9.36, the duration formula

$$\int \mathsf{In\_R} > R_{\mathrm{m}}$$

states that, relative to the (implicit) interval in which it is evaluated, the total time the train was in region R is more than R_{m}. The formula

$$\int (\mathsf{In\_I} \Rightarrow \mathsf{Closed}) = \ell$$

asserts that, for every implicit current interval over which we evaluate the integral, the duration over which $\mathsf{In\_I} \Rightarrow \mathsf{Closed}$ holds coincides with the length of the whole interval.   ∎

The "chop operator" " ; " (sometimes denoted by "$\cap$") is the fundamental modal operator of duration calculus. Just as the modal operators of LTL ($\Diamond$, $\square$, U, etc.) refer to time instants different from the current (implicit) one, the chop operator in a formula $D_1 \; ; \; D_2$ interprets the two subformulae D_1 and D_2 over two adjacent subintervals of the current (implicit) one, referred to by the whole formula $D_1 \; ; \; D_2$. This is expressed formally by

$$I, [b, e] \models D_1 \; ; \; D_2 \quad \text{if and only if} \quad \text{there exists } m \in [b, e] \text{ such that}$$
$$I, [b, m] \models D_1 \text{ and } I, [m, e] \models D_2 \,. \tag{9.17}$$

From (9.17), it is clear that the chop operator expresses a sort of existential quantification: $D_1 \; ; \; D_2$ states the existence of a partition of the original interval $[b, e]$ into two contiguous subintervals $[b, m]$ and $[m, e]$ in which the two subformulae D_1 and D_2 are respectively satisfied.

Finally, a duration formula is satisfied in an interpretation if it holds, according to the previously provided definitions, in every initial interval of the time domain:

$$I \models D \quad \text{if and only if} \quad \text{for all } t \in \mathbb{T} : I, [0, t] \models D . \qquad (9.18)$$

As we expected from the extensive use of intervals, duration calculus is well suited for describing sequences of system states: for instance, with reference to the kernel railroad crossing example, the following formula states that a train first approaches region R, then goes through it, then crosses the critical region I, and finally exits it:

$$(\smallint \neg \mathsf{In\_R} = \ell \wedge \ell > 0) \,; \, (\smallint \mathsf{In\_R} = \ell \wedge \ell > 0) \,;$$
$$(\smallint \mathsf{In\_I} = \ell \wedge \ell > 0) \,; \, (\smallint \neg \mathsf{In\_I} = \ell \wedge \ell > 0) . \quad (9.19)$$

The various occurrences of ℓ in (9.19) refer to the distinct subintervals that compose the overall interval on which the entire formula is interpreted; hence they have distinct values in general. Formula (9.19) also suggests that the formula schema $\pi = \ell \wedge \ell > 0$ is very common, and in fact it is customary to introduce it as the derived operator:

$$\lceil \pi \rceil := \quad \smallint \pi = \ell \wedge \ell > 0 .$$

As a side remark, let us note that duration calculus (like many other temporal logics such as TRIO) is based on a minimal set of basic operators and systematically introduces derived ones to raise the level of abstraction in the modeling and to provide concise, intuitive notations for specifying properties. The following ones are among the most frequently used:

$$\begin{array}{llll} \lceil \pi \rceil^t & := & \smallint \pi = \ell \wedge \ell = t & \text{exact duration of a property} \\ \Diamond D & := & \textbf{True} \,; \, D \,; \, \textbf{True} & \text{sometimes} \\ \Box D & := & \neg \Diamond \neg D & \text{always} \end{array}$$

where the correspondence in meaning with the same LTL operators is evident.

Real-time features can be expressed in duration calculus by predicating on occurrences of the ℓ special symbol. For instance, the lower and upper bounds on the time taken by the train in the regions R and I can be expressed as

$$\lceil \neg \mathsf{In\_R} \rceil \,; \, \lceil \mathsf{In\_R} \rceil \wedge R_m \leq \ell \leq R_M \,; \, \lceil \mathsf{In\_I} \rceil \wedge I_m \leq \ell \leq I_M \,; \, \lceil \neg \mathsf{In\_I} \rceil .$$

Since the value of ℓ depends on the subformula where it occurs, to compare lengths of different intervals we must use global variables to link the various values. For instance, to state that the time to cross region I is strictly shorter than that needed to cross region R, we write

$$\lceil \neg \mathsf{In\_R} \rceil \,; \, \lceil \mathsf{In\_R} \rceil \wedge \ell = d \,; \, \lceil \mathsf{In\_I} \rceil \wedge \ell = h \,; \, \lceil \neg \mathsf{In\_I} \rceil \quad \Rightarrow \quad d < h .$$

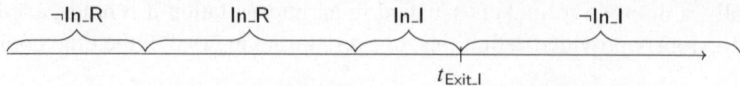

Fig. 9.17 Interpretation I_{run} representing the run of a train through regions R and I

All the examples of duration calculus formulae seen so far have a clear, intuitive meaning when interpreted with reference to given intervals. Intricacies may arise, however, when formalizing *requirements*, that is, properties that must be satisfied by all intended interpretations. In this case, it is necessary to consider the implicit quantification on the observation interval that is embedded in clause (9.18) of the satisfaction relation of duration calculus. Let us consider, for instance, the formula describing the sequence of regions traversed by the train

$$\lceil \neg \mathsf{In\_R} \rceil \ ; \ \lceil \mathsf{In\_R} \rceil \ ; \ \lceil \mathsf{In\_I} \rceil \ ; \ \lceil \neg \mathsf{In\_I} \rceil \qquad (9.20)$$

and, as a possible model for it, the interpretation I_{run} in Fig. 9.17, where proposition $\mathsf{In\_I}$ is false at every time point greater than $t_{\mathsf{Exit\_I}}$, the time when the train exits region I.

Formula (9.20) is not satisfied in the interpretation I_{run} of Fig. 9.17: in fact, $I_{\text{run}}, [0, t] \models \lceil \neg \mathsf{In\_R} \rceil \ ; \ \lceil \mathsf{In\_R} \rceil \ ; \ \lceil \mathsf{In\_I} \rceil \ ; \ \lceil \neg \mathsf{In\_I} \rceil$ for every $t > t_{\mathsf{Exit\_I}}$, but, for $t' < t_{\mathsf{Exit\_I}}$, it is not the case that $I_{\text{run}}, [0, t'] \models \lceil \neg \mathsf{In\_R} \rceil \ ; \ \lceil \mathsf{In\_R} \rceil \ ; \ \lceil \mathsf{In\_I} \rceil \ ; \ \lceil \neg \mathsf{In\_I} \rceil$, because the observation interval $[0, t']$ does not have a suffix subinterval in which $\neg \mathsf{In\_I}$ holds.

Since duration formulae must hold in all initial intervals of the temporal domain, it may be impossible to specify some properties simply in positive form as sequences of intervals where some predicates hold. In fact, requirements are often specified in duration calculus in negative form by describing the undesired behavior. As a simple example, the fact that the train exits region R to enter region I is not expressed by the formula $\lceil \mathsf{In\_R} \rceil \ ; \ \lceil \mathsf{In\_I} \rceil$, but by the slightly more involved

$$\neg \diamond (\lceil \mathsf{In\_R} \rceil \ ; \ \lceil \neg \mathsf{In\_I} \rceil) \ .$$

The property that the train, after entering region R, enters region I with a delay d, with $R_{\mathrm{m}} \le d \le R_{\mathrm{M}}$, is similarly specified by saying that the length of an interval where $\mathsf{In\_R}$ holds (denoted as usual by ℓ) cannot be less that R_{m} or more that R_{M}:

$$\neg \diamond (\lceil \neg \mathsf{In\_R} \rceil \ ; \ (\lceil \mathsf{In\_R} \rceil \wedge (\ell < R_{\mathrm{m}} \vee \ell > R_{\mathrm{M}})) \ ; \ \lceil \neg \mathsf{In\_R} \rceil) \ .$$

On the other hand, invariance properties are naturally specified with the "always" operator $\square$: for instance, the safety property 9.36 of the railroad crossing (whenever the train is in the critical region I, the bar is closed) is expressed by the following clear, concise specification formula:

$$\square(\mathsf{In\_I} \Rightarrow \mathsf{Closed}) \ .$$

Exercise 9.42. An intrusion detection system in a building includes a sensor of movement that continuously outputs a signal indicating, for every time instant, if a movement is being perceived. An alarm is raised if the sensor movement signal is on for at least two consecutive time units, or if the duration in which a movement is detected accumulates to three time units (possibly scattered over many disjoint intervals) within any interval of 30 time units. Provide a duration calculus specification of such an intrusion detection system.   ■

Exercise 9.43. Consider a safe equipped with two different keys: the safe opens if and only if either key is inserted in the lock and kept inserted for at least 2 s and no more than 4 s; then, within the following 3 s, the other key should be inserted and kept in for least 2 s and no more than 4 s; at that point the safe opens after exactly 1 s and remains open for exactly 1 min. Formalize such a safe by means of duration calculus formulae.   ■

9.5.1   Interval-Based and Point-Based Predicates

Using duration calculus, we can easily state properties of durations, that is, of the intervals over which Boolean-valued functions hold a value, under the assumption of finite variability of observable entities (non-Zenoness). The calculus can therefore specify *interval-based* properties (intended as system configurations) that remain stable for a non-null time intervals.

In contrast, the integral operator of duration calculus cannot model point-based predicates, that is, properties that hold at isolated time points. This may seem surprising given that the syntax of duration calculus can represent intervals with null duration through the standard abbreviation

$$\lceil \rceil \quad := \quad \ell = 0,$$

which characterizes any interval with null duration. However, isolated values of the observable entities, though admissible in the interpretations, cannot be discerned by the integral operator – the cornerstone of the duration calculus – because the integral of any function over an interval consisting of a single point is zero regardless of the function value at that point.

As a consequence, the original version of duration calculus cannot formalize events associated with state transitions, a fairly common idiom in system requirements specification which point-based linear temporal logics support quite naturally. For instance, in the railroad crossing example, it is natural to introduce the proposition Enter_I representing the event of the train exiting region R (and hence the state In_R becoming false) to enter region I (state In_I becoming true).

To overcome this limitation, two extensions to duration calculus have been proposed. In one approach, a Boolean-valued so-called δ-function, having non-null values at isolated time points, is introduced to model instantaneous actions and events, and, to denote such instantaneous actions, the integral of a function that is non-null at isolated time points is replaced by a suitably defined "mean value". The other approach retains the same basic calculus for the integrals of Boolean-valued functions, but the syntax is extended directly at the higher level of duration formulae by including predicates representing events. The literature in the bibliographic remarks provides complete descriptions of these two approaches.

9.6 Dealing with Heterogeneous Time Granularities

Once suitable constructs are available to denote, in a quantitatively precise way, time distances among events, new challenges arise in the modeling of systems that include several components that evolve on different time scales, possibly independently.

Many systems, in fact, combine components whose dynamics are regulated by time constants that differ by orders of magnitude, as we already discussed in Sect. 3.2.2. For instance, a hydroelectric power plant encompasses a lake (whose water level varies over many days or months), turbines and electric generators (whose states change in a few seconds), and control machines, possibly implemented with digital hardware (whose evolutions occur within milliseconds). All these cooperate in an integrated system. Another example is an office organization, where there are actions (such as writing a letter) performed in a few minutes, while paying salaries or completing projects are matters of months or years. We say that systems of such types have different "time GRANULARITIES".

As a consequence, a specification language for heterogeneous systems with different time granularities should offer references to different time scales, so that the requirements can always be stated in terms of their "natural" time scales. For instance, the requirements for a hydroelectric power plant should not include statements such as "the filling of the water basin must be completed within $4 \cdot 10^{12}$ microseconds" instead of "the filling of the water basin must be completed within two months".

A trivial solution to this problem uses different time units – say, months and minutes – to measure the "time variable" in the same dynamic model. Syntactically, we can label temporal expressions with a reference to their time domains: for instance, 30_D means 30 days, whereas 3_h denotes 3 h. The combined uses of various time scales are expressible within a model using the finest-grained time unit. We can obtain such an underlying model by means of suitable translations that apply appropriate multiplication factors; for example, a factor of $24 \cdot 60 \cdot 60$ translates values from days into seconds.

There are cases, however, where the use of different time scales in the description of dynamic system properties leads to subtle semantic problems. We now present two typical specification issues related to (a) the mapping, between time domains of different granularity, of time instants were predicates hold, and (b) the synchronization of events in the presence of boundaries between time units, such as the

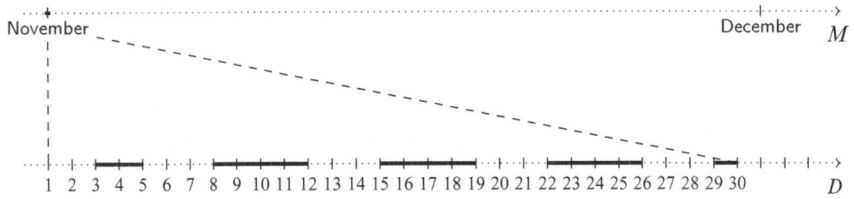

Fig. 9.18 Interpretation of a predicate in an upper-level time granularity in terms of a lower-level granularity. *Thick solid lines* denote the intervals where the predicate holds

time instant "midnight" separating two consecutive days. We briefly illustrate the notation introduced in TRIO to address such specification issues, and refer to the bibliographic remarks for other notations and frameworks addressing these and other issues related to time granularity.

Consider the requirement mentioned in Sect. 3.2.2 (applied in the context of the administrative office of a business organization):

Every month, if an employee works, then she gets her salary.

As we already observed, the sentence does not imply that an employee must work each and every day of a month to get her salary, nor that the salary is paid every day of a month in which an employee works. This example highlights the necessity of specifying the way in which a predicate is mapped from an interpretation based on the time domain of months to one based on that of days.

The notation adopted to specify such a property must support the definition of a mapping between time domains of different granularities; hence, the truth of a predicate at a given time value in a higher (coarser) level of granularity can be defined in terms of an *interval* in a lower (finer) level. For instance, Fig. 9.18 specifies that working during the month of November means working from the 3rd through the 5th, from the 8th through the 12th, and so on. Predefined but customizable predicate mappings are used to specify standard patterns. For instance, given two temporal domains T_1 and T_2, such that T_1 is coarser than T_2,

$$p \text{ event in } T_1 \rightarrow T_2$$

means that predicate p is true in any $t \in T_1$ if and only if it is true in just one instant of the interval of T_2 corresponding to t;

$$p \text{ complete in } T_1 \rightarrow T_2$$

means that p is true in any $t \in T_1$ if and only if it is true in the whole corresponding interval of T_2;

$$p \text{ intervalSequence in } T_1 \rightarrow T_2$$

means that p is true in any $t \in T_1$ if and only if it is true in a sequence of subintervals of the corresponding interval of T_2.

Therefore the TRIO formula

$$\text{Alw}_M \, (\forall emp(\text{work}(emp) \Rightarrow \text{getSalary}(emp)))$$

formalizes the previously introduced sentence "every month, if an employee works, then she gets her salary" using the mappings

$$\text{getSalary } \textbf{event in } M \rightarrow D \, ,$$

$$\text{work } \textbf{intervalSequence in } M \rightarrow D \, ,$$

defined as in Fig. 9.18.

As another example of an issue related to the meaning of predicates expressed at various time granularity levels let us consider, in the context of an educational institution, the sentence

> (E1) During every written exam, once the instructor has completed the explanation of the exercise, the students must solve it within exactly three hours; then, the instructor will collect the solutions and will publish and register the grades after two days.

The two parts of the sentence (separated by the semicolon) refer to different synchronization criteria among events with respect to the time units of hours and days: the former part of the sentence means "students have to complete their jobs within 180 minutes starting from the minute the explanation ended"; the latter part says that "the results will be published by midnight (or another predefined time) of the second calendar day following the one when the exam was held".

More generally, in application domains having administrative, business, or financial implications, changes in time granularity are often paired with references to a global time calendar that evolves in a *synchronous* way. For instance, time units such as days, weeks, months, and years change synchronizing at certain predefined time instants (e.g., midnight or New Year) that are conventionally established in a global fashion. In contrast, when some process events are related directly to one another on a given time scale (minutes, in the example) but independently of the coarser time scales (e.g., hours), the distances between events in the finer-grained scale can be translated "verbatim" to the coarser scale, ignoring the global synchronization references in that scale. In such cases, we say that time granularity is managed *asynchronously*. Quite often, the distinction between synchronous and asynchronous is implicit in natural language and depends on the context and custom knowledge, but it must be made explicit when formalizing. For instance, in the above example (E1) the interpretation of "the students must solve it within exactly three hours" is of asynchronous type; hence it is equivalent to "within 180 minutes", whereas the interpretation of "will publish and register the grades after two days" is synchronous. Under asynchronous interpretations fractional values are perfectly normal; in a similar example, an exam running for 2.5 h is the same as one running for 150 min. With synchronous interpretation, in contrast, it is unnatural and

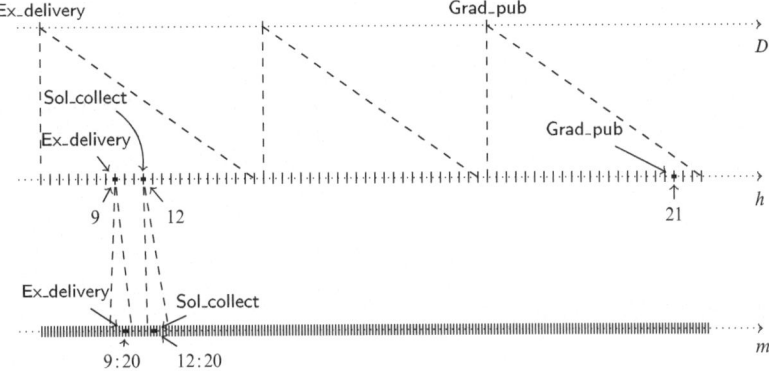

Fig. 9.19 Synchronous and asynchronous mapping of events

ambiguous to use fractional values, such as in "the instructor will publish the grades after 1.3 weeks".

We make explicit these notions of synchronous and asynchronous interpretations by labeling temporal operators "S" for synchronous and "A" for asynchronous, denoting the intended mode of granularity refinement. The description of the written examination is formalized by the following formula ("h" stands for hours, and "D" for days):

$$\mathrm{Alw}_{h,A}(\mathsf{Ex\_delivery} \Rightarrow \mathrm{Futr}(\mathsf{Sol\_collect}, 3)) \ \wedge$$

$$\mathrm{Alw}_{D,S}(\mathsf{Ex\_delivery} \Rightarrow \mathrm{Futr}(\mathsf{Grad\_pub}, 2)) \quad (9.21)$$

Figure 9.19 shows an interpretation satisfying (9.21); the requirement on the distance, in days, between Ex_delivery and Grad_pub would not be satisfied by the depicted model if the interpretation were asynchronous, because the distance in hours between the occurrence of the Ex_delivery predicate and that of Grad_pub is 60, which is greater than the value of 48 that corresponds to 2 days in an asynchronous interpretation.

9.7 Explicit-Time Logics

Explicit-time logics are another category of descriptive formalisms, still based on mathematical logic but not quite in the same way as (modal) temporal logics, as they adopt a "timestamp" **EXPLICIT** view of time. This is typically done by introducing some ad hoc feature (such as a variable that represents the current time, or a time-valued function providing a timestamp for every event occurrence). The family of explicit time logics is, if possible, even richer and more varied than that of temporal logics, as every notation focuses on different language features that are considered more relevant.

The present section discusses three distinguished approaches: the "real-time logic" (RTL), the "temporal logic of actions" (TLA), and the "timed propositional temporal logic" (TPTL). Another prominent example of explicit-time logic is the "real-time temporal logic" (RTTL) presented in the context of the dual language approach in Sect. 11.2.

9.7.1 Real-Time Logic

Real-time logic (RTL) formulae explicitly refer to the timing of events and actions. RTL's time model is **LINEAR** and **DISCRETE**, with a start point and no upper bound (in practice, the temporal domain is the set $\mathbb{N}$ of naturals). Timing constraints among event and action occurrences are expressed as formulae including first-order predicate calculus with arithmetic and relations over the natural numbers. The configuration at any given time instant is characterized by a set of time-dependent Boolean predicates. There are five basic kinds of events: the start and end of an action A (respectively, $\uparrow A$ and $\downarrow A$), the change of value of a predicate S from false to true ($S := \textbf{True}$) and from true to false ($S := \textbf{False}$), and external events E (ΩE denotes an occurrence of E). All events may occur multiple times during a computation: to explicitly represent the various event occurrence times and to distinguish different occurrences of the same event, RTL uses the occurrence function "@": for each event E, @(E, i) denotes the time of E's ith occurrence. The occurrence function makes RTL an **EXPLICIT-TIME** logic, since it explicitly denotes the absolute time at which events occur. With reference to the example of the real-time resource allocator, a low-priority request is an external event in RTL – denoted by Ωlpr – while the resource being occupied is a time-dependent predicate occ. The property that any low-priority resource request is satisfied within two time units is then expressed in RTL as

$$\forall i \exists j (@(\Omega\text{lpr}, i) + 2 \geq @(\text{occ} := \textbf{True}, j) \quad \wedge \quad @(\text{occ} := \textbf{True}, j) \geq @(\Omega\text{lpr}, i)).$$

The semantics of RTL specifications is captured by an efficiently decidable subclass of quantifier-free Presburger[2] formulae; their satisfiability is determined by computing the occurrence function @, if there is one, that satisfies them.

Exercise 9.44 (Simultaneous events). Is it possible that @$(E, i) = $ @(E, j) for some $i \neq j$? Which constraints would make it impossible? Which system behaviors and operational conditions require it to be possible?   ∎

---

[2] Presburger arithmetic is the first-order theory of the natural numbers with only addition.

9.7.2 Temporal Logic of Actions

The "temporal logic of actions" (TLA) interprets formulae over LINEAR, DIS-
CRETE state sequences, and supports variables, first-order quantification, predicates, and the usual modal operators $\Diamond$ and $\Box$ for referring to some or all future
states. While basic TLA is not natively endowed with a QUANTITATIVE treatment
of time, we can still model real-time features by introducing a distinguished
state variable *now* with CONTINUOUS domain $\mathbb{R}$. The value of *now* represents
the current time, so the specification of temporal properties consists of formulae
explicitly predicating on the values of *now* in different states, thus introducing real-
time constraints.

With reference to the resource manager example, let us describe behavior in the
case of low-priority requests: an action lpr describes the occurrence of the request,
independently of timing information. "Actions" are predicates over two states,
where the values of variables at the current and next state are respectively denoted by
unprimed and primed symbols. Therefore, the untimed behavior of an accepted low-
priority request is simply changing the value of the state of the resource (indicated
by variable res) from free to occupied, as in the following action:

$$\mathsf{lpr} \quad := \quad \mathsf{res = free} \wedge \mathsf{res' = occ} \,.$$

Then, the timed behavior associated with an action is specified by lower and
upper bounds on the time it takes. For lpr, we specify that the action must happen
within two time units whenever it is continuously enabled. Following TLA's idiom,
we define a *timer* by means of two formulae: the first one defines predicate
MaxTime(t), which holds in all states whose timestamp (represented by the value
of variable *now*) is less than or equal to the absolute time t; the second one defines
predicate VTimer(t, A, δ, v), where A is an action, δ is a delay, v is the set of all
variables, and t is a state variable representing a timer; VTimer(t, A, δ, v) holds if
and only if either action A is not currently enabled and t is ∞, or A is enabled and
t is $now + \delta$ and will remain so until either A occurs or A is disabled.

Finally, we define the complete timed behavior of low-priority requests by action
lpr', where T_{gr} is a state variable representing the maximum time within which
action lpr must occur:

$$\mathsf{lpr'} \quad := \quad \mathsf{lpr} \wedge \mathrm{VTimer}(T_{\mathsf{gr}}, \mathsf{lpr}, 2, v) \wedge \mathrm{MaxTime}(T_{\mathsf{gr}}) \,.$$

More precisely, the formula above states that after action lpr is enabled, it must
occur before time surpasses the value $now + 2$.

ZENO behaviors are possible in TLA because formulae involving time may be
satisfied by interpretations where the variable *now*, which is a normal state variable,
never increases. To ensure non-Zenoness, we can a priori incorporate into system
descriptions the requirement that time always advances, via formula NZ.

$$NZ \quad := \quad \forall t \in \mathbb{R} \, (\Diamond(now > t)) \,.$$

In TLA, as in other temporal logics previously discussed, two consecutive states may refer to the same time instant (if *now* does not change value), so the logic departs from the notion of time inherited from traditional dynamical system theory (see Chap. 4). In fact, in every timed TLA specification it is customary to explicitly state the separation of time-advancing steps from ordinary program steps. This approach is similar in spirit to that adopted in TTM/RTTL, presented in Sect. 11.2.

9.7.3   Timed Propositional Temporal Logic

The "timed propositional temporal logic" (TPTL) represents a quite interesting example of how a careful choice of operators can make a great difference in terms of expressiveness, decidability, and complexity of the verification procedures. TPTL may be informally described as a "half-order" logic, in that it extends propositional **LINEAR**-time logic with variables that refer to time, but with only a restricted form of quantification called "freeze" quantification. For a variable x, the freeze quantifier "x." binds x to the time when the formula in the scope of the quantification is evaluated. We can think of it as the analogue for logic languages of clock resets in timed automata (see Sect. 7.4.1). Freeze quantifiers are combined with the usual temporal operators such as $\Diamond$ and $\Box$: if $\phi(x)$ is a formula in which variable x occurs free, formula $\Diamond x.\phi(x)$ asserts that there is some future instant, with some absolute time k, such that $\phi(k)$ will hold at that instant; similarly, $\Box x.\phi(x)$ asserts that $\phi(h)$ will hold at every future instant, h being the absolute time of that instant.

The property of the resource allocator that low-priority requests are satisfied within two time units is expressed in TPTL as

$$\Box x.(\mathsf{lpr} \Rightarrow \Diamond y.(\mathsf{occ} \wedge y < x + 2)) .$$

TPTL is decidable over **DISCRETE** time with a doubly exponential **DECISION PROCEDURE**; furthermore, adding unrestricted first-order quantification on variables representing the current time or adding past operators makes TPTL of non-elementary complexity. Therefore, TPTL constitutes an interesting trade-off between expressiveness and complexity for a temporal logic with **METRIC** on time.

9.8   Probabilistic Temporal Logics

Section 7.5.1 showed how to extend transition systems with probabilities to model random phenomena. In particular, discrete-time Markov chains model systems where choices among transitions are probabilistic, whereas continuous-time Markov chains are transition systems in which the timing of transitions is stochastic. This section presents two natural extensions of the branching-time temporal logic CTL (Sect. 9.1.2) that can express timing properties with probabilistic features: desired or undesired features are described not only qualitatively – holding in some or all of the possible executions of the modeled system – but also *quantitatively* – holding with a certain *probability* over all possible system executions. We first discuss "probabilistic computation tree logic" (PCTL) with reference to discrete-time Markov chains, and then introduce "continuous stochastic logic" (CSL) to describe properties of continuous-time Markov chains.

9.8.1 Probabilistic Computation Tree Logic

The "probabilistic computation tree logic" (PCTL) is a branching-time temporal logic based on CTL, from which it inherits the fundamental syntactic features: state formulae are interpreted in states and path formulae over state sequences, and the fundamental modal timed operators are the "next time" operator $\circ$ and the "until" operator U, which retain very much the same meaning as in CTL. The major novelty concerns path quantification in state formulae: in CTL the state formulae $\forall \phi$ and $\exists \phi$ mean that property ϕ is satisfied in all (or some) of the state sequences starting from the current implicit state. PCTL state formulae offer, instead of two path quantifiers, a single parametric probabilistic operator $\mathscr{P}_J$, where J is any interval of the reals contained in $[0, 1]$. $\mathscr{P}_J(\phi)$ asserts that the probability of the set of paths departing from the current state s that satisfy the path formula ϕ is within interval J. For example, $\mathscr{P}_{[0.5,1]}(\Diamond \mathsf{gr})$ specifies that a grant will occur with probability greater than or equal to 0.5 or, equivalently, a grant will eventually occur in all the paths of a set whose probability is between 0.5 and 1.

The probabilistic operator $\mathscr{P}_J(\phi)$ assigns a quantitative measure to the set of paths satisfying path formula ϕ, whereas the path operators $\exists \phi$ and $\forall \phi$ assert the existence (or the absence) of a path that satisfies (or does not satisfy) ϕ. Therefore, the probabilistic operator $\mathscr{P}_J$ is considered the **QUANTITATIVE** counterpart of the **QUALITATIVE** path operators $\forall \phi$ and $\exists \phi$.

In addition to CTL's until operator, PCTL includes a *bounded* version of the until, which puts a limit on the number of states, following the current one, where the until's arguments are evaluated. Therefore PCTL is a **METRIC** temporal logic, with the usual proviso of considering the sequences of states as a bijection of the discrete time domain.

The syntax of PCTL, much like that of CTL, is defined in two steps. State formulae are formed as

$$\Phi := p \mid \neg \Phi \mid \Phi_1 \wedge \Phi_2 \mid \mathscr{P}_J(\phi),$$

where $p \in AP$, the set of atomic propositions, Φ, Φ_1, and Φ_2 are state formulae, ϕ is a path formula, and J an interval in $[0,1]$. PCTL path formulae are defined as

$$\phi := \circ \Phi \mid \Phi_1 \cup \Phi_2 \mid \Phi_1 \cup^{\leq n} \Phi_2,$$

where Φ, Φ_1, and Φ_2 are state formulae and $n \in \mathbb{N}$. Thus, every PCTL path formula is immediately preceded by a probability operator $\mathscr{P}_J$, just as every CTL path formula is immediately preceded by a path quantifier $\exists$ or $\forall$.

We introduce useful abbreviations for the probability bounds J in occurrences of the operator $\mathscr{P}_J$, such as "> 0" for the interval $(0, 1]$, and "$= 1$" for the singleton $[1, 1]$. Derived time operators are also customary, with the novelty that PCTL also incorporates time-bounded versions, and that duality must be applied also to probability intervals, as in the following examples:

$$\diamond \Phi \quad := \quad \textbf{True } \mathsf{U} \ \Phi \,,$$

$$\diamond^{\leq n} \Phi \quad := \quad \textbf{True } \mathsf{U}^{\leq n} \ \Phi \,,$$

$$\mathscr{P}_{\leq p}(\Box \Phi) \quad := \quad \mathscr{P}_{\geq 1-p}(\diamond \neg \Phi) \,.$$

The semantics of PCTL is defined with reference to discrete-time Markov chains, in a way similar to the semantics of a CTL formula with respect to transition systems (see Sect. 9.1.2.1), with the notable addition of clauses for the probability operator $\mathscr{P}_J$ and the bounded until $\mathsf{U}^{\leq n}$. The satisfaction relation for state formulae is defined by the following clauses (for $p \in AP$):

$s \models p$	if and only if	$p \in s$;
$s \models \neg \Phi$	if and only if	it is not the case that $s \models \Phi$;
$s \models \Phi_1 \wedge \Phi_2$	if and only if	both $s \models \Phi_1$ and $s \models \Phi_2$ hold;
$s \models \mathscr{P}_J(\phi)$	if and only if	$\Pr(s \models \phi) \in J$;

where

$$\Pr(s \models \phi) \ = \ \Pr_s\{\pi \in paths(s) \mid \pi \models \phi\}$$

is the probability of the set of paths departing from state s that satisfy the path formula ϕ. The satisfaction relation between a path $\pi = s_0 s_1 s_2 \ldots$ and a path PCTL formula is defined as follows:

$\pi \models \bigcirc \Phi$	if and only if	$s_1 \models \Phi$;
$\pi \models \Phi_1 \mathsf{U} \Phi_2$	if and only if	there exists $j \geq 0$ such that $s_j \models \Phi_2$
		and, for all i such that $0 \leq i < j$, $s_i \models \Phi_1$;
$\pi \models \Phi_1 \mathsf{U}^{\leq n} \Phi_2$	if and only if	there exists j such that $0 \leq j \leq n$, $s_j \models \Phi_2$,
		and, for all i such that $0 \leq i < j$, $s_i \models \Phi_1$;

and the semantics of a PCTL formula with respect to a discrete-time Markov chain M is defined by means of the relation between a *state* formula and the initial states of M:

$M \models \Phi$	if and only if	$s_0 \models \Phi$ for every initial state s_0 of M.

PCTL constitutes a concise notation for specifying **QUANTITATIVE** requirements of probabilistic systems formalized as discrete-time Markov chains. It can characterize quantitatively both timing requirements and the probability of events.

Fig. 9.20 A Markov chain
with an unfair execution

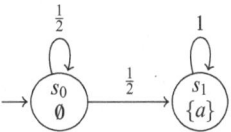

For instance, with respect to the discrete-time Markov chain of Example 7.44
modeling a faulty resource allocator, the formula

$$\mathscr{P}_{\geq 0.1}(\Diamond^{\leq 10}\text{pendh}) \tag{9.22}$$

asserts that the state with a pending high-priority request is reached within ten
computation steps with a probability greater than or equal to 0.1. The formula

$$\mathscr{P}_{\leq 0.01}(\neg\text{pendh } \mathsf{U}^{\leq 15} \text{ pendl})$$

states that the probability of reaching within 15 computation steps a state with a
pending low-priority request and no high-priority request pending meanwhile is less
than or equal to 0.01.

Remark 9.45. We noted that the operator $\mathscr{P}_J$ in state formulae replaces the path
quantifiers $\forall$ and $\exists$. Correspondingly, the PCTL formula $\mathscr{P}_{>0}(\phi)$ (stating that the
probability of the set of executions satisfying ϕ is non-null) *should* have a similar
meaning to that of the CTL formula $\exists\phi$ (stating the *existence* of an execution
satisfying ϕ); and the PCTL formula $\mathscr{P}_{=1}(\phi)$ (stating that the probability of the
set of executions satisfying ϕ is 1) *should* have a meaning similar to that of the CTL
formula $\forall\phi$ (stating that *all* executions satisfy property ϕ).

The above-mentioned PCTL and CTL formulae are, however, not really equivalent in general, as the following simple example shows. Consider the Markov
chain in Fig. 9.20. State s_1, where proposition a holds, is eventually reached with
probability 1; hence the PCTL formula $\text{Pr}_{=1}(\Diamond a)$ holds; however, the "intuitively
equivalent" CTL formula $\forall\Diamond a$ does not hold for the considered Markov chain,
because the transition system underlying the Markov chain admits the computation
s_0^ω, where a never holds.

We can state, however, suitable conditions under which PCTL formulae with
probability > 0 or $= 1$ and CTL formulae with path quantification are equivalent.
In the above example, the execution s_0^ω of the Markov chain of Fig. 9.20 can be
considered *unfair* as the system, being infinitely often in state s_0, always chooses
the self-loop transition over the other one to s_1. We can introduce the constraint of
strong fairness, as in Sect. 9.1.1.1: a transition system is strongly fair if, for every
state s and every successor state t of s, if state s is reached infinitely often, state t is
also reached infinitely often.

Under the strong fairness hypothesis, one can prove that, for every finite-state
Markov chain, for any state s and propositions a and b,

$$s \models \mathscr{P}_{=1}(a \mathsf{U} b) \qquad \text{if and only if } s \models \forall(a \mathsf{U} b).$$

In the example of Fig. 9.20, the path s_0^ω is not admissible under strong fairness, and the formulae $\mathscr{P}_{=1}(\Diamond a)$ and $\forall \Diamond a$ are in fact equivalent.

Similarly, in Example 7.44 (the faulty resource allocator) we have that the PCTL formula $\mathscr{P}_{=1}(\Diamond \mathsf{pendh})$ holds (the state pendh is eventually reached with probability 1), and so does the corresponding CTL formula $\forall \Diamond \mathsf{pendh}$, provided strong fairness is guaranteed; otherwise, if the system is not strongly fair, there exist (unfair) paths that never reach the pendh state; hence the CTL formula does not hold.   ∎

9.8.2 Continuous Stochastic Logic

In continuous-time Markov chains, the time domain is the set of nonnegative reals $R_{\geq 0}$, and execution paths are functions from the time domain to the set S of states; transitions are labeled by rates $\lambda_{j,k}$ of exponential distributions, as described in Chap. 6.

"Continuous stochastic logic" (CSL) is a natural adaptation of PCTL to this continuous-time setting: it maintains, with minor modifications, the same syntactic structure based on state and path formulae; the semantics of the logic, however, interprets bounds on the temporal operators as the *length of real time intervals*: for instance, the until operator is a path formula with the syntax

$$\Phi_1 \; \mathsf{U}^{[d_1, d_2]} \; \Phi_2 \,,$$

where Φ_1 and Φ_2 are state formulae and d_1 and d_2 define the lower and upper bounds of the continuous time interval, in the path over which the formula is interpreted, in which Φ_2 is required to hold. Thus, the clause defining the semantics of the until is

$$\pi \models \Phi_1 \; \mathsf{U}^{[d_1,d_2]} \; \Phi_2 \quad \text{if and only if} \quad \text{there exists } t \geq 0 \text{ such that } d_1 \leq t \leq d_2$$
$$\text{and } \pi(t) \models \Phi_2 \text{ and, for all } t' \text{ such that}$$
$$0 \leq t' < t, \pi(t') \models \Phi_1.$$

CSL can state QUANTITATIVE time properties of executions of continuous-time Markov chains and their probabilities. For instance, with reference to the continuous-time model of the resource allocator in Example 7.57, the following formula asserts that there is a probability less than or equal to 0.2 that a low-priority request occurs that is not satisfied (so that the system goes to state pendl) within 20.7 and 27.3 time units; the low-priority request was not preceded by any other similar one, since the state wg has never been reached from the current time:

$$\mathscr{P}_{\leq 0.2}(\mathsf{wg} \; \mathsf{U}^{[20.7,27.3]} \; \mathsf{pendl}) \,.$$

As another example on the same allocator system, the following formula specifies that a state with an unsatisfied high-priority request is reached within 15 time units with probability greater than 0.1:

$$\mathcal{P}_{\geq 0.1}(\diamondsuit^{\leq 15}\mathsf{pendh})\,.$$

Notice the difference in meaning with respect to formula (9.22): the latter is similar in structure, but its time parameter indicates the number of computation steps, not the length of a time interval within which the state pendh is reached.

After its first introduction as a continuous-time variant of PCTL, CSL has been enriched with various operators, thus enlarging the set of properties expressible (and verifiable by means of suitable model checking tools). We mention the "steady-state" operator $\mathcal{S}_{\sim p}(\Phi)$, which states that, in the long run, the state formula Φ holds with probability $\sim p$, with $\sim$ one of $<, \leq, >$, and $\geq$, and $p \in [0, 1]$.

For example, again with reference to the continuous-time Markov chain in Example 7.57, the formula $\mathcal{S}_{\geq 0.5}\mathsf{occ}$ states that, in the long run, the shared resource is used for at least half of the time, while the formula

$$\mathcal{S}_{\leq 0.01}(\mathsf{pendl} \vee \mathsf{pendh})$$

specifies that unsatisfied (low- or high-priority) requests occur only for 1 % of the overall system running time.

9.9 Methods and Tools Based on Temporal Logics

The very nature of formal notations like temporal logics suggests that their analysis can exploit deductive approaches. Tools based on deductions are normally called "theorem provers", in that they provide the means to derive properties as consequences of given assumptions.

Theorem provers targeting undecidable logics cannot be completely automatic; hence they are normally called "interactive": users are required to possess a working knowledge of the analyzed specification and be familiar with the logic and the tool itself to provide guidelines for a proof in the form of a sequence of intermediate steps (such as a series of lemmas). The tool can still handle automatically some steps of the proof, and it can check that the overall proof interactively constructed is completely correct. This way, proofs assisted by an interactive theorem prover are simpler and less error-prone than manual proofs. The main advantage of interactive deductive techniques is their generality, as they can deal with very expressive undecidable logics; on the other hand, they are not fully automated and therefore may require significant ingenuity and effort from users.

A second approach to the formal analysis of logics is still based on deductive techniques, but targets decidable logics for which reasonably efficient procedures exist for deciding satisfiability and validity. Tools implementing such fully automatic deductive techniques are called "satisfiability solvers". A satisfiability solver inputs a formula F and returns an interpretation that satisfies F, if it is satisfiable, or states that F is unsatisfiable. The simplest satisfiability solvers are called "SAT-solvers", because they target propositional satisfiability (referred to as "SAT"). So-called "SMT-solvers" (Satisfiability Modulo Theory solvers) build on top of SAT-solvers to deal with the satisfiability problem for decidable fragments of first-order theories, including non-Boolean variables and possibly interpreted and uninterpreted functions and predicates. Advantages and limitations of satisfiability solvers are symmetrical to those of interactive theorem provers: they are *push-button* tools which do not require user interaction other than for writing the formulae to be analyzed and interpreting the tool's output; the restriction to propositional or decidable first-order logics limits, however, their applicability. In fact, the results provided by SAT or SMT solvers may be partial or inconclusive, sharing some of the features of program analysis carried out by means of simulation and testing.

Theorem provers (both interactive and automatic) are general-purpose tools adaptable to any language based on mathematical logic. Therefore, tools of these kinds for temporal logics are typically not developed from scratch, but rather they rely on encodings of the target temporal logics into the input language of the chosen back-end general-purpose theorem prover.

A third approach to the analysis of temporal logics relies on their intimate connection with transition systems, which was underlined several times in the present chapter. The most prominent techniques in this area are referred to as "model-checking", an approach to verification in which a transition system TS and a temporal logic formula ϕ are analyzed to determine if all executions of TS satisfy ϕ (that is, if $TS \models \phi$). Model checking will be discussed in Chap. 11 as a "dual-language approach". Another area where the connection between temporal logic and transition systems yields useful techniques is in the synthesis of "runtime monitors": finite-state automata that can efficiently monitor whether a property, formalized in temporal logic, holds during a system execution.

In the remainder of this chapter, we give an overview of three illustrative examples of techniques for each of the previously described tool categories: Sect. 9.9.1 presents an approach for analyzing TRIO specifications based on the PVS interactive theorem prover, Sect. 9.9.2 an approach for propositional MTL based on the use of SAT-solvers, and Sect. 9.9.3 a translation of LTL formula into finite-state automata for runtime monitoring.

9.9.1 A Deductive Approach to the Analysis of TRIO Specifications

This section describes an approach supporting deductive analysis of TRIO specifications based on an encoding into the proof checker PVS. The language of PVS is a classical typed higher-order logic; its type system supports user-defined types and standard primitive types and type constructors, such as the Boolean, integer, real, and function types.

The encoding of TRIO in PVS uses two user-defined types: time and duration; they respectively represent the time domain and lengths of time intervals or distances between events. Under a dense-time interpretation, they are defined in terms of the real predefined PVS type:

> *time* : TYPE = *real*
> *duration* : TYPE = { t : *time* | $t \geq 0$ }

The encoding of TRIO formulae extensively exploits the higher-order PVS features: each time-dependent entity in TRIO (such as variable, predicate, or formula) maps to a PVS function whose domain is extended with an additional argument representing time. For instance, a time-dependent integer variable is encoded as a function from the time domain to the integers, and a time-dependent binary predicate on natural numbers is encoded as a Boolean function with domain $\mathbb{N} \times \mathbb{N} \times time$. This encoding relies on a PVS *parametric theory TD_Terms*:

> *TD_Terms* [D : TYPE] : THEORY
> BEGIN
> $\cdots$
>   *TD_Term* : TYPE = [*time* $\rightarrow D$]
> $\cdots$

Logical connectives are also defined accordingly as higher-order functions; for example, logical conjunction between TRIO formulae is defined as

> *AND* (A, B : *TD_Term* [**bool**]) : *TD_Term* [**bool**] =
>   LAMBDA (t : *time*) : $A(t)$ AND $B(t)$.

Similar encodings translate temporal operators; for instance, the fundamental operator Dist is encoded as

> *Dist*(A : *TD_Term* [**bool**], d : *time*) : *TD_Term* [**bool**] =
>   LAMBDA (t : *time*) : $A(t+d)$

while the derived operators Lasts, Until, and Alw are encoded as

> *Lasts*(A : *TD_Term* [**bool**], *dur* : *duration*) : *TD_Term* [**bool**] =
>   FORALL (t : *time* | $0 < t$ AND $t < dur$) : *Dist*(A,t),
> *Until*(A, B : *TD_Term* [**bool**]) : *TD_Term* [**bool**] =
>   EXISTS (pt : *duration*) : *Futr*(B, pt) AND *Lasts*(A, pt),
> *Alw*(A : *TD_Term* [**bool**]) : **bool** =
>   FORALL (t : *time*) : $A(t)$.

Besides the proof strategies incorporated into PVS (such as basic axioms and inference rules of first-order logic, and a rich collection of arithmetic properties on real intervals), further properties peculiar to the TRIO notation can become domain-specific rules of inference, or, in more complex cases, be used as strategies for carrying out the more involved proofs. For instance, the following formula, asserting a sort of "temporal induction", is a valid TRIO formula,

$$(A \wedge \text{Lasts}(A \Rightarrow \text{NowOn}(A), d)) \Rightarrow \text{Lasts}(A, d),$$

and can be useful in the derivation of other properties. A component of the PVS tool supports the visualization of formulae and derivations in the syntax of TRIO, without showing details of the PVS encoding (such as the explicit argument for current time). For instance, the following is a part of the proof of the utility property of the generalized railroad crossing (Exercise 9.39).

```
{-1}  UpToNow (Closed)
{-2}  Go_up
{-3}  Lasts (not Go_down, gamma)
|-------
{1}   Futr (gate = Open, gamma)
```

It asserts that if the gate is closed, a command Go_up is issued, and no command is issued to lower the bar in the next γ time units, then the gate will be open after γ time units.

9.9.2   An Approach to Discrete-Time MTL Verification Based on Satisfiability Solvers

This section briefly presents a technique for the verification of temporal logic specifications based on SAT-solving tools; the presentation is based on the features of the Zot toolsuite, which supports the analysis of a variety of notations, including LTL, MTL, and TRIO, over discrete time.

In a SAT-based verification environment, users submit a temporal logic specification and obtain, if the specification is satisfiable, an interpretation that satisfies it, or else the result "Unsatisfiable". Let us go back to the example of the "time reset lamp" of Exercise 9.23, and take $\Delta = 8$. Based on the fact that the light goes off automatically Δ time units after the pressing of the on button if no other button is pressed, we consider the conjecture that the light never remains on for more than Δ time units. To check the conjecture, we submit to the tool the complete system specification in conjunction with the property

$$\forall d(d > 8 \Rightarrow \Box(\neg\Box_{=d}\mathsf{L})), \tag{9.23}$$

stating that the lamp never remains lighted for more than $\Delta = 8$ time units (correspondingly, d ranges over the finite integer interval $[0..10]$).

Zot considers the negation of the submitted formula and looks for an interpretation where it is satisfied, such as the one in Fig. 9.21, where the lamp is lighted

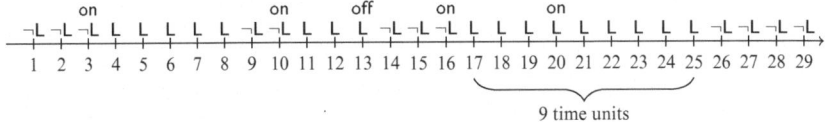

Fig. 9.21 A run of the "time reset lamp" violating property (9.23)

for nine consecutive time units, as a consequence of the fact that the on button was pressed twice four time units apart. This output disproves property (9.23).

Since a formula is unsatisfiable if and only if its negation is valid, satisfiability solvers can also verify properties. To prove that a given property ϕ holds for a system specified by a formula Σ under the assumption of another property ψ (say, a hypothesis on the input coming from the environment), we check the satisfiability of the formula

$$\Sigma \wedge \psi \wedge \neg\phi.$$

If the tool reports "Unsatisfiable", we have proved the validity of $\Sigma \wedge \psi \Rightarrow \phi$, that is, ϕ is a consequence of Σ and ψ. Otherwise, the tool shows an interpretation satisfying $\Sigma \wedge \psi \wedge \neg\phi$, which we can inspect to understand why the property does not hold.

SAT solvers can manage only *finite* alphabets of propositional letters. It is then quite natural to encode LTL or MTL over finite intervals of a discrete time domain, typically of the form $[0..k]$ for some $k \in \mathbb{N}$. The encoding uses a set of propositional letters $p_{\phi,t}$ for every subformula ϕ of the formula to be analyzed and for every time $t \in [0..k]$. For instance, if the MTL formula $\diamond_{=3}a$ holds at time instant 5, it determines the encoding

$$p_{\diamond_{=3}a,5} \wedge \left(p_{\diamond_{=3}a,5} \Longleftrightarrow p_{a,8}\right).$$

The fact that SAT solvers can only deal with finite formulae over a finite alphabet does not correspond to the standard interpretation of LTL and MTL over infinite (or bi-infinite) state sequences. This difficulty is overcome by taking advantage of a property enjoyed by propositional MTL over discrete time and equally expressive fragments: if a formula is satisfiable in an infinite discrete-time structure, then it is also satisfiable in an *ultimately periodic* infinite discrete-time structure, that is, a structure that, from some time instant on, repeats itself indefinitely. The encoding deployed in SAT-based tools takes advantage of this property by encoding ultimately periodic infinite state sequences as finite cyclic structures, as shown in Fig. 9.22: the encoded sequences "loop back" from a last state s_k to a previous state s_h; for consistency, the two states s_{h-1} and s_k must assign identical values to all elements of the specification alphabet.

As a very simple example of this kind of finite encoding for an ultimately periodic infinite state sequence, consider the LTL formula $\Box\diamond a$ which requires proposition a to hold infinitely often: the finite structure in Fig. 9.23 encodes a state sequence

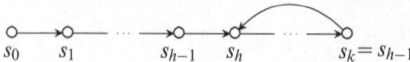

Fig. 9.22 An ultimately periodic infinite state sequence encoded as a finite cyclic structure

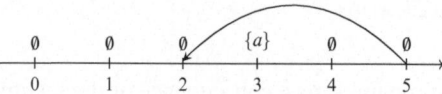

Fig. 9.23 A structure encoding a sequence satisfying formula $\Box\Diamond a$

σ such that $\sigma \models \Box\Diamond a$: the propositional letter a holds at times $3, 7, 11, \ldots$ and in general $\forall k \in \mathbb{N}(a \in s_{3+4k})$. Tools applying techniques based on encoding infinite state sequences into cyclic finite structures are called "*bounded* satisfiability checkers", where the "bound" is given by the length of the finite time domain assumed in the encoding.

A critical issue in bounded satisfiability checking is the number of states that are included in the cyclic finite structure encoding ultimately periodic infinite state sequences: for any given number of states, there exist formulae that are satisfiable only in structures having more states. For instance, no structure with six states (like the one shown in Fig. 9.23) satisfies the formula $\Box(\Diamond(\Box_{<5}a)) \wedge \Box(\Diamond(\Box_{<5}\neg a))$, while there exist longer cyclic structures (for example, with 12 states) that satisfy it.

Suitable techniques have been devised that, under certain hypotheses, can determine the *completeness bound* of a formula: if a formula ϕ has completeness bound K, then ϕ is satisfiable if and only if it is satisfiable by some cyclic structure of length at most K.

If the completeness bound is unknown or too large (hence, applying it is prohibitively expensive), bounded satisfiability checking is inconclusive: a formula that is unsatisfiable for a certain bound K might still be satisfiable over structures with more than K states. Even under this limitation, the analysis carried out by means of satisfiability checking can still increase the confidence that the conjectured properties hold (or can help find errors in the specification), especially in cases in which the bound is "large enough" with respect to the time constants appearing in the analyzed specification, even if without absolute guarantee.

Tools based on SAT-solvers can also be the base for simulation tools: they can provide system executions given a system specification in conjunction with a temporal logic formula characterizing the class of system executions we are interested in investigating. A further usage of satisfiability solvers is to perform "sanity-checks" of specifications, which are a sort of initial validation: if a system specification is unsatisfiable, there is a serious problem, such as contradictory requirements or excessively severe constraints that therefore cannot be implemented.

Fig. 9.24 A Büchi
automaton equivalent to the
LTL formula (9.24)

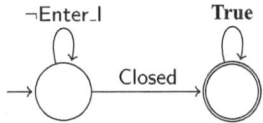

SAT-based technologies have been successfully applied to the design and
verification of safety-critical computer-based embedded systems, such as software-
intensive systems for signaling in railroad transportation.

9.9.3   Runtime Verification Based on Temporal Logic

This section outlines the basic ideas to build runtime monitors of system properties
specified by LTL formulae. As pointed out in Sect. 9.1.1.2, each LTL formula
admits a nondeterministic Büchi automaton that accepts exactly the same set of
infinite state sequences satisfying the formula. For runtime monitoring, however,
we must remember that both the notion of acceptance for Büchi automata and that
of satisfaction of LTL formulae refer to *infinite* state sequences, while a runtime
monitor can only check executions, which correspond to finite initial portions
of infinite state sequences. Therefore, plain Büchi automata are unfit to perform
runtime monitoring. Furthermore, the analysis of a monitored execution up to the
currently reached point may be inconclusive for determining whether the complete
execution will comply with the specification.

Consider, for instance, the system of Example 9.36, and the specification

$$\neg \mathsf{Enter\_I} \ \mathsf{U} \ \mathsf{Closed}, \tag{9.24}$$

which requires that the train not enter the critical region I until the gate is closed.
Formula (9.24) is equivalent to the Büchi automaton in Fig. 9.24.

The automaton structure highlights the fact that the specification formula incorporates two basic conditions: first, **Closed** must hold *eventually* (with no specific
deadline); second, **Enter_I** cannot occur before then. Therefore, a monitor that
checks an execution can be in three possible states.

The property fails: the monitor raises an alarm if an **Enter_I** event occurs before
the proposition **Closed** becomes true, because it is a clear violation of (9.24).

The property succeeds: if **Closed** takes place before any occurrence of **Enter_I**,
(9.24) is satisfied by the monitored execution, irrespective of any future event.

Don't know yet: as long as neither **Closed** nor **Enter_I** takes place, the monitor
cannot conclude whether (9.24) holds or not, and can only proceed with further
observations.

This state of affairs, generalized from the example, leads to the definition of the logic
LTL_3 using three values that correspond to the possible values output by the monitor

Fig. 9.25 An automaton
formalizing the monitor for
the LTL formula (9.24)

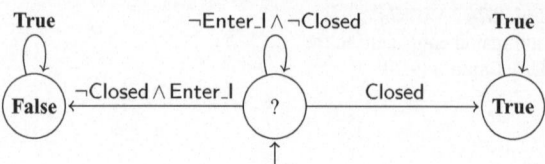

during the check of the execution: "**True**" for "property succeeds", "**False**" for "property fails", and " **?** " for "don't know yet". The semantics of LTL_3 formulae is defined over finite state sequences, which makes it suitable for formalizing runtime monitoring. LTL_3 formulae can be translated into Moore automata (see Sect. 5.2.3) that in real time, while reading the variables of the system being monitored, output the current evaluation of the LTL_3 formula on the execution read so far. Fig. 9.25 shows this automaton for the specification formula (9.24).

Moore automata that monitor LTL_3 formulae are deterministic, so the monitoring takes place in real time. The construction of monitors from temporal logic specifications has also been extended to a metric, discrete-time temporal logic called TLTL ("Timed LTL"), thus extending the application range of the method to real-time systems.

9.10 Bibliographic Remarks

LTL was introduced in computer science by Pnueli [65] as a development of modal and tense logics [48, 52, 68] mainly intended for reasoning about executions of programs.

Several kinds of branching-time temporal logics have been introduced; going from the least to the most expressive, we have the Hennessy-Milner logic [43] CTL and CTL* [20], and the modal μ-calculus [51]. Gabbay et al. [34] and Emerson [24] provide a comprehensive presentation of the classic temporal logics.

A long-standing dispute on the superiority of linear- to branching-time logics dates back to the early 1980s. Pnueli [66] was one of the first to point out the substantial difference in the notions of time adopted by linear- and branching-time logics. Clarke and Draghicescu [19] show that the expressiveness of LTL and CTL is incomparable. Vardi [76] thoroughly compares the relative merits of the linear and branching frameworks considering expressiveness, complexity, compositionality, and other pragmatic issues such as usability, intuitiveness, support of analysis and verification.

Past operators and bi-infinite time were considered natural features in the very first definitions of temporal logic [69], and widely adopted by philosophers [70], as they ensure symmetrical definitions and no border effects. In the first applications of temporal logic to computer science, however, mono-infinite time was customary [65]. Past operators were introduced later [55], because of their convenience and conciseness, but still over mono-infinite time domains. While PLTL (LTL with past operators) is not more expressive than LTL over mono-infinite time domains [33], it can be significantly more concise: Gabbay's "separation" algorithm to translate PLTL formulae into equivalent LTL ones may introduce nonelementary blowup in size [32], and there exist some PLTL formulae such that every equivalent LTL formula is exponentially larger [54].

Expressiveness changes significantly when temporal logics are interpreted over dense time domains; for instance, Bouyer et al. [11] shows that, over the reals, propositional MTL with past operators is strictly more expressive than its future-only version. Alur and Henzinger first investigated [2] the question of the expressiveness of past operators over dense time domains, and highlighted the differences with respect to discrete time. The expressiveness of temporal logic over dense time is also affected in subtle ways by the inclusion or exclusion of interval endpoints in the definitions of basic operators [28].

When temporal logic is used to specify not only program properties but also systems, it becomes more natural to adopt bi-infinite time, as we argued in previous works [18, 35, 37, 59].

In his paper introducing MTL [49, 50], Koymans provides a very general framework for metric temporal logic, accommodating a wide variety of metric temporal structures. Alur and Henzinger studied the expressiveness and decidability of propositional MTL over discrete and dense time [2] and showed, in particular, that MTL is highly undecidable in the latter model. In a later, somehow unexpected, turn of events, Ouaknine and Worrell showed [62] that the undecidability of MTL over dense time depends on the details of the semantic model considered; in the special case of weakly monotonic timed words of finite (but arbitrarily large) length, propositional MTL is decidable, albeit with nonelementary complexity. Propositional MTL is also decidable over dense-time interval-based behaviors of finite length [63] or with bounded variability [29].

We introduced TRIO in a previous work of ours [37] with a semantics parametric with respect to the adopted temporal domain (and in particular supporting both dense and discrete time domains [60]). This chapter focused the presentation of TRIO on dense time.

The notion of non-Zenoness was introduced by Abadi and Lamport [1]. Gargantini and Morzenti formalized non-Zenoness and point- and interval-based entities using TRIO [35]; Hirshfeld and Rabinovich [44] and Lutz et al. [56] formalized similar notions for other metric temporal logics over dense time. Gargantini and Morzenti showed [36] how the adoption of a small set of predefined categories of specification items (such as point- and interval-based predicates) can make the modeling of real-time hybrid systems quite systematic and amenable to automated verification.

Furia and Rossi introduced the notion of "sampling" and verification techniques based on it with reference to TRIO [27] and MTL [30]. Fainekos and Pappas [25] discussed a different notion of robust satisfiability of MTL over dense time.

The duration calculus was originally introduced by Chaochen et al. [16] with a formal semantics [40]. A book provides an extensive presentation of the language and its applications [14]. A few works extended duration calculus to support modeling of point-based events [15, 77].

From among the several extensions and variants of duration calculus, let us mention the "interval duration logic" [64, 74], a decidable variant of duration calculus, and the "interval-based temporal logic" (ITL) [61, 73], where formulae are asserted with reference to explicitly denoted intervals, using modal operators to state that they hold in some or in every instant of the interval. A set of operators is provided to construct intervals from primitive intervals associated with event occurrences or to derive intervals by composing other intervals. The fundamental time entity of ITL is still the time instant, since intervals are defined as segments of points satisfying some given property. The expressiveness and decidability of ITL have been extensively studied [12, 38, 39].

The "metric interval temporal logic" (MITL) [4] is a syntactic subset of MTL, where bounded temporal operators can only include intervals of positive length. This simply expressible restriction makes MITL fully decidable over non-Zeno dense-time behaviors, and every MITL formula admits an equivalent timed automaton [58]. Hirshfeld and Rabinovich introduced a class of decidable metric dense-time logics that subsume MITL and that are obtained as straightforward generalizations of classic LTL interpreted over dense time [44, 45].

Corsetti et al. introduced [22] the TRIO-based notation dealing with time granularities presented in Sect. 9.6. Among the few other approaches to the problem of time granularity, Burns and Hayes [13] present a comprehensive model and a language which relate granularity with a variety of issues arising in the specification of complex real-time systems, such as the notions of event and state, the precision of observations, simultaneity, and rate of change. Franceschet and Montanari [26] focus on the relation between a type of automata extended with time and a temporal logic incorporating (possibly unbounded) time granularities; they provide results concerning expressiveness of the notations and complexity of their analysis. Analysis and verification of systems modeled in LTL that accommodate a notion of time granularity is treated by Furia and Spoletini [31].

RTL was first introduced by Jahanian and Mok [46] as a purely descriptive, logic-based formalism for specifying and analyzing quantitative properties of timed systems; later, the same authors developed the notation into Modecharts [47], a language that combines descriptive and operational elements, and constructs supporting modularization for structuring the specification of complex systems.

TLA was introduced by Lamport [53] as a logic for specifying and analyzing program properties, and then extended [1] by means of additional constructs, sometimes quite technical, for specifying quantitative timing features.

TPTL was introduced by Alur and Henzinger [3] with the goal of combining expressiveness, readability, decidability, and an acceptable complexity of analysis for temporal logics with a metric on time.

Hansson and Jonsson introduced PCTL [41]. Later, CSL was explicitly introduced [7] as a continuous-time variant of PCTL, and extended with steady-state operators [8] and with features for expressing properties by means of single-clocked timed automata [23].

The presentation of the PVS-based environment for analyzing TRIO specifications in Sect. 9.9.1 has been simplified with respect to the actual implementation [35], which also supports the constrained elements discussed in Sect. 9.4.1 for carrying out requirements analysis of embedded time-critical systems. The original development of the TRIO support in PVS was after the TAME toolset for the analysis of a type of automaton with time [5, 6], which is also based on an encoding in PVS.

Skakkebæk and Shankar report [75] on an implementation over PVS of an environment supporting analysis of duration calculus specifications.

TLA enjoys a rich toolset for verification [17]; it incorporates a proof manager for interactive theorem proving, and support for various back-end verification engines, including the interactive theorem prover Isabelle, the first-order logic theorem prover Zenon, and standard SMT solvers.

Bounded LTL satisfiability checking was introduced by Biere et al. [10, 21] and it was used in the Zot tool [67] as a way to support verification of time-critical systems based exclusively on their descriptive specification in metric temporal logic, independently of their implementation in terms of operational formalisms. Rozier and Vardi [71, 72] document the use of satisfiability checking to perform sanity checks of requirements specifications.

The approach to runtime monitoring based on LTL_3 and deterministic finite-state automata is presented by Bauer et al. [9]. Maler et al. introduced automata-based approaches to runtime monitoring over continuous time [42, 57].

References

1. Abadi, M., Lamport, L.: An old-fashioned recipe for real time. ACM Trans. Program. Lang. Syst. **16**(5), 1543–1571 (1994)
2. Alur, R., Henzinger, T.A.: Real-time logics: complexity and expressiveness. Inf. Comput. **104**(1), 35–77 (1993)
3. Alur, R., Henzinger, T.A.: A really temporal logic. J. ACM **41**(1), 181–204 (1994)
4. Alur, R., Feder, T., Henzinger, T.A.: The benefits of relaxing punctuality. J. ACM **43**(1), 116–146 (1996)
5. Archer, M., Heitmeyer, C.L.: Human-style theorem proving using PVS. In: Gunter, E.L., Felty, A.P. (eds.) Theorem Proving in Higher Order Logics, Proceedings of the 10th International Conference, TPHOLs'97, Murray Hill, 19–22 August 1997. Lecture Notes in Computer Science, vol. 1275, pp. 33–48. Springer, Berlin/Heidelberg (1997)
6. Archer, M., Heitmeyer, C.L., Riccobene, E.: Proving invariants of I/O automata with TAME. Autom. Softw. Eng. **9**(3), 201–232 (2002)
7. Aziz, A., Sanwal, K., Singhal, V., Brayton, R.K.: Model-checking continous-time Markov chains. ACM Trans. Comput. Log. **1**(1), 162–170 (2000)
8. Baier, C., Haverkort, B., Hermanns, H., Katoen, J.P.: Model-checking algorithms for continuous-time Markov chains. IEEE Trans. Softw. Eng. **29**(6), 524–541 (2003)
9. Bauer, A., Leucker, M., Schallhart, C.: Runtime verification for LTL and TLTL. ACM Trans. Softw. Eng. Methodol. **20**(4), 14 (2011)
10. Biere, A., Heljanko, K., Junttila, T.A., Latvala, T., Schuppan, V.: Linear encodings of bounded LTL model checking. Log. Methods Comput. Sci. **2**(5) (2006)

11. Bouyer, P., Chevalier, F., Markey, N.: On the expressiveness of TPTL and MTL. In: Ramanujam, R., Sen, S. (eds.) Proceedings of the 25th International Conference on Foundations of Software Technology and Theoretical Computer Science (FSTTCS'05). Lecture Notes in Computer Science, vol. 3821, pp. 432–443. Springer, Berlin/New York/Heidelberg (2005)

12. Bresolin, D., Della Monica, D., Goranko, V., Montanari, A., Sciavicco, G.: The dark side of interval temporal logic: sharpening the undecidability border. In: Combi, C., Leucker, M., Wolter, F. (eds.) Eighteenth International Symposium on Temporal Representation and Reasoning, TIME 2011, Lübeck, 12–14 September 2011, pp. 131–138. IEEE Computer Society (2011)

13. Burns, A., Hayes, I.J.: A timeband framework for modelling real-time systems. Real-Time Syst. **45**(1–2), 106–142 (2010)

14. Chaochen, Z., Hansen, M.R.: Duration Calculus: A Formal Approach to Real-Time Systems. Springer, Berlin/New York (2004)

15. Chaochen, Z., Xiaoshan, L.: A mean value calculus of durations, pp. 431–451. Prentice Hall, Hertfordshire (1994)

16. Chaochen, Z., Hoare, C.A.R., Ravn, A.P.: A calculus of duration. Inf. Process. Lett. **40**(5), 269–276 (1991)

17. Chaudhuri, K., Doligez, D., Lamport, L., Merz, S.: The TLA$^+$ proof system: building a heterogeneous verification platform. In: Cavalcanti, A., Déharbe, D., Gaudel, M.C., Woodcock, J. (eds.) ICTAC. Lecture Notes in Computer Science, vol. 6255, p. 44. Springer, Berlin (2010)

18. Ciapessoni, E., Mirandola, P., Coen-Porisini, A., Mandrioli, D., Morzenti, A.: From formal models to formally based methods: an industrial experience. ACM Trans. Softw. Eng. Methodol. **8**(1), 79–113 (1999)

19. Clarke, E.M., Draghicescu, I.A.: Expressibility results for linear-time and branching-time logics. In: de Bakker, J.W., de Roever, W.P., Rozenberg, G. (eds.) REX Workshop. Lecture Notes in Computer Science, vol. 354, pp. 428–437. Springer, Berlin/New York (1988)

20. Clarke, E.M., Emerson, E.A.: Design and synthesis of synchronization skeletons using branching-time temporal logic. In: Logic of Programs, Workshop, pp. 52–71. Springer, London (1982)

21. Clarke, E.M., Biere, A., Raimi, R., Zhu, Y.: Bounded model checking using satisfiability solving. Form. Methods Syst. Des. **19**(1), 7–34 (2001)

22. Corsetti, E., Crivelli, E., Mandrioli, D., Morzenti, A., Montanari, A., San Pietro, P., Ratto, E.: Dealing with different time scales in formal specifications. In: Proceedings of the 6th International Workshop on Software Specification and Design, pp. 92–101. IEEE Computer Society, Los Alamitos (1991)

23. Donatelli, S., Haddad, S., Sproston, J.: Model checking timed and stochastic properties with CSL^{TA}. IEEE Trans. Softw. Eng. **35**(2), 224–240 (2009)

24. Emerson, E.A.: Temporal and modal logic. In: van Leeuwen, J. (ed.) Handbook of Theoretical Computer Science, vol. B, pp. 996–1072. Elsevier, Amsterdam/New York (1990)

25. Fainekos, G.E., Pappas, G.J.: Robustness of temporal logic specifications for continuous-time signals. Theor. Comput. Sci. **410**(42), 4262–4291 (2009)

26. Franceschet, M., Montanari, A.: Temporalized logics and automata for time granularity. TPLP **4**(5–6), 621–658 (2004)

27. Furia, C.A., Rossi, M.: Integrating discrete- and continuous-time metric temporal logics through sampling. In: Asarin, E., Bouyer, P. (eds.) Proceedings of the 4th International Conference on Formal Modelling and Analysis of Timed Systems (FORMATS'06). Lecture Notes in Computer Science, vol. 4202, pp. 215–229. Springer, Berlin/New York (2006)

28. Furia, C.A., Rossi, M.: On the expressiveness of MTL variants over dense time. In: Raskin, J.F., Thiagarajan, P.S. (eds.) Proceedings of the 5th International Conference on Formal Modelling and Analysis of Timed Systems (FORMATS'07). Lecture Notes in Computer Science, vol. 4763, pp. 163–178. Springer, Berlin/New York (2007)

29. Furia, C.A., Rossi, M.: MTL with bounded variability: decidability and complexity. In: Cassez, F., Jard, C. (eds.) Proceedings of the 6th International Conference on Formal Modelling and Analysis of Timed Systems (FORMATS'08). Lecture Notes in Computer Science, vol. 5215, pp. 109–123. Springer, Berlin (2008)

30. Furia, C.A., Rossi, M.: A theory of sampling for continuous-time metric temporal logic. ACM Trans. Comput. Log. **12**(1), 1–40 (2010). Article 8

31. Furia, C.A., Spoletini, P.: On relaxing metric information in linear temporal logic. In: Combi, C., Leucker, M., Wolter, F. (eds.) Proceedings of the 18th International Symposium on Temporal Representation and Reasoning (TIME'11), pp. 72–79. IEEE Computer Society, Los Alamitos (2011)

32. Gabbay, D.M.: The declarative past and imperative future. In: Banieqbal, B., Barringer, H., Pnueli, A. (eds.) Proceeding of Temporal Logic in Specification (TLS'87). Lecture Notes in Computer Science, vol. 398, pp. 409–448. Springer, Altrincham (1987)

33. Gabbay, D.M., Pnueli, A., Shelah, S., Stavi, J.: On the temporal basis of fairness. In: Conference Record of the 7th Annual ACM Symposium on Principles of Programming Languages (POPL'80), pp. 163–173. ACM, New York (1980)

34. Gabbay, D.M., Hodkinson, I., Reynolds, M.: Temporal logic (vol. 1): Mathematical foundations and computational aspects. Oxford Logic Guides, vol. 28. Oxford University Press, New York (1994)

35. Gargantini, A., Morzenti, A.: Automated deductive requirements analysis of critical systems. ACM Trans. Softw. Eng. Methodol. **10**(3), 255–307 (2001)

36. Gargantini, A., Morzenti, A.: Automated verification of continuous time systems by discrete temporal induction. In: Proceedings of the 13th International Symposium on Temporal Representation and Reasoning (TIME'06). IEEE Computer Society Press, Los Alamitos (2006)

37. Ghezzi, C., Mandrioli, D., Morzenti, A.: TRIO: a logic language for executable specifications of real-time systems. J. Syst. Softw. **12**(2), 107–123 (1990)

38. Goranko, V., Montanari, A., Sciavicco, G.: A road map of interval temporal logics and duration calculi. J. Appl. Non-Class. Log. **14**(1–2), 9–54 (2004)

39. Goranko, V., Montanari, A., Sala, P., Sciavicco, G.: A general tableau method for propositional interval temporal logics: Theory and implementation. J. Appl. Log. **4**(3), 305–330 (2006)

40. Hansen, M.R., Chaochen, Z.: Duration calculus: logical foundations. Form. Asp. Comput. **9**(3), 283–330 (1997)

41. Hansson, H., Jonsson, B.: A logic for reasoning about time and reliability. Form. Asp. Comput. **6**(5), 512–535 (1994)

42. Havlicek, J., Little, S., Maler, O., Nickovic, D.: Property-based monitoring of analog and mixed-signal systems. In: Chatterjee, K., Henzinger, T.A. (eds.) Formal Modeling and Analysis of Timed Systems—8th International Conference, FORMATS 2010, Klosterneuburg, 8–10 September 2010. Proceedings. Lecture Notes in Computer Science, vol. 6246, pp. 23–24. Springer, Berlin/Heidelberg (2010)

43. Hennessy, M., Milner, R.: Algebraic laws for nondeterminism and concurrency. J. ACM **32**(1), 137–161 (1985)

44. Hirshfeld, Y., Rabinovich, A.M.: Logics for real time: decidability and complexity. Fundam. Informormaticae **62**(1), 1–28 (2004)

45. Hirshfeld, Y., Rabinovich, A.M.: Decidable metric logics. Inf. Comput. **206**(12), 1425–1442 (2008)

46. Jahanian, F., Mok, A.K.: Safety analysis of timing properties in real-time systems. IEEE Trans. Softw. Eng. **12**(9), 890–904 (1986)

47. Jahanian, F., Mok, A.K.: Modechart: a specification language for real-time systems. IEEE Trans. Softw. Eng. **20**(12), 933–947 (1994)

48. Kamp, J.A.W.: Tense logic and the theory of linear order. Ph.D. thesis, University of California at Los Angeles (1968)

49. Koymans, R.: Specifying real-time properties with metric temporal logic. Real-Time Syst. **2**(4), 255–299 (1990)

50. Koymans, R.: (Real) time: a philosophical perspective. In: de Bakker, J.W., Huizing, C., de Roever, W.P., Rozenberg, G. (eds.) Proceedings of the REX Workshop: "Real-Time: Theory in Practice". Lecture Notes in Computer Science, vol. 600, pp. 353–370. Springer, Berlin/New York (1992)

51. Kozen, D.: Results on the propositional mu-calculus. Theor. Comput. Sci. **27**, 333–354 (1983)

52. Kripke, S.A.: Semantical analysis of modal logic I. Z. Math. Log. Grundl. Math. **9**, 67–96 (1963)
53. Lamport, L.: The temporal logic of actions. ACM Trans. Program. Lang. Syst. **16**(3), 872–923 (1994)
54. Laroussinie, F., Markey, N., Schnoebelen, P.: Temporal logic with forgettable past. In: Proceedings of the 17th Annual IEEE Symposium on Logic in Computer Science (LICS'02), pp. 383–392. IEEE Computer Society Press, Los Alamitos (2002)
55. Lichtenstein, O., Pnueli, A., Zuck, L.D.: The glory of the past. In: Proceedings of 3rd Workshop on Logic of Programs, Brooklyn. Lecture Notes in Computer Science, vol. 193, pp. 196–218. Springer (1985)
56. Lutz, C., Walther, D., Wolter, F.: Quantitative temporal logics over the reals: PSPACE and below. Inf. Comput. **205**(1), 99–123 (2007)
57. Maler, O., Nickovic, D.: Monitoring temporal properties of continuous signals. In: Lakhnech, Y., Yovine, S. (eds.) Formal Techniques, Modelling and Analysis of Timed and Fault-Tolerant Systems, Proceedings of the Joint International Conferences on Formal Modelling and Analysis of Timed Systems, FORMATS 2004 and Formal Techniques in Real-Time and Fault-Tolerant Systems, FTRTFT 2004, Grenoble, 22–24 September 2004. Lecture Notes in Computer Science, vol. 3253, pp. 152–166. Springer, Berlin/New York (2004)
58. Maler, O., Nickovic, D., Pnueli, A.: From MITL to timed automata. In: Asarin, E., Bouyer, P. (eds.) Formal Modeling and Analysis of Timed Systems, Proceedings of the 4th International Conference, FORMATS 2006, Paris, 25–27 September 2006. Lecture Notes in Computer Science, vol. 4202, pp. 274–289. Springer, Berlin/New York (2006)
59. Morzenti, A., Pietro, P.S.: Object-oriented logical specification of time-critical systems. ACM Trans. Softw. Eng. Methodol. **3**(1), 56–98 (1994)
60. Morzenti, A., Mandrioli, D., Ghezzi, C.: A model parametric real-time logic. ACM Trans. Program. Lang. Syst. **14**(4), 521–573 (1992)
61. Moszkowski, B.: Executing Temporal Logic Programs. Cambridge University Press, New York/Cambridge (1986)
62. Ouaknine, J., Worrell, J.: On the decidability and complexity of metric temporal logic over finite words. Log. Methods Comput. Sci. **3**(1), 8 (2007)
63. Ouaknine, J., Rabinovich, A., Worrell, J.: Time-bounded verification. In: Bravetti, M., Zavattaro, G. (eds.) CONCUR 2009—Concurrency Theory, Proceedings of the 20th International Conference, CONCUR 2009, Bologna, 1–4 September 2009. Lecture Notes in Computer Science, vol. 5710, pp. 496–510. Springer, Berlin/New York/Heidelberg (2009)
64. Pandya, P.K.: Interval duration logic: expressiveness and decidability. Electron. Notes Theor. Comput. Sci. **65**(6), 254–272 (2002)
65. Pnueli, A.: The temporal logic of programs. In: Proceedings of the 18th IEEE Symposium on Foundations of Computer Science (FOCS'77), pp. 46–67. IEEE Computer Society, New York (1977)
66. Pnueli, A.: Linear and branching structures in the semantics and logics of reactive systems. In: Brauer, W. (ed.) Proceedings of the 12th Colloquium on Automata, Languages and Programming (ICALP'85). Lecture Notes in Computer Science, vol. 194, pp. 15–32. Springer, Berlin/Heidelberg (1985)
67. Pradella, M., Morzenti, A., San Pietro, P.: Refining real-time system specifications through bounded model- and satisfiability-checking. In: 23rd IEEE/ACM International Conference on Automated Software Engineering (ASE 2008), 15–19 September 2008, pp. 119–127. IEEE Computer Society, Piscataway (2008)
68. Prior, A.: Time and Modality. Oxford University Press, Oxford (1957)
69. Prior, A.: Past. Present and Future. Oxford University Press, London (1967). Reprinted 2002
70. Rescher, N., Urquhart, A.: Temporal Logic. Springer, New York (1971)
71. Rozier, K.Y., Vardi, M.Y.: LTL satisfiability checking. STTT **12**(2), 123–137 (2010)
72. Rozier, K.Y., Vardi, M.Y.: A multi-encoding approach for LTL symbolic satisfiability checking. In: Butler, M., Schulte, W. (eds.) FM 2011: Formal Methods. Proceedings of the 17th International Symposium on Formal Methods, Limerick, 20–24 June 2011. Lecture Notes in Computer Science, vol. 6664, pp. 417–431. Springer, Berlin/New York/Heidelberg (2011)

73. Schwartz, R.L., Melliar-Smith, P.M., Vogt, F.H.: An interval logic for higher-level temporal reasoning. In: Proceedings of the 2nd Annual ACM Symposium on Principles of Distributed Computing (PODC'83), pp. 198–212. ACM, New York (1983)
74. Sharma, B., Pandya, P.K., Chakraborty, S.: Bounded validity checking of interval duration logic. In: Halbwachs, N., Zuck, L.D. (eds.) Tools and Algorithms for the Construction and Analysis of Systems, 11th International Conference, TACAS 2005, Held as part of the Proceedings of the Joint European Conferences on Theory and Practice of Software, ETAPS 2005, Edinburgh, 4–8 April 2005. Lecture Notes in Computer Science, vol. 3440, pp. 301–316. Springer, Berlin/New York (2005)
75. Skakkebæk, J.U., Shankar, N.: Towards a duration calculus proof assistant in PVS. In: Langmaack, H., de Roever, W.P., Vytopil, J. (eds.) FTRTFT. Lecture Notes in Computer Science, vol. 863, pp. 660–679. Springer, Berlin (1994)
76. Vardi, M.Y.: Branching vs. linear time: final showdown. In: Margaria, T., Yi, W. (eds.) Proceedings of the 7th International Conference on Tools and Algorithms for the Construction and Analysis of Systems (TACAS'01). Lecture Notes in Computer Science, vol. 2031, pp. 1–22. Springer, Berlin/Heidelberg/New York (2001)
77. Zhou, C., Hansen, M.: Chopping a point. In: Cooke, J., Jifeng, H., Wallis, P. (eds.) BCS-FACS 7th Refinement Workshop, Electronic Workshops in Computing. Springer, Berlin/Heidelberg (1996)

Chapter 10
Algebraic Formalisms

In mathematics, the term *algebra* has two closely related meanings, one general and one specific. "Abstract algebra" is the branch of mathematics that studies the abstract properties of symbolic operations and transformations. It encompasses, among other things, the formal axiomatic description of structures and their classification according to the properties of operations defined on them. For example, group theory is a part of algebra concerned with the properties of every operation that is associative, invertible, and admits a neutral element, such as addition on integers or reals, as well as the permutations of every finite set of elements. The other meaning of "algebra" denotes specific instances of axiomatic structures and their usage in symbolic manipulations. "Boolean algebra" and "linear algebra" are named according to this more specific meaning of the word and are formal systems based on axioms and symbolic calculations.

This chapter presents some formalisms for the description of systems collectively known as *process algebras* or *process calculi*. Process algebras follow the operational paradigm (Chap. 2) similarly to the automata of Chap. 7, because they describe system behavior in terms of abstract processes, their states, and the state transformations over time in response to transitions. To some extent, process algebras are just convenient notations for describing transition systems – with finite or infinite sets of atomic states – and their composition; hence they embody the first definition of algebra mentioned above, with its focus on abstract operations and transformations applicable to large classes of systems.

By virtue of being "algebras" also according to the second meaning illustrated above, process calculi include some characteristics of descriptive formalisms, which were among the main historical motivations for their introduction; in particular, they support verification as a calculation from a set of axioms, similarly to how symbolic manipulations are used to deduce formulae from axioms in various flavors of mathematical logic (see Chaps. 2 and 9).

From the perspective of this book, the traditional process algebras do not put much emphasis on *time*. Rather, they focus on the *interactions* of computational processes that work in parallel and occasionally synchronize, and much of the

C.A. Furia et al., *Modeling Time in Computing*, Monographs in Theoretical Computer Science. An EATCS Series, DOI 10.1007/978-3-642-32332-4_10,
© Springer-Verlag Berlin Heidelberg 2012

research on process algebras has insisted on the abstract study of concurrent systems and phenomena such as deadlocks (see Chap. 3). While presenting a few significant examples of process algebras, this chapter will try to make the connection with time modeling and its dimensions explicit.

The rest of the chapter starts with the presentation of the most relevant features of a classic process algebra: Communicating Sequential Processes (CSP). Section 10.1 introduces them through examples and an informal presentation of their semantics; then Sect. 10.2 tackles the problem of providing a rigorous and formal semantics. As we have experienced with other notations, formalizing the semantics of CSP requires unraveling a number of subtleties which are not apparent in the intuitive presentation, and which lend themselves to different rigorous solutions. These solutions summarize historically distinct approaches, which have often been the object of intense debate. The presentation of this chapter specially emphasizes the semantics based on transition systems (defined in Chap. 7), because it is the most general and comprehensive – hence our suggestion that process algebras are "notations for describing transition systems"; Sect. 10.2, however, also discusses other semantic approaches and illustrates their differences and similarities with that based on transition systems. Subsequently, Sect. 10.3 discusses extensions of CSP that accommodate quantitative time, while Sect. 10.4 shows how to model probabilistic behavior with process algebras. The selection of topics is limited to a few significant instances demonstrating fundamental features. As usual, Sect. 10.5 and the bibliographic remarks at the end of the chapter mention other widely used notations and verification techniques and tools based thereon, and refer you to the literature for additional details.

10.1   Communicating Sequential Processes

Introduced by Tony Hoare in the late 1970s, "Communicating Sequential Processes" (CSP) is a process algebra which established several paradigms used in modeling concurrency and is still widely used – directly or indirectly – as a basis for more sophisticated notations. This section introduces the most significant aspects of CSP on an intuitive basis, through a series of simple examples.

10.1.1   Processes and Events

As mentioned in the introduction of this chapter, and as Sect. 10.2 will illuminate in greater depth, the core of CSP, and of every other similar process algebra, is a notation to describe transition systems (discussed in Sect. 7.1) with "algebraic flavor". Take, for example, the two-state automaton of Fig. 7.1: CSP represents each transition with the *prefix* operator "→". The transition that goes from state 0 to state 1 upon reading s is then described by the equation:

$$\mathsf{ZERO} \;=\; s \to \mathsf{ONE}\,. \tag{10.1}$$

ZERO and ONE are *processes* and correspond to states of the transition system; s is called *event* and denotes the input to the process. Equation 10.1 specifies that process ZERO *becomes* process ONE after inputting s. In all the examples, we stick to CSP's convention of naming processes with uppercase alphabetical strings and events with lowercase alphabetical strings.

The complete CSP model of the transition system TS_{sr} of Fig. 7.1 includes two possible events for every process (separated by "|") and defines process TS_{sr} to be initially ZERO (the labeling of states is not represented in CSP, and the prefix operator conventionally has higher precedence than any other operator):

$$
\begin{aligned}
\text{ZERO} &= r \rightarrow \text{ZERO} \mid s \rightarrow \text{ONE} \\
\text{ONE} &= s \rightarrow \text{ONE} \quad \mid r \rightarrow \text{ZERO} \\
TS_{sr} &= \text{ZERO} .
\end{aligned}
\tag{10.2}
$$

This simple example shows how CSP consists, at its core, of sequences of atomic events. Intuitively, process ZERO becomes process ONE if it receives event s, whereas it remains process ZERO – more precisely, another instance of process ZERO – if it receives event r, and so on; thus, $r\,r\,s\,r\,s$ is a sequence of events compatible with the system specification (10.2). Of course, $r\,r\,s\,r\,s$ and all other sequences of events defined by (10.2) are also valid inputs to the automaton of Fig. 7.1. Viewing CSP as defining sequences of events entails a notion of time that is **DISCRETE** and **IMPLICIT**, as will be apparent in the definition of CSP semantics in Sect. 10.2.

CSP offers a compact indexed notation to model systems with a large, possibly infinite, number of events and processes. For example, the equations

$$
\begin{aligned}
\text{COUNTER} &= \text{AT}(0) \\
\text{AT}(k) &= \text{next} \rightarrow \text{AT}(k+1)
\end{aligned}
\tag{10.3}
$$

describe a process COUNTER which counts up from zero through a denumerable sequence of parametric processes $\text{AT}(k)$, with k ranging over the natural numbers.

10.1.2 Internal and External Choice

CSP, and process algebras in general, focuses on process *interaction*. All the processes share an implicit global *environment*, which provides their inputs and interacts with them: by selecting events, the environment drives the dynamics of processes and chooses from among alternatives. We have already seen how processes can "offer to receive" multiple events from the environment with the prefix operator; for example, Eq. (10.2) says that the environment can repeatedly choose between producing event r and producing event s. For this reason, "|" is called "external choice" operator, because it is the external environment which resolves the choice by producing an event. Unlike in other formalisms, the environment does not play the role of a special process; on the contrary, it is the source of inputs to the overall system that consists of all the available processes.

CSP also provides an "internal choice" operator, denoted by "⊓". Consider a variant of (10.2) where internal choice replaces external choice:

$$\mathsf{ZEROND} = r \to \mathsf{ZEROND} \sqcap s \to \mathsf{ONEND}. \qquad (10.4)$$

The informal semantics of internal choice is the following: process ZEROND nondeterministically chooses, independently of the environment, between behaving as $r \to$ ZEROND or behaving as $s \to$ ONEND; the environment can only proceed with event r as input if the former is chosen, and with event s if the latter is. Internal choice introduces a form of **NONDETERMINISTIC** process behavior, where system evolution is not completely determined by an arbitrary input sequence. From the point of view of the environment, this internal nondeterminism is *demonic* and *universal* (according to the terminology introduced in Chap. 6), in that the environment must be able to deal with any possible outcome of internal choice in order to control the system.

We can obtain a more explicit characterization of internal choice by means of special *silent events* τ, which are generated internally and are invisible to the environment. Then, process ZEROND of (10.4) can be equivalently described as first executing a silent event that transforms it into one of two processes, each of them offering (that is, being compatible with) only one of r and s:

$$
\begin{aligned}
\mathsf{ZEROND} &= \tau \to \mathsf{R} \mid \tau \to \mathsf{S} \\
\mathsf{R} &= r \to \mathsf{ZEROND} \qquad\qquad (10.5) \\
\mathsf{S} &= s \to \mathsf{ONEND}.
\end{aligned}
$$

We can already see a subtle difference between internal and external choice: even if both (10.2) and (10.5) define systems that can read arbitrary sequences of events r and s, the former really accepts all such sequences as delivered by the environment, whereas the latter only accepts events r if process ZEROND "chooses" to become R *before* reading input from the environment (and symmetrically for events s).

In CSP, it is customary to use silent events only in the formalization of the semantics (see Sect. 10.2); hence Eq. (10.5) would not be normally written. This restriction is however merely a convention, and in fact other process algebras, such as CCS, allow for silent events in system descriptions.

If we allow for explicit silent events in CSP definitions, internal and external choice can be presented as special cases of a "general" choice, denoted by "□". A process P offering a general choice between processes X and Y,

$$P = X \,\square\, Y,$$

behaves deterministically – as with external choice – if X and Y offer distinct events to the environment, for example, $X = x \to X$ and $Y = y \to Y$ with $x \neq y$.

If, instead, X and Y offer the same event (possibly τ), P behaves nondeterministically, as with internal choice. For example, if $X = \tau \to A$ and $Y = \tau \to B$, P internally chooses between A and B and the environment has no control over which process gets executed next.

10.1.3   Hiding of Events

The environment is common to all processes in the system, and every event other than τ is visible to every process by default. However, CSP offers a "hiding" (concealment) operator "\" that restricts the set of events that are visible to the environment. Take, for example, process TS_{sr} described above, and hide event r from it:

$$TS_s = TS_{sr} \setminus \{r\}.$$

Process TS_s replaces r events by silent τ events; hence it introduces internal nondeterministic behavior outside the control of the environment. In fact, TS_s is equivalently described with internal choice:

$$
\begin{aligned}
\mathsf{ZERO} &= \mathsf{ZERO} \quad \sqcap \ s \to \mathsf{ONE} \\
\mathsf{ONE} &= s \to \mathsf{ONE} \ \sqcap \ \mathsf{ZERO} \\
TS_s &= \mathsf{ZERO}.
\end{aligned}
\tag{10.6}
$$

Exercise 10.1. Write a transition system with explicit silent events equivalent to TS_s.   ◼

The presence of silent events or, equivalently, of hiding, and the notion of time as discrete and implicit pose a critical problem to modeling time in CSP: what happens if we assume that time elapses whenever processes make transitions? If we follow the same approach used in other formalisms – such as finite state automata – and associate one time unit (or any other finite positive amount) with every transition made, we have the problem of silent events: do they take time as well? If they do, time has an implicit metric based on counting events, but it may be inappropriate according to what the real "nature" of internal nondeterministic choices is (for example, they may simply be a modeling artifact to account for different deterministic implementations; hence they would be immaterial). On the other hand, if we posit that silent events take no time, we may have Zeno behaviors of processes that do not **ADVANCE** in time because they engage infinite sequences of internal transitions and stop reading input from the environment.

For example, process TS_s *may* stop progressing if its internal choices always select process **ZERO**. A more extreme example is a process that never interacts with the environment in any situation, conventionally called "STOP" in CSP.

Exercise 10.2. Define **STOP** from another process using hiding.

(*Hint*: you can, for example, start from process TS_s or process **COUNTER**).

◼

In summary, it seems appropriate that the notion of time used in CSP (and most process algebras) be purely **QUALITATIVE**, because the environment may be unable to count (or even detect) hidden transitions to measure time elapsing precisely. The CSP trace semantics, presented in Sect. 10.2.1, follows this intuition and abstracts away from the representation of silent transitions τ. Under different semantics, however, it is possible to introduce a rigorous **METRIC** on time for CSP, as Sect. 10.3 will demonstrate.

10.1.4   Concurrent Process Composition

Multiple processes can run concurrently and synchronize on shared events produced by the environment. CSP offers two operators to define the concurrent composition of processes; they essentially capture the same semantics as the composition of transition systems (described Sect. 7.1), but they offer some more flexibility in the definition of synchronization events.

We illustrate CSP concurrency operators with a variant of the producer/consumer example of Fig. 7.13 where the buffer is unbounded and hence is an infinite-state process,

$$\text{BUFFER} \;=\; \text{HAS}(0)$$

$$\text{HAS}(k) \;=\; \begin{cases} \text{put} \to \text{HAS}(k+1) \mid \text{get} \to \text{HAS}(k-1) & k > 0 \\ \text{put} \to \text{HAS}(k+1) & k = 0, \end{cases}$$

and k ranges over the natural numbers. Consider now a **PRODUCER** process which puts elements in the buffer:

$$\text{PRODUCER} \;=\; \text{produce} \to \text{READY}$$

$$\text{READY} \;=\; \text{put} \to \text{PRODUCER}.$$

The *parallel composition* of **PRODUCER** and **BUFFER** that synchronizes on event put is written

$$\text{BUFFER}\,_{\{\text{put,get}\}}\|_{\{\text{produce,put}\}}\,\text{PRODUCER}.$$

The semantics of this composition is the same as that of the parallel composition of transition systems defined in Sect. 7.1: **BUFFER** and **PRODUCER** **SYNCHRONIZE** on event put, whereas they proceed **ASYNCHRONOUSLY** on the other events (get for **BUFFER** and produce for **PRODUCER**). In this setting, the **PRODUCER** can produce an unlimited number of items, and every item can be put in the **BUFFER**.

The sets of events used in parallel composition may be subsets of all possible events the composed processes offer. Namely, the composition

$$P \ _{A}\|_{B}\ Q$$

of processes P and Q is a process where P can only engage in events in A, Q can only engage in events in B, and P and Q must share the events in $A \cap B$ but proceed independently on the other events. It is customary to choose A and B such that they respectively include all events mentioned in P's and Q's definitions. Otherwise, the transitions corresponding to events not included in A (or B) are permanently disabled in P (or Q) when P (or Q) executes in $P \ _{A}\|_{B}\ Q$. For example,

$$(a \rightarrow \mathsf{STOP}) \mid (b \rightarrow \mathsf{STOP}) \ _{\{a\}}\|_{\{a,c\}} \ (a \rightarrow \mathsf{STOP}) \mid (c \rightarrow \mathsf{STOP})$$

can proceed with events a (synchronously) or c (asynchronously), but not with event b. The definition of parallel composition implies that, if A is the set of all events offered by both P and Q, the composition $P \ _{A}\|_{A}\ Q$ has the same semantics as the *synchronous composition* of transition systems (see Sect. 7.1).

Exercise 10.3. Let A be the set of all events offered by process P. What is the behavior of – that is, the set of event sequences accepted by – the composition $P \ _{A}\|_{\emptyset}\ Q$, where Q is any other process composed with an empty set of events? ∎

CSP also offers an "interleaving" composition operator "|||". Processes composed with interleaving proceed completely independently; whenever the environment produces an event accepted by more than one process, exactly one process – chosen nondeterministically – inputs the event and proceeds while the others wait. For example, the interleaving composition

BUFFER ||| PRODUCER

defines a composite process where **BUFFER** and **PRODUCER** do not share put events: whenever the producer is **READY** and a put event is triggered, either the **BUFFER** increments its counter, or process **READY** changes to **PRODUCER**, but not both.

Exercise 10.4 (♣). Interleaving composition seems inappropriate for modeling synchronization between processes such as **PRODUCER** and **BUFFER** that must share certain events. Find different types of systems where coordination among processes is more naturally captured by interleaving than by synchronous composition. ∎

The term "interleaving" refers to the standard CSP semantics of the "|||" operator formalized in Sect. 10.2. In the particular case where A is the set of all events offered by process P, B the set of all events offered by process Q, and $A \cap B = \emptyset$, the interleaving $P \ ||| \ Q$ is equivalent to the parallel composition $P \ _{A}\|_{B}\ Q$, as in the case of transition systems (Sect. 7.1).

Exercise 10.5 (♣). Discuss how to represent interleaving composition as parallel composition by means of internal choice.   ∎

Exercise 10.6. Replicate the composite transition systems of Figs. 7.7, 7.8c, 7.10, and 7.11 using CSP processes and the appropriate composition operators.   ∎

10.1.5   Inter-process Communication

Processes running in parallel communicate via events shared in the environment, as we have seen in Sect. 10.1.4; this is a form of **SYNCHRONIZATION** via **MESSAGE PASSING**. Section 3.7.2 discussed message passing in general terms, and observed that it normally is an *asynchronous* communication mechanism. In the CSP language, however, as in the channel-based composition of transition systems described in Sect. 7.1, "message passing" defines synchronous communication where the "messages" are shared input events. Event sharing is indeed the only synchronization primitive available in CSP; the language offers two additional constructs to facilitate the definition of complex shared events: parametric and input/output events. Both are syntactic sugar which does not increase the expressive power of CSP but make formalizations more compact and readable.

CSP offers *parametric* (compound) events analogous to parametric processes; for example, we can extend Eq. (10.3) to define a counter that prints the current value before incrementing it:

$$\text{COUNTER\_PRINT} \;=\; \text{AT}(0)$$
$$\text{AT}(k) \;=\; \text{print}.k \to \text{AT}(k+1)\,.$$

$\text{print}.k$ (with k ranging over the natural numbers) is a parametric event, which can be thought as the appearance of value k over a *channel* with name print and of *type* $\mathbb{N}$. The terminology suggests that parametric events are tailored for inter-process communication.

Input/output events are defined on top of parametric events, and they make the roles of the communicating processes explicit. They have essentially the same semantics and syntax as those of the channels of transition systems (discussed in Sect. 7.1): for a channel c of type T and a value $t \in T$, the expressions $c?t$ and $c!t$ respectively define the input and output of message t on channel c; hence, when the environment sends message t on channel c, two processes composed in parallel and sharing the channel c will synchronize if one accepts $c?t$ and the other accepts $c!t$. The behavior is the very same as that obtained with just a parametric event $c.t$ for both processes, but the syntax of input/output events helps make communication explicit and restricts each synchronization event to happening between exactly two processes. Notice that this "internal input/output" among processes is logically distinct from the input generated by the environment to the system of processes: in the example above, the environment technically only sends one event $c.t$, but a process interprets it as internal input event $c?t$ from another process that interprets $c.t$ as internal output event $c!t$ through channel c.

Exercise 10.7. Complete the producer/consumer example introduced in Sect. 10.1.4 as follows:

- Introduce a **CONSUMER** process and formalize its behavior;
- Introduce input/output events to synchronize the **BUFFER** and its clients;
- Define an overall process **SYSTEM** with the appropriate concurrent composition. ∎

Example 10.8. Example 8.15 modeled with Petri nets an allocator that accepts low-priority requests only when it is free, and that enqueues an arbitrary number of high-priority requests and serves all of them before considering other types of requests. Let us formalize the same system as a CSP process **ALLOCATOR**. The **ALLOCATOR** is initially free:

$$\text{ALLOCATOR} = \text{FREE}.$$

A free allocator can receive low-priority or high-priority requests sent by the environment:

$$\text{FREE} = \text{hpr} \rightarrow \text{HIGH} \mid \text{lpr} \rightarrow \text{LOW}.$$

The allocator grants a free resource immediately to high-priority requests, and after some time to low-priority requests; we model time elapsing by means of a dummy **wait** event. Since time modeling is purely **QUALITATIVE** with the CSP semantics illustrated so far, we cannot specify the precise duration of the waiting event:

$$\text{HIGH} = \text{GRANT}$$
$$\text{LOW} = \text{wait} \rightarrow \text{GRANT}.$$

When the allocator becomes occupied, it spawns a process to count the number of pending requests, initialized to zero; the counter will synchronize on an event **done** when the last request is served:

$$\text{GRANT} = \text{OCC}_{\{\text{done}\}} \|_{\{\text{hpr,rel,done}\}} \text{PENDING}(0).$$

The number of pending requests is increased with every new **hpr** event and decreased with every **rel** event (k ranges over the natural numbers):

$$\text{PENDING}(k) = \begin{cases} \text{hpr} \rightarrow \text{PENDING}(k+1) \mid \text{rel} \rightarrow \text{PENDING}(k-1) & k > 0 \\ \text{hpr} \rightarrow \text{PENDING}(k+1) & k = 0. \end{cases}$$

When there are no pending requests, a release event releases the resource, which becomes free again:

$$\text{PENDING}(0) = \text{rel} \to \text{RELEASE}$$
$$\text{RELEASE} = \text{done} \to \text{SKIP}$$
$$\text{OCC} = \text{done} \to \text{FREE}.$$

The special CSP process **SKIP** simply terminates; hence it kills the current instance of the counter process. If we do not want to expose the synchronization event **done** to the environment, which would only interact with the allocator by issuing requests and by releasing resources, we can hide it:

$$\text{ALLOCATOR}' = \text{ALLOCATOR} \setminus \{\text{done}\}. \qquad \blacksquare$$

Exercise 10.9. CSP also offers a "sequential composition" of processes: $S = P \; ; \; Q$ is a process which behaves as P until it terminates; if P does terminate, S will then behave as Q. Sequential composition is derivable from other CSP operators (but we do not delve into these details, which are rather technical). Consider the special processes **SKIP** (defined in Example 10.8) and **STOP** (defined in Sect. 10.1.3). For a generic process P, describe the behavior of

- SKIP ; P
- P ; SKIP
- STOP ; P
- P ; STOP   ■

10.2 Formal Semantics of Process Algebras

Section 10.1 presented the semantics of CSP only informally, but it uncovered many subtle points which call for rigorous definitions. Formal semantics is, indeed, one of the primary foci of process algebras, and in fact there is a vast literature that deals with relations and equivalences between the various operators of process algebras. This section gives a succinct account of the main approaches to the semantics of process algebras – CSP primarily, but also some aspects of other algebras; as usual, the discussion concentrates on the dimensions of time modeling.

10.2.1 Trace Semantics

CSP is equipped with a semantics based on *traces* (also called "denotational"), which are FINITE sequences of *visible* events: even if CSP processes can engage in indefinitely long interactions with the environment, the trace model only records observations of finite length. As in the semantics of finite state machines (Definition 7.14), every CSP process P defines a set *traces*(P) of all the traces it can generate. CSP processes may or may not terminate. The trace t of a terminating process conventionally always ends with a unique instance of the special visible event $\checkmark$; predicate $term(t)$ holds for *terminating* traces t. The notion of termination is somehow similar to the notion of acceptance for finite state machines. The set $traces(P)$ contains the finite prefixes of every execution of P, both terminating and nonterminating ones. Notice that process SKIP only has finite executions, while the STOP process may have nonterminating executions. Partial executions are included as well, because an execution can stop at any point if the environment does not produce events; hence, the set $traces(P)$ is closed under prefix: if $e_1 e_2 \ldots e_n$ is a trace in $traces(P)$, then so is $e_1 e_2 \ldots e_k$ for every $1 \le k < n$.

The set $traces(P)$ admits an inductive definition on the structure of process P. Let us see a few significant examples of this definition for the operators introduced in Sect. 10.1. Unsurprisingly, the prefix operator populates *traces* by prefixing single events to the traces of the prefixed process (ε denotes the empty sequence):

$$traces(e \to P) \quad = \quad \{\varepsilon\} \cup \{e \cdot t \mid t \in traces(P)\}.$$

Notice that the empty trace is always possible if the environment does not supply e.

The trace semantics completely ignores silent events. Therefore, the various choice operators have the same trace semantics: the choice between processes P and Q includes all of P's traces as well as all of Q's traces:

$$traces(P \mid Q) = traces(P \sqcap Q) = traces(P \,\square\, Q) = traces(P) \cup traces(Q).$$

Hiding removes events from traces: if $s|_A$ denotes sequence s with all occurrences of events *not* in the set A removed, and Σ is the alphabet of all possible events (including $\checkmark$, which cannot be hidden), the trace semantics of hiding is

$$traces(P \setminus A) \quad = \quad \{t|_{\Sigma \setminus A} \mid t \in traces(P)\}.$$

The traces of the parallel composition $P \;_A\|_B\; Q$ include all traces t such that t restricted to only events in A is a trace of P, t restricted to only events in B is a trace of Q, and t contains only events in A or B (other than possibly a trailing $\checkmark$):

$$traces(P \;_A\|_B\; Q) \quad = \quad \left\{ t \;\middle|\; \begin{array}{l} t|_{A \cup \{\checkmark\}} \in traces(P), \\ t|_{B \cup \{\checkmark\}} \in traces(Q), \\ t \in (A \cup B)^* \cup ((A \cup B)^* \cdot \checkmark) \end{array} \right\}.$$

The above definition entails that the parallel composition terminates if and only if both P and Q terminate.

The trace semantics of interleaving composition relies on the definition of $interleave(t_1, t_2)$ as the set of all sequences obtained by interleaving the sequences t_1 and t_2; for example, $interleave(a\,b, c\,d)$ is $\{a\,b\,c\,d, a\,c\,b\,d, a\,c\,d\,b, c\,a\,b\,d, c\,d\,a\,b\}$. Similarly to parallel composition, the trace semantics of interleaving composition ensures that two interleaved processes terminate if and only if both processes terminate:

$$
traces(P \;|||\; Q) \;=\; \left\{ t \;\middle|\; \begin{array}{l} \text{there exist } t_1, t_2, t_3 \text{ such that :} \\ t_1 \in traces(P), \; t_2 \in traces(Q), \\ t_3 \in interleave(t_1|_{\Sigma \backslash \{\checkmark\}}, t_2|_{\Sigma \backslash \{\checkmark\}}), \text{ and} \\ \text{either } term(t_1) \wedge term(t_2), \text{ and } t = t_3 \cdot \checkmark \; ; \\ \text{or } \neg term(t_1) \vee \neg term(t_2), \text{ and } t = t_3 \end{array} \right\} .
$$

Exercise 10.10. What is the trace semantics of the special processes SKIP (defined in Example 10.8) and STOP (defined in Sect. 10.1.3)?   ∎

Exercise 10.11. Define the trace semantics of the sequential composition operator of Exercise 10.9.   ∎

Notice that process definitions are in general recursive. For example, process ZERO in Sect. 10.1 includes the recursive definition ZERO $= r \to$ ZERO; hence $traces(\text{ZERO})$ also has a recursive definition according to the rules above. Well-formed recursive definitions implicitly determine the sets $traces$ of traces; for example, ZERO $= r \to$ ZERO determines the set $traces(\text{ZERO}) = r^*$ of all traces with an arbitrary number of repeated r's.

Exercise 10.12 (♦). Section 10.1.3 gives two definitions of a process TS_s, first in terms of hiding, and then with internal choice. Show that the two definitions have equivalent trace semantics.

As observed in Sect. 10.1.3, traces ignore silent events; hence, time is QUALI-TATIVE in the trace semantics because it is impossible to measure time elapsing by counting events. Another characteristic of time in the model based on traces is its being a LINEAR sequence of events which ignores the branching structure induced by nondeterministic (internal) choice. Finally, concurrent interactions are explicitly represented as interleavings where an arbitrary TOTAL ORDER is assigned to asynchronous events. The preference for this model of concurrency is already apparent in term for the $|||$ composition operator, which refers explicitly to a total order semantics even if, in principle, concurrent processes might be given a semantics based on partial orders, as in untimed Petri nets.

Exercise 10.13. Derive the trace semantics of the following processes PROC$_1$ and PROC$_2$ and show that they coincide. How would you characterize the difference between PROC$_1$ and PROC$_2$ that is not captured by the trace semantics? (See also Exercise 10.14).

$$\text{PROC}_1 \;=\; a \to A$$

$$A \;=\; b \to \text{SKIP} \;|\; b \to \text{SKIP}$$

$$\begin{aligned}
\mathsf{PROC_2} &= \quad a \to H \mid a \to K \\
K &= \quad b \to \mathsf{SKIP} \\
H &= \quad b \to \mathsf{SKIP} \qquad\qquad \blacksquare
\end{aligned}$$

10.2.2   Transition System Semantics

The semantics based on traces, introduced in the previous section, has the advantage of being simple to present and analyze, as it offers the intuitiveness of linear time and provides suitable abstractions that focus on visible interactions and are independent of the detailed internal system behavior.

This level of abstraction may, however, be too imprecise in some modeling situations, and in fact there exist more detailed process algebra semantics. While CSP generalizes trace semantics with the failure model, described in Sect. 10.2.3, other process algebras – most notably, CCS – offer semantics based on *transition systems*, which uniformly represent system and environment events and hence support arbitrarily detailed analyses. The rest of this section briefly discusses a transition system semantics for CSP (often called "operational semantics").

The informal presentation of Sect. 10.1 has already pointed out the close connection between CSP and transition systems; it is therefore natural to formalize the semantics of CSP models as transition systems: given a CSP process P, $TS(P)$ denotes a transition system whose nodes are labeled with CSP processes and whose edges are labeled with events (including silent events τ). Similarly to $traces(P)$, $TS(P)$ has an inductive definition on P's structure; for example:

- For every process P of the form $e \to Q$, $TS(P)$ includes an edge $P \xrightarrow{e} Q$;
- For every process P of the form $a \to A \mid b \to B$, $TS(P)$ includes two edges $P \xrightarrow{a} A$ and $P \xrightarrow{b} B$;
- For every process P of the form $a \to A \sqcap b \to B$, $TS(P)$ includes two edges $P \xrightarrow{\tau} (a \to A)$ and $P \xrightarrow{\tau} (b \to B)$;
- For every process P of the form $Q \setminus \{e\}$, $TS(P)$ is the same as $TS(Q)$, with every transition $\xrightarrow{e}$ changed into a silent transition $\xrightarrow{\tau}$;
- For every process redefinition of the form $P = Q$, $TS(P)$ includes an edge with silent transition $P \xrightarrow{\tau} Q$;
- For concurrent processes, $TS(P)$ explicitly includes all possible interleaving and synchronizing transitions of the processes running in parallel (see also Exercise 10.15).

$TS(P)$ may be infinite if P involves an unbounded number of processes or events.

Using the above rules, it is easy to see that process TS_{sr} (presented in Sect. 10.1.1) determines the transition system shown in Fig. 10.1 – corresponding to that of Fig. 7.1 (except for the labeling of states and the marking of the initial

Fig. 10.1 Transition system
of process TS_{sr}

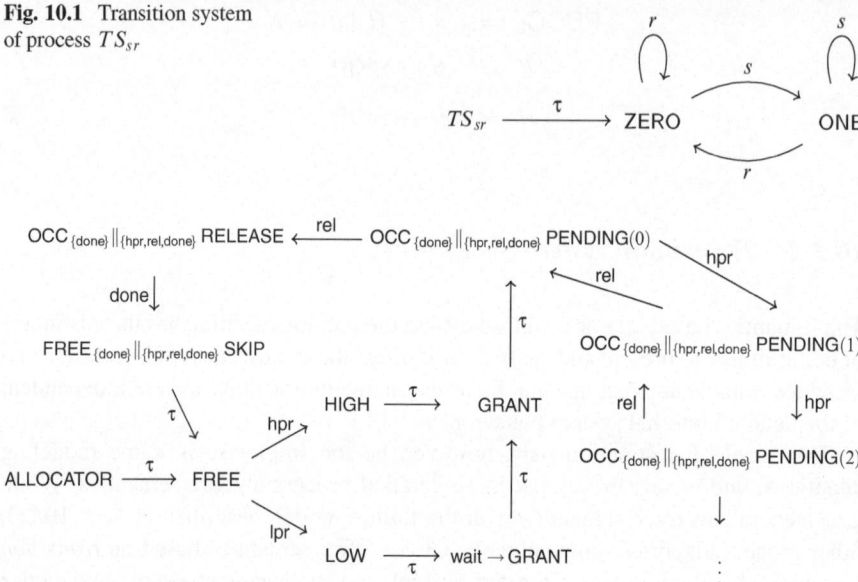

Fig. 10.2 Transition system of process ALLOCATOR

state with a silent event),[1] from which it was derived in the first place. A less trivial
example is presented in Fig. 10.2, showing a significant portion of the transition
system $TS(\mathsf{ALLOCATOR})$ for the allocator of Example 10.8; $TS(\mathsf{ALLOCATOR})$
is infinite because it can count arbitrarily many pending high-priority requests.

Exercise 10.14. Determine the transition system semantics of the processes
$\mathsf{PROC_1}$ and $\mathsf{PROC_2}$ defined in Exercise 10.13.   ■

Exercise 10.15. Present the transition system semantics of the CSP composition
operators "$\|$" and "$\|\|\|$" and demonstrate that they are the same as those of the
corresponding operators for transition systems, discussed in Sect. 7.1.   ■

The transition system semantics of process algebras emphasizes the **BRANCHING**
structure of time, as opposed to linear traces: it is straightforward to derive
trees that define all possible branching-time models of a transition system.
It also leads to notions of process *equivalence* that are finer grained than those
obtained by merely comparing sets of traces, and in particular takes into account the
NONDETERMINISM introduced by internal choice, as suggested by the comparison
of Exercises 10.13 and 10.14. By representing explicitly both internal and external
events, the transition system semantics allows, in principle, for a simple **METRIC**
model of time where one time unit is associated with every transition; the same

---

[1]The states of the transition systems in the current chapter are not enclosed in circles to
accommodate wide labels with process definitions.

Fig. 10.3 "Nice" and "rude"
interactions

$$
\begin{aligned}
\text{NICE} &= \quad \text{knock} \rightarrow (\text{E} \mid \text{L}) \\
\text{RUDE} &= \quad \text{knock} \rightarrow (\text{E} \sqcap \text{L}) \\
\text{E} &= \quad \text{enter} \rightarrow \text{SKIP} \\
\text{L} &= \quad \text{leave} \rightarrow \text{SKIP}
\end{aligned}
$$

model was inconsistent with the trace semantics. Section 10.3 discusses, in greater detail, more general approaches to introducing metric time in CSP.

10.2.3   Failure Semantics

The previous sections already pointed out that the trace semantics ignores silent transitions; hence, in particular, it does not distinguish scenarios where the interaction between system and environment cannot proceed because it is blocked from other scenarios where the environment has simply finished supplying events.

Consider, for example, a person wanting to enter another person's office; the person knocks on the door, and then she may enter or leave the office. If the office owner is "nice", he will let the knocker decide whether to enter or leave after knocking; if he is "rude", he will impose his decision on the knocker. The processes NICE and RUDE in Fig. 10.3 model the two situations, with the environment in the role of the knocker. It is clear that NICE and RUDE define the same set of *traces*. However, process RUDE may block if the process internally chooses to behave as E but the environment is only willing to leave or, conversely, the environment wants to enter but the process chooses to behave as L.

These situations define a form of DEADLOCK in the interaction between environment and process. A slightly different form of deadlock can also occur between multiple processes that make irreconcilable internal choices where progress on shared events is impossible. The trace semantics makes it impossible to reason about deadlock and other similar *liveness* properties because it fails to characterize internal choice (Sect. 9.1.1.1 introduced the notion of liveness with reference to temporal logic). More precisely, *absence of deadlock* is the liveness property of interest: "the process will always complete its interaction with the environment" holds for NICE but not for RUDE.

Exercise 10.16. Modify the system described in Fig. 10.3 to model the following situation using parallel composition of processes. The knocker is seeking access to a high security area; to this end, she issues an access_request to two independent guardians; each guardian can either grant or deny access; if both grant, the knocker manages to enter; if both deny she has to leave; but if the guardians disagree a DEADLOCK occurs and the whole system hangs. Show how this type of deadlock differs from the one mentioned in the previous paragraph, in that the latter is a consequence of process RUDE's direct interaction with the environment (representing the knocker), whereas the former follows from lack of synchronization between two parallel processes internal to the system.   ∎

Exercise 10.17 (♣). Consider a situation in which a system is composed of different elements, and a computation in which some of these components make irreconcilable choices and stop interacting with the environment, while others keep executing; how would you define the semantics for such a situation?   ∎

The transition system semantics of Sect. 10.2.2 includes enough information for reasoning about liveness, but it is not a standard approach in CSP. Instead, CSP defines the "stable failure" semantics, an extension of the trace semantics that keeps track of the effects of internal choice and hence supports reasoning about liveness properties such as absence of deadlock. The stable failure semantics is another form of *denotational* semantics.

Similarly as the trace semantics, the stable failure semantics defines a set *failures*(P) of *failures* for every CSP process P. A "failure" is a pair $\langle t, R \rangle$, where t is a trace in *traces*(P) and R is the set of events *refused* by P right after executing trace t. For example, process **RUDE** includes the failure

$$\langle \text{knock}, \{\text{leave}, \text{knock}\}\rangle ,$$

which record the fact that, after executing **knock**, **RUDE** may refuse to accept **leave** or **knock** again if it has internally chosen to behave as **E**. Another failure $\langle \text{knock}, \{\text{enter}, \text{knock}\}\rangle$ models the opposite situation where **RUDE** selects **L** and hence refuses **enter**.

The standard definition of the stable failure semantics uses inductive characterizations that extend the one for the trace semantics seen in Sect. 10.2.1. In this chapter, however, we prefer to give a definition of stable failure semantics that is based on the transition system semantics of Sect. 10.2.2; this presentation has the advantage of being simpler to describe and understand once the transition system semantics is available.

Given the *transition system* $TS(P)$ of a generic process P, we define failures for every node N of $TS(P)$ that has no outgoing silent transitions τ to a different state; such nodes are called "stable" (hence the name *stable* failures) because the system has made its possible internal choice and proceeds only when the environment issues an event. In particular, **STOP** is stable too, as it certainly ("stably") does not accept any event from the environment. If $t = e_1 e_2 \ldots e_k$ is the sequence of visible events along some path reaching N from an initial node, and if S is the set of all events outgoing node N, then

$$f = \langle t, \Sigma \setminus (S \cup \{\checkmark\})\rangle$$

is a stable failure of P. f includes all visible events (other than termination) that cannot be performed at N and hence are *refused*.

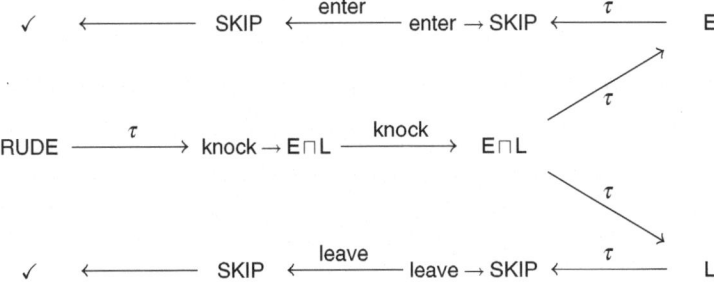

Fig. 10.4 Transition system of process RUDE

Figure 10.4 shows the transition system of process RUDE, from which it is clear that $\langle knock, \{leave, knock\}\rangle$ and $\langle knock, \{enter, knock\}\rangle$ belong to *failures*(RUDE) as mentioned above. *failures*(RUDE) also includes the failures $\langle \varepsilon, \{enter, leave\}\rangle$, $\langle knock\ enter, \{knock, enter, leave\}\rangle$, and $\langle knock\ leave, \{knock, enter, leave\}\rangle$.

Exercise 10.18. Derive the set *failures*(NICE) and show that it does not include a refusal of enter or leave after knock. ∎

Exercise 10.19 (♦). For generic processes P, Q, define *failures*$(P \mid Q)$ and *failures*$(P \sqcap Q)$ inductively from *failures*(P) and *failures*(Q) and show that, in general, they differ. ∎

The stable failure semantics can be seen as an attempt to augment the purely linear-time model of traces with some **BRANCHING** information about internal behavior. The additional information is needed to reason about liveness properties such as absence of **DEADLOCKS** among interacting processes and the environment. The stable failure semantics still relies on **FINITE** traces; hence it does not fully capture nonterminating interactions between the environment and the processes. On the other hand, refusals can describe processes that stop their interaction with the environment. Such processes, whose simplest instance is process STOP (discussed in Sect. 10.1.3), are nonterminating from their own internal perspective, but are deadlocked from the viewpoint of the environment.

While the stable failure semantics records more information than the simpler trace semantics, it is still coarser grained than the transition system semantics because it focuses on stable states without outgoing silent transitions. The following exercise shows a simple example of two processes with equivalent stable failure semantics but structurally different transition system semantics.

Exercise 10.20. Consider processes $X = $ SKIP $\sqcap$ SKIP and $Y = X \sqcap X$. Show that *failures*$(X) = $ *failures*$(Y) = \{\langle\varepsilon, \emptyset\rangle, \langle\checkmark, \emptyset\rangle\}$, but that the transition systems $TS(X)$ and $TS(Y)$ are structurally different (i.e., *not* isomorphic). ∎

10.2.4   Algebraic Laws

The semantics introduced in the previous sections do not take advantage of the *algebraic* nature of process definitions as a collection of symbolic *equations*. This section exploits this *calculational* aspect of process algebras and demonstrates how to reason formally about CSP processes by purely syntactic manipulation according to well-defined rules, called "algebraic laws". Historically, the calculational approach was one of the main original motivations for the very introduction of process algebras.

The algebraic laws of CSP define the semantics of operators indirectly, by means of equations that state their formal properties (e.g., commutativity, associativity, etc.) and the relations among different operators (e.g., when general choice is equivalent to external choice). The equations are valid with respect to a specific semantics (trace, stable failure, transition system), whose essential properties they capture via axiomatization. The following is a sample of CSP algebraic laws reflecting the stable failure semantics (and the simpler trace semantics as well).

(L1)   Internal choice is commutative:

$$P \sqcap Q = Q \sqcap P$$

(L2)   General choice reduces to internal choice:

$$(p \to P) \ \square \ (p \to Q) = p \to (P \sqcap Q)$$

(L3)   General choice reduces to external choice (for $p \neq q$):

$$(p \to P) \ \square \ (q \to Q) = p \to P \mid q \to Q$$

(L4)   Hiding removes prefix:

$$(p \to P) \setminus \{p\} = P \setminus \{p\}$$

(L5)   Hiding an event q does not affect a process NO_Q that never offers q:

$$NO_Q \setminus \{q\} = NO_Q$$

(L6)   Prefix distributes over internal choice:

$$p \to (P \sqcap Q) = (p \to P) \sqcap (p \to Q)$$

Exercise 10.21 (♦). Prove the consistency with trace semantics (Sect. 10.2.1) of the laws (L1)–(L6) as follows: for each law of the form $LHS = RHS$, compute the recursive definitions of the sets $traces(LHS)$ and $traces(RHS)$, and show that they are equivalent.   ∎

$$\text{IMPOLITE} = \text{CHOOSE} \setminus \{\text{ch}\}$$
$$\text{CHOOSE} = \text{ch} \to \text{Y} \quad \square \quad \text{ch} \to \text{Y}$$
$$\text{X} = \text{knock} \to \text{X}_2$$
$$\text{X}_2 = \text{enter} \to \text{DONE}$$
$$\text{Y} = \text{knock} \to \text{Y}_2$$
$$\text{Y}_2 = \text{leave} \to \text{DONE}$$
$$\text{DONE} = \text{SKIP}$$

Fig. 10.5 Process IMPOLITE

Exercise 10.22 (♦). Which of the laws (L1)–(L6) are *not* valid in the transition system semantics? That is, which equations $LHS = RHS$ are such that the transition systems $TS(LHS)$ and $TS(RHS)$ are *not* isomorphic?   ∎

With the systematic application of algebraic laws, we can formally prove properties and equivalences of CSP processes by deductive **SYNTACTIC TRANS-FORMATIONS**, in the same way as we prove theorems in mathematical logic (as demonstrated in Chap. 2).

Consider, for example, process IMPOLITE defined in Fig. 10.5. The process is very similar to process RUDE (Fig. 10.3), up to renaming of processes and a different layout of the definitions. In fact, we can prove that IMPOLITE and RUDE are equivalent (with respect to stable failure semantics) by transforming the former into the latter with the application of laws (L1)–(L6) plus the principle of *congruence*[2] (implicitly used whenever necessary).

First, we prove that X_2 is equivalent to E and Y_2 is equivalent to L by congruence:

$$
\begin{aligned}
\text{X}_2 &= \text{enter} \to \text{DONE} && \text{(definition of } \text{X}_2) \\
&= \text{enter} \to \text{SKIP} && \text{(definition of DONE)} \\
&= \text{E} && \text{(definition of E)} .
\end{aligned}
$$

The proof of $\text{Y}_2 = \text{L}$ is similar and it is left as an exercise. Then, we rewrite CHOOSE in terms of processes L and E as follows:

$$
\begin{aligned}
\text{CHOOSE} &= \text{ch} \to \text{Y} \;\square\; \text{ch} \to \text{X} && \text{(definition of CHOOSE)} \\
&= \text{ch} \to (\text{Y} \sqcap \text{X}) && \text{(L2)} \\
&= \text{ch} \to (\text{knock} \to \text{Y}_2 \sqcap \text{knock} \to \text{X}_2) && \text{(definitions of X, Y)} \\
&= \text{ch} \to (\text{knock} \to \text{L} \sqcap \text{knock} \to \text{E}) && \text{(equivalences } \text{X}_2 = \text{E}, \text{Y}_2 = \text{L}) .
\end{aligned}
$$

Finally, we derive RUDE from IMPOLITE by applying laws (L1)–(L6) and by rewriting CHOOSE according to the equivalence we just derived:

---

[2]A congruence is an equivalence relation that is preserved by some operations; for example, if X and Y are equivalent ($X \equiv Y$), then the maps of X and Y under any function f are also equivalent ($f(X) \equiv f(Y)$).

$$
\begin{aligned}
\text{IMPOLITE} & = \text{CHOOSE} \setminus \{\text{ch}\} & & \text{(definition of IMPOLITE)} \\
& = (\text{ch} \to (\text{knock} \to \text{L} \sqcap \text{knock} \to \text{E})) \setminus \{\text{ch}\} \\
& & & \text{(definition of CHOOSE)} \\
& = (\text{knock} \to \text{L} \sqcap \text{knock} \to \text{E}) \setminus \{\text{ch}\} & & \text{(L4)} \\
& = (\text{knock} \to \text{L} \sqcap \text{knock} \to \text{E}) & & \text{(L5)} \\
& = \text{knock} \to (\text{L} \sqcap \text{E}) & & \text{(L6)} \\
& = \text{knock} \to (\text{E} \sqcap \text{L}) & & \text{(L1)} \\
& = \text{RUDE} & & \text{(definition of RUDE)} \\
& & & \text{QED}
\end{aligned}
$$

Exercise 10.23. Determine which condition (if any) on the sets of events A and B makes the law:

$$
p \to P \ {}_A\|_B \ q \to Q \ = \ \text{STOP}
$$

valid with respect to the stable failure semantics when

- $p \neq q$,
- $p = q$. ∎

Exercise 10.24. Find a process P for which the following equivalence holds in the stable failure semantics:

$$
P \ ||| \ \text{STOP} \ = \ \text{STOP}.
$$
∎

Exercise 10.25 (♣). List other laws for the CSP operators introduced in this chapter. ∎

In summary, the transition system semantics offers the most general and uniform approach to formal CSP analysis. Even if the other semantics have historically been more prominent, the transition system semantics prevails in recent approaches to CSP analysis: by offering connections with other process algebras and operational formalisms, it has allowed for the reuse of powerful automated analysis techniques such as model checking (presented in Chap. 11).

10.3   Process Algebras with Metric Time

There are two fundamental approaches to introducing a **QUANTITATIVE** notion of time in untimed process algebras such as CSP. In the first approach, described in Sect. 10.3.1, time is discrete, whereas in the second one it can be dense, as described in Sect. 10.3.2.

10.3.1   Time Advancement as an Explicit Event

Section 10.1.3 pointed out the difficulties of introducing a simple discrete metric on time in CSP by assuming every event to take one time unit (or any fixed positive amount of time) due to the somehow special nature of silent events τ. We do not delve into further detail about this unsophisticated approach.

Let us discuss, instead, another simple and fairly conventional way to model the passing of time in CSP, based on the usage of special events that occur whenever time advances by one time unit. The approach is essentially the same as that used in timed transition modules and described in Sect. 7.3.1. In CSP, however, the event "time advances" is called tock instead of "tick", to avoid the potential confusion due to the fact that the latter is the way the special termination event "√" is usually addressed in English.

With an **EXPLICIT** event tock pacing time advancement, time ranges over a **DISCRETE** domain isomorphic to the natural numbers. The approach has the obvious advantages of being straightforward to present and not requiring extensions to the usual CSP semantics, since the special meaning of tock is only a convention: whenever a process model includes constraints on time advancements, it deploys event tock consistently with the convention. This also has the advantage that the analysis methods and techniques for standard CSP are applicable, without modifications, to processes with tock events as well.

There are, however, a number of semantic subtleties that arise with this approach to modeling time. Consider the very simple example of a unidirectional communication channel, which reads input events and writes them on the output after some time ("→" is right-associative):

$$\text{CHANNEL} = \text{read} \to \text{write} \to \text{CHANNEL} .$$

Suppose that the timed behavior is such that at least one time unit and at most two elapse between reading and writing the message. Using the tock event, we model this behavior as

$$\text{TCHANNEL} = (\text{read} \to \text{tock} \to \text{write} \to \text{TCHANNEL})$$

$$\square \ (\text{read} \to \text{tock} \to \text{tock} \to \text{write} \to \text{TCHANNEL}) .$$

What happens when **TCHANNEL** operates concurrently with other processes (in particular, clients using the channel)? Obviously, **TCHANNEL** must synchronize on the tock event with every other process in the system; otherwise time would not be *global* and the very notion of timed synchronization would become immaterial. If **TCHANNEL** synchronizes *only* on event tock, the other processes can proceed asynchronously except when tock is taken. This constraint allows for processes that never offer the tock event and hence perform an unbounded number of other events without advancing time; this may generate a problem of **ZENO BEHAVIOR**. If, instead, **TCHANNEL** synchronizes on tock as well as on other events, a different problem – still related with **TIME ADVANCEMENT** – may also occur: process **TCHANNEL** insists that two time steps elapse consecutively; hence it may block other processes that want instead to perform some activity in that time lapse. Both types of problem are typically avoided by **A POSTERIORI** analysis of the complete system model, to ensure that no interacting process imposes unacceptable constraints on the progress of time. Similar problems may occur whenever processes synchronize on shared events, but if the shared events do not represent explicit time advancement (as tock does), generic progress – rather than time advancement – is affected.

Example 10.26. With a lengthier specification, we can modify TCHANNEL so that it allows at most three events e between two consecutive occurrences of tock:

$$TCHANNEL = read \rightarrow tock \rightarrow THREE \ \Box \ read \rightarrow tock \rightarrow DONE$$
$$THREE = e \rightarrow TWO \ \Box \ STEP$$
$$TWO = e \rightarrow ONE \ \Box \ STEP$$
$$ONE = e \rightarrow STEP \ \Box \ STEP$$
$$STEP = tock \rightarrow DONE$$
$$DONE = write \rightarrow TCHANNEL . \qquad \blacksquare$$

Other subtleties related with the semantics of explicit progress events occur in the presence of *hiding* in process definitions: hiding replaces visible events with silent ones; hence it may violate quantitative constraints – think, for example, of properties such as "at least two events occur between any two consecutive instants". In fact, Sect. 10.2 already emphasized that the trace and stable failure semantic models are intrinsically qualitative.

10.3.2   Timed Process Algebras

When the simple model of time based on tock transitions is too coarse-grained, process algebras are extended with a finer notion of QUANTITATIVE time based on the separation of delays and instantaneous events. This notion describes the evolution of processes as an alternation of "evolution" phases and "event" phases. During an *evolution* phase, time elapses while process interaction blocks; from a dynamical system point of view, the time variable changes value, while every other global state component holds. During an *event* phase, a sequence of instantaneous events takes place while time is frozen; therefore, system state changes but time does not. For example, the beating of a clock every second is the process

$$beat \quad delay \text{ of } 1\,s \quad beat \quad delay \text{ of } 1\,s \quad beat \quad \dots .$$

This notion of alternated evolution produces a model of time that is "abstract" with respect to real physical time, where processes evolve concurrently as time flows. It is consistent with the notion of time as a global uniform flow shared by every process and synchronized with them, but whose progress is modeled explicitly and separately from the processes' – unlike what happens in classical physics and engineering (as discussed in Chap. 4), where system evolution is formalized as a function of the independent time variable t.

Another feature of the process algebraic model of time is the central notion of *event* as an action consuming no time. As we have seen in several other chapters – and in particular in the synchronous models of Chap. 5 – events are useful abstractions for actions whose duration is negligible with respect to global system dynamics. At the same time, if every non-delay action is zero-time, episodes of significant duration must be modeled by events explicitly marking their beginning and end (using an approach similar to that of timed Petri nets, shown, among other places, in Example 8.26). For example, a meal that takes thirty minutes is the process

$$\text{begin\_meal} \qquad \text{delay of 30 min.} \qquad \text{end\_meal}\,.$$

The process algebraic notion of metric time based on evolution and event phases lends itself, once again, to formalization by means of transition systems equipped with two distinct types of transition: time delays (for evolution phases) and events (for event phases). In fact, the formal semantics of timed automata (described in Definition 7.34) relies on similar transition systems, and hence adopts a similar model of abstract real-time. Time is typically **DENSE** (possibly **CONTINUOUS**) in this view to allow for the description of fully asynchronous processes, although the same time model can also support a discrete notion of time.

Timed process algebras that follow this general model of time provide specific language constructs to describe timed processes by means of **EXPLICIT** reference to the global time during evolution. The syntactic constructs vary, in general, from algebra to algebra; the rest of the current section describes an extension of CSP that supports real-time modeling along these lines.

10.3.2.1   Timed Communicating Sequential Processes

There have been a number of proposals to extend CSP with timed constructs. The extension we consider in this section only defines two new fundamental operators: the timed interrupt and the timed prefix. Every constraint on the timed behavior is expressed through them.

The "timed interrupt" "$\overset{d}{\triangleright}$" (also called "time-out") is a parametric binary operator that introduces upper bounds on the processing time of its first argument. Namely, the process

$$P \overset{d}{\triangleright} Q$$

initially behaves as P for up to d time units (included). If P accepts any event from the environment during this time, Q is not considered at all. If, instead, P does not engage in any interaction with the environment for d time units, P is stopped and the process immediately continues behaving as Q. More precisely, after exactly d time units have elapsed, if an event offered by P occurs, the process must continue as P; otherwise, it instantaneously turns into Q with a silent event. This means that, if no event is accepted from the environment by process P before the time-out d, the interrupt process may associate two "states" (P and Q) with the same relative time instant d: the choice between process P and Q is determined by the presence, at time d, of an event from the environment accepted by P. d is any arithmetic expression that evaluates to a nonnegative real number.

While the timed interrupt essentially introduces upper bounds on the duration of processes, we can also delay the beginning of processes by combining the timed interrupt with the special process STOP, which never interacts with the environment. For example, a process DELAY(P, d) that waits for d time units and then behaves as P is defined as

$$\text{DELAY}(P, d) \;=\; \text{STOP} \overset{d}{\rhd} P \,;$$

DELAY(P, d) defines a delayed start for P, which is enabled only in the interval $[d, \infty)$.

Exercise 10.27. Define a process BETWEEN(e, l, u) that accepts event e only in the interval $[l, u]$, and then terminates. ∎

The other fundamental timed operator introduced in timed CSP is the "timed prefix" "$@d \;\rightarrow$", which annotates the standard prefix operator "$\rightarrow$" with a reference d to time. The timed prefix does not directly constrain the time of process occurrences; rather, it introduces an explicit reference to time, which process definitions can refer to and use within other timed operators. Indeed, the timed process $e @ d \;\rightarrow\; P$ behaves exactly as its untimed counterpart $e \;\rightarrow\; P$. The only difference is that variable d becomes bound, in P, to the time at which event e takes place, relative to the time it is initially offered. For example, a digital display that shows the time of the latest thunderstorm is incremented whenever a new storm occurs:

$$\text{LAST\_STORM}(t) \;=\; \text{storm} @ d \rightarrow \text{LAST\_STORM}(t + d)\,.$$

Process LAST_STORM also demonstrates that time variables may be used as **EXPLICIT TIME** parameters of processes.

Exercise 10.28. Combine timed interrupt and timed prefix to define a process DOUBLE_WAIT(e, P) that initially offers event e. Let x be the time elapsed when e occurs since the beginning; then DOUBLE_WAIT blocks for another x time units and then behaves as P. ∎

Example 10.29. Let us introduce timed behavior in the allocator of Example 10.8, a behavior similar to that of the timed automaton model of Fig. 7.31: when used, the resource must be released within 100 time units. We first rewrite the definition of process PENDING in a way that makes explicit the external choice between events hpr and rel:

$$\text{PENDING}(k + 1) = \text{hpr} \rightarrow \text{PENDING}(k + 2) \mid \text{rel} \rightarrow \text{PENDING}(k)$$

$$\text{PENDING}(0) = \text{hpr} \rightarrow \text{PENDING}(1) \mid \text{rel} \rightarrow \text{RELEASE}\,. \qquad (10.7)$$

We cannot simply add a time-out to the definition of **PENDING** in (10.7) with the interrupt operator (for example: $\text{PENDING}(k+1) \overset{100}{\vartriangleright} \text{STOP}$), because each new high-priority request would reset the interrupt timer. Instead, we add a parameter *tout* to process **PENDING** that denotes the remaining time before the system stops unless a release occurs (see Exercise 10.30 for a variant of this behavior where the system does not block). High-priority requests increment the counter k but they also decrement *tout*; release events **rel** reset the timer to 100. The resulting process blocks if no event occurs before the time-out:

$\text{PENDING}(k+1, tout) =$

  $(\text{hpr@}d \rightarrow \text{PENDING}(k+2, tout-d) \mid \text{rel} \rightarrow \text{PENDING}(k, 100)) \overset{tout}{\vartriangleright} \text{STOP}$
$\text{PENDING}(0, tout) =$

  $(\text{hpr@}d \rightarrow \text{PENDING}(1, tout-d) \mid \text{rel} \rightarrow \text{RELEASE}) \overset{tout}{\vartriangleright} \text{STOP}.$

Finally, **PENDING** is initially $\text{PENDING}(0, 100)$ when process **GRANT** is spawned.   ∎

10.3.2.2 Transition System Semantics of Timed CSP

As anticipated earlier in this section, the notion of timed process as an alternation of event and evolution phases, described at the beginning of Sect. 10.3.2, naturally refers to a semantics based on transition systems. This extends the semantics of untimed CSP (Sect. 10.2.2) as follows. First, nodes include timestamps that store the value of global time since system initialization. Event transitions, which correspond to event phases, do not change the values of timestamps. Evolution transitions, which correspond to evolution phases, only increase the values of timestamps.

Based on these assumptions, the transition system semantics inductively defines a transition system $TT(P)$ for every timed CSP process P. The usage of a dense set for the timestamps entails that timed CSP processes define, in general, transition systems with uncountably many nodes and transitions. Let us sketch a few significant samples of the inductive definition.

A timed interrupt $P = (a \rightarrow A) \overset{d}{\vartriangleright} B$ introduces the following transitions.

Event occurs: from every state $\langle P, t \rangle$, an event transition $\overset{a}{\rightarrowtail}$ to state $\langle A, t \rangle$.
Time elapses: from every state $\langle P, t \rangle$, for every $0 < d' \leq d$, an evolution transition $\overset{d'}{\rightarrowtail}$ to state $\langle (a \rightarrow A) \overset{d-d'}{\vartriangleright} B, t + d' \rangle$.
Interrupt occurs: from every state $\langle (a \rightarrow A) \overset{0}{\vartriangleright} B, t \rangle$, a (silent) event transition $\overset{\tau}{\rightarrowtail}$ to state $\langle B, t \rangle$.

An important characteristic of event transitions (including silent events) is that they always have precedence over delay transitions; that is, if a state $\langle P, t \rangle$ has an event transition $\overset{e}{\rightarrowtail}$ to state $\langle Q, t \rangle$ and an evolution transition $\overset{d}{\rightarrowtail}$ to state $\langle P, t + d \rangle$, whenever $\langle P, t \rangle$ is the current state and an event e occurs, the event transition must be chosen.

External choice behaves as in the untimed case; hence it defines the same event transitions without delays, but time can elapse while events are offered. For example, $P = a \to A \mid b \to B$ introduces the transitions:

- From every state $\langle P, t \rangle$, an event transition $\overset{a}{\rightarrowtail}$ to state $\langle A, t \rangle$.
- From every state $\langle P, t \rangle$, an event transition $\overset{b}{\rightarrowtail}$ to state $\langle B, t \rangle$.
- From every state $\langle P, t \rangle$, for every $d > 0$, an evolution transition $\overset{d}{\rightarrowtail}$ to state $\langle P, t + d \rangle$.

Internal choices are in contrast made immediately, as soon as they become enabled. For example, $P = a \to A \sqcap b \to B$ introduces, from every state $\langle P, t \rangle$, two silent event transitions $\overset{\tau}{\rightarrowtail}$ to state $\langle a \to A, t \rangle$ and state $\langle b \to B, t \rangle$, but no evolution transitions.

Concurrent processes behave as in the untimed case with respect to event transitions. They must, however, synchronize on evolution transitions, because time is global and shared by all processes in the systems. Hence, an evolution transition

$$\langle P \,_A\|_B\, Q, t \rangle \overset{d}{\rightarrowtail} \langle P \,_A\|_B\, Q, t + d \rangle$$

of two parallel processes P and Q is defined if and only if evolution transitions $\langle P, t \rangle \overset{d}{\rightarrowtail} \langle P, t + d \rangle$ and $\langle Q, t \rangle \overset{d}{\rightarrowtail} \langle Q, t + d \rangle$ are both defined. The rule for evolution transitions with interleaving composition is the same, because even independent processes must synchronize on time.

The transition system semantics of timed interrupt suggests that the timed behavior of CSP is such that process definitions cannot directly constrain the environment to produce some events; they can only interrupt processes if the environment refuses to interact. Once events occur, however, the processes cannot ignore them. The only partial exceptions are transitions for internal silent events τ, whose execution does not depend on the environment: a silent transition is always possible whenever it is enabled.

Exercise 10.30. Example 10.29 features a semantics where process **PENDING** does not forcefully decrement the counter, but simply inhibits progress if the environment does not produce a release event **rel** within the time-out. Modify the example to have process **PENDING** releas the resource with silent transitions – independently of the environment – when a grant lasts longer than 100 time units. ∎

Another aspect of the timed behavior that becomes apparent in the transition system semantics is that the possible delays of timed CSP processes define intervals that are right-closed – with the only exception of unbounded delays whose intervals are right-open to infinity. This is a consequence of the definition of the timed interrupt, on which every timed behavior is ultimately based, which still allows events to happen at the very instant the time-out expires.

10.3.2.3   Time Advancement of Timed CSP

Given the formal semantics of timed CSP, it is interesting to classify the possible behavior of processes with respect to TIME ADVANCEMENT and, more generally, progress. Whether the behaviors we consider are acceptable or not depends, in general, on the application domain of the modeled systems. For example, executions that do not terminate are unacceptable for traditional application software, but they are appropriate in reactive systems. The semantics should then be able to represent appropriately all the acceptable behaviors. For example, to model infinite nonterminating executions under the trace semantics, we will use infinite traces. Let us now examine a few cases of lack of progress independently of timing, followed by examples where time cannot advance as well.

Finite execution: a process interacts with the environment with finitely many events, and then terminates. What happens to time after termination is irrelevant for the environment; we may assume that time stops or diverges, but no more visible events take place. For example,

$$\mathsf{FINITE} = a \rightarrow b \rightarrow c \rightarrow \mathsf{SKIP}$$

processes at most three events before terminating.

Infinite execution: a process interacts with the environment with infinitely many events over an unbounded amount of time. For example, process

$$\mathsf{INFINITE} = p \rightarrow \mathsf{INFINITE}$$

allows infinite executions with infinitely many p events.

Zero-time execution: an infinite sequence of environment events occurs, which prevents time advancement. For example, process INFINITE above may engage a zero-time sequence of p events without ever advancing time.

Blocked execution: time does not advance, nor does any other transition take place. In timed CSP, the special process TIMESTOP is postulated to have precisely this behavior. From the point of view of interaction with the environment, the only difference between TIMESTOP and STOP is that only the latter allows time to pass.

Divergent execution: time advances unboundedly; interaction with the environment may or may not progress. For example, process

$$\mathsf{FOREVER} = \mathsf{STOP} \overset{1}{\triangleright} \mathsf{FOREVER}$$

guarantees time advancement but does not otherwise interact with the environment.

Zeno execution: time advances by infinitesimal amounts and converges to a finite value. Interaction with the environment may or may not occur. For example, process

$$\text{ZENO} = \text{ZENO}(1)$$

$$\text{ZENO}(k) = e \to (\text{STOP} \overset{k}{\triangleright} \text{ZENO}(k/2))$$

includes the behavior where event e occurs at times $0, 1, 1 + 1/2, 1 + 1/2 + 1/4,$ $\ldots$; hence infinitely many events occur with time never going past instant 2.

Process algebras usually deal with problems of Zenoness and progress by A POSTERIORI analysis, possibly by automated means. Alternatively, in some cases it is possible to introduce syntactic restrictions that guarantee absence of non-progressing behavior A PRIORI by construction. For example, a simple approach constrains every transition to take a minimum finite amount of time δ; this ensures that time diverges as long as the environment indefinitely prolongs its interaction with the system.

Exercise 10.31. Devise sufficient syntactic conditions that guarantee a priori the absence of unstable executions (infinite sequences of internal events without time advancement) in processes.   ∎

10.3.2.4   Timed Trace Semantics of Timed CSP

The presentation of timed CSP focused on the transition system semantics, which is apt to describe the model of time underlying this formalism. Indeed, most formal presentations have the same focus, and tend to overlook algebraic laws – which are studied in detail for untimed CSP – in the timed version.

Similarly, the trace and failure semantics of timed CSP are based on the transition system semantics, from which they follow along the same lines as the semantics of timed automata (Definition 7.34). The rest of this section briefly outlines these timed semantics with a few simple examples.

Section 10.3.2.2 mentioned how every timed CSP process definition P induces a timed transition system $TT(P)$. In turn, $TT(P)$ defines a set $runs(TT(P))$ of runs or executions. Executions are finite or infinite sequences of transitions of the form

$$N_1 \overset{x_1}{\rightarrowtail} N_2 \overset{x_2}{\rightarrowtail} N_3 \overset{x_3}{\rightarrowtail} \ldots,$$

where each N_k is a state of $TT(P)$ and each x_k is a visible event, a silent event τ, or a positive real number denoting a delay. Each execution determines a timed trace (defined in Sect. 7.4.1), that is, a sequence of pairs $\langle ev, t \rangle$ of events and absolute times at which they take place; the latter are called *timestamps* and are always monotonically nondecreasing in accordance with the assumption that time is globally shared. Consistently with the trace semantics of untimed CSP, timed traces of timed CSP processes only record *visible* events. For example, the execution

$$P \overset{e}{\rightarrowtail} Q \overset{e}{\rightarrowtail} P \overset{2}{\rightarrowtail} Q \overset{\tau}{\rightarrowtail} Q \overset{1.3}{\rightarrowtail} R \overset{b}{\rightarrowtail} P$$

determines the timed trace

$$\langle e, 0 \rangle \ \langle e, 0 \rangle \ \langle b, 3.3 \rangle$$

which ignores the silent transition occurring at absolute time 2. Infinite traces are more significant in timed CSP than in their untimed counterparts, because it is natural to think about time as a divergent unbounded quantity; therefore, the timed trace semantics considers timed traces of both **FINITE** and **INFINITE LENGTH**. *timedTraces*(P) denotes the set of all such timed traces – induced by *runs*($TT(P)$) – for process P. The trace semantics features a notion of **LINEAR** time, but the underlying timed CSP processes may be **DETERMINISTIC** or **NONDETERMINISTIC**.

Example 10.32. Consider the processes defined in Sect. 10.3.2.3.

- *timedTraces*(FINITE) includes $\langle a, 0 \rangle \ \langle b, 1 \rangle \ \langle c, 2 \rangle \ \langle \checkmark, 3 \rangle$;
- *timedTraces*(INFINITE) includes $\langle p, 0 \rangle \ \langle p, 1 \rangle \ \langle p, 2 \rangle \ \langle p, 3 \rangle \ \langle p, 4 \rangle \ \dots$;
- *timedTraces*(TIMESTOP) and *timedTraces*(FOREVER) are empty;
- *timedTraces*(ZENO) includes $\langle e, 0 \rangle \ \langle e, 1 \rangle \ \langle e, 1.5 \rangle \ \langle e, 1.75 \rangle \ \langle e, 1.875 \rangle \ \dots$.   ∎

Exercise 10.33. Consider process $P = (e \rightarrow \mathsf{STOP}) \overset{d}{\triangleright} \mathsf{TIMESTOP}$, for some $d > 0$. Define the complete set *timedTraces*(P).   ∎

The notions of *refusals* and *failures*, which introduce some information about **LIVENESS** in untimed CSP, can be extended to timed processes. "Timed refusals" are pairs $\langle re, I \rangle$, where re is a set of refused visible events (exactly as in the untimed case), and I is a nonempty interval of the form $[t_1, t_2)$. A refusal $\langle re, I \rangle$ denotes that all the events in re are refused throughout interval I of absolute time. Refusal intervals are left-closed and right-open; this is a consequence of the fact that delays do not impose any refusal at their right endpoint, where a new event may occur. For example, process $(a \rightarrow \mathsf{STOP}) \overset{3}{\triangleright} (b \rightarrow \mathsf{STOP})$ refuses event b over the interval $[0, 3)$; at time 3, both b and a may occur; if a does not occur at 3, b is enabled by a silent transition and remains so indefinitely.

"Timed failures" combine timed traces and timed refusals as in the untimed case: a timed failure $\langle tr, \langle rf, I \rangle \rangle$ corresponds to an execution with timed trace tr and timed refusal $\langle rf, I \rangle$. The bibliographic remarks point to several references that contain detailed and rigorous treatments of the timed failure semantics of CSP.

10.4   Probabilistic Process Algebras

This section shows how to extend CSP with probabilities: Sect. 10.4.1 associates probabilities with internal choice; Sect. 10.4.2 discusses how to accommodate external choice; Sect. 10.4.3 shows the difficulties of including pure nondeterminism in probabilistic CSP.

10.4.1   Probabilistic CSP

The basic idea behind introducing stochastic behavior in CSP is replacing nondeterministic internal choice with *probabilistic choice*. To this end, we annotate the internal choice operator with real values $0 \leq p \leq 1$ that denote probabilities of the choices; namely,

$$P \;_p\sqcap\; Q$$

is a process that behaves as P with probability p, and as Q with probability $1 - p$. For example, the flipping of an unbiased coin is the process

$$\mathsf{COIN} = (\mathsf{head} \to \mathsf{COIN}) \;_{0.5}\sqcap\; (\mathsf{tail} \to \mathsf{COIN}) . \tag{10.8}$$

We can adjust the algebraic laws of basic CSP to accommodate the semantics of probabilistic internal choice. Some laws directly generalize the basic case; for example, the commutativity of nondeterministic internal choice ($P \sqcap Q = Q \sqcap P$) becomes

$$P \;_p\sqcap\; Q = Q \;_{1-p}\sqcap\; P$$

in the probabilistic case where Q has probability $1 - p$ of being chosen. Other laws capture specific properties of probabilistic behavior, such as the fact that processes with null probability are never chosen:

$$P \;_0\sqcap\; Q = Q .$$

We can also define a *trace semantics* for probabilistic internal choice. Let us outline its essential traits by examples, without delving into the technical details required to reason rigorously about probability. Unlike with untimed CSP, it is customary to consider a semantics based on **INFINITE** traces, where terminating processes determine traces with a finite sequence of events followed by an infinite tail of padding characters "ε" denoting "absence of events". Whereas the trace semantics of standard CSP focuses on finite traces, infinite traces are preferable for probabilistic CSP because "steady-state" behavior is especially significant for probabilistic processes (see Sect. 6.4 for some examples), and it is only defined for executions of unbounded length.

The trace semantics of probabilistic CSP defines a probability $prob^P$ over infinite traces for every process definition P using probabilistic internal choice. $prob^P$ is a function from *sets* of infinite traces to a probability value in the range $[0, 1]$. A few examples illustrate $prob^P$ for simple process definitions.

Take the trivial example of process **STOP**, which does not accept any visible event. Its only infinite trace is ε^ω, and therefore **STOP** induces a probability that assigns 1 to the single trace ε^ω, and 0 to any other set of traces:

$$prob^{\mathsf{STOP}}(T) = \begin{cases} 1 & \{\varepsilon^\omega\} = T \\ 0 & \text{otherwise} . \end{cases}$$

The laws of probability imply that the sequences in T must share a finite prefix (possibly empty); furthermore, the padding event ε can only appear in tails of infinite length; hence T contains ε^ω if and only if it contains *only* ε^ω.

Consider now two processes P and Q, and suppose we know that their trace semantics induce probabilities $prob^P$ and $prob^Q$. Process $\mathsf{IP} = P \;_p\sqcap\; Q$ proceeds with a probabilistic choice, after which it behaves as P – with probability p – or as Q – with probability $1-p$. Therefore, the trace semantics $prob^{\mathsf{IP}}(T)$ of probabilistic internal choice weights the probabilities given by the possible processes P and Q by the choice probabilities:

$$prob^{\mathsf{IP}}(T) \;=\; p \cdot prob^P(T) + (1 - p) \cdot prob^Q(T)\,.$$

Exercise 10.34 (♦). Determine the probability of process $a \to P$ that offers only event a recursively from the probability $prob^P$ of process P.

(*Hint*: consider the two cases of sets of traces that all begin with a and that all begin with an event other than a). ∎

10.4.2   External Choice and Concurrency in Probabilistic CSP

The simple semantics of probabilistic internal choice – introduced in Sect. 10.4.1 – cannot be applied directly to accommodate the semantics of *external* choice. This is because the notion of probabilistic internal choice entails an extremely simplified model of interaction between processes and the environment: processes make internal choices with assigned probabilities; every choice results in a *unique* event offered to the environment, which can either agree on it – so that the process can proceed – or disagree – so that the process blocks. If we consider again the process of coin flipping in (10.8), we see that it is the process that "flips" the coin, whereas the environment only provides a sequence of events that make the outcome of the flipping visible.

This cannot accommodate the case where the process offers different events to the environment, which can draw one from among them or block, because there are no probabilities associated with external choice. To illustrate with an example, consider two biased coins CT and CH: CT lands on tails with 99 % probability, whereas CH lands on heads with 99 % probability:

$$\mathsf{CT} = (\mathsf{tail} \to \mathsf{STOP})\;_{0.99}\sqcap\;(\mathsf{head} \to \mathsf{STOP})$$

$$\mathsf{CH} = (\mathsf{head} \to \mathsf{STOP})\;_{0.99}\sqcap\;(\mathsf{tail} \to \mathsf{STOP})\,.$$

Process BIASED lets the environment choose one of the two coins, and then flips it:

$$\mathsf{BIASED} \;=\; \mathsf{t\_bias} \to \mathsf{CT} \mid \mathsf{h\_bias} \to \mathsf{CH}\,.$$

What is the probability that process **BIASED** induces on the trace $\text{tail}\,\varepsilon^{\omega}$? It depends on the external choice: if the environment produces t_bias, then the probability is 99 %; if the environment produces h_bias, it is 1 %; if the environment produces any other event, the probability is 0. This suggests that probabilistic processes with external choice determine *probabilities* over traces that depend on the sequences of events provided by the environment. This notion of probability is called "conditional probability", because it expresses probabilities *depending on* an assumption of environment behavior. The environment then corresponds to the notion of *scheduler* of Markov decision processes (discussed in Sect. 6.4.1.2), which also provides sequences of input events that affect the probabilistic behavior of the process reading them.

Another problem with the semantics of probabilistic CSP from Sect. 10.4.1 occurs when trying to define the probability of parallel composition. More precisely, there is no problem when concurrent processes synchronize on shared events: since parallel events are *independent*, the probability of the composition is the *product* of the individual probabilities. For example, the probability that two synchronized unbiased coins (i.e., process $\text{C2} = \text{COIN}\,{}_{\{head,tail\}}\|{}_{\{head,tail\}}\,\text{COIN}$) both flip heads at the first round is

$$prob^{\text{C2}} = prob^{\text{COIN}}(\text{head} \cdot \{\text{head, tail}\}^{\omega}) \cdot prob^{\text{COIN}}(\text{head} \cdot \{\text{head, tail}\}^{\omega})$$

$$= 0.5 \cdot 0.5 = 0.25\,.$$

If, however, concurrent processes can proceed asynchronously on some events, parallel composition behaves as external choice at some steps; hence it determines conditional probabilities. For example, process **AB**,

$$AB = a \rightarrow \text{STOP}\,{}_{\{a,c\}}\|{}_{\{b,c\}}\,b \rightarrow \text{STOP}\,,$$

becomes

$$\text{STOP}\,{}_{\{a,c\}}\|{}_{\{b,c\}}\,b \rightarrow \text{STOP}$$

if the environment produces a, and becomes

$$a \rightarrow \text{STOP}\,{}_{\{a,c\}}\|{}_{\{b,c\}}\,\text{STOP}$$

if the environment produces b.

To model these behaviors appropriately, the trace semantics of probabilistic CSP with external choice defines a *conditional* probability $cprob^{P}$ over infinite traces, for every process definition P using probabilistic internal choice, depending on a trace of visible environment events. $cprob^{P}$ is a function of the two arguments T and ρ to probability values in the range $[0, 1]$: T is a set of infinite traces, and ρ a single infinite trace produced by the environment; hence $cprob^{P}(T \mid \rho)$ denotes the probability that traces in T have to occur when the environment produces the sequence of events ρ.

For example, the semantics of process **STOP** does not change with conditional probabilities, because the process's behavior is independent of the environment choice:

$$cprob^{\mathsf{STOP}}(T \mid \rho) = \begin{cases} 1 & \{\varepsilon^{\omega}\} = T \\ 0 & \text{otherwise} . \end{cases}$$

The behavior of probabilistic internal choice is also independent of the environment choice; more precisely, $\mathsf{IP} = P \;_p\!\sqcap\; Q$ continues as P or Q without reading any external events. Therefore,

$$cprob^{\mathsf{IP}}(T \mid \rho) = p \cdot cprob^{P}(T \mid \rho) + (1 - \rho) \cdot cprob^{Q}(T \mid q) .$$

The probability of external choice does depend, however, on the next event produced by the environment. To illustrate, consider process $\mathsf{SEL} = a \to A \mid b \to B$. If the environment produces a next, SEL inputs it and continues as process A. Thus, for a set T_a of traces that all start with a, $cprob^{\mathsf{SEL}}(T_a \mid a \cdot t)$ is the same as the probability of process A on the set T_a' obtained by removing the leading a – read by SEL – from all traces in T_a, conditionally on the tail t of the trace produced by the environment:

$$cprob^{\mathsf{SEL}}(T_a \mid a \cdot t) = cprob^{A}(T_a' \mid t) .$$

Symmetrically, the environment may produce b first; hence

$$cprob^{\mathsf{SEL}}(T_b \mid b \cdot t) = cprob^{B}(T_b' \mid t) .$$

If, instead, the environment produces some event c other than a or b, SEL blocks and behaves as STOP without reading c:

$$cprob^{\mathsf{SEL}}(T_c \mid c \cdot t) = cprob^{\mathsf{STOP}}(T_c \mid c \cdot t) = 0 .$$

Exercise 10.35. How would you define the probability $cprob^{\mathsf{SEL}}(T_a \mid b \cdot t)$ for a set T_a of traces that all begin with a, with respect to an environment that produces event b instead? ∎

Exercise 10.36 (♦). Sketch the semantics of parallel composition for process ABC recursively from the conditional probabilities of processes A, B, and C:

$$ABC = (a \to A \mid c \to C) \;_{\{a,c\}}\|_{\{b,c\}}\; (b \to B \mid c \to C) .$$ ∎

10.4.3   Nondeterminism and Hiding in Probabilistic CSP

Probabilistic choice replaces pure nondeterminism in probabilistic CSP. A consequence of this fact is that probabilistic CSP can offer external choice only for processes with distinct initial events. Consider, for example, process X:

$$X = a \to P \mid a \to Q .$$

If we interpret X as a nondeterministic CSP process, its behavior is expressible with an internal nondeterministic choice between behaving as $a \to P$ and behaving as $a \to Q$. In contrast, X's behavior as a probabilistic CSP process is undefined, because the environment cannot choose between the two options (behaving as $a \to P$ and behaving as $a \to Q$) as they correspond to the same initial event, and the system can only make *random* choices based on fixed probabilities, but no probability is defined in this case. Therefore, processes such as X are not well formed in probabilistic CSP.

A different manifestation of the same problem occurs with *hiding*. The intuition is that hiding turns visible events into silent events that represent nondeterministic choices, which are inexpressible in probabilistic CSP. To illustrate, take process H, which hides event a from process V:

$$B = b \to B$$
$$V = a \to B \mid b \to \text{STOP}$$
$$H = V \setminus \{a\}.$$

What is the probability $cprob^H(\{b^\omega\} \mid b^\omega)$ associated with the single trace b^ω if the same trace is provided as input? Unfortunately, the value is undefined because hiding is not injective; hence there is no way of reconstructing the trace that became b^ω after hiding event a. If b^ω comes from hiding a in $a\,b^\omega$, then b^ω is a valid trace of the system; hence the conditional probability is 1. If, instead, b^ω comes from hiding a in b^ω, b^ω is not a valid trace of the system, because if b is the first event, H blocks afterward; hence the conditional probability is 0. For these reasons, hiding is not allowed in probabilistic CSP. Renaming is, however, well defined – as long as it is injective – because it simply changes the name of visible events without making them invisible.

Exercise 10.37. Discuss why hiding is well defined in the absence of external choice (with the semantics introduced in Sect. 10.4.1). In particular, define the semantics of process $P = (q \to Q) \setminus \{q\}$ recursively from the probability $prob^Q$ of process Q.   ∎

10.5   Methods and Tools Based on Process Algebras

Many expressive languages and methods based on process algebras have been developed over the years; several of them are also supported by verification techniques and tools. This section gives a short overview of some representative examples.

LOTOS ("Language Of Temporal Ordering Specification") is an expressive notation that combines a process algebraic kernel with features, such as data types and genericity, for writing formal specifications in-the-large. Its semantics combines elements from both CCS and CSP. LOTOS is an ISO standard and supports different methods and styles of specification writing.

Occam is a full-fledged concurrent programming language that incorporates most of CSP's features to describe synchronization among concurrent processes. It was developed in the 1980s under the direct supervision of Tony Hoare.

The traditional CSP approach to verification is distinctively logic. Specifications are expressed in first-order predicate logic with arithmetic and interpreted functions on traces that directly express properties of generic traces of a process. For example, a specification of the **BUFFER** process described in Sect. 10.1.4 is, "in every trace, the number of **rel** events never exceeds the number of **hpr** events". This specification is expressed in CSP notation as

$$\textbf{BUFFER sat } \left| tr \right|_{\text{rel}} \leq \left| tr \right|_{\text{hpr}},$$

where P **sat** S states that any trace tr of process P satisfies specification S. The axiomatic semantics of CSP, together with general first-order theory deduction, allow us to prove such specifications directly on the process definitions. This approach can handle very expressive specifications, but it is neither very practical nor automated in general.

More recent approaches to process algebra verification take advantage of the great advances in model checking and automata-based verification (discussed in Chap. 11) and exploit the operational nature of process algebras as transition systems.

LTSA ("Labelled Transition System Analyser") uses an explicit finite-state machine notation to define both systems and specifications. A process algebra notation is also available as syntactic sugar for defining specifications concisely, but the verification algorithms implemented in LTSA rely on the underlying transition systems.

FDR ("Failure-Divergence Refinement") is a model checker for state machines defined with CSP notation; its latest version, FDR2, supports operators that extend CSP and allow for complex multi-process synchronization rules. As its name indicates, FDR also includes functionality to check the refinement between processes, according to the failure semantics discussed in Sect. 10.2.3. FDR can also be adapted to verify timed CSP through discretization techniques such as digitization (described in Sect. 7.4.3). Other approaches to the verification of timed CSP exploit equivalences with certain classes of timed automata, and take advantage of the corresponding tools (mentioned in Sect. 11.4).

PAT ("Process Analysis Toolkit") is an extensible model checking framework which implements several automated analysis techniques and algorithms (reachability, fairness, divergence analysis, refinement checking, partial-order reduction, and so on). It includes modules for several input languages, including CSP and a flavor of probabilistic CSP, and also other notations such as timed automata and transition systems including notions of time and probability (along the same lines as those presented in Sect. 7.5.3).

CADP ("Construction and Analysis of Distributed Processes") is a collection of tools for the design and analysis of communication protocols and distributed systems. It features a LOTOS front end, which can create executable C programs from LOTOS specifications or convert them to transition system representations. These are input to the other tools in the CADP collection, which includes model-checkers, visualizers, and testing drivers.

10.6   Bibliographic Remarks

In his brief history of process algebras, Baeten [2, 3] credits Bekič [7, 8] with the introduction of process calculi to model parallel programs. The two most influential fathers of the process algebraic approach are, however, Milner and Hoare, who respectively developed the Calculus of Communicating Systems (CCS) and the Communicating Sequential Processes (CSP) in the late 1970s. In general, the process algebraic approach was initiated by the desire to reason about concurrent processes more abstractly than is possible with operational models such as transition systems, through algebraic manipulations of abstract terms representing execution traces. Nonetheless, the chapter has discussed how specific semantic models – such as those based on traces and refusals – may not adequately represent every property of interest, so that, in practice, a semantic approach based on transition systems is still preferred for achieving the greatest generality, and process algebras can be presented as convenient notations for formalizing transition systems.

Hoare's first official article [28] introduced CSP as a language for parallel programs, generalizing Dijkstra's guarded commands [19], without formal semantics. Inspired by the work on CCS, Hoare later introduced the trace and failure semantics [15] for only a subset of the original CSP, excluding actual programming language constructs and focusing on the process algebraic kernel. Within successive publications, this subset became known as CSP, and it is the algebra presented in this chapter. General and authoritative references on CSP include Hoare's book [29] – which focuses on the traditional CSP failure and algebraic law semantics – Roscoe [43] – whose extensive presentation of CSP includes the transition system semantics and the tock timed semantics discussed in Sect. 10.3.1 – and Schneider [45] – who covers basic as well as timed CSP; the presentation of timed CSP in this chapter mostly follows his approach.

Some presentation details of CSP may differ in different sources. For example, some articles and books (including Schneider [44]) refer to $\Box$ as "external choice", and present it as a variant of | with fewer syntactic restrictions. In the present chapter, instead, we preferred to always use | for external choice and to refer to $\Box$ as "general choice" (as in Hoare [29]); anyway, this is merely a presentation detail that does not affect the expressiveness of the CSP language. The chapter also ignored a technical subtlety that arises in the presence of a recursive process definition, such as $P = a \to P$: a recursive definition determines an unambiguous set of traces only if the recursion has a *unique fixed point*, but unconstrained CSP process definitions may have zero or multiple fixed points. The aforementioned textbooks on CSP include discussions of this issue, and introduce sufficient syntactic conditions to guarantee well-founded recursive definitions.

The chapter omitted two other technical considerations that are prominent in the specialized literature on process algebra, but that would have steered the discussion into specialized, complicated technical areas and away from the main issue of time modeling. The first consideration is whether the parallel composition operators are expressible in terms of purely sequential operators [1]; semantics where this is *not* the case are called "truly asynchronous", as parallel processes are primitive notions. The other notions not discussed in the chapter are precise equivalence relations for the transition system semantics, and most notably *bisimulation equivalence* and its variants; see, for example, Roscoe [43] and Milner [33] for discussions of the subject.

Ouaknine and Schneider [38] offer a short history of timed CSP. Ouaknine and Worrell established equivalences between timed CSP and timed automata [39, 40] that make it possible to reuse the verification tools of timed automata for timed CSP. A different approach to dense-timed CSP verification [37] exploits some of the discretization techniques discussed in Sect. 7.4.3. Nicollin and Sifakis [35] and Ulidowski and Yuen [48] present general approaches for extending process algebras with time.

Other popular process algebras include the already mentioned CCS [32, 33], ACP by Bergrstra and Klop [6, 10] (who are also credited [3] with introducing the expression "process algebra" [9]), LOTOS [12, 14, 21, 50], and the π calculus [34]. In addition to the already cited monographs that focus on specific process algebras, there are several books featuring the more general approaches of untimed process algebras [4, 11, 13, 24, 26] and describing process algebras with time [5].

The probabilistic CSP of Sect. 10.4 are after Seidel [46]; other work introducing probabilities in process algebra includes Clark et al. [17], Hansson and Jonsson [25], Hillston [27] (whose process algebra PEPA has a semantics based on continuous-time Markov chains), and D'Argenio and Katoen [18] (whose process algebra "spades" $\mathbb{Q}$ includes quantitative time and probabilities, and is the process algebraic counterpart to the stochastic automata presented in Sect. 7.5.3). Baeten [3] has many additional references to process algebras and their variants or extensions.

Abstract algebra is an important topic in advanced mathematics, covered by many textbooks such as Pinter [42] and Dummit and Foote [20]. Some details in the presentation of probabilistic CSP refer to basic notions of probability theory (such as probability measures and conditional probability) that can be acquired from many sources and textbooks [23, 47, 49].

The languages and tools discussed in Sect. 10.5 are often complemented by documentation on their official websites. This is the case with Occam [36], LTSA [30, 31], FDR [22], PAT [41], and CADP [16].

References

1. Baeten, J.C.M.: The total order assumption. In: Purushothaman, S., Zwarico, A.E. (eds.) Proceedings of the First North American Process Algebra Workshop (NAPAW'92), Workshops in Computing, pp. 231–240. Springer, London/New York (1993)

2. Baeten, J.C.M.: Over thirty years of process algebra: past, present and future. In: Aceto, L., Ésik, Z., Fokkink, W.J., Ingólfsdóttir, A. (eds.) Process Algebra: Open Problems and Future Directions. BRICS Notes Series, vol. NS–03–3, pp. 7–12 (2003)
3. Baeten, J.C.M.: A brief history of process algebra. Tech. Rep. CSR 04–02, Department of Mathematics and Computer Science, Technische Universiteit Eindhoven (2004)
4. Baeten, J.C.M., Basten, T., Reniers, M.A. (eds.): Process Algebra: Equational Theories of Communicating Processes. Cambridge University Press, Cambridge (2009)
5. Baeten, J.C.M., Middelburg, C.A.: Process Algebra with Timing. Springer, Berlin/New York (2010)
6. Baeten, J.C.M., Weijland, W.P.: Process Algebra. Cambridge Tracts in Theoretical Computer Science, vol. 18. Cambridge University Press, Cambridge/New York (1990)
7. Bekič, H.: Towards a mathematical theory of processes. Tech. Rep. 25.125, IBM Laboratory, Wien (1971). Published in [8]
8. Bekič, H.: Programming languages and their definition. In: Jones, C.B. (ed.) Selected Papers by Hans Bekič. Lecture Notes in Computer Science, vol. 177. Springer (1984)
9. Bergstra, J.A., Klop, J.W.: Fixed point semantics in process algebra. Tech. Rep. IW 208, Mathematical Center, Amsterdam (1982)
10. Bergstra, J.A., Klop, J.W.: Process algebra for synchronous communication. Inf. Control **60**(1–3), 109–137 (1984)
11. Bergstra, J.A., Ponse, A., Smolka, S.A. (eds.): Handbook of Process Algebra. Elsevier, Amsterdam/New York (2001)
12. Bolognesi, T., Lucidi, F.: Timed process algebras with urgent interactions and a unique powerful binary operator. In: de Bakker, J.W., Huizing, C., de Roever, W.P., Rozenberg, G. (eds.) Real-Time: Theory in Practice, REX Workshop. Lecture Notes in Computer Science, vol. 600, pp. 124–148. Springer, Berlin/New York (1991)
13. Bowman, H., Gomez, R. (eds.): Concurrency Theory: Calculi and Automata for Modelling Untimed and Timed Concurrent Systems. Springer, London (2010)
14. Brinksma, E. (ed.): Information Processing Systems – Open Systems Interconnection – LOTOS – A Formal Description Technique Based on the Temporal Ordering of Observational Behaviour. ISO (1989). ISO 8807:1989
15. Brookes, S.D., Hoare, C.A.R., Roscoe, A.W.: A theory of communicating sequential processes. J. ACM **31**(3), 560–599 (1984)
16. Cadp: Construction and analysis of distributed processes. http://www.inrialpes.fr/vasy/cadp/
17. Clark, A., Gilmore, S., Hillston, J., Tribastone, M.: Stochastic process algebras. In: Bernardo, M., Hillston, J. (eds.) Formal Methods for Performance Evaluation. Lecture Notes in Computer Science, vol. 4486, pp. 132–179. Springer, Berlin/Heidelberg/New York (2007)
18. D'Argenio, P.R., Katoen, J.P.: A theory of stochastic systems. part II: process algebra. Inf. Comput. **203**(1), 39–74 (2005)
19. Dijkstra, E.W.: Guarded commands, nondeterminacy and formal derivation of programs. Commun. ACM **18**(8), 453–457 (1975)
20. Dummit, D.S., Foote, R.M.: Abstract Algebra, 3rd edn. Wiley, New York (2003)
21. van Eijk, P.H.J., Vissers, C.A., Diaz, M. (eds.): The Formal Description Technique LOTOS. Elsevier, Amsterdam (1989)
22. FDR2 by formal systems (Europe). http://www.fsel.com
23. Feller, W.: An Introduction to Probability Theory and Its Applications, 3rd edn. Wiley, New York (1968)
24. Fokkink, W.: Introduction to Process Algebra. Springer, Berlin/New York (2000)
25. Hansson, H., Jonsson, B.: A calculus for communicating systems with time and probabitilies. In: IEEE Real-Time Systems Symposium, Lake Buena Vista, pp. 278–287 (1990)
26. Hennessy, M.: Algebraic Theory of Processes. MIT, Cambridge (1988)
27. Hillston, J.: A Compositional Approach to Performance Modelling. Cambridge University Press, Cambridge/New York (1996). Available online at http://www.dcs.ed.ac.uk/pepa/book.pdf

28. Hoare, C.A.R.: Communicating sequential processes. Commun. ACM **21**(8), 666—677 (1978)
29. Hoare, C.A.R.: Communicating Sequential Processes. Prentice Hall, Englewood Cliffs (1985). Updated version available online at http://www.usingcsp.com/
30. LTSA: Labelled transition system analyser. http://www.doc.ic.ac.uk/ltsa/
31. Magee, J., Kramer, J.: Concurrency: State Models and Java Programs. Wiley, Chichester/ New York (1999)
32. Milner, R.: A Calculus of Communicating Systems. Lecture Notes in Computer Science, vol. 92. Springer, Berlin/New York (1980)
33. Milner, R.: Communication and Concurrency. Prentice Hall, New York (1989)
34. Milner, R.: Communicating and Mobile Systems: The π-Calculus. Cambridge University Press, Cambridge (1999)
35. Nicollin, X., Sifakis, J.: An overview and synthesis of timed process algebras. In: Larsen, K.G., Skou, A. (eds.) Proceedings of the 3rd International Workshop on Computer Aided Verification (CAV'91). Lecture Notes in Computer Science, vol. 575, pp. 376–398. Springer, Berlin (1991)
36. The Occam language. http://www.wotug.org/occam/
37. Ouaknine, J.: Digitisation and full abstraction for dense-time model checking. In: Katoen, J.P., Stevens, P. (eds.) Proceedings of the 8th International Conference on Tools and Algorithms for the Construction and Analysis of Systems (TACAS'02), Grenoble. Lecture Notes in Computer Science, vol. 2280, pp. 37–51. Springer (2002)
38. Ouaknine, J., Schneider, S.: Timed CSP: a retrospective. Electron. Notes Theor. Comput. Sci. **162**, 273–276 (2006)
39. Ouaknine, J., Worrell, J.: Timed CSP = closed timed safety automata. Electron. Notes Theor. Comput. Sci. **68**(2), 142–159 (2002)
40. Ouaknine, J., Worrell, J.: Timed CSP = closed timed epsilon-automata. Nord. J. Comput. **10**(2), 99–133 (2003)
41. PAT: process analysis toolkit. http://www.comp.nus.edu.sg/~pat/
42. Pinter, C.C.: A Book of Abstract Algebra, 2nd edn. Dover, Mineola (2010)
43. Roscoe, A.W.: The Theory and Practice of Concurrency. Prentice-Hall, New York (1997)
44. Schneider, S.: Concurrent and Real-time Systems: The CSP Approach. Wiley, New York/ Chichester (1999). Available online at http://www.cs.rhul.ac.uk/books/concurrency/
45. Schneider, S.: Concurrent and Real-Time Systems: The CSP Approach. Wiley, Chichester/ New York (2000)
46. Seidel, K.: Probabilistic communicating processes. Theor. Comput. Sci. **152**(2), 219–249 (1995)
47. Shiryaev, A.N.: Probability, 2nd edn. Springer, Heidelberg (1995)
48. Ulidowski, I., Yuen, S.: Extending process languages with time. In: Johnson, M. (ed.) Proceedings of the 6th International Conference on Algebraic Methodology and Software Technology (AMAST'97), Sydney. Lecture Notes in Computer Science, vol. 1349, pp. 524–538. Springer (1997)
49. Varadhan, S.R.S.: Probability Theory. Courant Institute of Mathematical Sciences, New York (2001)
50. Vissers, C.A., Scollo, G., van Sinderen, M., Brinksma, E.: Specification styles in distributed systems design and verification. Theor. Comput. Sci. **89**(1), 179–206 (1991)

Chapter 11
Dual-Language Approaches

At this point, we have presented many formalisms of various types that deal with timing aspects. We examined them singularly, although we devoted much attention to comparing them according to the dimensions of Chap. 3. We have also emphasized from the beginning that most of them are clearly attributed to the two major categories of operational and descriptive formalisms. Given the complementarity of the two categories, it is natural to exploit it to create frameworks in which operational and descriptive languages coexist for describing different aspects of the problems at hand. In the most typical setting, an operational language describes the system under design, whereas a descriptive (usually logic) language formalizes the desired system properties. Then, suitable methods – preferably supported by tools – are used to establish that the modeled system satisfies the stated properties. This approach is called the "dual-language approach".

Oftentimes, verification in dual-language approaches relies on some sort of transformation of models expressed in one of the two languages into the other. For example, if we can express the semantics of an operational formalism in terms of logic axioms, then verification techniques for the logic (e.g., deductive techniques such as those shown in Chaps. 2 and 9) can prove properties of models that mix operational and logic components. Conversely, if we can translate temporal logic formulae into equivalent (for a suitable notion of equivalence) automata, dual-language verification relies on state-space search techniques.

Historically, the first example of a dual-language approach is Floyd-Hoare's axiomatic semantics of sequential programs, which essentially consists of *axiomatizing* the behavior of an abstract machine (described by flowcharts in Floyd's approach, and by imperative program statements in Hoare's), that is, of formally defining the semantics of the abstract machine's behavior (for example, the effect of an assignment to a state variable) as transformations of *predicates* in first-order logic. The axioms and inference rules of axiomatic semantics can be used to derive the correctness of a given program with respect to a given specification as a *theorem*. As correctness is in general undecidable for Turing-complete languages, the Floyd-Hoare axiomatic approach cannot be fully automated.

C.A. Furia et al., *Modeling Time in Computing*, Monographs in Theoretical Computer Science. An EATCS Series, DOI 10.1007/978-3-642-32332-4_11,
© Springer-Verlag Berlin Heidelberg 2012

In the same vein as the axiomatic semantics of sequential programs, pioneering work by Pnueli and Ostroff applied a dual-language approach to reactive and timed systems. In particular, Ostroff combined the operational formalism of Timed Transition Models (treated in Chap. 7) with the "real-time logic" we will present in Sect. 11.2.

Model checking techniques are arguably the most popular recent example of the dual-language approach. Model checking – introduced by Clarke, Emerson, and Sifakis in the 1980s – combines system models formalized through finite transition systems and property specifications expressed using propositional temporal logics such as CTL or LTL. Model checking expressly targets combinations of notations with a decidable verification problem; hence fully automated "push-button" tools are a fundamental ingredient in every model checking framework. Decidability has, however, a price in terms of restricted expressiveness of the modeling languages. Model checking frameworks have been defined for many of the languages presented in Chaps. 7–10.

The rest of this chapter presents the essentials of dual-language approaches within the field of timing analysis. Section 11.1 first describes the principles on which model checking is based. Then, Sect. 11.2 outlines Ostroff's dual-language approach to the verification of timed systems. Finally, Sect. 11.3 presents an application of the axiomatic approach à la Floyd-Hoare to timed system analysis, which combines timed Petri nets as abstract machine and TRIO as logic language. The final Sect. 11.4 is devoted to the tools supporting dual-language verification techniques.

11.1 Model Checking Frameworks

In the last few years, model checking techniques have attracted considerable interest from both the academic and the industrial worlds as a means for rigorously verifying properties of system models. An important part of the appeal of such techniques is the possibility of applying them in fully automated tools: after building the system model and formalizing the properties to be verified, all the user has to do is "push a button", and the model checker will provide a positive or negative answer to the question of whether the modeled system satisfies the stated property. In addition, if the system does not satisfy the property, the model checker will return a counterexample, that is, an execution of the system violating the property. Due to the intrinsically high complexity of verification, model checking techniques suffer from the so-called "state explosion problem": as the size of the model to be analyzed grows, the time and memory required for verification increase exponentially. As a consequence, model checkers analyzing models of nontrivial size may fail to terminate in an acceptable amount of time or exhaust all the available memory before reaching a conclusion. Much of the research on model checking techniques has tried to mitigate state explosion, a direct consequence of the computational complexity of the verification problems, which are typically PSPACE-complete (see Chap. 6), and often even harder.

Every model checking framework has three main ingredients:

1. An operational formalism, used to describe the systems to be analyzed, which ultimately corresponds to a transition system.
2. A descriptive formalism, used to describe the properties to be verified, which is typically a flavor of temporal logic.
3. An algorithm that can check whether a given system model S (expressed in the operational formalism) satisfies a given property P (expressed in the temporal logic).

Correspondingly, we can roughly identify three main categories of model checking algorithms:

1. Those in which the property P is transformed into an equivalent transition system, which is then combined with the transition system S; the algorithm performs an exploration of the state space of this combination.
2. Those in which the transition system S is transformed into an equivalent logic formula, which is then combined with P; the algorithm analyzes this combined formula (typically, it checks its satisfiability).
3. Those in which both S and P are unmodified, and the algorithm performs an exploration of the state space of S guided by P's structure.

Throughout this book, the focus has been on modeling languages and their features, rather than on analysis algorithms. In keeping with this approach, this chapter does not delve into the details of the various checking algorithms, for which an extensive literature exists (see the bibliographic remarks). Rather, the presentation focuses on the relationships among the languages used to represent systems and properties, and it highlights the features of these languages that are exploited by the verification algorithms.

11.1.1 Automata-Theoretic Model Checking

Automata-theoretic model checking follows the approach of translating properties into equivalent transition systems. Given an automaton S representing the system to be analyzed, and a temporal logic formula P formalizing the property to be checked:

A1. Build an automaton A_P that accepts exactly all traces that satisfy formula P (that is, A_P accepts σ if and only if $\sigma \models P$).
A2. Build the *complement* automaton $\overline{A_P}$, which accepts exactly all traces that A_P does *not* accept.
A3. Build the *intersection* automaton $S \cap \overline{A_P}$, which accepts exactly all traces accepted by both S and $\overline{A_P}$,

Fig. 11.1 The Büchi
automaton S, model of a safe

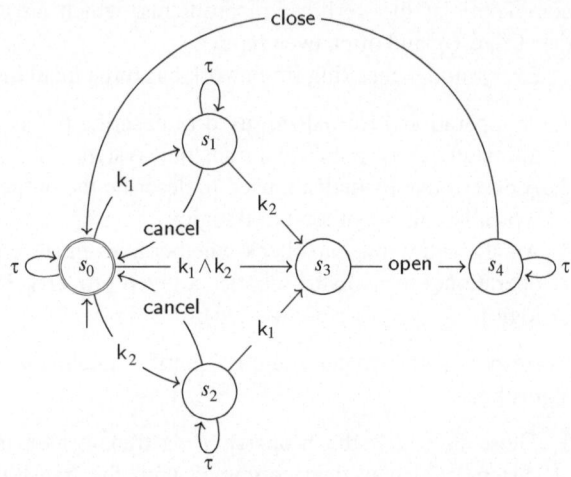

A4. Check whether $S \cap \overline{A_P}$ accepts *any* trace (that is, whether the set of traces accepted by both S and $\overline{A_P}$ is *empty*). If it does, return one of such traces and report unsuccessful verification; if it does not, report successful verification.

If the automaton $S \cap \overline{A_P}$ does not accept any trace, there is no input sequence compatible with system S that violates property P. In other words, all traces of system S also satisfy property P; hence P holds for S. If, on the other hand, the automaton $S \cap \overline{A_P}$ accepts some trace, this is a *counterexample*: a valid execution of S that violates P.

For finite-state models, all four steps above can be automated. The details of the various algorithms implementing the steps A1–A4 are described in the bibliographic references listed at the end of the chapter. The rest of the section provides some insight into the key features of automata-theoretic model checking through some examples.

Consider the Büchi automaton S of Fig. 11.1, which represents a safe that opens only after two keys, k_1 and k_2, have been inserted. The insertion order of the keys is irrelevant; in particular, label $k_1 \wedge k_2$ represents the simultaneous insertion of both keys. Then, a close command, given at any time after the safe opens, closes the safe again.

We would like to verify the following property for the model of Fig. 11.1: after key k_1 is inserted, the safe opens only if key k_2 is later inserted. We formalize such a property as the LTL formula

$$P_1 \quad \equiv \quad \Box(k_1 \Rightarrow \neg(\neg k_2 \ \mathsf{U} \ \text{open})) . \tag{11.1}$$

Exercise 11.1. (♣) By inspecting the automaton in Fig. 11.1 and the formula P_1, determine whether P_1 is a property of every run of the automaton. Compare your conclusion with the outcome of the model checking algorithm, discussed in the rest of the section. ■

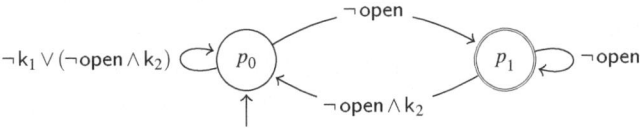

Fig. 11.2 Büchi automaton A_{P_1} equivalent to LTL formula P_1

The property formalized by formula P_1 is equivalently represented by the Büchi automaton A_{P_1} in Fig. 11.2. The labels on the transitions of the automaton have the following meanings: $\neg k_1$ stands for "any combination of symbols that does not include k_1"; $\neg$open $\wedge k_2$ stands for "any combination of symbols that does not include open but includes k_2".[1] The automaton of Fig. 11.2 is nondeterministic: if, for example, k_2 occurs in state p_0, the automaton has the choice of staying in p_0 or of moving to p_1. To see that A_{P_1} accepts the same traces that satisfy formula P_1, notice that if k_1 occurs while in state p_0 (which entails that the automaton goes to state p_1 if there is no simultaneous k_2), then either the safe never opens (which would satisfy formula P_1), or at some point in the future k_2 occurs before open occurs (which is forbidden as long as the automaton is in state p_1).

Exercise 11.2.   Build a *deterministic* Büchi automaton equivalent to the automaton in Fig. 11.2.   ∎

The creation of automaton A_{P_1} completes step A1 above. To perform step A2, we could complement A_{P_1} to obtain $\overline{A_{P_1}}$. This is possible in principle, but the construction of the complement of a generic Büchi automaton is quite complex in practice. Since $\overline{A_{P_1}}$ represents, in the form of an automaton, the "violating traces" that do not satisfy property P_1, or, equivalently, that satisfy $\neg P_1$, we can avoid the complementation of A_{P_1} and directly build an automaton equivalent to $\neg P_1$. This leads to the following revised sequence of steps:

A1'.   Build an automaton $A_{\neg P}$ that accepts exactly all traces that do *not* satisfy formula P.

A2'.   Build the *intersection* automaton $S \cap A_{\neg P}$, which accepts exactly all traces accepted by both S and $A_{\neg P}$,

A3'.   Check whether $S \cap A_{\neg P}$ accepts *any* trace.

In our example, $\neg P_1$ is $\neg\square(k_1 \Rightarrow \neg(\neg k_2 \text{ U open}))$. Using customary logic equivalences (see Chaps. 2 and 9), we equivalently rewrite $\neg P_1$ as

$$\neg P_1 \quad \equiv \quad \Diamond(k_1 \wedge (\neg k_2 \text{ U open})). \tag{11.2}$$

---

[1] This is a notational convenience that is common when depicting automata corresponding to LTL formulae; the automaton in Fig. 11.1 uses the more common convention of indicating on the transitions all and only the events that hold.

Fig. 11.3 Büchi automaton
$A_{\neg P_1}$ equivalent to LTL
formula $\neg P_1$

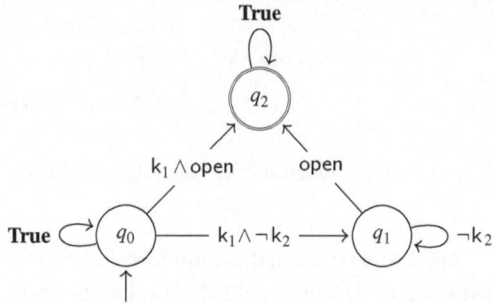

The corresponding Büchi automaton is depicted in Fig. 11.3, where **True** stands
for "any combination of symbols".

Figure 11.4 shows the intersection Büchi automaton $S \cap A_{\neg P_1}$. As mentioned
in Chap. 7, building the intersection of Büchi automata is slightly more complicated
than for finite-state automata over finite words because of the acceptance condition:
an execution is accepted by two Büchi automata BA_1 and BA_2 if it traverses
infinitely often some final states of both BA_1 and BA_2, but not necessarily at the
same time (for example, a trace that alternates between final states of BA_1 and
BA_2 is accepted). For this reason, the states of the intersection automaton include
a tag from $\{0, 1, 2\}$, with the following meaning: 0 denotes that no final state of
automaton BA_1 has been traversed, yet; 1 denotes that a final state of automaton
BA_1 has been traversed in the past, but no final state of BA_2 has been reached since;
the tag becomes 2 (from 1) when a final state of BA_2 is reached; transitions outgoing
from states with tag 2 "reset" the tag to 0, so that, to reach again a state with tag 2,
at least a final state of BA_1 and a final state of BA_2 must be traversed anew. The
final states of the intersection automaton are those with tag 2, so that an accepting
run of the intersection (which traverses states with tag 2 infinitely often) traverses
infinitely often both some final states of BA_1 and some final states of BA_2.

To determine whether property P_1 holds for automaton S, we check whether
automaton $S \cap A_{\neg P_1}$ accepts some trace. In practice, we need to determine whether
there is a loop in the graph of $S \cap A_{\neg P_1}$ that is reachable from the initial state and
that includes a final state. If such a loop exists, the trace that starts from the initial
state, reaches the loop, and then cycles along the loop is a trace of S that violates
property P_1. In the case of $S \cap A_{\neg P_1}$, we can see that execution

$$\langle s_0, q_0, 0 \rangle \; \mathsf{k_2} \; \langle s_2, q_0, 0 \rangle \quad \mathsf{k_1} \; \langle s_3, q_1, 0 \rangle \; \mathsf{open} \; \langle s_4, q_2, 1 \rangle$$

$$\mathsf{close} \; \langle s_0, q_2, 2 \rangle \; (\tau \; \langle s_0, q_2, 0 \rangle \; \tau \; \langle s_0, q_2, 1 \rangle \; \tau \; \langle s_0, q_2, 2 \rangle)^{\omega}$$

loops over final state $\langle s_0, q_2, 2 \rangle$; hence the corresponding input trace

$$\mathsf{k_2} \; \mathsf{k_1} \; \mathsf{open} \; \mathsf{close} \; \tau^{\omega}$$

is a counterexample for property P_1. In fact, P_1 requires that the safe open when
key $\mathsf{k_2}$ is inserted after key $\mathsf{k_1}$, but the system also allows for sequences where

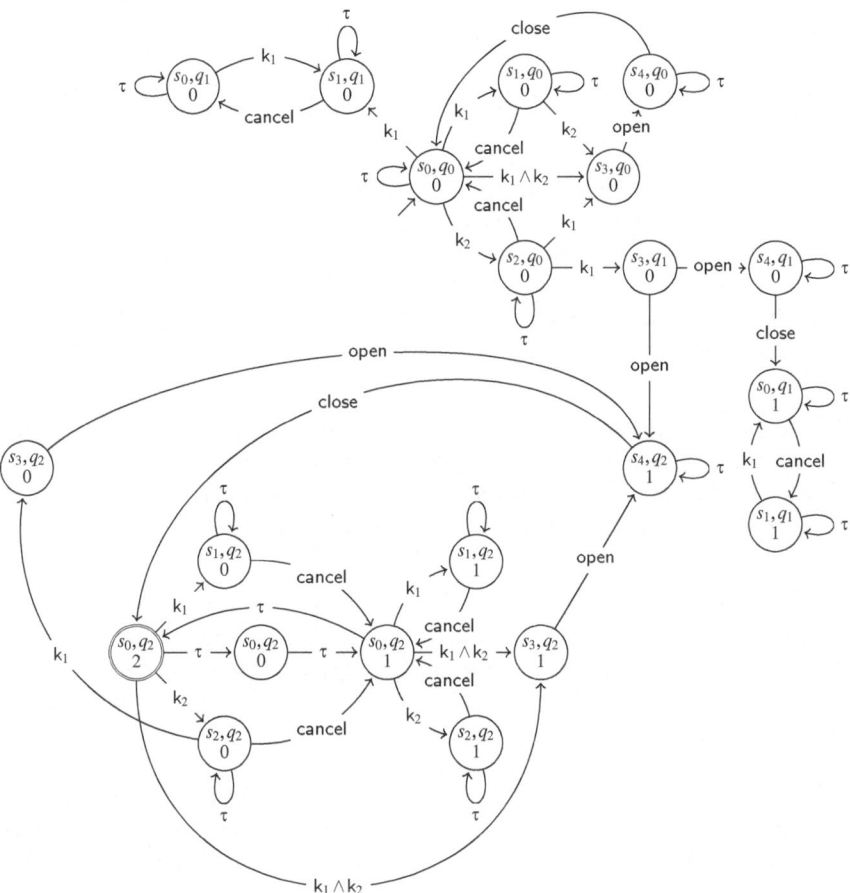

Fig. 11.4 Büchi automaton $S \cap A_{\neg P_1}$

k_2 is followed by k_1, which is precisely the kind of behavior represented in the counterexample.

Exercise 11.3. The safe model discussed in this section does not deal explicitly with time in a metric way. However, we can (as often done in Chap. 7) implicitly associate a time unit with the occurrence of every transition (τ included). Under this hypothesis, build a Büchi automaton model of a safe that opens if and only if both keys are inserted in any order no more than three time units apart. Then, formalize and prove (or disprove) the following properties for your automaton:

- If key k_1 is inserted two time units after key k_2, the safe opens.
- If key k_2 is inserted four time units after k_1, and k_1 is not inserted again in the meanwhile, the safe remains closed.   ∎

11.1.2   Satisfiability-Based Model Checking

Automata-theoretic model checking is based on the transformation of temporal logic formulae into automata. Satisfiability-based model checking follows the converse approach: it transforms the operational model S of the system into an equivalent logic formula F_S, and then applies logic-based algorithms to a suitable combination of F_S and P. The recent impressive practical improvements in the field of automatic satisfiability solvers have made satisfiability-based techniques practical and capable of handling verification problems of considerable size.

Given an automaton S representing the system to be analyzed, and a temporal logic formula P formalizing the property to be checked, satisfiability-based model checking works according to the following macro-steps.

B1. Build an LTL formula F_S such that the traces that satisfy F_S coincide with the accepting executions of automaton S.

B2. Check whether the LTL formula $F_S \land \neg P$ is satisfiable. If it is, return a trace that satisfies it and report unsuccessful verification; if it is not, report successful verification.

If $F_S \land \neg P$ is *not* satisfiable, there is no execution of automaton S (hence, which satisfies F_S) that also satisfies $\neg P$ or, equivalently, that does not satisfy P. In other words, all executions of system S also satisfy property P; hence P holds for S. If, on the other hand, $F_S \land \neg P$ is satisfiable, then the sequence of states that satisfies it is a *counterexample*: an execution of S violating P.

Let us demonstrate through the example presented in Sect. 11.1.1 how to transform an automaton S into a corresponding LTL formula F_S. First of all, we introduce symbols for the states and inputs of S. For each state s_i of S, we introduce a propositional letter (also written s_i): s_i is true at instant j if and only if the automaton is in state s_i at j. For each input symbol i_k, we introduce a propositional letter (also written i_k) that is true at instant j if and only if the automaton receives input i_k at j. Consider, for example, state s_0: if the automaton receives input k_1 (but not k_2) while in state s_0, it moves to state s_1. This is formalized by

$$\Box(s_0 \land k_1 \land \neg k_2 \Rightarrow \bigcirc s_1).$$

Similarly, if no input is received (in other words, τ holds) while in state s_0, the automaton remains in state s_0:

$$\Box(s_0 \land \tau \Rightarrow \bigcirc s_0).$$

In all, the following constraint describes the transitions from state s_0:

$$F_{S,s_0,1} \quad \equiv \quad \Box \begin{pmatrix} (s_0 \land k_1 \land \neg k_2 \Rightarrow \bigcirc s_1) \\ \land\ (s_0 \land k_2 \land \neg k_1 \Rightarrow \bigcirc s_2) \\ \land\ (s_0 \land k_2 \land k_1 \Rightarrow \bigcirc s_3) \\ \land\ (s_0 \land \tau \Rightarrow \bigcirc s_0) \end{pmatrix}. \tag{11.3}$$

Similar formulae can be written for all other states and transitions of the automaton of Fig. 11.1.

In addition to formulae such as $F_{S,s_0,1}$, we must also specify that in each instant the automaton is in exactly one state,

$$F_{S,2} \quad \equiv \quad \square \; (s_0 \vee s_1 \vee s_2 \vee s_3 \vee s_4) \,, \tag{11.4}$$

$$F_{S,3} \quad \equiv \quad \square \left(\begin{array}{l} (s_0 \Rightarrow \neg s_1 \wedge \neg s_2 \wedge \neg s_3 \wedge \neg s_4) \\ \wedge \, (s_1 \Rightarrow \neg s_2 \wedge \neg s_3 \wedge \neg s_4) \\ \wedge \, (s_2 \Rightarrow \neg s_3 \wedge \neg s_4) \\ \wedge \, (s_3 \Rightarrow s_4) \end{array} \right) \,, \tag{11.5}$$

and that only certain input symbols are acceptable in different states. For example, cancel, open, or close cannot happen in state s_0:

$$F_{S,s_0,4} \quad \equiv \quad \square(s_0 \Rightarrow \neg\text{cancel} \wedge \neg\text{open} \wedge \neg\text{close}) \,. \tag{11.6}$$

Since an execution is valid only if it traverses the accepting state s_0 infinitely often (a liveness condition, as described in Chap. 9), we must also add the constraint

$$F_{S,5} \quad \equiv \quad \square\lozenge s_0 \,. \tag{11.7}$$

Finally, the formula F_S below formalizes the automaton of Fig. 11.1 as the logic conjunction of all formulae of the types above, for each state $s_1, \ldots, s_n$ of the automaton, plus the condition that the automaton is initially in state s_0:

$$F_S \quad \equiv \quad s_0 \wedge \bigwedge_{i=1..n} F_{S,s_i,1} \wedge F_{S,2} \wedge F_{S,3} \wedge \bigwedge_{i=1..n} F_{S,s_i,4} \wedge F_{S,5} \,. \tag{11.8}$$

As expected from the results of Sect. 11.1.1, the sequence

$$\{s_0, \text{k}_2\} \, \{s_2, \text{k}_1\} \, \{s_3, \text{open}\} \, \{s_4, \text{close}\} \, \{s_0, \tau\}^\omega$$

satisfies formula $F_S \wedge \neg P_1$ and corresponds to the execution $s_0 \, \text{k}_2 \, s_2 \, \text{k}_1 \, s_3 \, \text{open} \, s_4$ close $(s_0 \, \tau)^\omega$ of the automaton in Fig. 11.1.

Off-the-shelf satisfiability solvers cannot directly deal with temporal logic formulae; hence we have to encode the semantics of $F_S \wedge \neg P$ in a simpler input logic (propositional or first-order fragment) supported by some solver, along the lines of the *bounded encodings* introduced in Sect. 9.9.2. Assuming that the back-end satisfiability solver inputs propositional logic, we have the following extended sequence of macro-steps for satisfiability-based model checking.

B1'. Build an LTL formula F_S such that the traces that satisfy F_S coincide with the accepting executions of automaton S.

B2'. Transform formula $F_S \wedge \neg P$ into an equi-satisfiable propositional formula $\Pi_{F_S \wedge \neg P}$ using the encoding of Sect. 9.9.2.

B3'. Using a SAT solver, check whether $\Pi_{FS \wedge \neg P}$ is satisfiable. If it is, return a trace that satisfies it and report unsuccessful verification; if it is not, report successful verification.

Even if the techniques presented in the current section and in Sect. 11.1.1 target LTL formulae, it is possible to reuse them to check some CTL properties that do not have equivalent LTL formulations, according to the discussion of Sect. 9.1.2.

Consider, for instance, with reference to the system of Fig. 11.1, the CTL formula

$$P_2 \quad \equiv \quad \exists\diamond(k_1 \wedge \exists(\neg k_2 \; U \; open)) \,. \tag{11.9}$$

P_2 states that there exists an execution path such that, at some point, key k_1 is inserted, and then, along some branch continuing from that point, the safe opens without key k_2 having been inserted in the meantime. CTL property P_2 is inexpressible in LTL: according to Theorem 9.11, if it were, then the LTL formula obtained by eliminating all path quantifiers from P_2 would be equivalent to it. However, if we remove path quantifiers from P_2 we obtain formula $\neg P_1$, which is not equivalent to P_2: $\neg P_1$ is valid for a system if and only if *all executions* are such that key k_1 is eventually inserted, and then, some time after that, the safe opens with no insertion of key k_2 in the meantime, which is clearly not the meaning of P_2.

If however, we do not check the *validity* of formula $\neg P_1$ for system S, but rather its *satisfiability*, we conclude that $\neg P_1$ is satisfiable, which entails that there exists an execution of the system S in which an insertion of k_1 is followed by the safe opening without further insertion of key k_2. This is precisely the meaning of CTL formula P_2, so the satisfiability of LTL formula $\neg P_1$ entails the validity of CTL formula P_2 for S.

This custom approach to model checking only works for CTL formulae with only existential path quantifiers without negation. The following section presents a more general approach to dealing with generic CTL formulae.

11.1.3   CTL Model Checking by Tableau

Some model checking algorithms are not founded on transformations of one formalism into another, but, rather, on the direct exploration of the state space of the operational model guided by the structure of the formula. This is, in particular, the typical approach to model checking properties expressed in CTL logic. As remarked at the beginning of this section, the focus in this book is on modeling issues rather than on specific algorithms. Nevertheless, since the exploration algorithm is *the* central issue in CTL model checking, this section briefly outlines its basic principles through an example.

Let us consider again the safe described in the previous sections. CTL-based algorithms work on transition systems with labels on states only. To this end, consider the transition system in Fig. 11.5, which models the safe discussed in

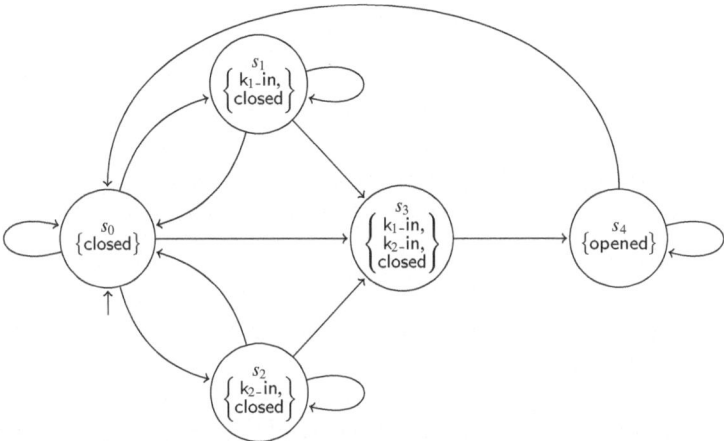

Fig. 11.5 Specification of the safe with labels on states

Sect. 11.1.1 with some simplified details. Each state in the transition system is labeled according to the state of the keys and of the safe; for example, state s_1 is labeled $\{k_1\_\text{in}, \text{closed}\}$ to represents the fact that, while in s_1, the key is inserted and the safe is still closed. Transitions are not labeled, nor final states defined, as requested by the CTL model checking algorithm.

Let us now consider the following CTL formula:

$$P_3 \equiv \exists\Diamond(k_1\_\text{in} \wedge \exists\Diamond(\neg k_2\_\text{in} \wedge \exists\bigcirc(k_2\_\text{in} \wedge \exists(k_2\_\text{in U opened})))). \quad (11.10)$$

It states that there is an execution from the initial state such that, eventually, key k_1 is inserted and then, sometime later, key k_2 is inserted (represented by the fact that there are consecutive instants in which the key is not in and then in), and it remains in until the safe opens.

To check whether property P_3 holds for the system of Fig. 11.5, we go through its *state subformulae* (see Sect. 9.1.2.1) to determine in which states they hold. For each subformula Ψ, we denote by *Sat* (Ψ) the set of states in which Ψ holds.

Let us start from subformula $\exists(k_2\_\text{in U opened})$. From its semantics (see Sect. 9.1.2.1), *Sat* ($\exists(k_2\_\text{in U opened})$) includes all states that are labeled with proposition opened, plus all states that can reach opened states along paths in which all states are tagged k_2_in. To determine such a set of states, we can perform a series of breadth-first searches in the transition system of Fig. 11.5, starting from states tagged opened and going backwards:

- In the first iteration, we add s_4 (the only state tagged opened) to *Sat* ($\exists(k_2\_\text{in U opened})$). Hence we write:

$$Sat_1 (\exists(k_2\_\text{in U opened})) = \{s_4\}.$$

- In the second iteration, we look for all predecessors of s_4 in which k_2_in holds. The only predecessor of s_4 is s_3, which is tagged with k_2_in; hence we write:

$$Sat_2\ (\ \exists(k_2\_in\ U\ opened)\ ) = \{s_4, s_3\}\ .$$

- In the third iteration, we explore the predecessors of s_3: s_0, s_1, and s_2. Of these, the only one in which k_2_in holds is s_2, hence we write:

$$Sat_3\ (\ \exists(k_2\_in\ U\ opened)\ ) = \{s_4, s_3, s_2\}\ .$$

- In the fourth iteration, we consider s_2, the only member of Sat_3 ($\exists(k_2\_in\ U\ opened)$) not explored yet. It has s_0 and itself as predecessors; since s_2 already belongs to Sat_3 ($\exists(k_2\_in\ U\ opened)$) and k_2_in does not hold in s_0, no further states are added to Sat ($\exists(k_2\_in\ U\ opened)$), which has then reached a stable state (more technically, a "fixed point"):

$$Sat\ (\ \exists(k_2\_in\ U\ opened)\ ) = Sat_4\ (\ \exists(k_2\_in\ U\ opened)\ )$$
$$= Sat_3\ (\ \exists(k_2\_in\ U\ opened)\ ) = \{s_4, s_3, s_2\}\ .$$

The algorithm continues on set Sat ($k_2\_in \wedge \exists(k_2\_in\ U\ opened)$), which corresponds to the intersection of Sat ($\exists(k_2\_in\ U\ opened)$) with Sat (k_2_in). Since the latter is simply the set $\{s_2, s_3\}$ of states in which k_2_in holds, we have

$$Sat\ (\ k_2\_in \wedge \exists(k_2\_in\ U\ opened)\ ) = \{s_2, s_3\}.$$

Next, Sat ($\exists\circ(k_2\_in \wedge \exists(k_2\_in\ U\ opened))$) is the set of states such that one of their successors belongs to set Sat ($k_2\_in \wedge \exists(k_2\_in\ U\ opened)$). It can be determined with a simple search of the state space; hence,

$$Sat\ (\ \exists\circ(k_2\_in \wedge \exists(k_2\_in\ U\ opened))\ ) = \{s_0, s_1, s_2\}.$$

Continuing with larger subformulae,

$$Sat\ (\ \neg k_2\_in \wedge \exists\circ(k_2\_in \wedge \exists(k_2\_in\ U\ opened))\ )$$

is the intersection of Sat ($\exists\circ(k_2\_in \wedge \exists(k_2\_in\ U\ opened))$) with Sat ($\neg k_2\_in$),

$$Sat\ (\ \neg k_2\_in \wedge \exists\circ(k_2\_in \wedge \exists(k_2\_in\ U\ opened))\ ) = \{s_0, s_1\},$$

and Sat ($\exists\diamond(\neg k_2\_in \wedge \exists\circ(k_2\_in \wedge \exists(k_2\_in\ U\ opened)))$) is the set of states from which $\{s_0, s_1\}$ can be reached, and can be computed in a similar way as Sat ($\exists(k_2\_in\ U\ opened)$), thanks to the definition $\exists\diamond\Psi \equiv \exists(\textbf{True}\ U\ \Psi)$:

$$Sat\ (\ \exists\diamond(\neg k_2\_in \wedge \exists\circ(k_2\_in \wedge \exists(k_2\_in\ U\ opened)))\ ) = \{s_0, s_1, s_2, s_3, s_4\}.$$

Along the same lines, we compute

$$Sat (\ k_1\_in \wedge \Diamond(\neg k_2\_in \wedge \exists\bigcirc(k_2\_in \wedge \exists(k_2\_in \ \mathsf{U} \ opened)))) \ ,$$

and finally

$$Sat (\ P_3 \ ) = \{s_0, s_1, s_2, s_3, s_4\}.$$

Since formula P_3 holds in the initial state ($s_0 \in Sat (\ P_3 \ )$), we conclude that P_3 is a property of the system of Fig. 11.5.

The example showed how to compute $Sat (\Psi)$ when Ψ is one of the following:

1. A simple proposition p,
2. The negation of a formula $\neg\Phi$,
3. A conjunction $\Phi_1 \wedge \Phi_2$,
4. An "exists next state" $\exists\bigcirc\Phi$,
5. An "exists until" $\exists(\Phi_1 \ \mathsf{U} \ \Phi_2)$.

It can be shown that every CTL formula can be written in "existential normal form", in which only propositions, the connectives $\neg$ and $\wedge$, and the temporal operators $\exists\bigcirc$, $\exists \ \mathsf{U}$ and $\exists\square$ appear. Formula P_3 already is in existential normal form. We conclude that, if we are able to compute $Sat (\Psi)$ when Ψ has one of the forms p, $\neg\Phi$, $\Phi_1 \wedge \Phi_2$, $\exists\bigcirc\Phi$, $\exists(\Phi_1 \ \mathsf{U} \ \Phi_2)$, and $\exists\square\Phi$, we can model check every CTL formula.

Exercise 11.4. The only case of existential normal form we have not dealt with in the example is $\exists\square\Phi$. Complete the description by discussing how to compute $Sat (\exists\square\Phi)$.

(*Hint*: $\exists\square\Phi$ holds in a state s if and only if one can reach from s a loop where Φ holds in all states leading to and in the loop). ∎

Exercise 11.5. (♦) Consider the following variation of the rules regulating the opening of the safe defined by the automaton of Fig. 11.5: the safe opens if and only if the two keys are inserted in any order and then extracted in the reverse order of insertion. Formalize and prove by model checking the property that, even if the keys are extracted in the wrong order, by mistake, it is always possible to open the safe. ∎

11.1.4 Model Checking Timed Automata

All the model checking approaches presented so far in this chapter target models with discrete time and do not specifically address metric timing properties. Model checking techniques, however, exist also for continuous-time formalisms with metric constructs, such as timed automata and timed Petri nets. This section presents one such technique to model check timed automata models.

Model checking techniques for continuous-time metric operational models typically use the CTL logic (possibly in some metric extension) to express the properties to verify, and algorithms similar to the one outlined in Sect. 11.1.3 to carry out the actual verification. The continuous notion of time leads to an infinite state space which cannot be exhaustively explored without suitable abstractions.

Model checking techniques for timed automata rely on an abstraction that captures the semantics of finite-state automata by means of finite transition systems on which CTL properties can be verified, as done in Sect. 11.1.3. The abstraction of timed automata is based on the notion of *clock region*. A clock region is a set of clock valuations that have common features in regard to the behavior of timed automata (for example, they satisfy the same *guards*).[2]

Let us consider a clock constraint of the form $x < c$, with c a natural number. All values for x that have the same integral part (all values v_{x_1}, v_{x_2} such that $\lfloor v_{x_1} \rfloor = \lfloor v_{x_2} \rfloor$) behave in the same way in regard to the satisfaction of the constraint: if $\lfloor v_{x_1} \rfloor = \lfloor v_{x_2} \rfloor$, then either both $v_{x_1} < c$ and $v_{x_2} < c$, or both $v_{x_1} \geq c$ and $v_{x_2} \geq c$. For constraints of the kind $x \leq c$, however, $\lfloor v_{x_1} \rfloor = \lfloor v_{x_2} \rfloor$ is not enough to guarantee the same behavior: if one of v_{x_1}, v_{x_2} has fractional part equal to 0 ($frac(v_{x_1}) = 0$), then the other must too, in addition to having the same integral part, for them to be equivalent with respect to constraint $x \leq c$.

Let us now consider two clocks, x and y, and two valuations v_1 and v_2 such that $\lfloor v_1(x) \rfloor = \lfloor v_2(x) \rfloor$ and $frac(v_1(x)) = frac(v_2(x)) = 0$, or both $frac(v_1(x))$, $frac(v_2(x)) \neq 0$; and similar constraints hold for clock y. If the fractional parts of x and y have the same ordering in v_1 and v_2 (i.e., $frac(v_1(x)) \leq frac(v_1(y))$ if and only if $frac(v_2(x)) \leq frac(v_2(y))$), then, when time advances by a quantity d_1 from v_1 there is a positive real value d_2 such that $v_2 + d_2$ "behaves" like $v_1 + d_1$, in the sense that all clock constraints that hold for $v_1 + d_1$ also hold for $v_2 + d_2$. This is depicted in Fig. 11.6: as time advances from the two valuations v_1 and v_2 (along diagonal lines), it reaches the same regions (line segments, their intersections, and open bounded or unbounded regions), where each region corresponds to a clock constraint. For example, the first vertical line segment encountered corresponds to constraint $x = c_x \wedge c_y - 1 < y < c_y$.

If $c_{x,\max}$ is the maximum constant with which clock x is compared in an automaton (for instance, through guard $x < c_{x,\max}$), all valuations where the value of x is above $c_{x,\max}$ are equivalent, in that they all satisfy the same set of constraints in the automaton.

These informal considerations lead to the following rigorous definition of clock equivalence and region.

Definition 11.6 (Clock equivalence). Given a timed automaton TA with a set of clocks C, let $c_{x,\max}$ denote the maximum constant with which clock x is compared

---

[2]Recall that guards are combinations of clock constraints, where a clock constraint has the form $x \sim c$, with x a clock, c a natural number, and $\sim \in \{<, \leq, =, \geq, >\}$.

Fig. 11.6 Advancement of
clocks with the same ordering
of decimal parts

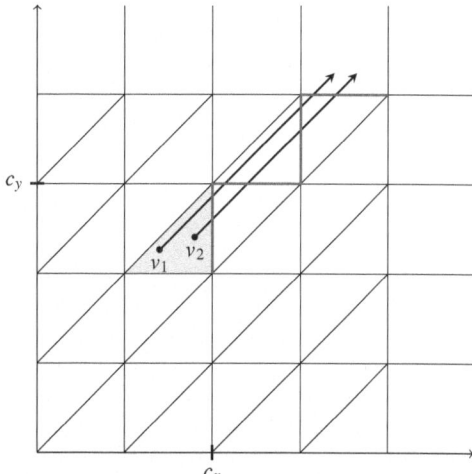

in TA, for $x \in C$. Two clock valuations v_1 and v_2 are equivalent, written $v_1 \simeq v_2$, if
and only if all the following conditions hold:

(i)   For each $x \in C$, $v_1(x) > c_{x,\max}$ if and only if $v_2(x) > c_{x,\max}$.
(ii) For each $x \in C$, if $v_1(x) \leq c_{x,\max}$ (hence $v_2(x) \leq c_{x,\max}$), then $\lfloor v_1(x) \rfloor = \lfloor v_2(x) \rfloor$, and $frac(v_1(x)) = 0$ if and only if $frac(v_2(x)) = 0$.
(iii) For each pair $x, y \in C$, $frac(v_1(x)) \leq frac(v_1(y))$ if and only if $frac(v_2(x)) \leq frac(v_2(y))$.   ∎

Each set of equivalent clock valuations defines a *clock region*. Since $\simeq$ is an
equivalence relation, the set of all clock regions is the quotient set $V_C / \simeq$, where
V_C is the set of all clock valuations for clocks C. Then, if v is a valuation of clock
region ρ, we have that $\rho = [v]$, where $[v]$ is the equivalence class of v.

Exercise 11.7. (♦) Prove that $\simeq$ is an equivalence relation, that is, it is reflexive,
symmetric, and transitive.   ∎

The equivalence relation of Definition 11.6 is defined differently for each timed
automaton, according to the constants $c_{x,\max}$ appearing in it; however, it *only*
depends on the $c_{x,\max}$'s, so different automata with the same clocks and values for
the $c_{x,\max}$'s produce the same equivalence relation and clock regions.

Example 11.8 (Clock regions). Let us consider an automaton with two clocks, x
and y, in which $c_{x,\max} = 2$ and $c_{y,\max} = 1$. The clock regions of the automaton
are depicted in Fig. 11.7, where each region is marked by a symbol: there are eight
open regions (marked by black stars), including four unbounded ones, six corner
points (marked by black squares), and 14 lines (marked by white circles), for a total
of 28 regions.   ∎

As Example 11.8 shows, the number of clock regions determined by any timed
automaton is *finite*. In addition, as time advances, if the clocks are in a certain

Fig. 11.7 Clock regions for
an automaton with
$C = \{x, y\}$, $c_{x,\max} = 2$ and
$c_{y,\max} = 1$

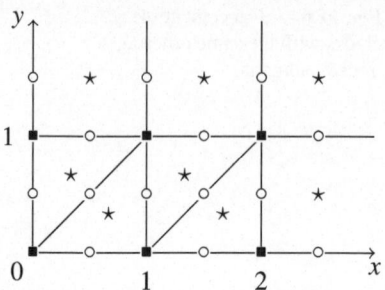

region, the *successor region* is well defined. Consider Fig. 11.6 again: given any clock valuation v that is in the same region as v_1 and v_2, as time advances the first clock region encountered is the vertical line segment (in which all valuations v' are such that $v'(x) = c_x \wedge c_y - 1 < v'(y) < c_y$, where $frac(v'(y)) > frac(v'(x)) = 0$).

Definition 11.9 (Successor region). Given a clock region ρ, its successor $succ(\rho)$ sssis

- ρ itself if ρ is the region in which all clocks have surpassed the maximum values with which they are compared (i.e., it is the set of valuations $\rho_{\max}$ such that $v \in \rho_{\max}$ if and only if, for all $x \in C$, $v(x) > c_{x,\max}$);
- Otherwise, it is the region $succ(\rho) \neq \rho$ such that $v' \in succ(\rho)$ if and only is there is a $v \in \rho$ and a positive real value d such that $v' = v+d$ and, for all $0 < d' < d$, $v + d'$ is either in ρ or in $succ(\rho)$ (i.e., $v + d' \in \rho \cup succ(\rho)$). ∎

Thanks to the notions of region and successor region, we can build, for any timed automaton TA, a corresponding finite transition system, called the *region transition system* RTS_{TA}, which precisely captures the semantics of TA abstracted by the region equivalence: every execution of TA has a corresponding execution in RTS_{TA}, and every execution of RTS_{TA} corresponds to a *set* of executions of TA.

Definition 11.10 (Region transition system). Given a timed automaton $TA = \langle L, A, C, E, l_0, I \rangle$, its corresponding region transition system $RTS_{TA} = \langle S, S_0, J, \rightarrowtail \rangle$ is such that:

(i) $S = L \times V_C / \simeq$;
(ii) $S_0 = \{\langle l_0, [v_0] \rangle\}$, where $[v_0]$ is the equivalence class of valuation v_0, in which all clocks have value 0 ($[v_0]$ contains only one valuation, as it corresponds to a corner point; see Fig. 11.7);
(iii) $J = I \cup \{\tau\}$, where $\tau \notin I$.

In addition, $\langle \langle l, [v] \rangle, e, \langle l', [v'] \rangle \rangle \in \rightarrowtail$ if and only if one of the following holds:

(iv) $e = \tau$, $l = l'$, $[v'] = succ([v])$, and both $[v]$ and $[v']$ satisfy the invariant $I(l)$ (recall that all valuations of the same clock region satisfy the same constraints, so the satisfaction relation is well defined for a region);
      or

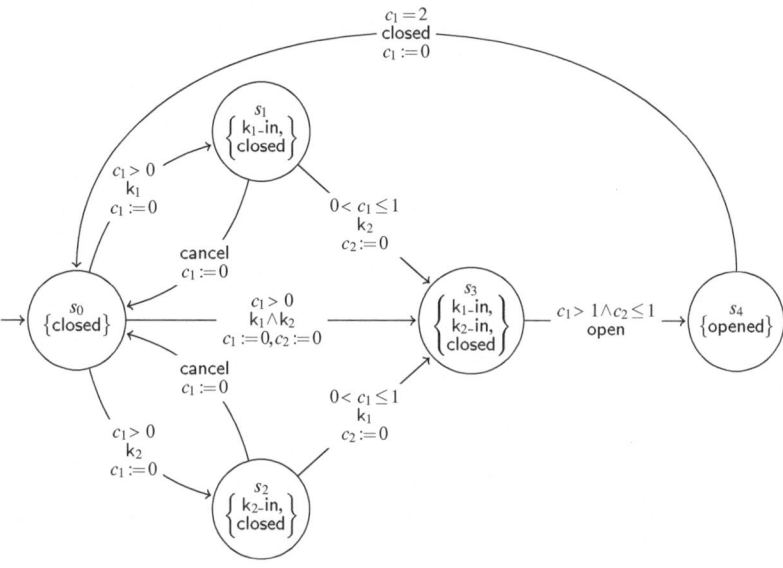

Fig. 11.8 Timed automaton specifying the behavior of the safe

(v) $e \in A$, and there is an edge $\langle l, g, e, r, l' \rangle$ such that $[v]$ satisfies guard g, $[v'] = [v[r := 0]]$ (recall that $v[r := 0]$ is the valuation obtained by resetting all clocks in set r to 0), and $[v']$ satisfies invariant $I(l')$. ∎

Example 11.11 (Region transition system). Consider the timed automaton in Fig. 11.8, which is a variation of the model of the safe in Fig. 11.5. The automaton uses two clocks, c_1 and c_2, to impose real-time constraints. For example, the automaton can stay in state s_3 for no more than one time unit; in addition, the constraint on clock c_1 (which is reset when entering s_1 or s_2) entails that, if the automaton moves from state s_1 to state s_3, then the total permanence in these two states must be at least one time unit (and similarly if the automaton moves from s_2 to s_3). Note that the timed automaton of Fig. 11.8 has labels associated with each state, similarly to the model of Fig. 11.5, though they are not considered in Definitions 11.10 and 7.32. Extending those definitions to cover state labels is straightforward.

Figure 11.9 shows a fragment of the region transition system corresponding to the automaton of Fig. 11.8 (we leave the completion of the region transition system as an exercise). Consider property P_3 of Sect. 11.1.3 with respect to the region transition system fragment; the "next" operator ○ in P_3 now refers to the "next state" in the transition system, instead of the "next time instant" (the latter is not well defined in a continuous-time model). Correspondingly, we can see that, in the fragment in Fig. 11.9, there is a path that satisfies CTL property P_3. The same result can also be mechanically obtained through the application of the CTL-based algorithm outlined in Sect. 11.1.3.

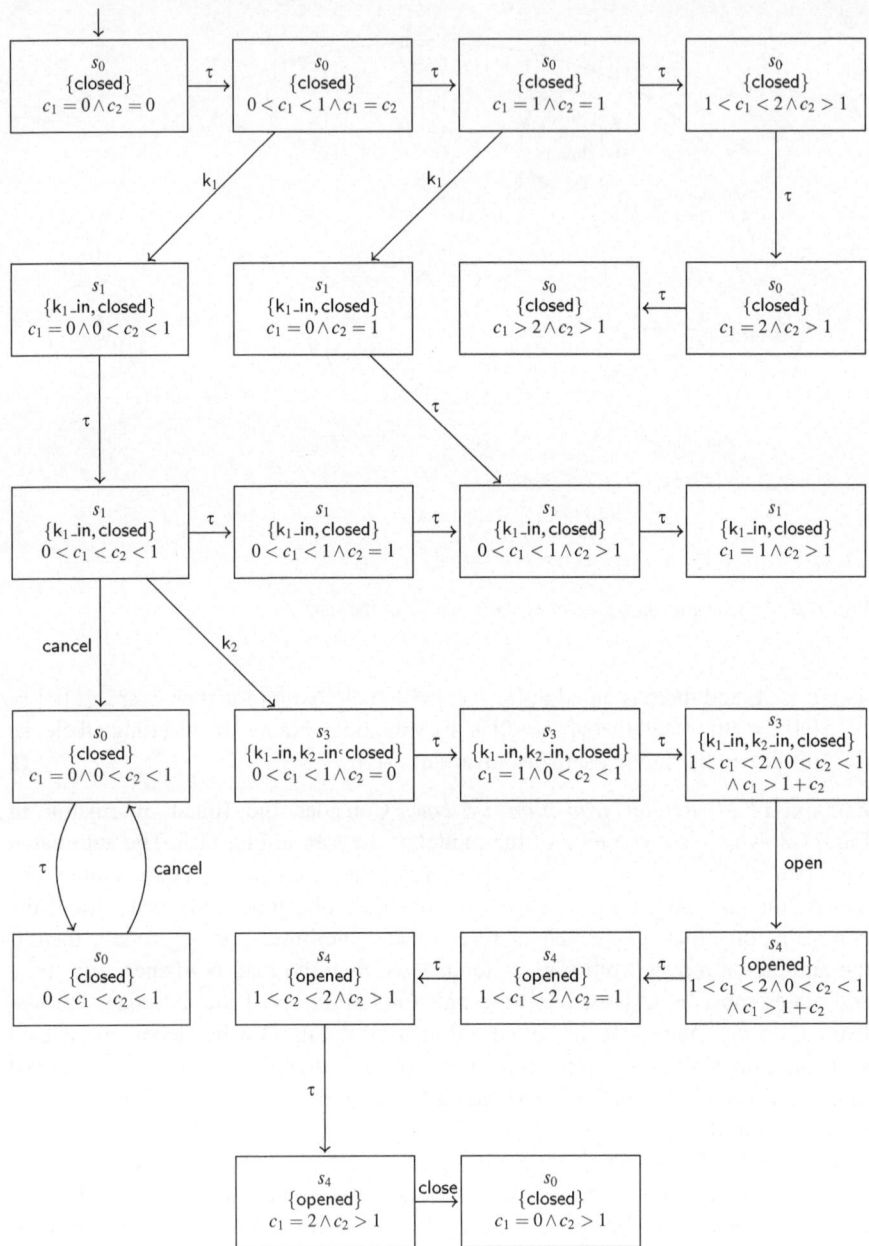

Fig. 11.9 Fragment of the region transition system of the automation in Fig. 11.8

In addition, the region transition system contains state information about the values of the clocks; hence metric properties such as "there exists a path such that until state opened is reached clock c_1 is less than 2" can also be verified using the CTL-based algorithm of Sect. 11.1.3. To express properties such as

$$P_4 \quad \equiv \quad \exists(c_1 < 2 \ \mathsf{U} \ \mathsf{opened}) \qquad (11.11)$$

we use "timed CTL" (TCTL), a metric extension of CTL that has been defined in the literature. The algorithm of Sect. 11.1.3 can be used, with suitable adjustments, to model check timed automata (through their region transition systems) against properties expressed in TCTL.

Let us finish our informal analysis of the fragment in Fig. 11.9 by discussing time and state progress conditions. There is a loop between states $\langle s_0, \{\mathsf{closed}\}, c_1 = 0 \wedge 0 < c_2 < 1 \rangle$ and $\langle s_0, \{\mathsf{closed}\}, 0 < c_1 < 1 \wedge 0 < c_2 < 1 \wedge c_1 < c_2 \rangle$. However, in any infinite path that loops through these two states, time does not progress, since clock c_2 remains always between 0 and 1, without ever being reset. As explained in Sect. 7.4.1, such a behavior is a priori excluded in the semantics of timed automata. In accordance with this assumption, a sufficient and necessary condition for the existence of a path where time progresses is that, for each clock $x \in C$, infinitely often either $x > c_{x,\max}$ or x is reset. For example, the loop between states $\langle s_0, \{\mathsf{closed}\}, c_1 = 0 \wedge 0 < c_2 < 1 \rangle$ and $\langle s_0, \{\mathsf{closed}\}, 0 < c_1 < 1 \wedge 0 < c_2 < 1 \wedge c_1 < c_2 \rangle$ does not satisfy the progress condition, as $c_{1,\max}$ and $c_{2,\max}$ are never surpassed, and clock c_2 is never reset. Every model checking algorithm must encode such progress constraints to exclude from their analysis all paths that do not conform the semantics of timed automata. ∎

Exercise 11.12. (♦) Describe how to modify the CTL algorithm illustrated in Sect. 11.1.3 to exclude paths where time does not progress. ∎

11.1.5 Concluding Remarks on Model Checking

As mentioned in the introduction, and highlighted by the examples throughout this section, model checking techniques suffer from the state explosion problem. For instance, in the example of Sect. 11.1.1, a system of five states and a property of three states require an automaton of 21 states to be explored. Similarly, the timed automaton of Fig. 11.8 determines a region transition system whose size is proportional to the product of number of regions and number of locations of the original timed automaton, which can be considerable, as shown in Example 11.11. In general, model checking techniques require the exhaustive exploration of a state space (in satisfiability-based techniques, such a state space is the set of possible interpretations of the formulae), which typically grows exponentially with the size of the elements involved.

For example, given a property P formalized in LTL, the size of its corresponding Büchi automaton is, in the worst case, exponential in the length of the formula, and the size of the intersection automaton to be searched is of the order of $|S| \cdot 2^{|P|}$, with $|S|$ the size of the system automaton and $|P|$ the size of the property. The properties to be verified rarely have significant sizes, so the limiting factor is usually the size of the system model, which is typically very large for nontrivial systems. In general, LTL model checking is PSPACE-complete.

Despite these limitations, which have been addressed through a number of optimizations and heuristics, model checking is a very appealing verification technique, if not for any other reason, because it is fully automated.

This Sect. 11.1 did not deal with techniques for probabilistic systems. Though they are a class of techniques of rapidly growing importance, supported by tools under constant improvement, their underlying principles are similar to those presented in Sects. 11.1.3 and 11.1.4. In fact, the logics used to express probabilistic properties (PCTL and CSL, discussed in Chap. 9) are variations of CTL; hence model checking of probabilistic models employs algorithms similar in nature to the one of Sect. 11.1.3: the state space is explored according to the structure of the property formula; during the exploration of the state space, suitable probability values are computed, and depending on these values new state labels are produced, which are then recursively used in the successive stages of the algorithm. The bibliographic remarks refer to specialized literature on this topic.

11.2   The TTM/RTTL Framework

Section 7.3.1 presented Timed Transition Models (TTMs) as abstract machines where time is modeled in a discrete, metric fashion: a special transition tick, which is implicitly combined with the various parts of the specification, has the effect of increasing the value of the integer variable t, expressly devoted to representing the global time. Variable t is a component of the state, and transitions other than tick cause a change of state without a change of time; therefore, the TTM notation departs from the traditional approach of dynamical systems where the state is a *function* of time and, as a consequence, for each time instant the modeled system is in a unique state. In the TTM/RTTL framework, TTMs are paired with a temporal logic, called "Real-Time Temporal Logic" (RTTL), to yield a dual-language approach. Historically, TTM/RTTL was one of the first approaches providing the means to model precisely quantitative time aspects, and hence was adequate for the specification and verification of real-time systems.

RTTL is a first-order, linear, discrete-time temporal logic which includes the temporal operators of LTL ($\circ$, $\cup$, etc.). The alphabet of RTTL formulae includes all the components of the state of TTMs (including the integer variable t for the current time) plus a special variable ε, whose value ranges over the set of TTM transitions, denoting the latest transition executed by the TTM, which just took the machine to

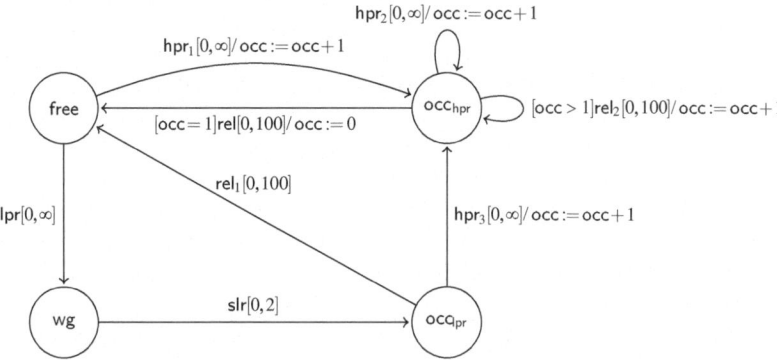

Fig. 11.10 A TTM representing a resource allocator

its current state (notice that the state includes the value of the time variable t; hence $\varepsilon = \textsf{tick}$ when the current time is the only component of the state that changes).

Customary arithmetic operators, plus first-order quantification on the value of variable t, can express quantitative time constraints. Consider, for instance, the resource allocator modeled with TTMs in Example 7.28, reproduced for convenience in Fig. 11.10. Recall that the $\textsf{hpr}$ and $\textsf{lpr}$ transitions respectively denote high-priority and low-priority requests, and the integer-valued state variable $\textsf{occ}$ counts the number of pending high-priority requests.

We next present two sample real-time properties of the allocator, specified in RTTL.

- It is always the case that, within 102 time units after a low-priority request, the resource is free:

$$\forall T \Box ((\varepsilon = \textsf{lpr} \wedge t = T) \Rightarrow \Diamond(\textsf{free} \wedge t \leq T + 102)) . \tag{11.12}$$

- The resource will be free infinitely often, provided no two high-priority requests $\textsf{hpr}_2$ can ever take place within 200 time units:

$$(\forall T \Box ((\varepsilon = \textsf{hpr}_2 \wedge t = T) \Rightarrow \neg\Diamond(t \leq T + 200 \wedge \varepsilon = \textsf{hpr}_2))) \Rightarrow \Box\Diamond\textsf{free}. \tag{11.13}$$

In the formulae (11.12) and (11.13), t is a *time-dependent* variable, in that it is a component of the state that may change from one instant to another, while T is, in the terminology of Sect. 9.3.3, a *global* or *rigid* variable, used as a placeholder in the universal quantification.

The same properties could be specified, more concisely, using the metric temporal logics MTL and TRIO – discussed in Sect. 9.3 – interpreted over discrete time: (11.12) is stated in TRIO as

$$\textsf{Alw}(\varepsilon = \textsf{lpr} \Rightarrow \textsf{WithinF}_{\textsf{ii}}(\textsf{free}, 102)) ,$$

while (11.13) in MTL is stated as

$$\Box(\varepsilon = \mathsf{hpr}_2 \Rightarrow \neg\Diamond_{\leq 200}(\varepsilon = \mathsf{hpr}_2)) \Rightarrow \Box\Diamond\mathsf{free}.$$

Exercise 11.13. Express in RTTL, in MTL, and in TRIO the following property of the resource allocator: at any time, if there is one pending high-priority request and no high-priority request hpr_2 is issued for 200 time units, then the resource will be released and free within 200 time units. ∎

One of the main advantages of the dual-language approach is that operational models can provide useful support and insight for reasoning about the proof of properties expressed using logic notations. The technique for analyzing an RTTL formula with reference to a TTM and the complexity of the proof vary depending on the structure of the stated property and the specific features of the TTM.

For instance, if the conjectured property specifies a feature that should be enjoyed by *every* execution, and the property does *not* hold, then the operational model can be used to produce a system run that does not possess the stated feature and thus constitutes a refutation of the property. This is the case, for instance, with property (11.12), which can be disproved by the following, informally described, state sequence: initially $t = 0$ and the TTM state is **free**, then the **lpr** transition is executed, then time advances by two units (i.e., the **tick** transition is executed twice), then **slr** is executed and the TTM goes into the $\mathsf{occ_{lpr}}$ state, then time advances by 99 units, then the transition hpr_3 is taken so that the $\mathsf{occ_{hpr}}$ state is reached, and finally the **tick** transition is executed; now the TTM is in state $\mathsf{occ_{hpr}}$ and $t = 102$; hence the property is violated.

If, instead, the RTTL formula analyzed specifies a feature that is enjoyed by *all* executions and it actually holds for the TTM at hand, then the proof may work by induction on the length of the execution, taking advantage of the topological structure of the TTM. As an example, we provide an informal outline of a proof of (11.13), based on the following lemmas:

- An invariant of all possible executions is that when the system is in the **free** or $\mathsf{occ_{lpr}}$ states, the state variable occ is 0.
- Based on this, whenever state $\mathsf{occ_{hpr}}$ is entered from some other state, occ is 1.
- After the $\mathsf{occ_{hpr}}$ state is entered from state **free** or $\mathsf{occ_{lpr}}$, either the transition rel_0 is taken and the **free** state is reached, or transition hpr_2 is executed: in this case occ becomes equal to 2 and, due to the hypothesis that transition hpr_2 will not be executed for the next 200 time units, transitions rel_2 and rel_0 will be executed next, taking the system to the **free** state.
- Thus, when the system is in the state **free**, it eventually returns to the same state.

These lemmas can be combined with the fact that, in the initial configuration, the TTM is in state **free** with $\mathsf{occ} = 0$ to derive that the **free** state is reached by the TTM infinitely often, as required by (11.13).

11.3   The TRIO-Petri Nets Approach

This section examines how the dual-language approach can combine the asynchronous operational formalism of timed Petri nets (introduced in Chap. 8) with the logic language TRIO (described in Sect. 9.3.2). We follow the classical approach that axiomatizes in logic (TRIO) the semantics of the abstract machine (timed Petri nets). The TRIO axioms are all implicitly universally quantified over time with an Alw operator.

Following a first intuition, it seems natural to express the semantics of an action as a pair of pre- and post-conditions on the state of the machine (the marking, for Petri nets). This follows the classic Hoare style which uses expressions such as

$$\{\Phi_x^{exp}\}\ x := exp\ \{\Phi\}$$

to express the fact that if $\{\Phi_x^{exp}\}$ holds before executing the statement $x := exp$, then $\{\Phi\}$ will hold after its execution, where $\{\Phi_x^{exp}\}$ denotes the result of replacing by exp every free occurrence of x in Φ.

Similarly, for Petri nets we write formulae such as

$$\{\,\mathsf{marked}(P) \wedge \mathsf{marked}(Q)\,\}\ \mathsf{fire}(tr)\ \{\,\mathsf{marked}(R) \wedge \mathsf{marked}(S)\,\}$$

to express that, for a Petri net fragment in the type of Fig. 11.11, the marking of input places P and Q is the precondition that guarantees that the firing of transition tr produces the effect of marking output places R and S.

Many subtle technicalities, however, surface when trying to write a complete formalization of Petri net semantics. Let us first consider untimed Petri nets, introduced in Sect. 8.1, where time is implicitly defined by the interleaving semantics, is discrete, and its unit corresponds to the time taken by the firing of a single transition. Thus, we can exploit LTL to express the semantics of the fragment of Fig. 11.11 as follows:

$$\mathsf{marked}(P) \wedge \mathsf{marked}(Q) \wedge \mathsf{fire}(tr) \Rightarrow$$

$$\bigcirc (\mathsf{marked}(R) \wedge \mathsf{marked}(S)\ \wedge \neg\mathsf{marked}(P) \wedge \neg\mathsf{marked}(Q))\,. \quad (11.14)$$

Apart from dealing with the expressiveness limits of LTL pointed out in Sect. 9.1.1, generalizing this approach requires dealing with the many facets of Petri net semantics. In particular:

- Formula (11.14) formalizes the marking of a place as a Boolean variable; this is a reasonable assumption for 1-bounded Petri nets, but unbounded nets require integer counters.
- Formula (11.14) describes the effect of firing tr when it is enabled due to the marking of its input places. According to the interleaving semantics, we should also formalize the fact that, in case several transitions are enabled in a given marking, one and only one enabled transition fires nondeterministically.

Fig. 11.11 A Petri net
fragment

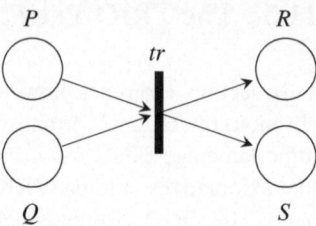

Exercise 11.14. (♦) Taking into account these issues, provide a complete LTL axiomatization of the Petri net in Fig. 8.6 according to the interleaving semantics.
■

As we have seen in Chap. 8, the semantics becomes considerably more complicated when we move to metric-timed Petri nets with a continuous time domain. With reference to the presentation of Chap. 8, we axiomatize, by means of the TRIO language, the semantics of timed Petri nets with the following features:

- Minimum and maximum firing times associated with transitions;
- Strong time semantics;
- Continuous time domain;
- No loops of transitions for which the sum of the lower bounds is 0 (hence Zeno behaviors are excluded a priori).

Before treating this general case, we start with additional restrictions that considerably simplify the model, and gradually relax them to deal with the general case in Sect. 11.3.2.

11.3.1 Axiomatization of a Subclass of Timed Petri Nets

Consider the following restriction hypotheses:

(BN) 1-bounded nets;
(CF) Input and output conflict-free nets, that is, every place is in the preset and postset of at most one transition;
(IO) No place is both input and output of the same transition;
(GT0) All lower bounds are strictly greater than 0.

These restrictions still allow for modeling nontrivial systems, as we will see in Example 11.16.

The first TRIO axiom specifies that the firing of a transition tr is instantaneous:

$$\mathsf{fire}(tr) \Rightarrow \mathsf{NowOn}(\mathsf{fire}(tr)) \wedge \mathsf{UpToNow}(\mathsf{fire}(tr)) \,.$$

Since we are dealing with 1-bounded nets (hypothesis BN), we can formalize the marking of a place P as a predicate $\mathsf{marked}(P)$. Also, hypothesis GT0 guarantees that every marking has non-null duration; this hypothesis is essential for formalizing

Fig. 11.12 Timed Petri nets fragments with one input place per transition (**a**) and two input places per transition (**b**)

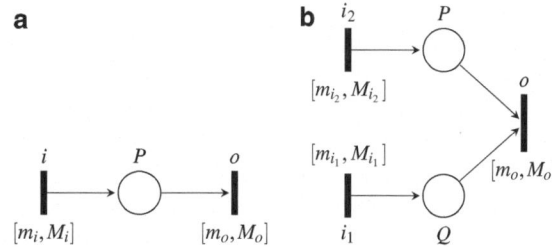

the relations between the net's state and the transition firings in a way that is close to the intuitive semantics. In fact, in Sect. 11.3.2, where we will drop hypothesis GT0, we will also have to abandon the approach based on representing the marking and pursue a more complicated one.

Let i and o denote the (unique, thanks to CF) input and output transitions of P. The following axiom specifies when P is marked:

$$\mathsf{marked}(P) \iff \mathsf{Since_{ii}}(\neg\mathsf{fire}(o), \mathsf{fire}(i)) \vee (\mathsf{fire}(i) \wedge \neg\mathsf{fire}(o)); \quad (11.15)$$

that is, P is marked if and only if at some time in the past its input transition fired and its output transition has not fired yet; or the input transition fires now, and the output transition does not fire simultaneously (which cannot happen under hypothesis GT0). Thus, marked is a predicate that holds over nonempty, left-closed, right-open intervals.

We now formalize the firing rules according to the strong time semantics. Thanks to hypotheses CF and IO, we only have to consider net fragments of the types in Fig. 11.12. (Their generalizations to the case of transitions with more than two input places is routine).

With reference to the fragment in Fig. 11.12a, we define the axioms

$$\mathsf{LowerBound(a)} \quad \equiv \quad \mathsf{fire}(o) \Rightarrow \mathsf{Lasted_{ie}}(\mathsf{marked}(P), m_o), \quad (11.16)$$

$$\mathsf{UpperBound(a)} \quad \equiv \quad \mathsf{Lasted_{ie}}(\mathsf{marked}(P), M_o) \Rightarrow \mathsf{fire}(o). \quad (11.17)$$

If o fires, P has been marked for at least m_o time units, and after M_o has elapsed, o *must* fire. We assume that, in the particular case of infinite upper bound, $\mathsf{Lasted_{ie}}(\mathsf{marked}(P), \infty)$ never holds, thus o is not forced to fire.

Exercise 11.15. Modify axiom (11.17) for the fragment of Fig. 11.12a if $M_o = \diamond$.
∎

Similarly, for the fragment in Fig. 11.12b, we define the axioms

$$\mathsf{LowerBound(b)} \equiv \mathsf{fire}(o) \Rightarrow \mathsf{Lasted_{ie}}(\mathsf{marked}(P) \wedge \mathsf{marked}(Q), m_o),$$
$$(11.18)$$

$$\mathsf{UpperBound(b)} \equiv \mathsf{Lasted_{ie}}(\mathsf{marked}(P) \wedge \mathsf{marked}(Q), M_o) \Rightarrow \mathsf{fire}(o).$$
$$(11.19)$$

We still have to formalize system initialization, that is, the initial marking. Axiom (11.15) states that a place can be marked exclusively as the consequence of one of its input transitions firing; hence we cannot just say that a marking holds initially (it would lead to contradictions). Instead, we elude the requirement of (11.15) and add a special initializing transition it$P$ as input to every place P that is initially marked. Let IT be the set of all initializing transitions and TR the set of all the other non-initializing transitions. Then

$$\text{InitialMarking} \equiv \text{Som}\left(\begin{array}{l} \forall \text{it}P \in IT \left(\text{fire(it}P) \wedge \neg\text{fire}(tr) \wedge \text{AlwF}_e(\neg\text{fire(it}P)) \wedge \\ \forall tr \in TR \left(\text{AlwP}_e(\neg\text{fire}(tr)) \wedge \neg\text{fire(it}P)) \right. \end{array} \right) \right) ;$$

(11.20)

that is, at some (initial) time all initializing transitions fire together, no other transition fires at the same time, and after that time initializing transitions never fire again.

The introduction of the initializing transitions may violate hypothesis CF that there are no input conflicts for places. However, axiom (11.20) excludes any conflict between the firing of initializing and non-initializing transitions, so axiom (11.15) can be easily extended to cover cases where both an initializing and a normal transition are input to the same place.

The next example shows how the collection of axioms written is sufficient to prove important properties of nontrivial systems.

Example 11.16. Let us go back to the "kernel railroad crossing" (KRC) problem introduced in Example 8.26. Figure 11.13 is a minor modification of Fig. 8.17 to make it compatible with hypotheses BN, CF, IO, and GT0.

Apart from the addition of transitions that initialize the net and purge tokens corresponding to trains which exited the critical region, the only significant change in Fig. 11.13 with respect to Fig. 8.17 is the replacement of 0-time transitions with ε-time ones, for some small positive constant ε whose exact value is irrelevant (a simplification often introduced, as pointed out in Chap. 3).

Let us use the axiomatization described before to prove the safety property of the KRC system,

$$\text{marked(In\_l)} \Rightarrow \text{marked(Closed)} ,$$

(11.21)

under the hypothesis that $R_m > \varepsilon + \gamma$, and that initially $\text{fire(itArrived)} \wedge \text{fire(itOpen)}$.

Let us first derive a few general properties from the axioms. With reference to the fragment of Fig. 11.12a,

$$\text{marked}(P) \Rightarrow \text{WithinP}_{ei}(\text{fire}(i), M_o) ;$$

(11.22)

that is, $\exists d (\text{Past(fire}(i), d) \wedge 0 < d \leq M_o)$.

Proof (of (11.22)). marked(P) implies by axiom (11.15) that P's input transition i fired in the past (or present) and that, since then, P's output transition o has not

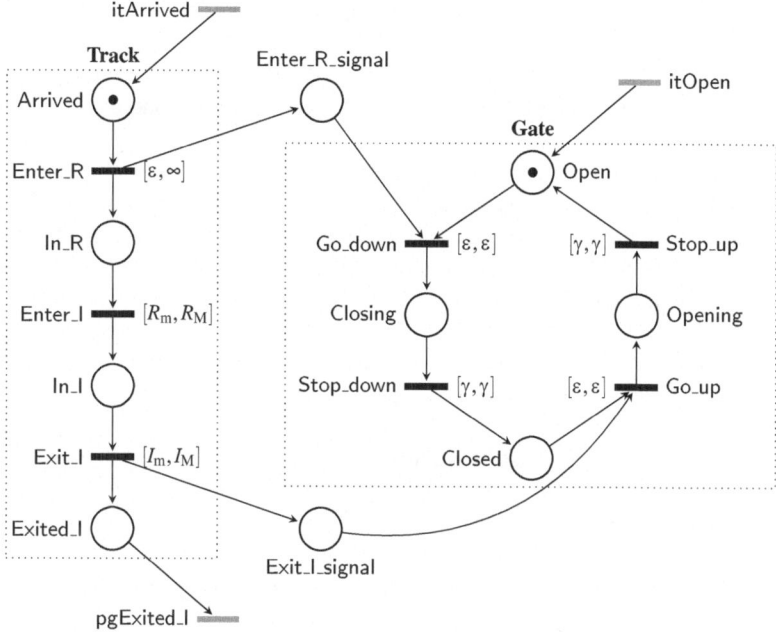

Fig. 11.13 A timed Petri net describing the KRC system

fired yet. Since the lasting of marked(P) for M_o would imply fire(o), the firing of i could not have occurred earlier than M_o time units ago (and not even exactly M_o time units ago).   ■

The proof of the following other properties, still referring to Fig. 11.12a, is left as an exercise.

$$\text{fire}(o) \;\Rightarrow\; \exists d\, \text{Past}(\text{fire}(i), d) \wedge m_o \le d \le M_o \tag{11.23}$$

$$\text{fire}(i) \;\Rightarrow\; \exists d\, \text{Futr}(\text{fire}(o), d) \wedge m_o \le d \le M_o \tag{11.24}$$

$$\text{fire}(o) \;\Rightarrow\; \text{UpToNow}(\text{marked}(P)) \wedge \text{NowOn}(\neg\text{marked}(P)) \tag{11.25}$$

With reference to the fragment of Fig. 11.12b, we have the properties (11.26) and (11.27).

$$\text{fire}(i_1) \wedge \text{UpToNow}(\text{marked}(P)) \;\Rightarrow\; \exists d\, \text{Futr}(\text{fire}(o), d) \,\wedge\, m_o \le d \le M_o\,, \tag{11.26}$$

that is, if i_1 fires at the current time and P is already marked, then o will fire not earlier than m_o and not later than M_o. (A symmetric property holds for place Q and transition i_2).

$$\text{fire}(o) \;\Rightarrow\; (\text{SomP}_i(\text{fire}(i_2)) \wedge \text{SomP}_i(\text{fire}(i_1))). \tag{11.27}$$

Finally, the following property is specific to the topology of the net in Fig. 11.13, and is stated only informally:

Every transition of the net of Fig. 11.13 can fire at most once.   (11.28)

Property (11.28) can be derived inductively by observing that the initializing transition itArrived fires exactly once. Thus, only when Enter_R fires does Arrived get unmarked; after that, it cannot be marked again since itArrived will not fire anew; consequently, Enter_R cannot fire anymore either; and so on for all other transitions, whose firing depends on input places that receive tokens from transitions ultimately enabled by the firing of Enter_R.

We are now ready for the proof of the safety property. Assume that at the current instant marked(In_l) holds. Then, marked(In_l) implies by (11.22)

$$\exists t \ \mathsf{Past}(\mathsf{fire}(\mathsf{Enter\_l}), t) \wedge t \leq I_{\mathrm{M}},$$

which implies by (11.23)

$$\exists t \ \mathsf{Past}(\mathsf{fire}(\mathsf{Enter\_l}) \wedge \exists p \ \mathsf{Past}(\mathsf{fire}(\mathsf{Enter\_R}), p), t) \wedge R_{\mathrm{m}} \leq p \leq R_{\mathrm{M}} \wedge t \leq I_{\mathrm{M}}$$

or, equivalently, by some basic properties of the Past operator,

$$\exists q \ \mathsf{Past}(\mathsf{fire}(\mathsf{Enter\_R}), q) \wedge R_{\mathrm{m}} \leq q \leq I_{\mathrm{M}} + R_{\mathrm{M}}.$$

Since (11.27) implies that Go_down cannot fire before Enter_R does, when Enter_R fired q instants in the past, place Open was still marked. By (11.26), this implies

$$\mathsf{Past}(\mathsf{fire}(\mathsf{Go\_down}), q - \varepsilon),$$

that is, by (11.24),

$$\mathsf{Past}(\mathsf{fire}(\mathsf{Stop\_down}), q - \varepsilon - \gamma).\qquad(11.29)$$

Notice that $q - R_{\mathrm{m}} \geq 0$ and $R_{\mathrm{m}} > \varepsilon + \gamma$; thus Stop_down indeed fired in the past. Next, recall the definition of marked:

$$\mathsf{marked}(\mathsf{In\_l}) \quad \Longleftrightarrow \quad \begin{pmatrix} (\mathsf{fire}(\mathsf{Enter\_l}) \wedge \neg\mathsf{fire}(\mathsf{Exit\_l})) \\ \vee \\ \mathsf{Since}_{\mathrm{ii}}(\neg\mathsf{fire}(\mathsf{Exit\_l}), \mathsf{fire}(\mathsf{Enter\_l})) \end{pmatrix}.$$

Since Past(fire(Enter_l), t) holds, with $t \leq I_{\mathrm{M}}$,

$$\mathsf{SomF}_{\mathrm{i}}(\mathsf{fire}(\mathsf{Exit\_l})),$$

and also

$$\neg\mathsf{fire}(\mathsf{Exit\_l})$$

because marked(In_l) holds now; hence $\text{AlwP}_i(\neg\text{fire}(\text{Exit\_l}))$ holds by (11.28). Thus, by (11.23),

$$\text{AlwP}_i(\neg\text{fire}(\text{Go\_up})) . \tag{11.30}$$

Finally, (11.29), (11.30), and $q - R_m \geq 0$ imply

$$\text{Since}_{ii}(\neg\text{fire}(\text{Go\_up}), \text{fire}(\text{Stop\_down})) ,$$

that is, marked(Closed) holds, which establishes the implication (11.21). ■

The safety proof of the KRC is "natural" in the sense that it parallels the normal informal reasoning based on the simulation of the graphical model. However, it is based on a simplified axiomatization which significantly restricts generality. To develop a more generic and powerful axiomatization, we must remove several of the restrictive hypotheses previously introduced. This is done in the next subsection.

11.3.2 Axiomatization of General Timed Petri Nets

Let us now progressively relax the restrictions assumed in Sect. 11.3.1, and adapt the axiomatization to work with general timed Petri nets. To simplify the presentation, we initially still retain the assumption BN of 1-bounded nets in Sect. 11.3.2.1; Sect. 11.3.2.2 will hint, however, at how the extended axioms can be adjusted to handle unbounded nets as well.

11.3.2.1 Axiomatization of 1-Bounded Timed Petri Nets

We now drop hypotheses GT0, CF, and IO; therefore:

- 0-time transition firings are possible. As we have seen in Chap. 8, this implies that different markings may occur in some order but at the same time; this exposes the risk of inconsistencies and contradictions.
- Input and output conflicts are allowed, that is, the same place can be in the postsets and presets of several transitions. This means that axioms such as (11.17) are not directly applicable, because they generate contradictions if applied to multiple output transitions for the same place P.
- The same place can be in the pre- and postset of the same transition.

We still assume restriction BN that nets are 1-bounded. To simplify the presentation, we also introduce a new minor topological assumption (often satisfied in practice):

(J1)   At most one place joins any pair of transitions.

To deal with these more general hypotheses, we turn around the approach used in Sect. 11.3.1, and instead of centering the axiomatization on the notion of marking, we use the sequences of transition firings as the basic building blocks of the axiomatization. In fact, 0-time transition firings make it impossible to describe a marking predicate that holds for zero time; instead, we rely on a predicate tokenF which formalizes the flowing of a token through a place.

Precisely, let i and o be two transitions connected through place P, which is output of i and input of o. tokenF(i, o, d) denotes that[3]:

- Transition i fires at the current instant.
- The token produced by i enters place P.
- The *same* token is consumed by a firing of transition o after d time units.

We proceed with the axiomatization of the semantics of timed Petri nets in TRIO. The first, fundamental, axiom formalizes the relation between tokenF and fire:

$$\text{FutureImplication} \quad \equiv \quad \text{tokenF}(i, o, d) \Rightarrow \text{fire}(i) \wedge \text{Futr}(\text{fire}(o), d). \quad (11.31)$$

For notational convenience, we also introduce a derived predicate tokenP(i, o, d), which has the same meaning as tokenF but refers to o's firing time (which consumes the token in P):

$$\text{PastToFuture} \quad \equiv \quad \text{tokenP}(i, o, d) \Longleftrightarrow \text{Past}(\text{tokenF}(i, o, d), d). \quad (11.32)$$

The following symmetric properties immediately follow from (11.31) and (11.32):

$$\text{PastImplication} \quad \equiv \quad \text{tokenP}(i, o, d) \Rightarrow \text{fire}(o) \wedge \text{Past}(\text{fire}(i), d), \quad (11.33)$$

$$\text{FutureToPast} \quad \equiv \quad \text{tokenF}(i, o, d) \Longleftrightarrow \text{Futr}(\text{tokenP}(i, o, d), d). \quad (11.34)$$

We are now ready to axiomatize the behavior of all the net fragments in Fig. 11.14, which cover most cases of normal use compatible with the restrictions BN and J1.

Lower Bound Axioms. Consider the net fragment of Fig. 11.14d. The axiom LowerBound(d) prescribes that, when transition o fires, it must consume a token produced either by a previous firing of transition i_1 or by one of transition i_2 no less than m_o time units earlier:

$$\text{LowerBound(d)} \quad \equiv \quad \text{fire}(o) \Rightarrow \exists d \ (d \geq m_o \wedge (\text{tokenP}(i_1, o, d) \vee \text{tokenP}(i_2, o, d))). \quad (11.35)$$

---

[3]Place P is not an argument of tokenF because hypothesis J1 guarantees that i and o uniquely determine P.

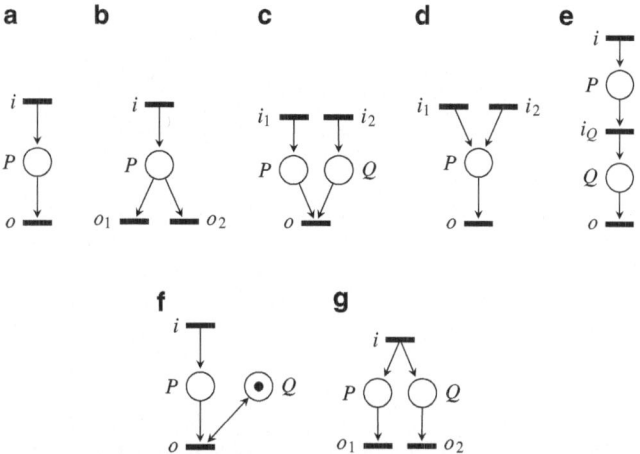

Fig. 11.14 A sample of Petri nets fragments (each transition t has a firing interval $[m_t, M_t]$)

LowerBound(c) is a similar axiom for the fragment of Fig. 11.14c:

$$\text{LowerBound(c)} \quad \equiv \quad \text{fire}(o) \Rightarrow \begin{pmatrix} \exists d_{i_1}\,(d_{i_1} \geq m_o \;\wedge\; \text{tokenP}(i_1, o, d_{i_1})) \\ \wedge \\ \exists d_{i_2}\,(d_{i_2} \geq m_o \;\wedge\; \text{tokenP}(i_2, o, d_{i_2})) \end{pmatrix}.$$

(11.36)

Writing the lower bound axioms for the other fragments in Fig. 11.14 is left as an exercise.

Finite Upper Bound Axioms. Let us illustrate axioms describing finite upper bounds for the fragments of Figs, 11.14b–d. The generalization to the case of upper bounds equal to ∞ or $\diamond$ is not complicated, and left as an exercise.

Consider first axiom UpperBound(c) for describing a finite upper bound on the firing of the transition o in Fig. 11.14c. If both i_2 and i_1 fired more than M_o time units ago, then *at least one* of the two tokens that they produced (respectively in P and Q) must have been consumed by a firing of o:

$$\text{UpperBound(c)} \quad \equiv \quad \begin{pmatrix} d_{i_1} \geq M_o \wedge \text{Past}(\text{fire}(i_1), d_{i_1}) \\ \wedge \\ d_{i_2} \geq M_o \wedge \text{Past}(\text{fire}(i_2), d_{i_2}) \end{pmatrix} \Rightarrow$$
$$\begin{pmatrix} \exists d_o\,(d_o \leq d_{i_1} \wedge \text{Past}(\text{tokenP}(i_1, o, d_{i_1} - d_o), d_o)) \\ \vee \\ \exists d_o\,(d_o \leq d_{i_2} \wedge \text{Past}(\text{tokenP}(i_2, o, d_{i_2} - d_o), d_o)) \end{pmatrix}. \quad (11.37)$$

To illustrate the use of axiom UpperBound(c), consider the firing sequence in Fig. 11.14a for the net in Fig. 11.14c. Places P and Q are initially empty;

Fig. 11.15 Firing sequences
for the net of Fig. 11.14c

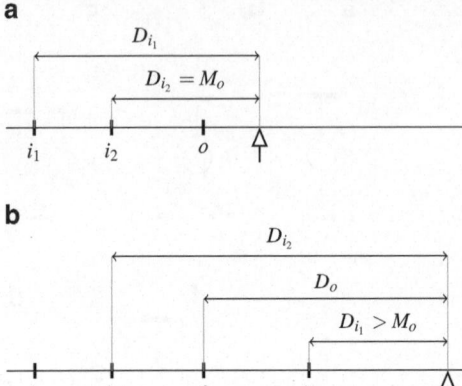

i_1 fires, followed by i_2. If axiom 11.37 is evaluated M_o time units later (in the position marked by the upward arrow), its antecedent holds for $d_{i_2} = D_{i_2} = M_o$ and $d_{i_1} = D_{i_1}$; hence the consequent implies that a firing of o occurred previously. Furthermore, axiom LowerBound(c) requires that o fire not earlier than m_o time units after i_1 and i_2, and o consume both tokens produced by i_1 and i_2.

As another example, consider the firing sequence in Fig. 11.15b, which contains the same events as Fig. 11.15a plus an additional firing of transition i_1. If axiom UpperBound(c) is evaluated at the time marked by the upward arrow, its premise holds for $d_{i_1} = D_{i_1}$ and $d_{i_2} = D_{i_2}$; then, the second disjunct of the consequence holds for $d_o = D_o$, consistently with the fact that the token produced by the latest firing of i_1 has not been consumed.

Axiom UpperBound(d) describes a finite upper bound on the firing of transition o in Fig. 11.14d – a fairly natural modification of UpperBound(c):

$$\text{UpperBound(d)} \quad \equiv \quad \text{UB(d)}(i_1, o) \ \wedge \ \text{UB(d)}(i_2, o), \tag{11.38}$$

where

$$\text{UB(d)}(i_1, o) \ \equiv \ \begin{pmatrix} d_{i_1} \geq M_o \ \wedge \\ \text{Past}(\text{fire}(i_1), d_{i_1}) \end{pmatrix} \Rightarrow \exists d_o \begin{pmatrix} d_o \leq d_{i_1} \ \wedge \\ \text{Past}(\text{tokenP}(i_1, o, d_{i_1} - d_o), d_o) \end{pmatrix} \tag{11.39}$$

and

$$\text{UB(d)}(i_2, o) \ \equiv \ \begin{pmatrix} d_{i_2} \geq M_o \ \wedge \\ \text{Past}(\text{fire}(i_2), d_{i_2}) \end{pmatrix} \Rightarrow \exists d_o \begin{pmatrix} d_o \leq d_{i_2} \ \wedge \\ \text{Past}(\text{tokenP}(i_2, o, d_{i_2} - d_o), d_o) \end{pmatrix}. \tag{11.40}$$

Each of the two main conjuncts in UpperBound(d) respectively relates the firings of transitions i_1 and o, and i_2 and o. As with UpperBound(c), they implicitly refer to a time different from the firing times of both transitions. It is sometimes simpler to assert $\text{UB(d)}(i_1, o)$ and $\text{UB(d)}(i_2, o)$ at the time of the firing of transitions i_1 and i_2, as in the following axioms.

$$\text{UB'}(d)(i_1, o) \quad \equiv \quad \text{fire}(i_1) \Rightarrow \exists d \ (d \leq M_o \wedge \text{tokenF}(i_1, o, d)) \tag{11.41}$$

$$\text{UB'}(d)(i_2, o) \quad \equiv \quad \text{fire}(i_2) \Rightarrow \exists d \ (d \leq M_o \wedge \text{tokenF}(i_2, o, d)) \tag{11.42}$$

The last upper bound axiom UpperBound(b) describes a finite upper bound on the firings of the transitions o_1 and o_2 in Fig. 11.14b. If M denotes the least upper bound $\min(M_{o_1}, M_{o_2})$, UpperBound(b) imposes the firing of *either* o_1 *or* o_2 within M time units.

$$\text{UpperBound(b)} \quad \equiv \quad \text{fire}(i) \Rightarrow \exists d \ (d \leq M \wedge (\text{tokenF}(i, o_1, d) \vee \text{tokenF}(i, o_2, d))) \tag{11.43}$$

Exercise 11.17. Write the finite upper bound axioms for the remaining fragments in Fig. 11.14. ■

Uniqueness Axioms. It is also necessary to explicitly formalize output and input uniqueness: every transition firing puts one and only one distinct token in each place of the postset, and consumes one and only one token from each place of the preset.

Axiom OutputUniqueness(b) formalizes output uniqueness for the net of Fig. 11.14b:

$$\text{OutputUniqueness(b)} \quad \equiv \quad \text{tokenF}(i, u, d) \wedge \text{tokenF}(i, w, e) \Rightarrow d = e \wedge u = w, \tag{11.44}$$

where u and w are variables ranging over the set $\{o_1, o_2\}$ of P's conflicting output transitions. OutputUniqueness(b) avoids inconsistencies in the presence of conflicts among transitions sharing an input place, as it avoids the same token being consumed by two distinct firings.

Axioms InputUniqueness(b)(o_1) and InputUniqueness(b)(o_2) formalize input uniqueness for the net of Fig. 11.14b.

$$\text{InputUniqueness(b)}(o_1) \equiv \text{tokenP}(i, o_1, d) \wedge \text{tokenP}(i, o_1, e) \Rightarrow d = e \tag{11.45}$$

$$\text{InputUniqueness(b)}(o_2) \equiv \text{tokenP}(i, o_2, d) \wedge \text{tokenP}(i, o_2, e) \Rightarrow d = e \tag{11.46}$$

The output and input uniqueness for the fragment in Fig. 11.14d are obtained from OutputUniqueness(b), InputUniqueness(b)(o_1), and InputUniqueness(b)(o_2) by symmetry.

Markings. The axiomatization of the marking of places and of the initial marking follows the same approach as that used in Sect. 11.3.1 under more restrictive assumptions; its extension to the fragments in Fig. 11.14 is also straightforward.

Example 11.18. Consider the net fragment of Fig. 11.14b and assume that $[m_{o_1}, M_{o_1}] = [1, 3]$ and $[m_{o_2}, M_{o_2}] = [4, 7]$. Let us show that transition o_2 never fires: $\text{Alw}(\neg\text{fire}(o_2))$.

The proof is by contradiction: assume that, at some time, $\text{fire}(o_2)$ holds. Then, the lower bound of o_2 implies that

$$\exists d \ (d \geq 4 \wedge \text{tokenP}(i, o_2, d)),$$

and therefore,

$$\exists d \ (d \geq 4 \wedge \text{Past}(\text{fire}(i), d)).$$

Axiom UpperBound(b) implies that

$$\exists d \ (d \geq 4 \wedge \text{Past}(\exists e \ (0 \leq e \leq 3 \wedge (\text{tokenF}(i, o_1, e) \vee \text{tokenF}(i, o_2, e))), d)),$$

but the output uniqueness axiom OutputUniqueness(b) implies that the disjunct $\text{tokenF}(i, o_1, e)$ cannot hold; hence it must be

$$\exists d \ (d \geq 4 \wedge \text{Past}(\exists e \ (0 \leq e \leq 3 \wedge (\text{tokenF}(i, o_2, e) \wedge d = e)), d)),$$

a clear contradiction.   ∎

Example 11.19 (Kernel railroad crossing proof revisited). Let us go back to the proof of the safety property for the KRC system in Example 11.16, and replace the ε-lower and upper bounds with the original 0-times.

The key difference of the new proof with respect to that of Example 11.16 lies in the way the lemmas are extended to cover the case of lower bound equal to 0 (and possibly also upper bound equal to 0). Consider, for instance, property (11.23); when $m_o = 0$, we cannot exploit the fact that place P, connecting transitions i and o, has been marked for some time. Instead, we combine the more general axioms for lower bounds and upper bounds with axiom (11.33) (which holds also for $d = 0$) to provide proofs of the same lemmas.

After having proved the lemmas in the new setting, the safety proof proceeds in parallel with the one in Example 11.16 by replacing ε with 0 and discussing the special case $R_\text{m} = \gamma$: if $q = R_\text{m}$, Stop_down may occur at the current instant. In this case, marked(In_l) implies fire(Exit_l); thus, tokenP(Exit_l, Go_up, 0), fire(Go_up), and therefore marked(Closed) hold as expected.   ∎

Exercise 11.20. The statements of properties (11.25) and (11.26) must be generalized to work with the axiomatization based on firing sequences. Provide such generalizations and prove them.   ∎

Exercise 11.21. A button that lights up after being pushed (for example, a button to call an elevator) can be described by the timed Petri net of Fig. 11.16. The firing of transition push represents the button being pressed. If the button's light is off (a token is in off) and push fires, then set immediately fires and sets the button to on (a token goes in on). Transition C acts as a token consumer to prevent accumulation

Fig. 11.16 A timed Petri net fragment describing button illumination

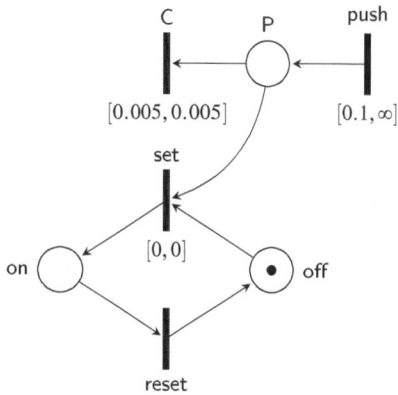

of redundant tokens in P. In such a way, the on button can be pushed multiple times (with a minimum delay of 0.1 time units between pressings) without any consequence. In other words, between any two firings of transition set, there is at least one of push. This informal analysis supports the property below, which you are required to prove.

$$\text{fire(set)} \wedge \text{Futr(fire(set)}, t) \quad \Rightarrow \quad \exists d \ (d \leq t \wedge \text{Futr(fire(push)}, d)) \quad \blacksquare$$

Example 11.22 (The dining philosophers). As a richer example, let us now apply the axiom system to derive a property of the dining philosophers problem. We refer to the timed version of the system in Fig. 11.17, which differs from those presented in Chap. 8 in that every philosopher has a maximum time he is allowed to hold a fork without using it. This time-out is represented by transitions $\text{rel}_i^{\text{F}_j}$ (the ith philosopher leaves the jth fork): if the ith philosopher picks up the jth fork (transition $\text{get}_i^{\text{F}_j}$) and in the following three time units he does not manage to pick up the second fork, he must release the jth fork. In contrast, if he obtains the second fork, he can start eating one time unit later (transition $\text{start}_i^{\text{Ea}}$).

With this scheme, we can prove deadlock freedom (which is not guaranteed in the untimed version and in the timed version of Example 8.20 without special assumptions on the time constants). Figure 11.17 also avoids an unnecessary explicit formalization of the thinking and hungry states.

The following formula formalizes deadlock freedom:

$$\text{InitialMarking} \Rightarrow \bigwedge_{i=1}^{5} \text{AlwF}\left(\text{WithinF}\left(\left(\begin{array}{c} \text{fire(get}_i^{\text{F}_i}) \\ \vee \\ \text{fire(get}_{i+1 \bmod 5}^{\text{F}_i}) \end{array}\right), 7\right)\right). \quad (11.47)$$

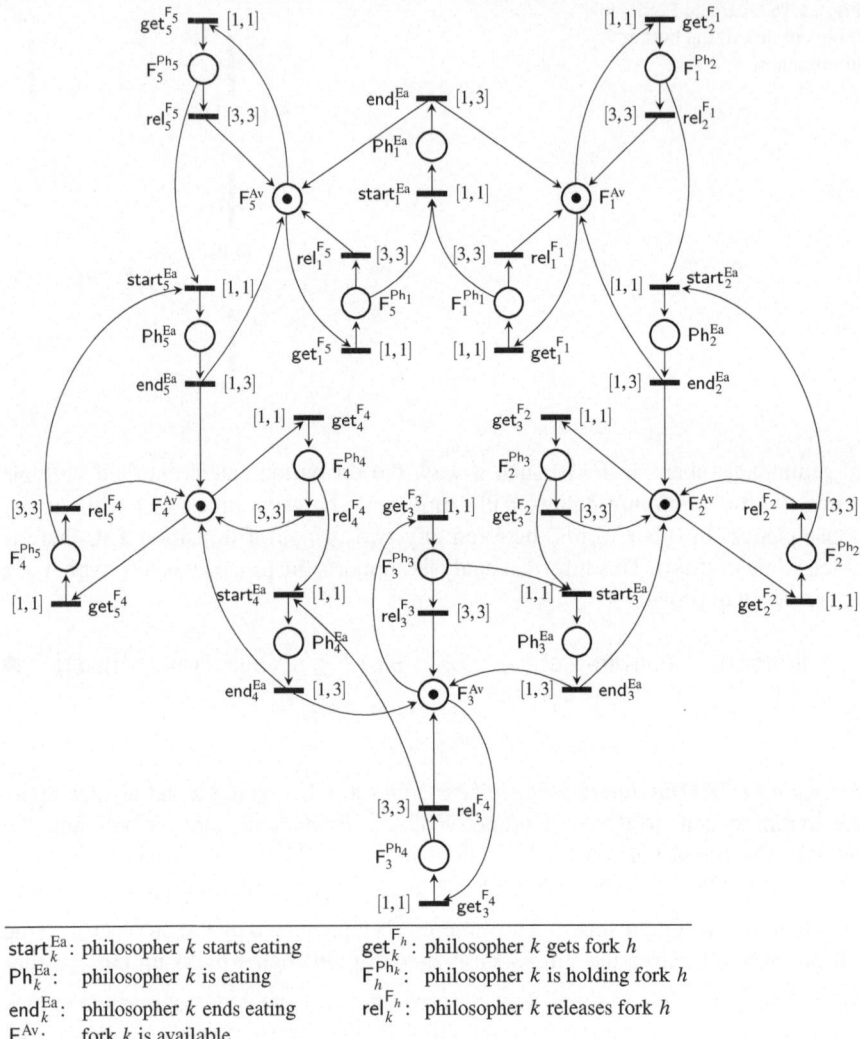

start$_k^{Ea}$: philosopher k starts eating   get$_k^{F_h}$: philosopher k gets fork h
Ph$_k^{Ea}$:   philosopher k is eating   F$_h^{Ph_k}$: philosopher k is holding fork h
end$_k^{Ea}$: philosopher k ends eating   rel$_k^{F_h}$: philosopher k releases fork h
F$_h^{Av}$:   fork k is available

Fig. 11.17   A real-time version of the dining philosophers problem

Formula (11.47) actually states a stronger property than deadlock freedom: *every fork* must be picked up within seven time units starting from any time. This does not mean that a philosopher will eventually eat: for instance, it could happen that every philosopher always picks up the left fork and never picks up the right one; thus (11.47) implies deadlock freedom but not starvation freedom.

We adopt two natural methods to master the complexity of the proof of (11.47). First, its formulation is *parametric* with respect to the number of philosophers and forks. Thanks to the obvious symmetry of the net, we do not need to prove (11.47)

separately *for every i*. In practice, such a parameterization could be embedded both in the net and in the TRIO syntax by introducing notions such as array constructors. Second, the proof itself is modularized along the following lines.

1. Prove a periodicity property of the net: if a fork is ever picked up, it will be picked up again no less than two and no more than seven time units later.

$$
\left(\begin{array}{c} \text{fire(get}_i^{F_i}) \\ \vee\ \text{fire(get}_{i+1\ \text{mod}\ 5}^{F_i}) \end{array} \right) \Rightarrow \text{Futr}\left(\text{WithinF}\left(\left(\begin{array}{c} \text{fire(get}_i^{F_i}) \\ \vee\ \text{fire(get}_{i+1\ \text{mod}\ 5}^{F_i}) \end{array} \right),5 \right),2 \right)
\tag{11.48}
$$

2. Prove that (11.48) and the initial marking, where each place F_h^{Av} stores one token, imply (11.47).

The proof of (11.48) is based on the following lemmas. If a fork is picked up, it will be either used for eating or released within three time units:

$$
\text{fire(get}_i^{F_i}) \quad \Rightarrow \quad \exists d\ \left(1 \leq d \leq 3 \wedge \left(\begin{array}{c} \text{Futr}\big(\text{fire(start}_i^{\text{Ea}}),d\big) \\ \vee \\ \text{Futr}\big(\text{fire(rel}_i^{F_i}),d\big) \end{array} \right) \right).
\tag{11.49}
$$

If a philosopher starts to eat, he will finish not before one time unit and not after three time units:

$$
\text{fire(start}_i^{\text{Ea}}) \quad \Rightarrow \quad \exists d\ (1 \leq d \leq 3 \wedge \text{Futr}\big(\text{fire(end}_i^{\text{Ea}}),d\big)).
\tag{11.50}
$$

Whenever a fork is released, either because a philosopher has finished eating (11.51) or because the time-out has expired (11.52), it will be picked up again after one time unit:

$$
\text{fire(end}_i^{\text{Ea}}) \Rightarrow \text{Futr}\big(\text{fire(get}_i^{F_i}) \vee \text{fire(get}_{i+1\ \text{mod}\ 5}^{F_i}),1\big),
\tag{11.51}
$$

$$
\text{fire(rel}_i^{F_i}) \Rightarrow \text{Futr}\big(\text{fire(get}_i^{F_i}) \vee \text{fire(get}_{i+1\ \text{mod}\ 5}^{F_i}),1\big).
\tag{11.52}
$$

At this point, the proof that a fork will not be picked up again sooner than two time units is obvious. Let us therefore focus on the proof that it will be picked up within seven units. If philosopher i picks up the fork, he may either start to eat or leave it because the time-out expires (11.49). In the former case, he will return the fork within six time units (11.50) and the fork will be picked up again after one more time unit either by philosopher i or by philosopher $i + 1$ (11.51). In the latter case, the fork will be available again within three time units. In both cases, the fork will be available again for philosophers i and $i + 1$ within six time units, and picked up by either philosopher within seven time units. A similar reasoning applies to the case where fork i is picked up by philosopher $i + 1$, so (11.48) holds.

Finally, (11.47) is derived from (11.48) as follows. Let γ denote

$$\gamma \;\; \equiv \;\; \mathsf{fire}(\mathsf{get}^{\mathsf{F}_i}_i) \vee \mathsf{fire}(\mathsf{get}^{\mathsf{F}_i}_{i+1 \bmod 5}) \, .$$

Then (11.48) is rewritten as

$$\mathrm{Alw}(\gamma \Rightarrow \mathrm{Futr}(\mathrm{WithinF}(\gamma, 5), 2)) \tag{11.53}$$

by also making explicit the universal temporal quantification. To prove (11.47), let us assume (11.20) holds. Then, $\mathsf{fire}(\mathsf{F}^{\mathrm{Av}}_i)$; hence

$$\exists d \; (d \leq 1 \wedge \mathrm{Futr}(\gamma, d)) \, .$$

Conservatively, let $\mathrm{Futr}(\gamma, 1)$ be the case; combined with (11.53), it implies

$$\mathrm{Futr}(\mathrm{AlwF}(\mathrm{WithinF}(\gamma, 7)), 1) \, . \tag{11.54}$$

Furthermore, $\mathrm{Futr}(\gamma, 1)$ also implies

$$\mathrm{Lasts}(\mathrm{WithinF}(\gamma, 1), 1) \, , \tag{11.55}$$

because, for any formula ψ, $\mathrm{Futr}(\psi, d)$ implies $\mathrm{Lasts}(\mathrm{WithinF}(\psi, d), d)$. Also notice that, for any formula ψ, $\mathrm{WithinF}(\psi, d)$ subsumes $\mathrm{WithinF}(\psi, d')$, for $d' > d$; hence, (11.54) and (11.55) can be combined to establish

$$\mathrm{Lasts}(\mathrm{WithinF}(\gamma, 7), 1) \, ,$$

which concludes the proof of (11.47). ∎

Exercise 11.23. (♦) (*For readers familiar with logical deduction systems*). Complete the lemmas and the proof of (11.48) in Example 11.22. ∎

11.3.2.2 Hints Towards the Axiomatization of Unbounded Timed Petri Nets

Let us briefly discuss how to extend the axiomatization if we drop the two remaining restrictions BN (1-bounded nets) and J1 (at most one place connects each pair of transitions). Dealing with k-bounded nets (for $k > 1$) or even unbounded ones is complicated because even the notion of a transition firing *at the current instant* becomes elusive, since a transition may fire *finitely many times* at the same instant (transitions cannot fire infinitely many times because loops of 0-time transitions are excluded, which, paired with the fact that minimum firing times are constant values, prevents Zeno behaviors). The axiomatization accommodates this new behavior by a natural extension of the predicate tokenF: since a transition may fire many times at the same instant, we explicitly introduce an argument h that denotes the hth firing of the transition at the current instant. Furthermore, since a pair of transitions i and o does not uniquely determine a place joining them, we also include a place identifier among tokenF's arguments. In conclusion, tokenF(i, h, P, o, j, d) denotes that

- The hth firing of transition i occurs,
- The token produced by i enters place P,
- The *same token* will be consumed by the jth firing of transition o after d time units.

Accordingly, we also enrich predicate fire into

- nFire(tr, n), to denote that transition tr fires exactly n times;
- fireth(tr, h), to denote that there is an hth firing of transition tr at the current time.

Markings must also be generalized with two arguments: marked(P, n) denotes that place P contains exactly n tokens.

With these conventions, the axioms discussed in Sect. 11.3.2.1 can be naturally generalized, at the only price of a more cumbersome notation. For instance, the axiom FutureImplication becomes

$$\text{GeneralFutureImplication} \equiv \text{tokenF}(i, h, P, o, j, d) \Rightarrow \begin{pmatrix} \text{fireth}(i, h) \\ \wedge \\ \text{Futr}(\text{fireth}(o, j), d) \end{pmatrix},$$
$$(11.56)$$

denoting that the token produced by the hth firing of i in P will be consumed by the jth firing of o at time distance d from the current time.

The literature mentioned in the bibliographic remarks presents more details about the general axiomatization.

Exercise 11.24. Axiomatize the weak time semantics of timed Petri nets. You can assume any of the simplifying hypotheses BN, CF, IO, and GT0. ∎

Exercise 11.25. Consider the following generalization of the KRC problem: many trains can reach the crossing; however, they run on a single track and therefore at a suitable distance from one another.

- Determine a minimum distance between trains that guarantees the safety property (and that also preserves the 1-boundedness hypothesis BN).
- Prove the safety property, both with and without hypothesis GT0. ∎

Exercise 11.26. Further generalize Exercise 11.25 as follows: when the gate is opening, a Go_down command reverses the bar's direction of motion and drives the gate back to a Closed state. Can you find a different minimum distance between trains than that in Exercise 11.25 that still guarantees the safety property? ∎

Exercise 11.27. With reference to the formalization of the dining philosophers in Fig. 11.17, find minimum and maximum firing times for the transitions such that:

1. At least one philosopher never starves.
2. All philosophers never starve.

Formally specify and prove Properties 1 and 2.   ■

11.4 Verification Tools for Dual-Language Approaches

The main focus of dual-language approaches is verification of properties (expressed in a descriptive formalism) of system models (expressed in an operational notation). Therefore, this chapter has provided a high-level description of verification techniques as an integral part of the pairs of the considered dual-language frameworks. The current section lists a few of the most relevant tools supporting verification in dual-language frameworks. Reflecting practical success, model checking tools are more prominent than those supporting the two other frameworks.

11.4.1 Model Checkers

As mentioned in Sect. 11.1, in the literature a wide range of model checking techniques has been defined; many of them are tailored to variants and subsets of general modeling languages and are supported by automated tools. We now provide a brief overview of some of the most commonly used model checkers, with an emphasis on those that are freely available. A number of commercial model checkers are also available, both as stand-alone applications and as part of larger CASE (Computer-Aided Software Engineering) tools, but we do not discuss them in this section.

The two best known tools dealing with finite-state and Büchi automata are SPIN and NuSMV, which is a successor of the SMV model checker. SPIN and NuSMV support model simulation and reachability analysis, and offer textual languages for defining finite-state automata. SPIN uses ProMeLa (Process Meta Language), a C-like notation for describing nondeterministic transitions, in which processes communicate via message passing. SPIN adopts the automata-based approach to model checking described in Sect. 11.1.1, and it can translate LTL formulae into Büchi automata. NuSMV supports both CTL-based and LTL model checking. For the latter, it can use both a technique not dissimilar to the automata-based approach of Sect. 11.1.3, and a SAT-based technique similar to the one of Sect. 11.1.2. The SAT-based technique of Sect. 11.1.2 is also supported by the Zot tool, which offers encodings of the model checking problem over discrete time for LTL, MTL, and fragments of first-order temporal logic.

As mentioned in Sect. 7.6, UPPAAL is the best-known tool for the verification of timed automata. It implements an optimized version of the technique of Sect. 11.1.4, based on a coarser and more compact representation of the state space that relies on the notion of *zone*, instead of region. UPPAAL can verify a subset of CTL properties, and in particular reachability properties of the form $\exists \Diamond \phi$ ("there is a path such that eventually a state in which ϕ holds is reached"); invariant properties of the form $\forall \Box \phi$ ("ϕ always holds along all paths") or $\exists \Box \phi$; liveness properties of the form $\forall \Diamond \phi$ ("along all paths, ϕ is eventually satisfied") or $\forall \Box (\phi \Rightarrow \forall \Diamond \psi)$ ("along all paths, if ϕ holds in some state, then in all paths from that state ψ eventually holds").

Finally, let us mention the PRISM tool and the MRMC (Markov Reward Model Checker) tool; they both are model checkers that support the analysis of probabilistic formalisms, including discrete-time and continuous-time Markov chains, and – in the case of PRISM – Markov decision processes and probabilistic timed automata.

11.4.2   Tools Supporting the TTL/RTTL Approach

Development, analysis, and verification of TTM/RTTL models are supported by the StateTime toolset. StateTime allows for a Statechart-like visual representation of hierarchically structured modular TTMs and includes an automatic translator of TTM/RTTL models into a form processable by the STeP model checker and theorem prover, supporting the treatment of both finite- and infinite-state systems. TTM/RTTL models can be tested by means of execution or checked for conformance with respect to the requirements described in RTTL.

11.4.3   Tools Supporting the TRIO/Petri Nets Approach

Section 11.3 showed that the verification of timed Petri net properties is reduced to the proof of TRIO theorems in the TRIO/Petri nets approach. Thus, the same tools associated with the TRIO language also support the TRIO/PN approach. In particular, we recall the PVS-based environment for analyzing TRIO specifications and the Zot toolsuite which implements SAT-solving techniques, described in Sects. 9.9.1 and 9.9.2 respectively.

11.5   Bibliographic Remarks

The pioneering work on the dual-language approach applied to sequential programming is due to Floyd [13] and Hoare [14]. Dijkstra extended Floyd and Hoare's seminal work [11] and introduced the notion of predicate transformer [10].

Clarke, Emerson, and Sistla [7, 8] and Queille and Sifakis [22] authored the seminal papers on model checking. Plenty of follow-up work has been devoted to optimizing, tailoring, enriching, and generalizing model checking techniques and tools; in this chapter we provided only a minimal overview. Clarke et al. [9] and

Baier and Katoen [3] are excellent books that go deep into the details of model checking techniques, including implementation and optimization issues. The idea of automata-based model checking was introduced by Vardi and Wolper [29]. Holzmann [15] describes in detail the principles behind the SPIN model checker [27] and its ProMeLa language for the description of concurrent systems. The NuSMV model checker [25] is described by Cimatti et al. [6]. Biere et al. introduced SAT-based model checking [4], and later presented different ways of encoding satisfiability of LTL formulae into propositional satisfiability [5]. The Zot tool [23] implements some of these encodings, as well as others for MTL satisfiability [21]. The first publications on model checking [7, 22] targeted CTL model checking by tableau, as described in Sect. 11.1.3. Alur and Dill [1] introduced timed automata and the region algorithm to check for emptiness, discussed in Sect. 11.1.4. The UPPAAL model checker [28] and its underlying principles are described by Bengtsson and Wang. Baier and Katoen present [3] PCTL model checking of discrete-time Markov chains and of Markov decision processes. Kwiatkowska et al. [16] present the principles underlying model checking of probabilistic timed automata, while Baier et al. [2] introduce algorithms for model checking of continuous-time Markov chains. The PRISM model checker [26] and the MRMC model checker [24] are available online.

Pnueli [20] and Ostroff [17] introduced the notion of dual-language approach in the realms of concurrency and real time, and coined the expression "dual-language". Ostroff [18] presents a tool supporting his approach, and improves it with composition and refinement mechanisms in [19].

The dual-language approach based on timed Petri nets and TRIO is described by Felder et al. [12], who also develop Example 11.22 in great detail. For the tools supporting TRIO, see the bibliographic remarks in Chap. 9.

References

1. Alur, R., Dill, D.L.: A theory of timed automata. Theor. Comput. Sci. **126**(2), 183–235 (1994)
2. Baier, C., Haverkort, B., Hermanns, H., Katoen, J.P.: Model-checking algorithms for continuous-time Markov chains. IEEE Trans. Softw. Eng. **29**(6), 524–541 (2003)
3. Baier, C., Katoen, J.P.: Principles of Model Checking. MIT, Cambridge (2008)
4. Biere, A., Cimatti, A., Clarke, E.M., Zhu, Y.: Symbolic model checking without BDDs. In: Proceedings of the 5th International Conference on Tools and Algorithms for Construction and Analysis of Systems, TACAS '99, pp. 193–207. Springer, London (1999)
5. Biere, A., Heljanko, K., Junttila, T.A., Latvala, T., Schuppan, V.: Linear encodings of bounded LTL model checking. Log. Method Comput. Sci. **2**(5) (2006)
6. Cimatti, A., Clarke, E.M., Giunchiglia, E., Giunchiglia, F., Pistore, M., Roveri, M., Sebastiani, R., Tacchella, A.: NuSMV 2: An opensource tool for symbolic model checking. In: Proceedings of the 14th International Conference on Computer Aided Verification, CAV '02, pp. 359–364. Springer, London (2002)
7. Clarke, E.M., Emerson, E.A.: Design and synthesis of synchronization skeletons using branching-time temporal logic. In: Logic of Programs, Workshop, pp. 52–71. Springer, London (1982)
8. Clarke, E.M., Emerson, E.A., Sistla, A.P.: Automatic verification of finite-state concurrent systems using temporal logic specifications. ACM Trans. Program. Lang. Syst. **8**(2), 244–263 (1986)

9. Clarke, E.M., Grumberg, O., Peled, D.A.: Model Checking. MIT, Cambridge (2000)
10. Dijkstra, E.W.: Guarded commands, nondeterminacy and formal derivation of programs. Commun. ACM **18**(8), 453–457 (1975)
11. Dijkstra, E.W.: A Discipline of Programming. Prentice-Hall, Englewood Cliffs (1976)
12. Felder, M., Mandrioli, D., Morzenti, A.: Proving properties of real-time systems through logical specifications and Petri net models. IEEE Trans. Softw. Eng. **20**(2), 127–141 (1994)
13. Floyd, R.W.: Assigning meanings to programs. In: Schwartz, J.T. (ed.) Mathematical Aspects of Computer Science, Proceedings of Symposia in Applied Mathematics, vol. 19, pp. 19–32. American Mathematical Society, Providence (1967)
14. Hoare, C.A.R.: An axiomatic basis for computer programming. Commun. ACM **12**(10), 576–580 (1969)
15. Holzmann, G.J.: The SPIN Model Checker: Primer and Reference Manual. Addison-Wesley, Boston (2003)
16. Kwiatkowska, M., Norman, G., Parker, D., Sproston, J.: Verification of real-time probabilistic systems. In: S. Merz, N. Navet (eds.) Modeling and Verification of Real-Time Systems: Formalisms and Software Tools, pp. 249–288. Wiley, London (2008)
17. Ostroff, J.S.: Temporal Logic for Real Time Sytems. Advanced Software Development Series. Wiley, New York (1989)
18. Ostroff, J.S.: A visual toolset for the design of real-time discrete-event systems. IEEE Trans. Control Syst. Technol. **5**(3), 320–337 (1997)
19. Ostroff, J.S.: Composition and refinement of discrete real-time systems. ACM Trans. Softw. Eng. Methodol. **8**(1), 1–48 (1999)
20. Pnueli, A.: Specification and development of reactive systems (invited paper). In: IFIP Congress, pp. 845–858. North-Holland, Amsterdam (1986)
21. Pradella, M., Morzenti, A., San Pietro, P.: A metric encoding for bounded model checking. In: A. Cavalcanti, D. Dams (eds.) FM 2009: Formal Methods, Second World Congress, Eindhoven, 2–6 November, 2009. Proceedings, Lecture Notes in Computer Science, pp. 741–756. Springer, Berlin (2009)
22. Queille, J.P., Sifakis, J.: Specification and verification of concurrent systems in CESAR. In: Symposium on Programming, pp. 337–351. Springer, Berlin (1982)
23. The Zot bounded model/satisfiability checker. http://zot.googlecode.com
24. The MRMC model checker. http://www.mrmc-tool.org/
25. The NuSMV model checker. http://nusmv.fbk.eu/
26. The PRISM model checker. http://www.prismmodelchecker.org/
27. The SPIN model checker. http://spinroot.com
28. The UPPAAL model checker. http://www.uppaal.org
29. Vardi, M.Y., Wolper, P.: An automata-theoretic approach to automatic program verification (preliminary report). In: Proceedings, Symposium on Logic in Computer Science, 16–18 June 1986, Cambridge, pp. 332–344. IEEE Computer Society, Washington, DC (1986)

Chapter 12
Time Is Up

12.1 Modeling Time: Past, Present...

The notion of time has always been prominent in human life and in the descriptions
of the physical world, impacting – directly and indirectly – nearly every area of
knowledge and speculation. The classic scientific description of time has long relied
on a limited number of general mathematical models that embody a fairly intuitive
notion of time. A century ago, with the introduction of the radically new scientific
theories of relativity and quantum mechanics, the classical model of time had to
be reconsidered, revised, and generalized. Somehow independently, another push
towards a major rethinking of the conventional time models came from the modern
development of computer science: initially, it relied on very abstract notions of time
consisting of a purely functional view of computational processes; then – driven
by its own pervasive success in meeting with very many heterogeneous application
domains – it developed original models, notations, and tools to reason rigorously
about timed computational processes.

   The main goal of this book has been to survey critically and "put some order
to" such a rich – yet possibly tangled and occasionally redundant – collection of
models of time in computing. We pursued this goal through two main approaches.
First, by following a somehow historical perspective, we structured the presentation
in two main parts. Second, we defined and extensively used a repertoire of
"dimensions" that characterize the essential features of models of time and of their
usage. Chapter 3 presented the dimensions, and every following chapter extensively
used them to introduce and compare the notations and models. The dimensions
help us understand the essential features of the models and notations, as well as
evaluate them against design and formalization objectives, to compare, select, and
possibly also modify, integrate, and even invent new ones to address unprecedented
challenges and needs.

   You have seen many examples throughout the book, but let us sketch one more to
give a narrative of how concrete design goals can be approached. Suppose you are
envisaging an innovative unmanned vehicle that can operate in the traffic of a large

C.A. Furia et al., *Modeling Time in Computing*, Monographs in Theoretical Computer
Science. An EATCS Series, DOI 10.1007/978-3-642-32332-4_12,
© Springer-Verlag Berlin Heidelberg 2012

city. Before starting with its design, you perform a detailed feasibility study, which starts by collecting the *requirements* of safety, usefulness, and economic feasibility for the vehicle. The high risk of the project suggests *formalizing* the requirements, in order to detect any possible incompleteness and inconsistency, and to increase confidence in the feasibility of the project.

The need to formalize requirements flexibly may orient your choice towards a **DESCRIPTIVE**, perhaps logic, language. The whole system (including the vehicle under design embedded in the **ENVIRONMENT** where it will operate) is typically reactive, and time-invariant; hence **TEMPORAL LOGIC** could be a reasonable choice, from among all logic-based approaches since it allows you to specify your requirements in a parametric way with respect to a generic current instant. Furthermore, you soon realize that many requirements of the vehicle are *real time*; hence you need a temporal logic with a **METRIC** on time.

At this point, further choices surface that are somehow more controversial. Should you model using a **DISCRETE** or a **CONTINUOUS** time domain? Do you adopt simple decidable temporal logics (such as LTL), or more expressive ones at the expense of decidability? You should keep in mind that choosing an undecidable formalism entails having only limited supporting analysis tools, or no tools at all. How do you model the uncertainty in the behavior of the environment and the degrees of freedom allowed by your requirements? **DETERMINISTIC** models may be too constraining, but what about the choice between **NONDETERMINISTIC** and **PROBABILISTIC** notations for modeling unpredictable components such as the behavior of other vehicles? Probabilistic models may seem to be the most natural choice for reliability requirements, but they also entail being able to estimate with great accuracy the probabilities of component and system failures, which is often quite difficult [4, 5], and to guarantee extremely high standards when dealing with safety-critical systems. Following an explorative attitude, you may want to try different approaches to gain a better understanding of the problem. Your analysis may show that many requirements are difficult even to state precisely; hence at the end of your exploration you might be forced to conservatively conclude that the state of the art is not yet mature for the endeavor.

In another perspective, you may want to get a clear and concise picture of which features are possessed by a given notation, which ones can be added, and which ones are not present and incompatible with it. For example, a diagram such as the one in Fig. 12.1 concisely displays the main features of TRIO: it informally says that TRIO is a descriptive formalism with a metric on time, more oriented towards system-centric modeling than computer-centric, and normally used for formalizing nondeterministic specifications (although it is extensible in principle to probabilistic modeling). A similar diagram for Petri nets is shown in Fig. 12.2.

Based on these diagrams, you can investigate pros and cons of the various combinations of TRIO and Petri nets, possibly with the goal of pursuing a dual-language approach. Does it make sense to combine them in a probabilistic setting? Or without a metric on time? How should you evaluate the trade-off between decidability and expressiveness? You can combine Figs. 12.1 and 12.2 in a graphical

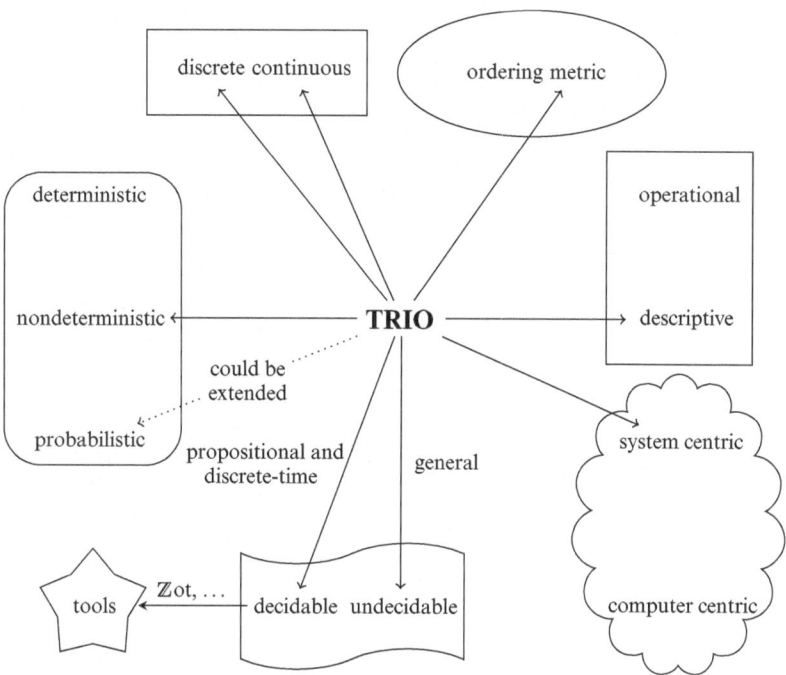

Fig. 12.1 A graphical summary of TRIO's features

summary diagram, which gives an overview of how an operational and a descriptive formalism can be combined.

Of course, diagrams such as those of Figs. 12.1 and 12.2 inevitably entail some, possibly rough, approximations and by no means should be interpreted as constraints in the application of the represented formalism: a personal "interpretation" of a model could lead to unexpected and original applications such as one applying an originally computer-centric notation to a system-centric view and vice versa.

Time and again we repeated that the dimensions of time modeling that guided us throughout this book are not intended to be "mathematically orthogonal"; consequently, the classification of models along the dimensions is often not univocal and occasionally leads to non-obvious, unexpected outcomes.

For instance, with reference to the various kinds of temporal logics discussed in Chap. 9, a very general version of branching-time temporal logic called CTL* generalizes LTL in that it is strictly more expressive, while a simpler kind of branching-time temporal logic, called CTL, is incomparable with LTL despite the availability of a richer set of modal operators; due to semantic subtleties of the way in which the operators are combined, each of LTL and CTL can express properties that cannot be expressed in the other. In addition, when features are introduced in temporal logics that can increase the modeling ability of the notations – for example, past-time operators (Sect. 9.2), metric operators (Sect. 9.3), and dense

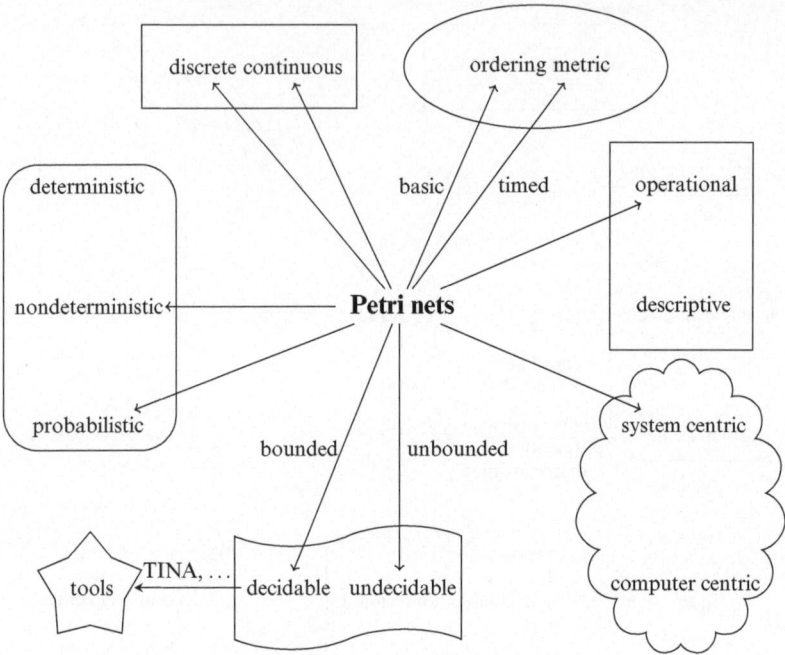

Fig. 12.2 A graphical summary of Petri nets' features

time domains (Sect. 9.4) – these features interact in a quite subtle and involved way in determining the qualities of the resulting logics, such as expressiveness, conciseness, and readability. For example, metric operators make temporal logic more concise but not more expressive in the presence of a discrete time domain, whereas metric temporal logic is strictly more expressive if the time domain is dense. Similarly, LTL with both past and future operators is only more concise (but not more expressive) than future-only LTL in the case of a mono-infinite time domain like the set of natural numbers, but it is more expressive in the presence of a bi-infinite temporal domain like the integer numbers.

Furthermore, it is often the case that the choice from among different formalisms with similar (but not equivalent) properties is also a matter of subjective preferences. This is a common situation, as we have seen many examples that can be modeled and analyzed through different formalisms, and perhaps from different perspectives.

12.2 … and Future

The pervasive impact of computer science in the most diverse application domains has fostered the rethinking, specialization, and generalization of the traditional models of time in science and engineering, and consequently the new notations and approaches have burgeoned. This has been beneficial as well as inevitable, given that diverse domains and needs do, in general, require different approaches and notations. On the other hand, the development has sometimes been disordered and has generated redundancies. A general conclusion of our journey through the chapters of the book may then be that it is now *time* to put some order into the current state of affairs, and distinguish between substantially equivalent features versus truly innovative approaches. Along the same lines, it is also quite important to have solid guidelines for *combining* and *integrating* complementary models and techniques.

An example of an "essential" criterion for classifying automata-based formalisms with a notion of time is whether they associate – implicitly or explicitly – a time unit with the occurrence of each transition (and therefore are bound to a discrete time domain) or whether they use ad hoc means – such as clocks – to measure time. Example of the difficulties that may arise when one combines features of different notations are those encountered when nondeterministic models must be integrated with probabilistic ones within the same system view.

The discussion of the book and these simple examples are only the start when addressing these challenges. A much more ambitious goal is the unification of a class of formalisms with similar characteristics in terms of a single "meta-formalism" suitable for arranging in a unique framework different problems and analyzing possible solutions – vaguely reminiscent of the efforts in physics to develop universal models and laws. Considering dimensions such as discrete vs. dense time domains, while the book illustrated how to choose between them, a new set of challenges involves finding ways to express, in a single model, evolutions that require both types of domains, such as when asynchronous processes operating on the scale of milliseconds cooperate with others synchronized by uniform clocks at much coarser time scales, such as seconds or minutes.

There have been a few significant attempts at such unifications. Let us mention, for example, the approach using nonstandard analysis with both standard and infinitesimal numbers, which offers a natural way for dealing simultaneously with discrete and continuous evolutions, and for reasoning about Zeno behaviors [1, 3]. As another example towards unification, some very general operational formalisms – such as transition systems and "abstract state machines" [2][1] – lend themselves to being used as a basis on top of which more specialized formalisms can be defined. We thoroughly applied this approach in Chap. 7 using transition systems; this showed how many automata-based models have very similar structures and semantics but also a few subtle differences (for example, how they can be composed or how they measure time).

Another field where a unifying approach has been actively pursued (at least in part) is the "dual-language" approach (treated in Chap. 11), which is based on the idea of exploiting the complementary features of operational and descriptive

---

[1] We did not discuss abstract state machines in the book because they have only rarely been applied in the context of time modeling. Ouimet and Lundqvist's work [6] is a noticeable exception.

notations in system modeling and requirements specification. The dual-language approach could also be extended, for example, by considering "triple-language" approaches where a third – more expressive – formalism provides an abstract and unifying base, on top of which the other two – descriptive and operational – simpler notations are formally defined and integrated.

We hope this book has offered a critical and comparative perspective on the fascinating topic of modeling time. The many research challenges that remain open will make for future exciting developments to which we look forward.

References

1. Bliudze, S., Krob, D.: Modelling of complex systems: systems as dataflow machines. Fundam. Inform. **91**(2), 251–274 (2009)
2. Börger, E., Stärk, R.F.: Abstract State Machines. A Method for High-Level System Design and Analysis. Springer, Berlin/New York (2003)
3. Gargantini, A., Mandrioli, D., Morzenti, A.: Dealing with zero-time transitions in axiom systems. Inf. Comput. **150**(2), 119–131 (1999)
4. Littlewood, B.: How to measure software reliability, and how not to. In: ICSE, pp. 37–45. IEEE Computer Society, Washington, DC/New York/Long Beach (1978)
5. Littlewood, B.: Dependability assessment of software-based systems: state of the art. In: Roman, G.C., Griswold, W.G., Nuseibeh, B. (eds.) Proceedings of the 27th International Conference on Software Engineering (ICSE 2005), 15–21 May 2005, St. Louis, pp. 6–7. ACM, New York (2005)
6. Ouimet, M., Lundqvist, K.: The timed abstract state machine language: abstract state machines for real-time system engineering. In: Proceedings of the 14th International Workshop on Abstract State Machines (ASM'07), Norway (2007)

Index

C.A. Furia et al., *Modeling Time in Computing*, Monographs in Theoretical Computer Science. An EATCS Series, DOI 10.1007/978-3-642-32332-4,
© Springer-Verlag Berlin Heidelberg 2012